General Biochemistry

J.H. WEIL
Professor of Biochemistry
Louis-Pasteur University
Strasbourg (France)

Contributors to this volume:
J. BONNET Y. BOULANGER P. CHAMBON
G. DUBERTRET D. GAUTHERON
C. KEDINGER M. LAZDUNSKI J. MONTREUIL
J.C. PATTE G. REBEL
J.M. ROSSIGNOL and M. WRIGHT

Preface by Professor J.P. Ebel

Translated by Dr. Jean-Edouard Mannar

SIXTH EDITION

PUBLISHING FOR ONE WORLD
NEW AGE INTERNATIONAL (P) LIMITED, PUBLISHERS
New Delhi • Bangalore • Chennai • Cochin • Guwahati • Hyderabad
Jalandhar • Kolkata • Lucknow • Mumbai • Ranchi
Visit us at **www.newagepublishers.com**

Published with the assistance of the Office of the Counsellor for Cultural, Scientific and Technical Cooperation, Embassy of France, New Delhi.
First Published in France under the title **Biochimie generale.**

Published by New Age International (P) Ltd., Publishers
Sixth Edition: 2005
Reprint: 2015

BRANCHES

- **Bangalore** 37/10, 8th Cross (Near Hanuman Temple), Azad Nagar, Chamarajpet, Bangalore-560 018, Tel.: (080) 26756823, Telefax: 26756820 **E-mail: bangalore@newagepublishers.com**
- **Chennai** 26, Damodaran Street, T. Nagar, Chennai-600 017 Tel.: (044) 24353401, Telefax: 24351463, **E-mail: chennai@newagepublishers.com**
- **Cochin** CC-39/1016, Carrier Station Road, Ernakulam South, Cochin-682 016 Tel.: (0484) 2377303, Telefax: 4051303, **E-mail: cochin@newagepublishers.com**
- **Guwahati** Hemsen Complex, Mohd. Shah Road, Paltan Bazar, Near Starline Hotel, Guwahati-781 008 Tel.: (0361) 2513881, Telefax: 2543669, **E-mail: guwahati@newagepublishers.com**
- **Hyderabad** 105, 1st Floor, Madhiray Kaveri Tower, 3-2-19, Azam Jahi Road, Near Kumar Theater Nimboliadda, Kachiguda, Hyderabad-500 027, Tel.: (040) 24652456, Telefax: 24652457 **E-mail: hyderabad@newagepublishers.com**
- **Kolkata** RDB Chambers (Formerly Lotus Cinema) 106A, 1st Floor, S N Banerjee Road, Kolkata-700 014 Tel.: (033) 22273773, Telefax: 22275247 **E-mail: kolkata@newagepublishers.com**
- **Lucknow** 16-A, Jopling Road, Lucknow-226 001, Tel.: (0522) 2209578, 4045297 Telefax: 2204098, **E-mail:lucknow@newagepublishers.com**
- **Mumbai** 142C, Victor House, Ground Floor, N.M. Joshi Marg, Lower Parel Mumbai-400 013, Tel.: (022) 24927869, Telefax: 24915415 **E-mail: mumbai@newagepublishers.com**
- **New Delhi** 22, Golden House, Daryaganj, New Delhi-110 002 Tel.: (011) 23262368, 23262370, Telefax: 43551305 **E-mail: sales@newagepublishers.com**

ISBN: 978-81-224-0641-2

₹ 235.00

C-15-04-8394

Printed in India at Kanishik Printing Press, Delhi.

PUBLISHING GLOBALLY
NEW AGE INTERNATIONAL (P) LIMITED, PUBLISHERS
7/30 A, Daryaganj, New Delhi - 110002
Visit us at **www.newagepublishers.com**

PREFACE

The study of the mechanisms of life at the molecular level constitutes one of the most exciting fields of modern science. The description of these mechanisms is called Biochemistry. This subject has evolved considerably in the recent past and the number of people who wish to specialize in it is substantial.

A knowledge of Biochemistry has become an indispensable tool for the biologist. No matter what his specialization, he should always keep in mind the fundamental molecular mechanisms of life in order to understand the phenomena governing complex biological systems, whether it be at the cellular level or those pertaining to the whole organism.

Moreover, it can be said that since we are at the stage where progress in the science of Biology can have important repercussions on human life, basic biochemical mechanisms should be a part of scientific study of even those scientists who do not wish to specialize in Biology.

The need was felt for an introductory work presenting the fundamental concepts of general biochemistry, neither too simplified nor too specialized in its scope. This book, by Jacques-Henry Weil, Professor of Biochemistry at the University Louis-Pasteur at Strasbourg, fulfills this need, and avoids both these pitfalls. The author has been able to present, in a simple and clear style, the essential structural and dynamic aspects of Biochemistry, and has closely integrated the two so as to avoid an unhealthy academic division between structural and metabolic Biochemistry. He is constantly aware of the need to put across the interrelations between the various metabolic systems. He ends the book with an exposé, remarkable for its clarity, of the regulatory mechanisms of biochemical phenomena. This book should be able to give its readers an all-encompassing view of Biochemistry. Perhaps it will inspire some of its readers to specialize further in this field.

J.P. EBEL

INTRODUCTION TO THE SIXTH EDITION

All my colleagues who had earlier participated in the preparation of the fifth edition kindly agreed, despite their multifarious occupations, to revise parts of the book pertaining to their specialization. I am very grateful to them.

In this sixth edition, while all chapters have been updated, the most important changes have been made in the pages devoted to the control of gene expression, in view of the major advances made in this domain in the past few years.

I also thank Mrs. M. Heinrich and Mrs. C. Schwoob who undertook with great care and a high degree of competence the task of typing the corrections and inserting them in the text.

INTRODUCTION TO THE FIRST EDITION

Biochemistry is the study of chemical reactions taking place in living organisms, notably reactions of degradation of food substances which provide the energy required by organisms, and transformation or biosynthesis reactions leading to the formation of compounds needed by the cells. In recent years, biochemistry has made gigantic strides and the results of major discoveries (like the elucidation of the genetic code, to cite just one example) largely surpassed the expectations of specialists. At present, in most domains of biology, one tries to comprehend the mechanisms of life at the molecular level. It is therefore understandable that many a student is attracted by biochemistry, no matter the subject of his specialization. Teachers of biochemistry therefore have to deal with youngsters interested in bacteriology, virology, zoology, animal or plant physiology, embryology, genetics etc. We have written this book for the benefit of these students who received different basic training and often pursue varied careers.

Today, one is beginning to understand in some cases, the differences existing between various types of living organisms, from the biochemical point of view, but this book has not been written from the angle of comparative biochemistry. On the contrary-and as indicated by the title of this work – we chose topics which cover the general aspects, so as to create within possible limits, a unified picture of biochemistry. It is indeed important that the student first assimilates the fundamental aspects of biochemistry; having acquired such basic knowledge, he will be able to concentrate on certain organisms, tissues or some particular domains of metabolisms, depending on the career of his choice. In short, he can become a specialist and our prayers will be answered if the present book can instill in him an interest in biochemistry and the desire to know more about it.

To maintain this book within reasonable limits, we often had to simplify or even schematize certain topics and some domains of biochemistry were not developed. Our colleagues can fill these gaps by dwelling at greater length, in their course, on certain aspects they judge important. Moreover, monographs which are brought out regularly contain studies and treatises on various aspects of modern biochemistry, written by specialists. It would be advantageous to consult these works to gain additional knowledge in a particular field of biochemistry. This book must therefore be regarded as an introduction to biochemistry enabling the student already equipped with fundamental notions of chemistry, physics and biology, to approach this discipline and gain an overall view of biological chemistry.

We did not wish to make the distinction between structural biochemistry and metabolic biochemistry, because one must always be able to establish the relationship between chemical structure and biological function; that is why we have every time grouped in the same chapter, structure and metabolism of carbohydrates, structure and metabolism of lipids, etc. Enzymatic catalysis and energy transfers are essential characteristics of biochemistry; enzymology and energetics are therefore treated at the beginning preceded only by a study on the structure of proteins because one cannot indeed speak of enzymes before having examined the proteins. We have then taken up carbohydrates, lipids, nucleic acids and amino acids. The study of metabolism is thus divided into chapters for convenience of presentation but we have underlined the relations between different groups of compounds to show that they are in reality a harmoniously integrated group. In the last chapter the reader will find an introduction to the mechanisms of regulation of metabolism. One is indeed beginning to understand how biosynthesis and enzyme activity can be controlled and we believe that having studied the important pathways of metabolism, the student – even a beginner – who has read through the book, will have acquired an idea of the possible process of regulation of metabolic activities in living organisms.

Our thanks are due to our colleagues, Professors Y. Boulanger, G. Dirheimer and J.-P. Ebel who kindly agreed to read our manuscript and offer their comments and suggestions. We also thank Professor H. Duranton who helped in revising the pages devoted to the light phase of photosynthesis. If some errors are still present in the text we request the reader to point them out to us so that they may be corrected in the event of a later edition. We are also thankful to Mrs. Issler who executed the typing work with much care and competence.

CONTENTS

CHAPTER 1[1]

AMINO ACIDS, PEPTIDES, PROTEINS, STRUCTURES AND IMPORTANT PROPERTIES

The term "*protein*" is derived from a Greek word meaning "*primacy*" or "*holding first place*". Proteins are indeed extremely important constituents of living cells: quantitatively, they generally represent more than half the dry weight of cells, and from a qualitative point of view, we find, besides the structural proteins, other proteins having a capital biological function, especially the enzymes. The latter are biological catalysts, indispensable for the reactions in cells of living organisms.

All proteins contain the four elements: C, H, O and N. Many contain sulphur. Phosphorus is present in some. The nitrogen content of proteins being around 16%, its approximate quantity can be estimated by a simple nitrogen titration after isolating the proteins.

Proteins are large molecules, i.e. *macromolecules*, formed by the condensation of a large number of units (50 to several thousands) called *amino acids*. There are two broad groups of proteins:

– Simple proteins (*holoproteins*), consisting exclusively of amino acid chains.

– *Conjugated proteins (heteroproteins)* which contain – in addition to one or several amino acid chains – a *prosthetic group* of carbohydrate, lipid, nucleic or other nature.

AMINO ACIDS

As indicated by their name, amino acids contain one acidic group and one primary amino group in α position with respect to the carboxyl group. Excepting proline and hydroxyproline, their general formula is:

$$R - \overset{\alpha}{C}H(NH_2) - COOH$$

(1)Revised by Y. BOULANGER.

Whatever their origin (viral, bacterial, plant or animal), we find in proteins about twenty amino acids which differ by their radical R and may be classified as follows:

I. – CLASSIFICATION OF AMINO ACIDS

1. – Aliphatic amino acids

A. – Hydrocarbon chain amino acids

1) *Glycine* (Gly) or glycocoll, or α-aminoacetic acid is the simplest of the amino acids because R = H:

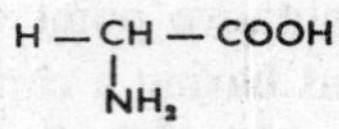

Glycine is very widely distributed in proteins; it is particularly abundant in gelatine and fibroin (silkworm protein) where it represents respectively 25% and 40% of the total amino acids.

As will be seen in the chapter on lipids, glycine is present in bile acids (see fig. 5-7).

On the other hand, glycine contributes to the detoxication processes in the liver; benzoic acid is thus eliminated in the form of a derivative where it is conjugated with glycine, giving N-benzoyl-glycine or *hippuric acid,* $C_6H_5-CO-NH-CH_2-COOH$ (see fig. 7-12).

The monomethyl derivative of glycine is called *sarcosine*. *Betaine* is the trimethyl derivative: $(CH_3)_3{\equiv}N-CH_2-COO^-$, which is an oxidation product of the corresponding alcohol, choline (see fig. 7-10).

The metabolism of glycine will be studied in Chapter 7.

2) *Alanine* (Ala), or α-aminopropionic acid is the higher homologue of glycine (it is also a widely distributed amino acid).

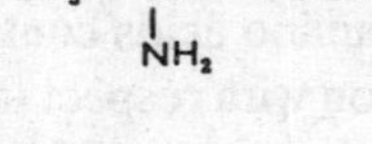

The structure of other amino acids is derived from that of alanine by substitution of various groups to one H of the methyl radical. Let us note that one sometimes finds *β-alanine*: $H_2N-CH_2-CH_2-COOH$ (particularly in coenzyme A, see fig. 2-18).

3) *Valine* (Val), or α-aminoisovaleric acid.

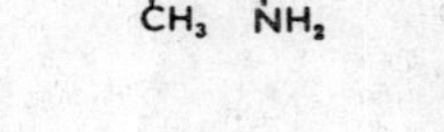

4) *Leucine* (Leu), or α-aminoisocaproic acid.

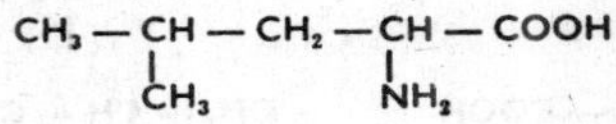

5) *Isoleucine* (Ile), or α-amino-β-methylvaleric acid.

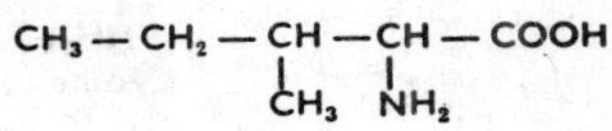

These three amino acids have a non-polar hydrocarbon chain which man and most animals cannot synthesize; they must be supplied through diet and are therefore called *indispensable* or *essential* amino acids.

B. – *Hydroxy amino acids*

1) *Serine* (Ser), or α-amino-β-hydroxypropionic acid, is found in most proteins.

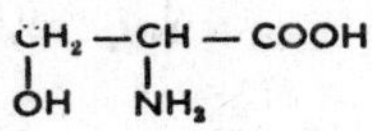

The hydroxyl group can be esterified by phosphoric acid; this gives phosphoserine, which is characteristic of *phosphoproteins* (like *vitellin* of egg yolk):

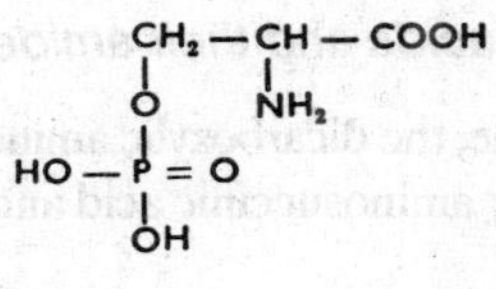

Certain antibiotics are derivatives of serine (azaserine, cycloserine, chloramphenicol). The metabolism of serine will be studied in Chapter 7.

2) *Threonine* (Thr), or α-amino-β-hydroxybutyric acid. This is an amino acid indispensable to man.

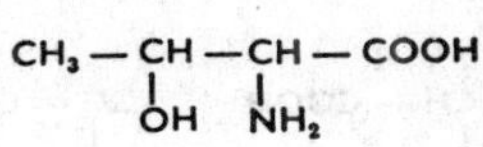

C. – *Sulphur-containing amino acids*

1) *Cysteine* (Cys), or α-amino-β-thiopropionic acid, differs from serine only by an SH in place of an OH. This thiol group has an important role in certain enzymes which lose their catalytic activity when the SH is blocked, or even

when it is simply oxidised according to the following reaction, with formation of a disulphide bridge:

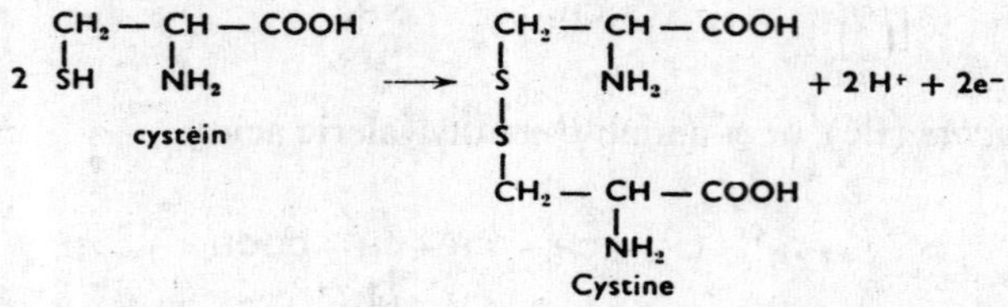

A more extensive oxidation of cysteine gives *cysteic acid.* The latter, by decarboxylation gives *taurine*, which – like glycine – çan be conjugated with cholic acid to produce the bile acids (see fig. 5-7).

The decarboxylation of cysteine gives *cysteamine* or *thioethanolamine*, which forms the reactive part of the molecule of coenzyme A (see fig. 2-18): $HS-CH_2-CH_2-NH_2$.

2) *Methionine* (Met), or α-amino-γ-methyl-mercapto-butyric acid:

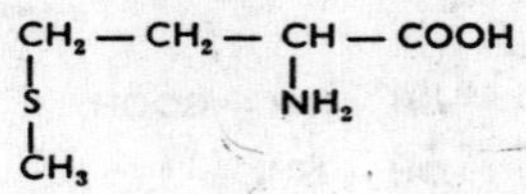

This amino acid is indispensable to man; it is present in numerous proteins, though in small quantities. It is also a *methyl group donor* in various biosynthetic reactions. We shall study in Chapter 7, the metabolism of cysteine and methionine.

D. – *Dicarboxylic amino acids and their amides*

As indicated by their name, the dicarboxylic amino acids have two acidic groups.

1) *Aspartic acid* (Asp), or aminosuccinic acid and its β-amide *asparagine* (Asn).

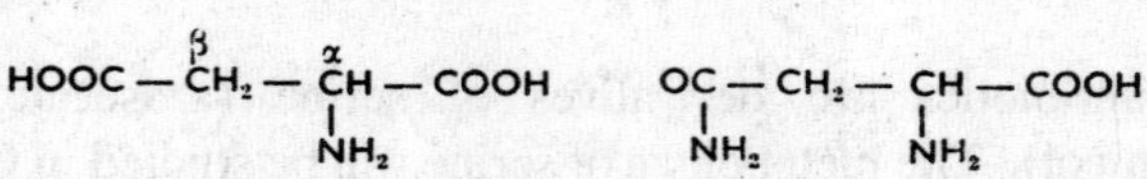

2) *Glutamic acid* (Glu) or α-aminoglutaric acid, and its γ-amide, *glutamine* (Gln).

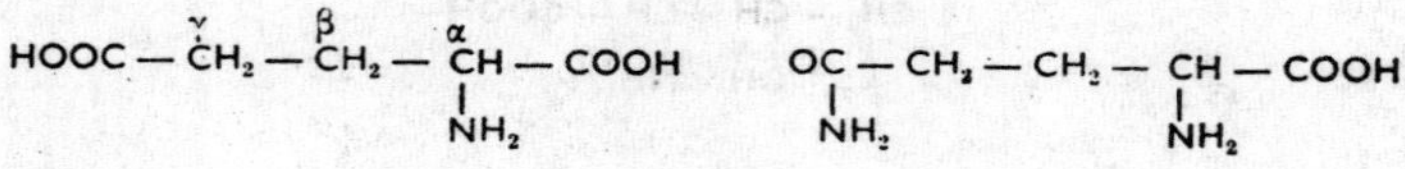

Aspartic and glutamic acids are very widely distributed in proteins. Besides, they play an important role in *transamination* processes as will be seen in connection with the metabolism of amino acids (see p. 392).

Both amides are also found in proteins. Furthermore, they play an important role in the *transport* and *storage of amino nitrogen* in plants (asparagine) and animals (glutamine).

We shall examine in Chapter 7, the metabolism of these two dicarboxylic amino acids and their amides.

E. – Basic amino acids

We find in proteins, three basic amino acids having an additional cationic group besides the α-amino group.

1) *Lysine* (Lys), or α-ε-diamino-caproic acid; it has a second NH_2 in ε. This amino acid is indispensable to man.

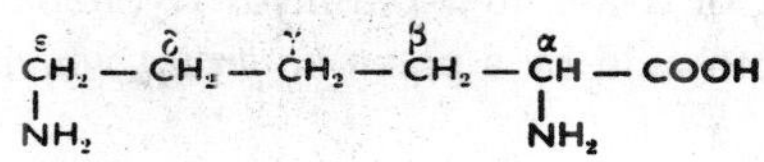

In certain proteins (collagen, gelatin) one finds δ-hydroxylysine, the δ-hydroxylated derivative of lysine.

2) *Arginine* (Arg), or α-amino-δ-guanidino-valeric acid is – like lysine – an amino acid particularly abundant in the basic proteins of cell nuclei: *histones* and *protamines* (certain protamines contain 80 to 90% arginine).

Arginine plays a very important role in the ureogenesis cycle (see fig. 7-30). Besides, *arginine-phosphate* plays an important role in the storage of energy in Invertebrates (the same role is played by *creatine-phosphate* in Vertebrates, see page 110).

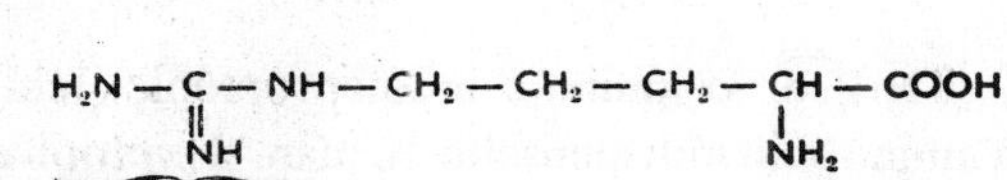

guanidinium group arginine

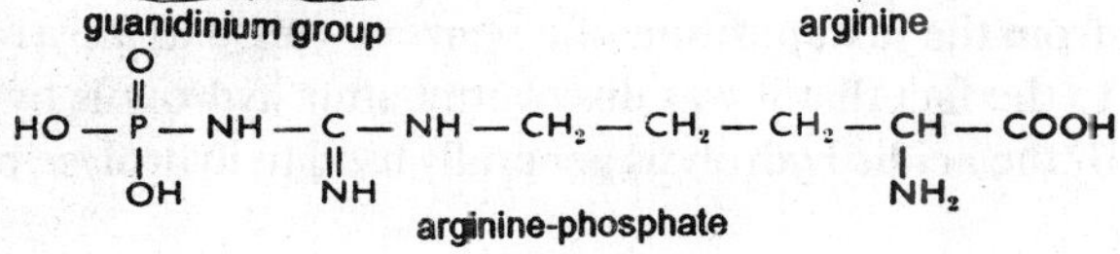

arginine-phosphate

3) *Histidine* (His), or α-amino-β-imidazole propionic acid, or imidazolyl-alanine, is not an aliphatic amino acid because it contains an imidazole ring; we shall however study it here together with the other two basic amino acids. Certain authors prefer to classify histidine with tryptophan and proline in the group of heterocyclic amino acids.

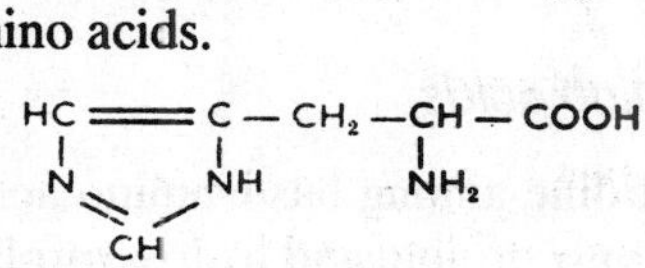

The weakly basic character of histidine is due to the imidazole ring. It is a widely distributed amino acid, particularly in globines.

2. – Cyclic amino acids

A. – Aromatic amino acids

1) *Phenylalanine* (Phe), or α-amino-β-phenyl propionic acid, is indispensable to man.

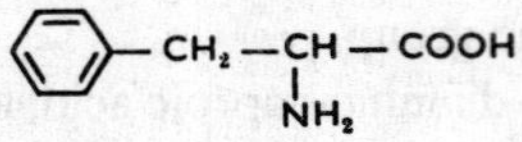

2) *Tyrosine* (Tyr), or α-amino-β-parahydroxyphenyl propionic acid, differs from the above amino acid by a *phenolic group* which gives it a weakly acid character.

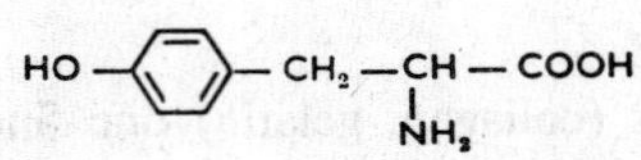

Tyrosine – together with tryptophan – is mainly responsible for the absorption in U.V. (between 260 and 290 nm) observed in proteins. Tyrosine is also responsible for the xanthoprotein reaction, i.e. yellow coloration observed on addition of HNO_3 and due to the formation of nitrophenol type compounds. As will be seen in Chapter 7, the metabolism of tyrosine leads to several important compounds like *hormones* (adrenaline, thyroxine) and *pigments* (melanines).

3) *Tryptophan* (Trp), or α-amino-β-indolepropionic acid, or β-indolylalanine. This is an amino acid indispensable to man. Tryptophan has an indole ring (resulting from the juxtaposition of a benzene ring and a pyrrole ring). It owes its name to the fact that it was discovered after hydrolysis by trypsine, as it is destroyed by the acidic hydrolysis generally used to hydrolyse proteins into amino acids.

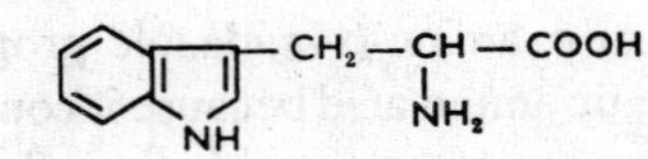

B. – Heterocyclic amino acids

As we included histidine among basic amino acids and tryptophan among aromatic amino acids, only proline and hydroxyproline are left in this group.

– *Proline* (Pro), or pyrrolidine-2-carboxylic acid and its 4-hydroxylated derivative, *hydroxyproline*, contain a heterocycle having 4 carbon atoms and one nitrogen atom.

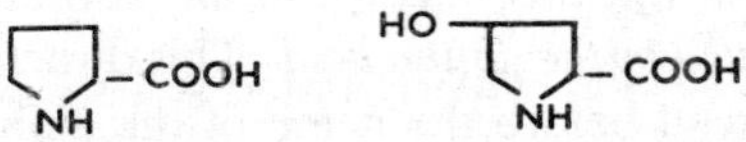

Proline Hydroxyproline

These two amino acids have a secondary amino group, not a primary one; they are substituted α-amino acids.[1]

Proline is very widely distributed; hydroxyproline less so, but it is found in appreciable quantities in collagen (13%).

Besides these amino acids constituting the proteins, one may find some less common amino acids, sometimes observed only in a small number of species, either in free state, or in peptides. For example in the toxic polypeptides of Amanita phalloides one finds hydroxyleucine and allothreonine; and in various cyclic peptides with antibiotic activity (tyrocidines, gramicidines, bacitracines) isolated from certain bacteria, one finds ornithine (see fig. 1-16); we will refer again to this amino acid because ornithine and citrulline are two amino acids which play an important role in ureogenesis (see fig. 7-30).

II. – IMPORTANT PHYSICAL PROPERTIES OF AMINO ACIDS

1. – Optical isomerism

All amino acids, except glycine which has two H atoms on the α-carbon, have an asymmetric α-carbon (the 4 valence bonds are attached to different atoms or groups). As may be seen in figure 1-1, one can therefore write – for example, in the case of alanine – the spatial formula of two *optical isomers* or *enantiomers* (which form a racemate when mixed), non-superimposable because one is a plane mirror image of the other, and whose physical and chemical properties are identical (except the rotatory power). One therefore distinguishes for each amino acid, an isomer D, and an isomer L, depending on whether the spatial structure is identical to that of D glyceraldehyde or L glyceraldehyde (whose absolute configuration is known). The spatial formulae being rather complicated, the plane projection formulae proposed by Fischer is preferred; by convention one writes the amino acids of the D-series with the amino group on the right, and those of

---

(1) Some authors use the term "imino-acids" which is not correct, because an imine is characterized by a $>CH=NH$ and not a $\genfrac{}{}{0pt}{}{C}{C}\!>NH$.

the L-series with the NH_2 on the left. It may be noted that the *prefixes D and L have absolutely no significance relative to the direction in which the plane of polarized light is rotated by the amino acid.* This direction may be specified by a sign (+) or (–) placed before the name of the amino acid [Example: L (–) leucine], keeping however in mind the fact that the rotatory power changes with the conditions of the medium, particularly the pH.

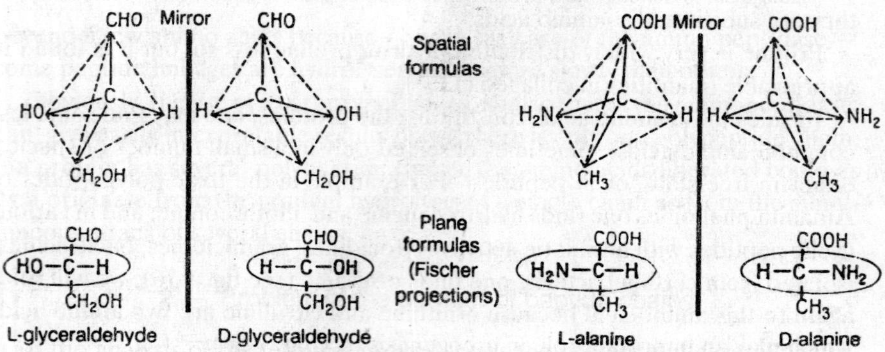

FIG. 1-1. — Structure of optical isomers of glyceraldehyde and alanine.

All amino acids constituting the proteins belong to the L-series. But in various bacterial peptides, particularly those present in the cell wall, or certain peptides having an antibiotic activity, one finds amino acids of the D-series (see fig. 1-16).

On the other hand, certain amino acids have an additional asymmetric carbon in the side chain, which increases the possibilities of isomerism; thus for threonine, one will have D and L threonine and D and L allothreonine.

Let us note that in organic chemistry, the designations L and D are now being replaced by the letters S (latin "Sinistrum": left) and R ("Rectum": right). Example: S-alanine (L) and R-alanine (D).

2. – Absorption in the ultraviolet region

Amino acids present an appreciable absorption at wavelengths below 230 nm (2 300 Å); furthermore, some of them absorb between 250 and 300 nm, due to the presence of chromophores like the phenyl radical (Tyr) or the indole ring (Trp) in their R chain, thus permitting the spectrophotometric titration of proteins.

3. – Ionization

All amino acids contain at least two ionizable groups, the carboxyl group and the amino group; they are amphoteric; the *carboxyl group* of an amino acid can yield one proton, and an *anion* appears:

The amino group can fix one proton and form a *cation*.

$$R-NH_2 + H^+ \rightleftharpoons R-NH_3^+$$

These two dissociation reactions correspond to equilibria governed by the law of mass action; the proportions of ionized and unionized amino acid present in solution will therefore depend on the concentration of H^+ ions. The two dissociation constants K_1 and K_2, corresponding to the two equilibria, may therefore be written as follows:

$$K_1 = \frac{[R{-}COO^-][H^+]}{[R{-}COOH]}, \quad K_2 = \frac{[R{-}NH_2][H^+]}{[R{-}NH_3^+]}$$

The exact values of K_1 and K_2 vary for different substances but it may be said that K_1 for a carboxylic acid is of the order of 10^{-4} to 10^{-6} (1) and K_2 for an amine, between 10^{-8} and 10^{-10}. Knowing the values of K_1 and K_2 we can calculate for each H^+ ion concentration (i.e. for each pH), the percentage of ionized molecules. For example, at pH 7, one will have for the carboxyl group (taking for K_1 a mean value of 1×10^{-5}):

$$1 \times 10^{-5} = \frac{[R{-}COO^-]\,1 \times 10^{-7}}{[R{-}COOH]}, \quad \text{i.e.} \quad \frac{[R{-}COO^-]}{[R{-}COOH]} = 10^2$$

This means that at pH 7, there will be only one non-ionized molecule for 100 anion molecules, in other words, the percentage of ionization of the acid group is 99%. For the amino group, at the same pH 7, one will have (taking for K_2 a mean value of 1×10^{-9}):

$$1 \times 10^{-9} = \frac{[R{-}NH_2]\,1 \times 10^{-7}}{[R{-}NH_3^+]}, \quad \text{i.e.} \quad \frac{[R{-}NH_2]}{[R{-}NH_3^+]} = 10^{-2}$$

At this pH, the percentage of ionization of the amino group is therefore 99%.

When the H^+ ion concentration is equal to K_1, we have $\frac{[R-COOH]}{[R-COO^-]} = 1$, in other words, the number of anions present is equal to the number of non-ionized molecules. If, as in the example given above, $K_1 = 10^{-5}$, we also have $[H^+] = 10^{-5}$, and this situation is therefore reached at pH = 5, i.e. when the pH is equal to pK (–log K) of the ionizable group. *The pK corresponds to the pH of*

---

(1) We shall see later (page 11) while studying the results obtained from the titration of an amino acid, that the carboxyl group of an amino acid is actually 100 times more acidic than that of acetic acid, due to the proximity of the α-amino group and therefore K_1 is of the order of 10^{-2} to 10^{-3}.

half dissociation, it indicates the ease with which this dissociation takes place and it is a measure of the strength of the acid.

The same is true when the H^+ ion concentration is equal to K_2. One then has $\frac{[R-NH_2]}{[R-NH_3^+]} = 1$, i.e. cations and non-ionized molecules in equal number, and when $K_1 = 10^{-9}$ (above example) this situation is reached at pH = 9. Here again, the pH of half dissociation is called pK and is a measure of the strength of the base.

When the pH of a solution of an amino acid is raised from a low value to a high value, one has the following transformations:

$$\underset{\substack{\text{Cation (acid pH)}\\ \text{Net charge} = +1}}{^+H_3N-\underset{\substack{|\\ R}}{CH}-COOH} \underset{K_1}{\overset{-H^+}{\underset{+H^+}{\rightleftharpoons}}} \underset{\substack{\text{Zwitterion}\\ \text{Net charge} = 0}}{^+H_3N-\underset{\substack{|\\ R}}{CH}-COO^-} \underset{K_2}{\overset{-H^+}{\underset{+H^+}{\rightleftharpoons}}} \underset{\substack{\text{Anion (alkaline pH)}\\ \text{Net charge} = -1}}{H_2N-\underset{\substack{|\\ R}}{CH}-COO^-}$$

It may be noted that at a particular pH, the amino acid molecules are in the dipolar form (Zwitterion) and the net charge of the molecule is nil; this is the *isoionic* or *isoelectric*[(1)] *point* of the amino acid. At this pH, its solubility is minimal and it does not migrate when placed in an electric field (unlike the cation and the anion).

The dissociation of the various polar groups of an amino acid may be easily studied by adding HCl or NaOH to the solution and measuring the pH after each addition. We can thus plot *titration curves,* of varying shapes depending on whether the amino acid is neutral, acidic or basic.

A. – *Titration of a neutral amino acid*

When acid (HCl) is added, the base $R-COO^-$ is titrated (according to Bronsted's definition, a base is a compound capable of capturing one proton):

$$^+H_3N-CH_2-COO^- + H^+Cl^- \longrightarrow Cl^-\overset{+}{H_3N}-CH_2-COOH$$

When a base (NaOH) is added, the weak acid $R-NH_3^+$ is titrated (an acid, because it is capable of yielding one proton):

$$^+H_3N-CH_2-COO^- + Na^+OH^- \longrightarrow H_2N-CH_2-COO^-Na^+ + H_2O$$

---

(1) In general, the term *"isoionic point"* denotes the pH at which the net charge is nil when the substance is dissolved in pure water and has the possibility of fixing only protons. But in most cases there are other ions in solution, which can also be fixed by the substance, so that the molecule may still be in a charged state at the pH corresponding to the isoionic point; the term *"isoelectric point"* is therefore used to denote the pH at which the charge of a molecule having fixed foreign ions is nil.

It is observed (see fig. 1-2a) that there are 2 zones where the addition of HCl or NaOH produces only a very small variation of the pH; there is therefore a *buffer effect;* glycine is actually used to form a glycine-HCl buffer (where glycine is in the form of hydrochloride) and a glycine-NaOH buffer (where glycine is in the form of sodium salt). At the middle of each of these 2 buffer zones, one has the pH value corresponding to the pK of the 2 groups: pK_1 (half dissociation of the carboxyl group) = 2.3 approximately, and pK_2 (half dissociation of the amine) = 9.7 approximately. For comparison, the pK of acetic acid is 4.8: it therefore appears that the carboxyl group of glycine is an acid 100 times stronger than that of acetic acid, which is due to the effect of the α-amino group.

The isoelectric pH or $pH_i = \dfrac{pK_1 + pK_2}{2} = \dfrac{2.3 + 9.7}{2} = 6$. Figure 1-2a shows that this pH_i is at the centre of an isoelectric zone. We have seen earlier that for a pH = pK_1 + 2 (here 2.3 + 2 = 4.3), the acidic group is ionized to 99% and for a pH = pK_2 – 2 (here 9.7 – 2 = 7.7), the amino group is ionized to 99%; one may therefore consider that in the pH zone between 4.3 and 7.7, glycine is almost entirely in zwitterion state.

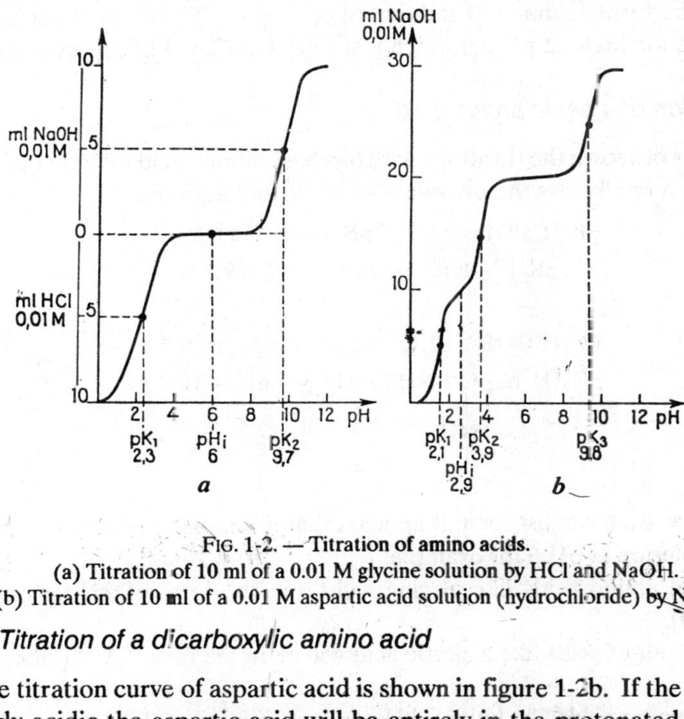

FIG. 1-2. — **Titration of amino acids.**

(a) Titration of 10 ml of a 0.01 M glycine solution by HCl and NaOH.

(b) Titration of 10 ml of a 0.01 M aspartic acid solution (hydrochloride) by NaOH.

B. — *Titration of a dicarboxylic amino acid*

The titration curve of aspartic acid is shown in figure 1-2b. If the solution is strongly acidic the aspartic acid will be entirely in the protonated form; one adds sodium hydroxide which will react with all the protons supplied by the amino acid (the 2 carboxyl groups and the protonated amino group $-NH_3^+$ are all acids according to Bronsted's definition, because they can all yield a proton). During the titration by sodium hydroxide one will therefore have the different forms represented in figure 1-3.

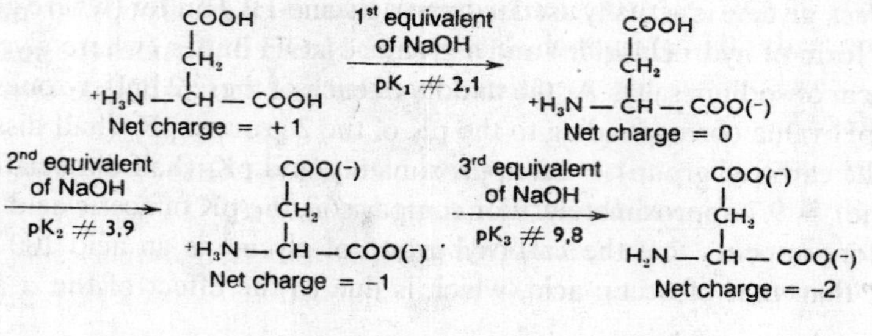

FIG. 1-3. — Different forms obtained upon addition of NaOH to an acidic solution of aspartic acid.

The pK values corresponding to the half dissociations of the 3 protons are respectively about 2.1, 3.9 and 9.8. It is seen that the most acidic carboxyl group is the one carried by the α-carbon because it is near the $-NH_3^+$ which, by repelling positive charges, promotes the dissociation of the α-carboxylic group.

At pH 2.1 one has, in equal quantities, the form having a net charge + 1 and the one having a net charge 0. Similarly, at pH 3.9, one will have in equal quantities the form of charge 0 and that of charge –1. The form of net charge 0 will be predominant at pH 2.9 which is the isoelectric pH of aspartic acid.

C. – *Titration of a basic amino acid*

We have not shown the titration curve of a basic amino acid but one can have an idea of it if one knows the pK values of lysine and arginine:

Lysine:

$$pK_1(COOH) = 2{,}2 \quad pK_2(\alpha\text{—}NH_3^+) = 9{,}0$$
$$pK_3(\varepsilon\text{—}NH_3^+) = 10{,}5 \quad pH_i = 9{,}7$$

Arginine:

$$pK_1(COOH) = 2{,}1 \quad pK_2(\alpha\text{—}NH_3^+) = 9{,}0$$
$$pK_3(H_2N\text{—}\underset{\underset{\displaystyle R}{|}}{\underset{\displaystyle NH}{\overset{}{C}}}{=}NH_2^+) = 12{,}5 \quad pH_i = 10{,}8$$

From what we have just seen, it appears that if one has a mixture of amino acids in a solution at pH 6 for example:

– neutral amino acids like glycine will be in the form of zwitterion $^+A^-$ (charge = 0),

– acidic amino acids like aspartic acid will be in the form $^+A^{2-}$ (charge = –1),

– basic amino acids like lysine will be in the form $^{2+}A^-$(charge = + 1).

This explains why *amino acids can be separated by fractionation methods based on charge differences,* especially electrophoresis and ion exchange chromatography.

So far, we have only referred to carboxylic and amino groups but one finds in amino acids other groups which also can be ionized. Besides the guanidium group of arginine we just mentioned, we may cite the imidazole nitrogen of histidine (pK = 6.7).

In proteins, except at each end, the α-NH_2 and α-COOH are all involved in the peptide bonds and therefore cannot be ionized. Hence, the *ionizable groups of the side chains mainly contribute towards conferring a charge to the macromolecule* (i.e. the groups we just mentioned and those of the basic and dicarboxylic amino acids). Depending on the pH, a given protein may have a positive charge, a zero charge (when the + and – charges balance each other), or a negative charge. For a given protein, one will therefore have an isoelectric point at which the protein will not migrate if it is placed in an electric field; this pH_i will be low if the protein has an excess of carboxylic groups; it will be high in the case of a protein having an excess of amino groups.

III. – IMPORTANT CHEMICAL PROPERTIES OF AMINO ACIDS

The chemical properties of amino acids are due to the carboxyl group, the amino group and the radical R. We shall mention only the important properties, particularly those referred to later while studying the peptides and proteins.

1. – Reactions due to the presence of the carboxyl group

A. – *Formation of salts*

This property may be used for titrating the amino acids, but due to the weak dissociation of the carboxyl group, high pH values (around 11 or 12) must be reached to obtain total saturation (except when titration is carried out in presence of formaldehyde as will be seen below).

B. – *Decarboxylation*

It can be carried out in the laboratory, but it takes place by enzyme action in living cells. We shall revert to this subject in the chapter on the metabolism of nitrogen-containing compounds (see page 391).

2. – Reaction due to the presence of the amino group

A. – *N-alkylation and N-arylation*

The two H of NH_2 can be replaced by:

– *aliphatic* radicals, the simplest of which is CH_3; N-methyl, N-dimethyl and N-trimethyl derivatives can be obtained (refer to glycine, for example; figure 7-10).

– *aromatic* radicals like the dinitrophenyl (DNP) derivatives used for the determination of the N-terminal amino acid of proteins (see page 20). For example, one can see in figure 1-4, the action of 1 fluoro -2-4-dinitro-benzene on an amino-group.

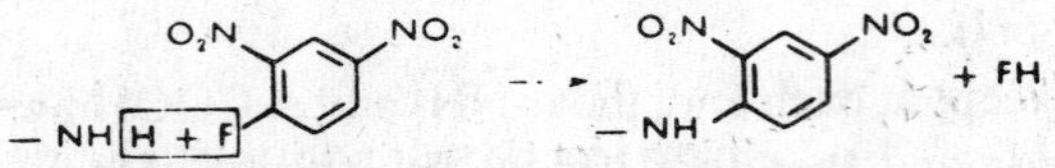

FIG. 1-4. — Action of 1-fluoro-2-4-dinitrobenzene on the amino group of an amino acid.

B. – N-acylation

Formyl, acetyl (easily hydrolyzable) or carbobenzoxyl radicals ($C_6H_5-CH_2-O-CO-$, easily eliminated by hydrogenation) are used in the laboratory to ensure a temporary blockage of the amino group.

In vivo, one knows the N-acetyl and N-benzoyl derivatives of glycine (see figure 7-12) formed during detoxication processes. But special mention must be made of N-formyl-methionine which plays an important part in the initiation of protein biosynthesis in bacteria and organelles (see pages 366-368).

C. – Action of formaldehyde

Formaldehyde (or methanal) reacts with the NH_2 of amino acids at room temperature and neutral pH to form dihydroxymethyl derivatives (figure 1-5) which are now tertiary amines and no longer primary amines; these are weaker bases, which have a lower pK, and one can neutralize the amino acids with a base in presence of phenolphtalein (i.e. without the necessity of reaching pH values as high as in the absence of formaldehyde). This is called *Sorensen's formaldehyde titration*.

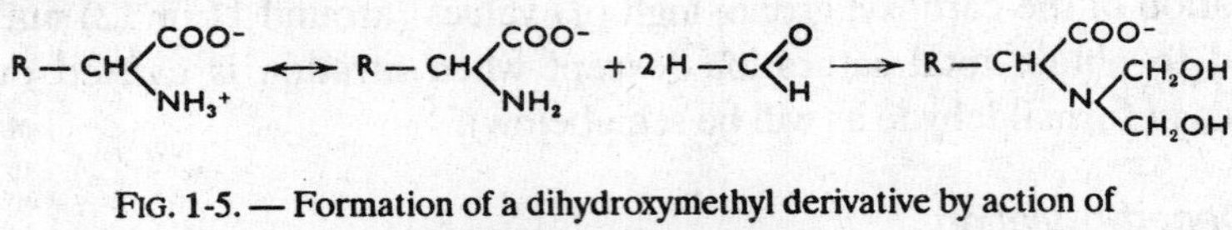

FIG. 1-5. — Formation of a dihydroxymethyl derivative by action of formaldehyde on an amino acid.

D. – Action of nitrous acid

Figure 1-6 shows that nitrous acid reacts with the NH_2 groups, liberating nitrogen which can be titrated. This is the principle of the Van Slyke method for the determination of the free NH_2 groups of amino acids, peptides or proteins. This method enables us to follow step by step the hydrolysis of a protein (because as hydrolysis progresses, the number of free NH_2 groups increases).

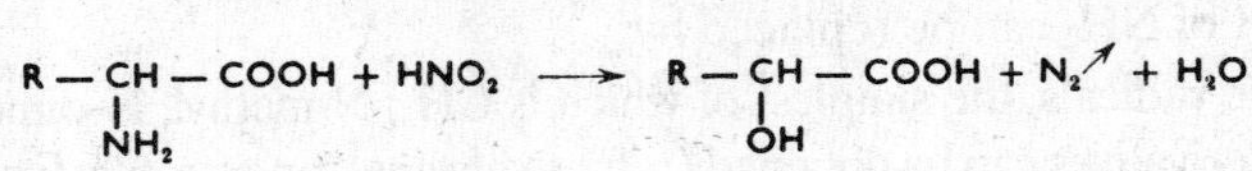

FIG. 1-6. — Formation of a hydroxylated acid and liberation of nitrogen by action of nitrous acid on an amino acid.

E. — Deamination

Deamination and transamination, causing the removal of the amino group, are two very important enzyme reactions which we shall study in the chapter devoted to the metabolism of nitrogen-containing compounds (see pages 392-396).

3. — Reactions necessitating the simultaneous presence of a carboxyl group and an amine in α position

A. Reaction with ninhydrin

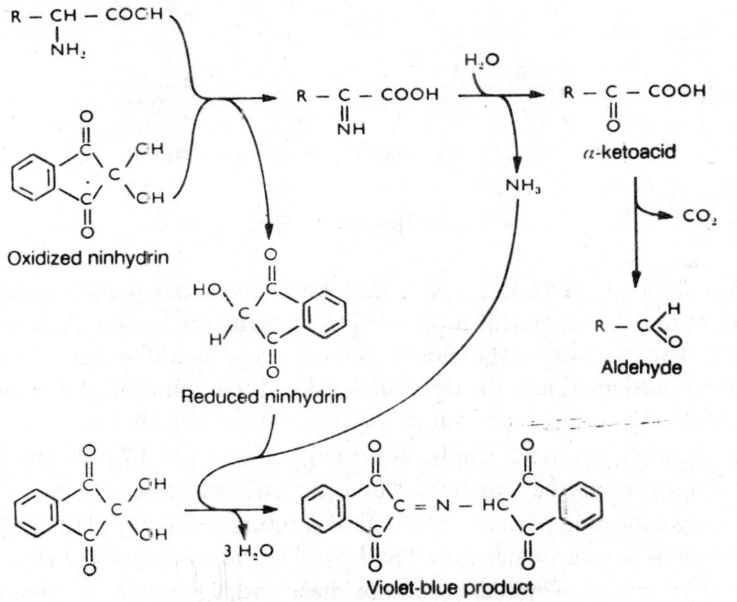

FIG. 1-7. — Reaction of amino acids with ninhydrin.

As can be seen in figure 1-7, this reaction consists of several steps: at first we observe the formation of reduced ninhydrin while the amino acid is transformed into an iminoacid; the latter is hydrolysed into an α-keto-acid which is decarboxylated into an aldehyde having one atom less than the starting amino acid; the amino acid can be determined by identification of the aldehyde obtained. On the other hand, Van Slyke developed a titration of free α-amino acids based on the measurement of CO_2 liberated. Then, NH_3 produced in the first part of the reaction reacts with one molecule of reduced ninhydrin (also formed during the first step) and one molecule of oxidized ninhydrin, to yield a product of violet-blue colour. This coloration may either be used to identify the amino acids after paper or plate chromatography or electrophoresis, or, it may be used in certain conditions for a colorimetric titration of amino acids, for example, at the exit of an ion exchange chromatographic column having fractionated a mixture of amino acids (see page 18).

B. Substituted amide linkage between 2 amino acids

A substituted amide linkage or *peptide bond* (–CO–NH–) can take place between 2 amino acids, the α-COOH of one of them being linked to the α-NH_2 of the other. It will be seen that this linkage joining the amino acids is found in peptides and proteins. If the linkage involves a COOH or a NH_2 in another position (β-COOH, γ-COOH, ε-NH_2) it is said to be a *peptidoid* linkage (for example, the linkage between glutamic acid and cysteine in glutathione, see fig. 1-12). One can obtain in the laboratory diketopiperazines (see fig 1-8) or cyclic dipeptides, where the α-NH_2 of amino acid 1 is involved in a peptide linkage with the α-COOH of amino acid 2 and the α-COOH of amino acid 1 is similarly linked with the α-NH_2 of amino acid 2.

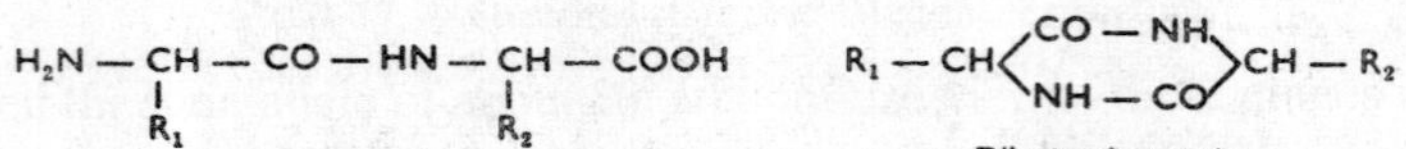

Dipeptide

Diketopiperazin (cyclic dipeptide)

FIG. 1-8. — Peptide linkages between two amino acids.

4. – Properties of the R radicals

Since most α-COOH and α-NH_2 groups are involved in peptide linkages, it is evident that the chemical properties of proteins are mainly due to the R radicals. The reactivity of these radicals is often attenuated by steric hindrance or by their participation in the three-dimensional configuration of proteins, but it is useful to know the important properties of these side chains.

The alcoholic *hydroxyls* can be esterified, for example, by phosphoric acid (phospho-serine) or they can be acylated (the OH of tyrosine also).

The *aromatic* ring permits substitution reactions. Example: the iodinated derivatives of tyrosine which form the thyroid hormones (see fig. 7-28).

The *thiol* groups of cysteine can be oxidized and the cysteine-cystine system ($2R-SH \leftrightarrow R-S-S-R$) can act as hydrogen donor or acceptor. The disulphide bridges are important for the three-dimensional structure of proteins. Besides, oxidation can proceed up to the stage $R-SO_3H$ (sulphonic acid, whose decarboxylation gives taurine).

Furthermore, the hydrogen of the SH group is a mobile H which can be substituted by acyl radicals, thus forming *thio-esters* of formula $R-S-CO-R'$, with high reactivity and of very great importance in biochemistry (example: the S-acylated derivatives of co-enzyme A).

STUDY OF THE PRIMARY STRUCTURE OF PEPTIDES AND PROTEINS

Peptides and proteins consist of chains of amino acids linked by *peptide bonds* (see fig. 1-8). We observe a structure analogy with the *biuret* molecule, formed by the condensation of 2 urea molecules (as indicated by its name) with elimination of one molecule of $NH_3(H_2N-CO-NH-CO-NH_2)$ and this is why polypeptides give the so-called biuret reaction characterized by a violet coloration observed upon addition of SO_4Cu in alkaline solution.

We distinguish, in general:

– the *oligopeptides*: dipeptides (formed by the binding of 2 amino acids), tripeptides (3 amino acids) which do not give the characteristic biuret reaction (some may give an atypical coloration).

– the *polypeptides*: from tetrapeptides onwards, which all give the biuret reaction.

Proteins are polypeptides, but we will see that in addition to peptide bonds linking the amino acids, there are other types of interactions giving the protein a three-dimensional structure. The size limit beyond which a polypeptide is called *protein* is not rigorously fixed; this designation is generally reserved to polypeptides of molecular weight > 10 000 and which do not dialyse through a cellophane membrane.

Having isolated a peptide or a protein, one generally determines first the nature of the constituent amino acids, and then, the order in which they are linked (i.e. the primary structure).

I. – AMINO ACID COMPOSITION

1. – Hydrolysis of proteins

To split the peptide bonds and liberate the amino acids constituting the proteins one may use acids, bases or proteolytic enzymes. As it is very difficult to obtain the complete enzymatic hydrolysis of a protein, we will not discuss here the hydrolysis using enzymes (we will study their specificity on page 92). As for the alkaline hydrolysis, it causes structure modifications and rearrangements of some amino acids. One therefore carries out *acidic hydrolysis* by HCl 6N, at 105°, in a tube sealed under vacuum or nitrogen, with a weak protein concentration (0.5 to 1%), generally for a period of 24 hours. This method has the disadvantage of converting glutamine and asparagine into the corresponding dicarboxylic acids ($+NH_3$) and destroying tryptophan which will have to be titrated separately after hydrolysis by another acid (methane-sulphonic acid)

2. – Amino acid analysis

Historically, it was first tried to use specific methods for each amino acid. These methods were either colorimetric or enzymatic. They were all abandoned due to their lack of specificity or low sensitivity, and more modern techniques of chromatographic separation of amino acids were adopted: these techniques combine excellent reproducibility and very high sensitivity. Besides, they are amenable to complete automation. Three types of chromatography are now proposed:

a) **Ion exchange chromatography.** – like electrophoresis, this method divides amino acids in 3 groups:

– acidic amino acids, retained by the *anion exchanger resins* (Dowex 2, Amberlite IRA-400) which are often *polyamino resins*:

$$\text{Resin}-NH_3^+OH^- + R-COO^- \longrightarrow \text{Resin}-NH_3^+ \ ^-OOC-R + OH^-$$

– basic amino acids, retained by *cation exchangers* which are often *polysulphonated* resins (Dowex 50) or *polycarboxylic* resins (Amberlite IRC 50)

$$\text{Resin}-SO_3^- - Na^+ + R - NH_3^+ \longrightarrow \text{Resin} - SO_3^- \ ^+H_3N - R + Na^+$$

– neutral amino acids which, in principle, are not retained by any of the above 2 categories of resins.

Stein and Moore carried out the complete fractionation of amino acids with a sulphonated polystyrene (SO_3^- Na^+) at a low pH where all the amino acids (even the dicarboxylic) are in the form of cations ($R-NH_3^+$) and therefore extensively retained at the top of the column where they are localized in a thin band even if the volume of the solution (hydrolysate) is large. One performs an elution gradient where the pH and ionic force increase gradually; the amino acids are eluted, according to their degree of ionization, in very distinct peaks which are titrated colorimetrically with ninhydrin (see page 15). The system can be automated and used for qualitative and quantitative analysis of a protein hydrolysate (see fig. 1-9).

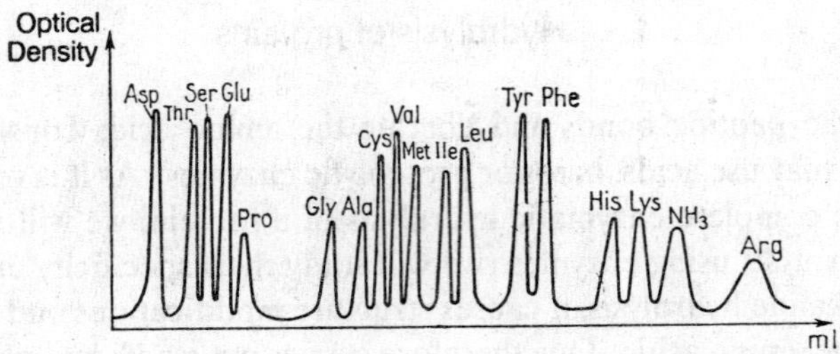

FIG. 1-9. — Profile obtained after chromatographic fractionation of a mixture of amino acids on sulphonated polystyrene column and colorimetric titration with ninhydrin. Elution is carried out using successively the following buffers:

pH 3.25 — Na^+ 0.2 N; pH 4.25 — Na^+ 0.2 N; pH 5.9 — Na^+ 1.4 N.

b) **Gas phase chromatography.** – This is a more recent method of complete separation of amino acids. These are first rendered volatile by chemical modification of their polar groups: methylation of the amine, alcohol or phenol groups, esterification of carboxylic groups. They are then separated by passage through a capillary column (2 m length, 0.1 mm diameter) with an inner coating of a liquid film. This column is heated at temperatures generally ranging between 100 and 200°C. The amino acids are partly soluble in this non-volatile stationary liquid phase and partly vaporized. They are then carried along by a vector gas (mobile phase) at different velocities depending on the respective solubilities in the stationary liquid film. This is therefore a liquid-gas partition chromatography. Different devices, now commercially available, apply this technique and allow the titration of very small quantities (of the order of one picomole).

c) **Adsorption chromatography.** – This is the latest method. It is a very high resolution, liquid-solid type adsorption chromatography called *high performance liquid chromatography* (HPLC). The amino acids are converted into aromatic derivatives by reaction with a chemical reagent, e.g. Edman's reagent (see later). They are adsorbed on a column of micro-granular silica (spherical grains of 5 μm diameter) whose OH groups are substituted by long hydrocarbon chains of 18 carbon atoms. This highly hydrophobic phase retains the amino acid derivatives depending on the hydrophobicity of the radical R. The elution of the column is performed by gradients of water and acetonitril based solvents and titration is carried out by simple spectrophotometry in UV. Quantities of the order of one picomole can be detected.

It must be noted that the commercial devices available for these three methods are now entirely automated thanks to the systematic introduction of microprocessors and micro-computers which control the different operations including the calculation of the quantities titrated and the printing of results (on paper printer or magnetic tape).

3. – Results

Results of the analysis of amino acid content may be expressed in several ways:

– one can calculate percentages by weight (g amino acid/100 g protein), or express the results in number of residues[(1)] for 100 g protein or 100 amino acids,

– one can draw a nitrogen balance, i.e. calculate the quantity of nitrogen from each amino acid and refer it, either to 100 g protein (in this case, the sum must be equal to the nitrogen content in the protein, about 16%), or to 100 g total nitrogen (in this case, the sum must be near 100, if the titrations of different amino acids were carried out correctly),

– one can also – on the basis of weight percentages – calculate the number of residues of each amino acid per molecule of protein when the molecular weight of the latter has been determined (otherwise we can refer the number of residues to an arbitrary weight of 10^4 or 10^5 g). In this manner, one obtains, the gross formula of protein after rounding off the figures to the nearest

(1) The term "residue" denotes the amino acid integrated in a peptide chain and in which — except for terminal amino acids — neither the α-carboxylic group, nor the α-amino group is free.

digit. For example, the gross formula of sheep α-corticotrophine is: Ala 3, Arg 3, Asp 2, Glu 5, Gly 3. His 1, Leu 1, Lys 4, Met 1, Phe 3, Pro 4, Ser 3, Trp 1, Tyr 2, Val 3.

II. – STUDY OF THE SEQUENCE OF AMINO ACIDS

1. – Determination of terminal amino acids

Sequence studies generally begin with the determination of terminal amino acids because this determination is often comparatively easy, and also, because the number of end group gives information on the form of the polypeptide molecule and on the number of polypeptide chains which form the protein molecule: a cyclic chain has no end group; a semi-cyclic chain has one; a linear chain, two; two linear chains, four, etc.

A. – Determination of free terminal α-amino groups

A few methods have been used; only three of them will be presented here: two chemical and one enzymatic methods.

a) **Sanger's dinitrophenyl-amino acid method.** – While studying the chemical properties of amino acids, it was seen (figure 1-4) that the NH_2 group can condense with dinitro-fluorobenzene. If it is the NH_2 of the amino acid at the end of the polypeptide chain called NH_2-terminal end or N-terminal (precisely because of the presence of an amino acid having its amino group free), we obtain a dinitrophenyl-polypeptide. The linkage thus formed is more stable than the peptide linkage, so that after hydrolysis by HCl we obtain all the free amino acids and the N-terminal amino acid in the form of dinitrophenyl-amino acid (DNP amino acid), easily identifiable by paper chromatography.

One can also react 1-*d*imethyl-*a*mino-*n*aphthalene-5 *s*ulphonyl chloride with the NH_2-terminal to obtain – after hydrolysis – a *dansyl amino acid* detectable by its yellow fluorescence. The principle is the same, but this method is 100 times more sensitive than the previous one.

b) **Edman's phenylthiohydantoin method.** – The principle of this method is illustrated in figure 1-10. There is formation of phenylthiohydantoin of the N-terminal amino acid which can be separated and identified by chromatography. But contrary to the previous method where a total hydrolysis of the polypeptide chain is necessary, the chain here is simply severed of its N-terminal residue because in acid medium, the first peptide linkage becomes more fragile due to the substitution; it splits, liberating a phenylthiohydantoin which can be identified by chromatography, and it is the amino acid number 2 which is now in N-terminal position; another molecule of phenylthioisocyanate is added; it will condense on this amino acid and permit its detachment and identification, and so on: this is called a *recurring degradation.* This method can therefore allow the determination of the sequence of several amino acids

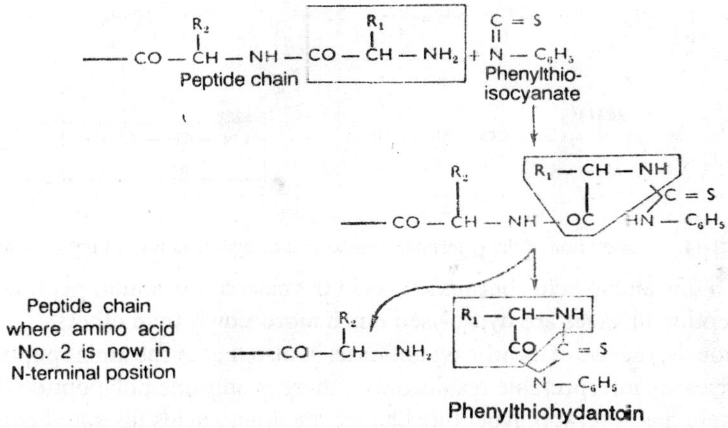

FIG. 1-10. — Reactions in Edman's method of phenylthiohydantoins.

starting from the N-terminal end; the method has even been automated in a device named "Sequenator".

c) **Enzymatic degradation by aminopeptidase.** – This enzyme hydrolyses the peptide linkage where the N-terminal amino acid is involved. After detaching the first amino acid it will liberate the second, which now carries the free NH_2, and so on; one can therefore deduce the sequence of amino acids at this end from the rate of liberation of amino acids. It was thought for some time that the enzyme was specific for leucine (hence its name: leucine-aminopeptidase), but it is not so; it also detaches other amino acids when they are in N-terminal position; however, in the case of some amino acids, its action is very slow which explains why this method does not represent a general approach to study sequences at the N-terminal ends.

B. – Determination of free terminal α-carboxylic groups

We shall present a chemical and an enzymatic method.

a) **Hydrazinolysis.** – When a protein is treated with *hydrazine* at 100°C, all the peptide linkages are split, and all the residues are converted into hydrazides, except the COOH-terminal (or C-terminal) residue which remains in the form of free amino acid, easily isolated and identified.

b) **Enzymatic degradation by a carboxypeptidase.** – One of the best methods of determination of the C-terminal amino acid consists in reacting a *carboxypeptidase*, a pancreatic enzyme which hydrolyses the peptide linkage involving the C-terminal amino acid. Once it is detached, the second amino acid which now carries the free COOH is freed, and so on. Here again, we have in principle, a method of *recurring degradation,* but actually one can rarely go

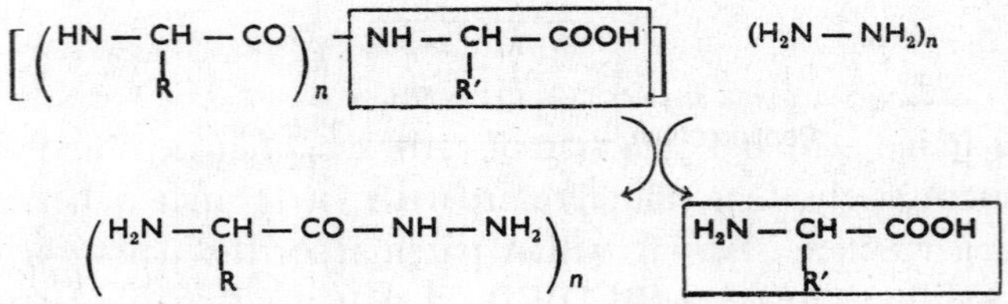

FIG. 1-11. — Liberation of the C-terminal amino acid by hydrazinolysis of the protein.

beyond a few amino acids because – as in the case of the amino peptidase – some peptide linkages are hydrolysed much more slowly than others.

Obviously, the use of a carboxypeptidase, as also that of the aminopeptidase, can give easily interpretable results only if there is only one polypeptide chain; if a protein has several polypeptide chains, the amino acids liberated could as well originate from the gradual hydrolysis of a single chain as from the simultaneous attack of several chains.

2. – Problem of the number of peptide chains

If the analysis of the terminal groups shows the presence of several chains, we may have two cases:

– the chains are identical (there is only one N-terminal amino acid and only one C-terminal amino acid). The determination of the sequence is then performed on the entire protein.

– the chains are different (there is more than one N-terminal amino acid or more than one C-terminal amino acid). In this case we must first separate the chains from one another and study the sequence of each of them.

This separation first requires a dissociation of the protein, i.e. a break of the inter-chain linkages. If these linkages are non-covalent (ionic, hydrogen, Van der Waals) they are split by a change of pH or ionic force, or by the action of certain agents like urea, guanidine or detergents. If these linkages are covalent (disulphide bridges between two residues of cysteine), they are split chemically, for example, by oxidation ($R-S-S-R' \longrightarrow R-SO_3H + R' - SO_3H$, see page 16).

As for the separation of dissociated chains, it uses the general methods of fractionation of mixtures of proteins (see page 42).

3. – Determination of the sequence of amino acids

We have just seen that certain methods permit a recurring degradation of polypeptide chains, but till now this type of approach has not taken us very far in the chain. In order to determine the complete sequence of a chain we must carry out partial hydrolyses, determine the sequence of the small peptides thus obtained and then – as in a Chinese puzzle – reconstitute the complete sequence by cross-check operations.

Let us take the example of a hexapeptide whose gross formula AB_2C_2D has been determined. Sanger's method of DNP-amino acids indicates that A is the N-terminal amino acid and by action of carboxypeptidase, B and then C are detached successively. One can therefore write the formula of hexapeptide in the following manner, showing within parentheses the amino acids whose sequence is still unknown: A(BCD)CB. A number of *partial hydrolyses* are then carried out, either *chemical* (by decreasing acid concentration, temperature or heating time), or *enzymatic* (using proteolytic enzymes like pepsine, trypsine, chymotrypsine, which – as will be seen later – hydrolyse preferentially certain peptide linkages, i.e. peptide linkages in which certain amino acids are involved). One thus obtains dipeptides and tripeptides which are fractionated by different techniques (electrophoresis, ion exchange chromatography, counter-current distribution etc.); their sequence is very easily determined; as can be seen below, from these dipeptides and tripeptides one can easily deduce by comparison, the sequence of the hexapeptide:

$$\text{dipeptides obtained}\quad \left\{\begin{array}{llllll} A-C & & & & & \\ & C-D & & & & \\ & & D-B & & & \\ & & & B-C & & \\ & & & & C-B & \end{array}\right.$$

$$\text{tripeptides obtained}\quad \left\{\begin{array}{l} C-D-B \\ \quad D-B-C \end{array}\right.$$

$$\text{sequence of the hexapeptide}\quad A-C-D-B-C-B$$

4. – Results

Since Sanger's determination of the complete sequence of *insulin* – the first polypeptide whose exact structure was established – numerous complete sequences were determined. This relates to other *hormones* of polypeptide nature (oxytocin, vasopressin, adrenocorticotropic hormone, glucagon, etc), *enzymes* (lysozyme, ribonuclease, etc.) *globins* or protein fractions of various hemoglobins and myoglobins (which are heteroproteins whose prosthetic group is heme), *electron transporters* like cytochrome *c* (also a heteroprotein), *coat proteins* of several phages and viruses, etc. Based on these studies, one may draw some general conclusions:

Amino acids can bind in any order, there is no incompatibility, so that a very large number of sequences is possible for a polypeptide chain of known amino acid content: for a tripeptide consisting of three different amino acids A, B, C, there are already as many as 6 possible sequences: ABC, ACB, BAC, BCA, CAB, CBA. We will see that *the sequence of amino acids in proteins is determined genetically.*

One always finds — except accidentally (mutation) — the same sequence of amino acids in the different molecules of the same protein originating from a given species; this suggests that during the biosynthesis of proteins, there is a control of the sequence of amino acids and that this control is hereditary. We will see in chapter 6, how this control is carried out.

On the contrary, some differences appear when one compares the sequences of similar proteins extracted from neighbouring or distant species (for example, while comparing the cytochromes *c* or the globins of different species), although these proteins have the same biological activity. This suggests that some amino acids are indispensable to the physiological role of the protein, while some others are not and can therefore be replaced. We will also see in chapter 6 that the substitution of a single amino acid by another, is sufficient in certain cases, to seriously impair the biological activity of the protein (see page 304). Resemblances in the primary structure of similar proteins originating from different species are not due to chance but are generally the reflection of a single ancestral structure; evolution indeed takes place through mutations, i.e. changes in the genetic heritage and the main effect of these mutations is a modification of the structure of proteins (see Chapter 6). The study of protein sequences is therefore of phylogenic interest because it offers the possibility of following the filiation of species in the course of evolution.

PEPTIDES

1. — CLASSIFICATION

As already mentioned *oligopeptides* and *polypeptides* are distinguished depending on whether the peptide is formed by the union of less than 4 or at least 4 amino acids. Natural peptides consist of different amino acids; but one can synthesize *homopeptides* (triglycine, polyphenylalanine etc.). Peptides can have different structures:

— *linear peptides* with a free NH_2 at one end and a free COOH at the other.

— *branched peptides* where the branching of one or several amino acids on a linear peptide chain takes place either by the ω-carboxylic group of a dicarboxylic amino acid, or by the ε-amino group of lysine.

— *cyclic peptides* which have neither N-terminal nor C-terminal end.

— *semi-cyclic peptides* having only one end; when there is only one N-terminal end, the α-COOH of the last amino acid is linked with the ε-NH_2 of an endopeptide lysine; inversely, when there is only one C-terminal end, the α-NH_2 of the first amino acid is linked with the ω-COOH of an endopeptide dicarboxylic amino acid.

II. – NOMENCLATURE

An amino acid incorporated in a peptide chain loses one H (of its NH_2) and one OH (of its COOH), or only one of the two if it is a terminal amino acid. This is called an amino acid *"residue"*; it is designated by adding the suffix "yl" to the root of the name (example: glycyl, seryl, tyrosyl etc.).

The amino acid of the N-terminal end is indicated first and then follow the others in their order of succession, all with the suffix "yl"; only the amino acid of the C-terminal end is designated by its unchanged name. Example: alanyl-valyl-phenylalanyl-isoleucine. One may also use abbreviations and write: Ala-Val-Phe-Ile.

III. – PREPARATION AND PURIFICATION

Peptides may be separated by different techniques: precipitation at the isoelectric point (minimal solubility), selective extraction by some solvents, etc.

Various methods are used for purification: crystallisation, counter-current distribution, adsorption chromatography, partition chromatography, ion exchange chromatography, electrophoresis, etc.

IV. – STUDY OF SOME BIOLOGICALLY IMPORTANT PEPTIDES

1. – Glutathione

This is a tripeptide (γ-L-glutamyl-L-cysteinyl-glycine); its structure is shown in figure 1-12.

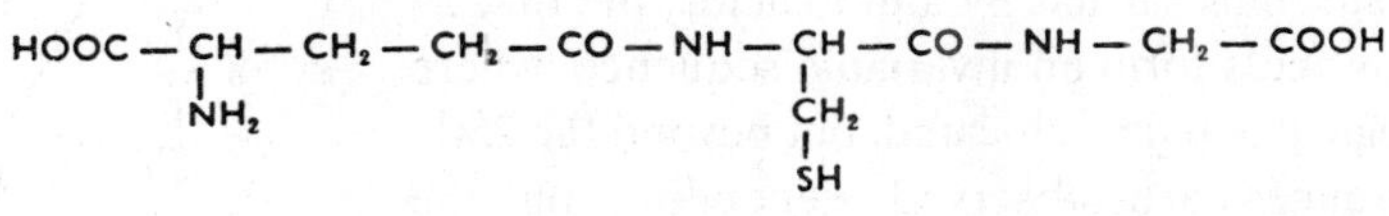

FIG. 1-12. — Structure of reduced glutathione.

It is seen that the glutamic acid is linked with cysteine by its γ-COOH (peptidoid linkage) and its α-COOH and α-NH_2 are therefore free.

Glutathione exists in reduced form (thiol) and oxidized form (where 2 molecules are linked by a disulphide bridge, as in cystine) and can thus play a role in some oxidation-reduction reactions.

2. – Peptide hormones

Certain hormones, especially those secreted by the hypophysis and the pancreas, are peptides. A few examples are given in the following:

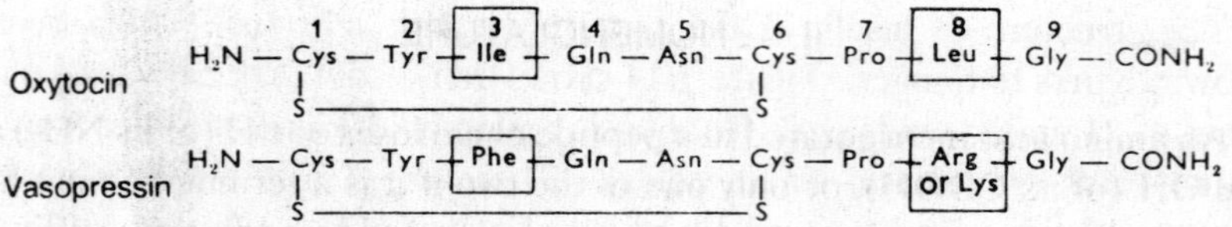

FIG. 1-13. — Structure of oxytocin and vasopressin.

A. – Pituitary hormones: oxytocin and vasopressin

These two peptides have a cycle due to a disulphide bridge, and a side chain. Figure 1-13 shows that their structures are very similar. The structure of vasopressin differs from that of oxytocin only by the substitution of Ile by Phe, and of Leu by a basic amino acid (Lys or Arg depending on the animal species). It is interesting to note that such minor structural differences cause great differences in biological activity; oxytocin stimulates the contraction of the uterine muscle whereas vasopressin increases blood pressure and has an antidiuretic action. It should be noted that in both cases there is no free terminal COOH, because the carboxyl group of the C-terminal glycine is amidified.

B. – Adrenocorticotropic hormone (ACTH)

It is a hormone of the ante-hypophysis, whose main role is to stimulate the synthesis and secretion of steroid hormones by the cortical part of the adrenals. It has 39 amino acids; the first 24 amino acids form an invariable sequence where biological activity is located, but beyond the 25th, differences are observed depending on the animal species.

C. – Insulin

This hormone, causing a decrease in the glucose concentration of the blood, is secreted by the pancreas. Its molecular weight is about 6 000 but dimers (M.W. = 12 000) are easily formed and can further combine to form polymers of molecular weight of 24 000 or 48 000. The distinction between peptides and proteins – with an arbitrary limit at 10 000 – is obviously very artificial (see page 17).

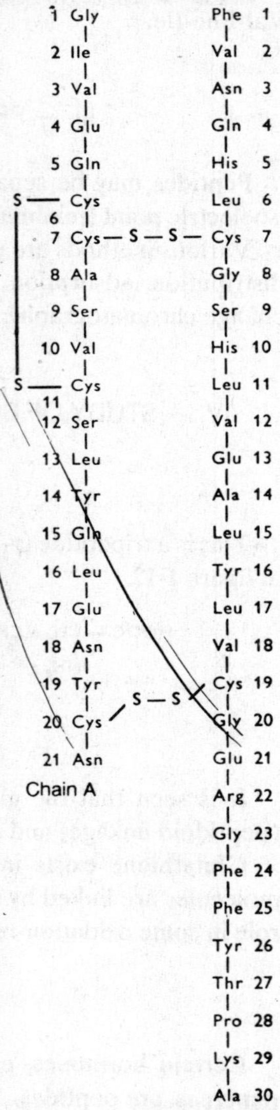

FIG. 1-14. — Structure of insulin.

The structure of insulin is the first polypeptide structure which became known thanks to Sanger. Figure 1-14 shows that insulin has 2 chains: one A chain of 21 amino acids (free from basic amino acids) and one B chain of 30 amino acids (with basic amino acids). There are 3 disulphide bridges: 2 interchain bridges and 1 intrachain bridge. The nature of amino acids A 8, A 9, and A 10, in privileged position in the region of the intrachain bridge, varies with the animal species.

3. – Peptides with antibiotic activity

A number of antibiotics produced by bacteria or other micro-organisms are of peptide nature. In the following, we are mentioning only a few of them as examples:

– *Penicillin*: valine and cysteine are found in penicillin (figure 1-15) but there are linkages other than the peptide linkage. The NH_2 of cysteine is acylated, in most cases by the benzyl radical.

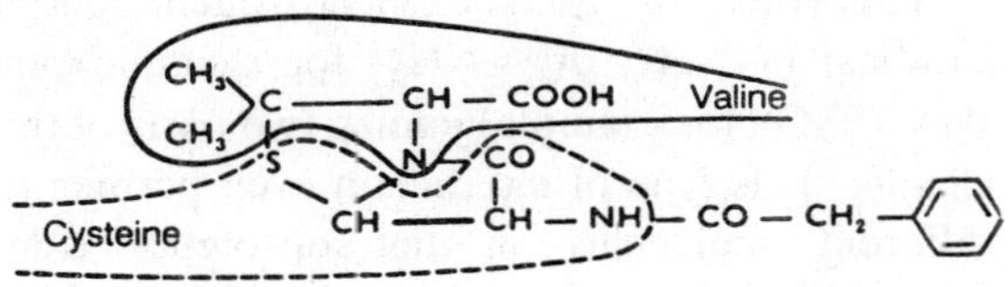

FIG. 1-15. — Structure of benzyl-penicillin.

– *tyrocidins* are cyclo-decapeptides. Tyrocidin B differs from tyrocidin A, merely by the replacement of L-Phe by L-Trp. Both have (figure 1-16):

– a residue of L-ornithine which is not an amino acid found in proteins.

– 2 residues of D-phenylalanine, an optical isomer never present in proteins.

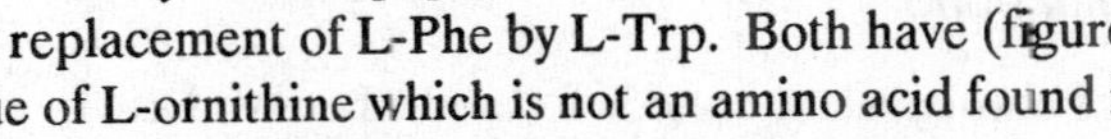

FIG. 1-16. — Structure of tyrocidin A.

Other antibiotics are also of peptide nature, or at least include a peptide part (gramicidins, polymyxins, bacitracins, actinomycins, etc).

PROTEINS

I. – THREE DIMENSIONAL CONFORMATION OF PROTEINS

Proteins consist of long polypeptide chains and we know how to determine the sequence of the amino acids of these chains. This constitutes the primary structure of proteins. But in reality, each protein has a three-dimensional structure of its own, established and maintained by types of linkages other than the peptide linkage. It is said that the "*native*" protein has a secondary, tertiary and even quaternary structure in certain cases, but before considering these different structures, one must understand the interactions which make them possible.

1. Bonds involved in the spatial structure of proteins

A. – Disulphide bond (or disulphide bridge)

We have already mentioned this covalent bond (strong bonding) between two residues of cysteine belonging to either the same peptide chain or two different chains (see formula of insulin, fig. 1-14).

B. – Ionic bond (or saline bond)

It is a non-covalent (therefore weaker), electrovalent bond between a positively charged radical ($-NH_3^+$, or $=NH_2^+$, for example) and a negatively charged radical ($-COO^-$, for example) joining two parts of the same chain or two different chains. This type of interaction even permits the bonding between two different molecules in heteroproteins (for example, in nucleoproteins, between the negatively charged nucleic acid and the positively charged basic proteins, especially the histones).

C. – Hydrogen bond

This type of non-covalent bond is formed when we have in close proximity, on the one hand, one hydrogen atom bonded to a nitrogen atom or to an oxygen atom and on the other hand, the unshared electronic doublet of another nitrogen atom or oxygen atom. These bonds can form:

– between the $>C=O$ and the $>N-H$ of peptide linkages;

– between the radicals of amino acid residues and involve for example, the phenolic group of tyrosine, the imidazole nucleus of histidine, the γ-carboxyl group of glutamic acid, the amide group of asparagine or glutamine.

We will see that the hydrogen bonds can either be intrachain and maintain the helicoidal conformation of the chain, for example, or can produce interchain interactions.

D. – Hydrophobic bond

Certain amino acids have a non-polar hydrophobic side chain (Ala, Val, Leu, Ile, Phe) which does not form any hydrogen bond with water molecules (which associate through hydrogen bonds): these side chains thus repelled tend to get closer to each other, leading to interactions between different parts of a peptide chain (these interactions are of the Van der Waals type).

2. – Secondary structure of proteins

X-ray diffraction patterns provided the data for this structural study. There are generally two main types of secondary structure: the stretched state and the helicoidal state.

Before describing these two states we will review the spatial properties of the peptide linkage.

A. – *Spatial properties of the peptide bond*

The peptide linkage can in fact, be written in two ways (mesomeric forms):

It has therefore a double bond character which implies that all atoms (C_α, C, O, N, H and $C_{\alpha'}$) are coplanar. Besides, there is a possibility of *cis-trans* isomerism for the two carbon atoms C_α and $C_{\alpha'}$ with respect to this $>C=\bar{N}<$ linkage. In natural peptides and proteins one mainly finds the plane *trans* configuration.

These properties give three possibilities of spatial arrangements for the planes formed by successive peptide linkages of a peptide chain:

– the successive planes alternate in two privileged spatial orientations forming a pleated sheet structure;

– the successive planes rotate regularly, always in the same direction, giving a structure of the helicoidal type;

– lastly, the planes can rotate at random. The resulting structure is said to be random coil.

The tetrahedral C_α atoms are always situated at the hinge of two consecutive planes.

B. – *Stretched state or pleated sheet structure*

The *fibrous proteins*, or scleroproteins, often found as structural proteins (β-keratin, silk fibroin) have this type of conformation, represented in figure 1-17. We see two antiparallel polypeptide chains ("moving" in opposite directions) united by *interchain hydrogen bonds.* The atoms of the peptide linkage are situated in the same plane, but the α carbon atoms belong simultaneously to two different planes. Periodicity is of the order of 7Å (7×10^{-7} mm).

C. – *Helicoidal state or α-helix*

The α-helix is shown in figure 1-18. It is seen that the peptide chain is maintained in this helicoidal configuration by *intrachain hydrogen bonds.* The helix has 3.7 amino acid residues per turn. The planes of peptide linkages form

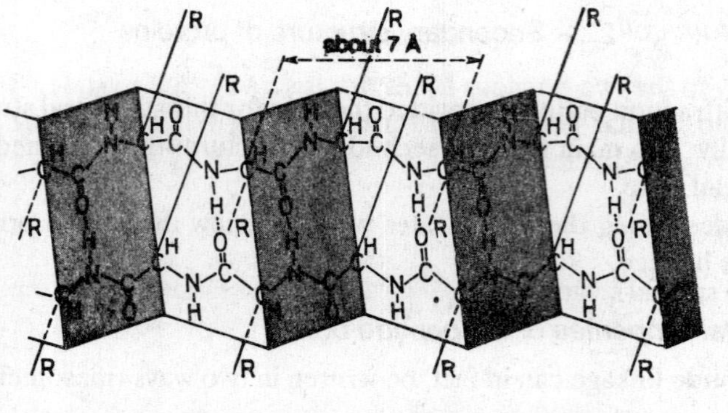

FIG. 1-17. — Stretched state or pleated sheet structure.

between them an angle of about 80° (the intersection of the 2 planes is always at the α carbon). The side chains are directed outwards and can react with each other or with the medium.

If a fiber of α-keratin of helicoidal structure is stretched, β-keratin in pleated sheets is obtained.

Several α-helices can coil round one another like a twisted, twined cable, forming fairly voluminous and strong fibers due to numerous disulphide and hydrogen bonds.

The figure shows a left-handed helix, but very often, natural proteins contain right-handed helices.

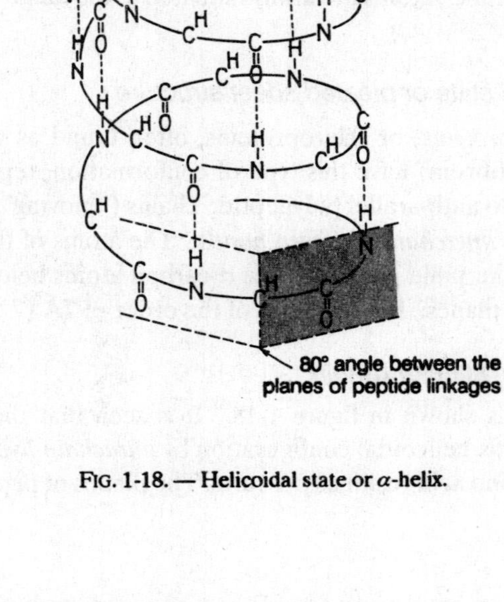

FIG. 1-18. — Helicoidal state or α-helix.

D. – Random coil

Contrary to the two previous states this is a non-ordered state. The peptide chain has no regular geometrical form in space, but the valency angles are however always respected.

E. – β turn

This is a special structure involving 4 residues of amino acids represented in the diagram below: the α-carbon atoms are numbered 1, 2, 3 and 4. A hydrogen bond is formed between the CO group of the first residue and the NH group of the fourth. In these conditions the sequence portions preceding and following this β turn are antiparallel.

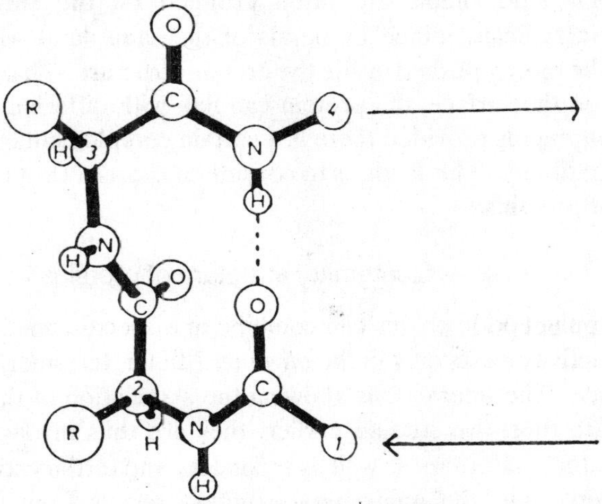

3. – Tertiary structure of proteins

Globular proteins or *spheroproteins* which form the majority of biologically active proteins (especially the enzymes) are, as indicated by their name, more compact than those mentioned above. They may consist of several chains; they may also consist of a single polypeptide chain having, at some places, zones in α-helix and at some others, zones of less regular curvatures and folds. The percentage of α-helix varies considerably among the proteins and depends on their primary structure: certain amino acids, especially proline, have a structure which breaks the regularity of the α-helix.

The study of high resolution (of the order of 2Å) X-ray diffraction patterns has revealed the tertiary structure of certain proteins. The first tertiary structures were determined by Kendrew and Perutz for myoglobin and hemoglobin. Figure 1-19 represents the tertiary structure of globin (protein part of myoglobin).

The tertiary structure is of capital importance for the biological activity of proteins. Residues of amino acids very distant from one another in the sequence, can be brought very near due to the folding and thus form regions

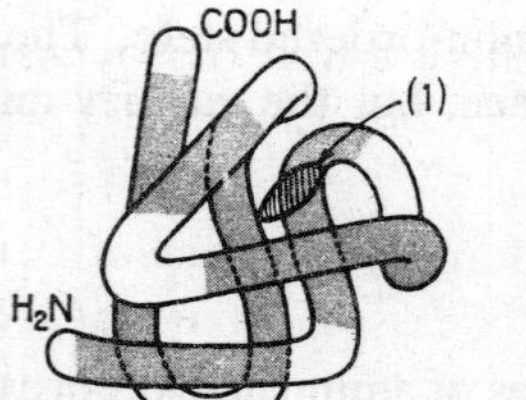

FIG. 1-19. — Tertiary structure of globin in myoglobin. The helicoidal regions are represented in grey.

(1) Prosthetic group of myoglobin = heme (tetrapyrrole ring containing one iron atom at the centre; see pages 50-52)

indispensable for the functioning of the protein, for example the *active site* (or catalytic site) of enzymes.

The polar side chains are often grouped on the surface, while the hydrophobic radicals, joined by bonds of the same kind which we studied above, can be mostly pushed inside the protein structure. Thanks to the groups accessible on the surface, the protein can link with different protein or non-protein compounds provided there is a certain complementarity, i.e. a certain steric compatibility. This leads us to consider a last aspect of complexity in the structure of proteins.

4. – Quaternary structure of proteins

Several polypeptide chains can combine in a specific manner. Very often, biological activity exists only in the *oligomer* (dimer, tetramer, etc.) and not in the *protomer*. The interactions allowing the association of these polypeptide sub-units are those we studied earlier; they are thus involved in the establishment of the quaternary as well as secondary and tertiary structures.

In certain cases, the quaternary structure results from the assembly of *identical* peptide units, and in other cases from the assembly of *different* sub-units. For example, hemoglobin which has to transport oxygen from the lungs to different tissues, consists of 4 sub-units, generally 2 α-chains and 2 β-chains; each of these chains has a tertiary structure similar to the one shown in figure 1-19. There are also many enzymes which consist of sub-units; in particular, we will see that enzymes whose activity is subjected to an allosteric regulation, consist of sub-units with a very different structure and role (see page 475).

Lastly, the notion of quaternary structure may be extended to hetero-proteins, formed by the association of protein and non-protein sub-units. Hence, a viral particle in which a nucleic acid and a protein are linked, may be considered as constituting a heterogeneous quaternary structure.

II. – DENATURATION OF PROTEINS

The three-dimensional conformation (the primary, secondary, tertiary and even in some cases, quaternary structure) is characteristic of a *native* protein.

This conformation can be upset and disorganized, without breakage of any peptide linkage, only by the rupture of linkages which enabled the structure to maintain its conformation in space; this is called *denaturation*. It can be triggered by diverse physical or chemical agents:

– heat (coagulation of ovalbumin of egg white is a well known example of denaturation),

– ultraviolet and ionizing radiations,

– pH variations. Certain acids (nitric, trichloroacetic, perchloric etc.) and certain acid solutions of heavy metals (Hg, Pb) are used in biological analysis to eliminate the proteins of the medium (blood for example); they are *deproteinizing agents*,

– detergents,

– organic solvents (except at a temperature less than 0°),

– urea or guanidine solutions,

– simple dilution or simple agitation can also cause the denaturation of proteins; this causes great inconvenience when purifying a protein.

Denaturation is sometimes *irreversible* (for example, in the coagulation of ovalbumin) and sometimes *reversible*. This process needs to be studied; we have here the passage from a highly ordered state to a less ordered state; there is an increase of *entropy* (entropy is a measure of the probability of existence of a state, and the least ordered states are the most probable). In the native state, the protein has the most stable conformation – from the energetic point of view – in intra-cellular conditions; if these conditions are changed, the conformation of the protein will be altered; there will be rupture of the hydrogen bonds and other secondary bonds studied earlier, disorganization of secondary and tertiary structures resulting in increased sensitivity to proteolytic enzymes, increase of reactivity of certain groups which were either involved in secondary bonds or inaccessible and, above all, a loss of biological properties (especially in the case of enzymes).

III. – DETERMINISM OF THE THREE-DIMENSIONAL CONFORMATION

The study of certain cases of reversible denaturation suggests that conformation is determined by the sequence of amino acids. We will take the example of *ribonuclease*. It is an enzyme. Its sequence is known (124 amino acids). Its three-dimensional conformation is maintained especially by 4 disulphide bridges resulting from the association in pairs of 8 cysteine residues of the chain. If the enzyme is reduced, the four S-S bridges are broken (generating 8 thiol groups), the conformation collapses and we have a denatured, inactive enzyme. But this denaturation is reversible, the four S-S bridges can form again by oxidation, the ribonuclease resumes a three-dimensional conformation and we have again an active enzyme. But the remarkable fact is that the same SH groups interact to form again the same four S-S bridges, although there is a large number of possible combinations (28). It therefore appears that the sequence of amino acids contains the information required to impose a given spatial structure in defined intra-cellular conditions. We will see however, that small molecules (substrates, activators, inhibitors) are capable of producing slight modifications of the three-dimensional conformation in the case of allosteric enzymes (see pages 472-477).

IV. – IMPORTANT PROPERTIES OF PROTEINS

1. – Solubility

Water solubility of proteins is so variable that it was proposed to classify them according to their solubility (see page 44). Certain proteins are soluble in pure water (albumins), others – especially the globulins – dissolve only in presence of neutral salts or when the medium is slightly acidic or weakly alkaline, and certain others (scleroproteins) are insoluble even in these diverse conditions.

A. – Effect of electrolytes

The action of neutral salts depends on their concentration and the charge of ions, i.e. the *ionic* strength[(1)]. When ionic strength is low a *dissolving effect* is observed. On the contrary when the ionic strength is high, *salting out* takes place, i.e. the precipitation of protein molecules. These two phases are distinctly observed for example, while studying the solubility variation of hemoglobin when ionic strength is increased by addition of ammonium sulphate (see fig. 1-20).

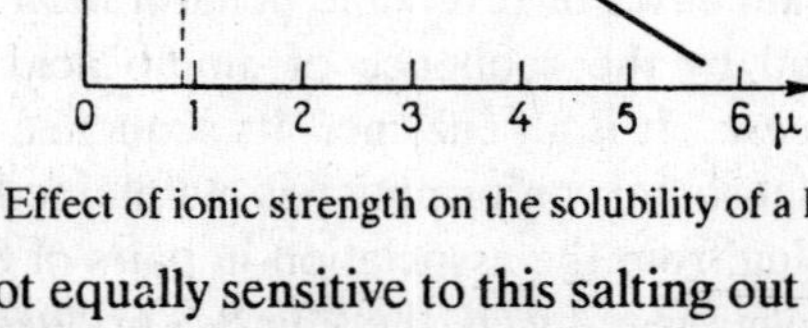

FIG. 1-20. — Effect of ionic strength on the solubility of a hemoglobin.

All proteins are not equally sensitive to this salting out effect by addition of a neutral salt; thus, certain proteins precipitate in a solution half saturated with $(NH_4)_2\ SO_4$, others only when the solution is 55% or 60% saturated etc. By

---

(1) By definition the molar ionic strength (μ) is $1/2\ \Sigma \times Z^2$ (C is the molar concentration of each ion and Z its valency). Thus, for a 0.1 M solution of a salt consisting of 2 monovalent ions (e.g., NaCl), $\mu = 1/2\ [(0.1 \times 1) + (0.1 \times 1)]$ i.e. 0.1 and for a 0.1 M solution of $(NH_4)_2SO_4$ (2 monovalent ions and one divalent ion) $\mu = 1/2\ [2(0.1 \times 1) + (0.1 \times 4)]$ i.e. 0.3.

gradual addition of a neutral salt (and centrifuging after each addition) it is therefore possible to fractionate a mixture of proteins, if not completely, at least partially.

B. – Effect of pH

Several methods of fractionation of proteins are based on the fact that under constant ionic strength and constant temperature, *the solubility of a protein is minimal near the isoelectric pH* (see fig. 1-21). In certain cases, crystallization of a protein may be achieved by increasing salt concentration in a solution maintained at a pH near the isoelectric point.

For the two values of ionic strength used, solubility is minimal near the isoelectric pH, but it is observed that whatever the pH, solubility is greater when μ = 0.01. This is therefore the range of dissolving effect as far as ionic strength is concerned. To obtain the precipitation of the protein, it is advantageous to act on both factors.

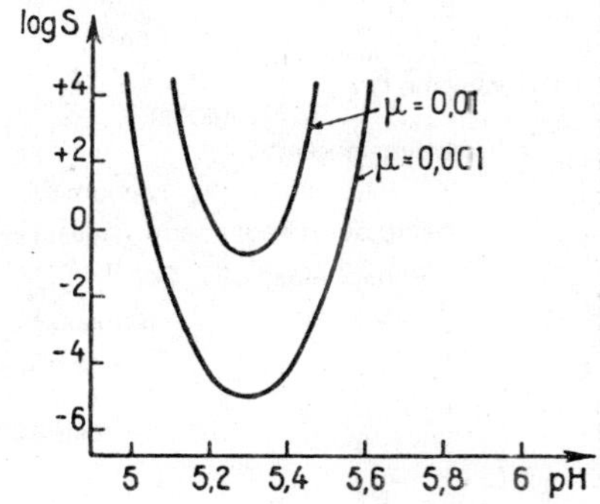

FIG. 1-21. — Effect of pH on the solubility of a globulin.

C. – Effect of organic solvents

Ethanol, methanol, acetone, can be used to precipitate the proteins, provided temperature is maintained around –5° C to avoid their denaturation by these solvents. A fractionation of plasma proteins can be achieved by increasing especially alcohol concentration.

2. – Molecular weight

Cryoscopy is not applicable because the lowering of the freezing point is too small to be measured accurately. The measurement of *osmotic pressure* is not very convenient, but it has been used to determine the molecular weight of some proteins through the relation:

Π (osmotic pressure) $= RT\dfrac{C}{M}$ (C is the concentration by weight and M, the molecular weight). Three methods are commonly used:

A. – Filtration through dextran gel

The dextran gels used (Sephadex type) are polyacids containing a varying number of cross-linkages which produce a certain degree of porosity; macromolecules penetrate these gels with varying ease depending on their size, hence the name "molecular sieve" given sometimes to these dextran gel columns: the largest molecules excluded from the gel will be the first to come out of the column, whereas the smallest molecules which can penetrate into the gel, will be delayed and will come out last. These dextran gel columns can be "calibrated" by passing proteins of known molecular weight; a protein of unknown molecular weight is then passed, and its molecular weight determined on the basis of the time it takes to come out of the column (see fig. 1-22). Columns of acrylamide gel (Biogel) can be used for the same purpose.

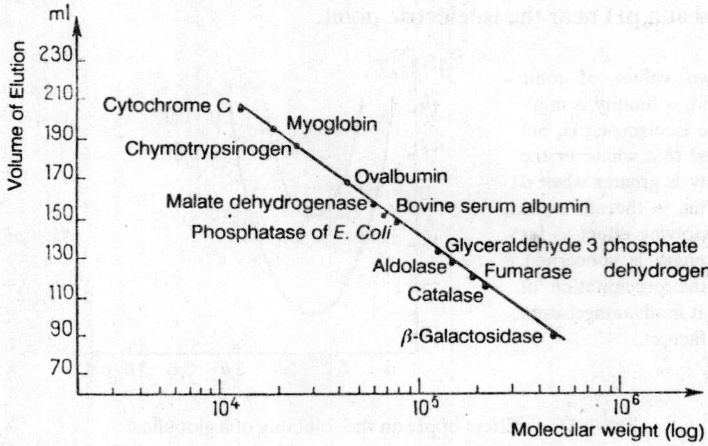

FIG. 1-22. — Elution of proteins from a Sephadex G.200 column at pH 7.5, as a function of their molecular weight.

B. – *Light scattering*

When a beam of light passes through a protein solution a part of the light is scattered (Tyndall effect); the proportion of scattered light increases (while the proportion of transmitted light obviously decreases) when the number and size of molecule in solution increase. The molecular weight of a protein can therefore be calculated from the ratio: incident light/transmitted light.

C. – *Ultracentrifugation*

We owe to Svedberg this high speed centrifugation method, carried out in devices rotating at 60 000 revolutions/minute, which corresponds to a gravitation field of 500 000 × g or even more (depending on the geometry of the rotor used). *The protein molecules in solution,* subjected to this ultracentrifugation, *sediment according to their density* (which is greater than that of the solvent) and concentration variations resulting from the sedimentation are observed by measuring the refractive index of the solution during the centrifugation and without interfering with the centrifugation operation (measurements are made with the help of an optical recording device).

a) **Sedimentation velocity.** – When a homogeneous, monodisperse solution (containing only one protein) is subjected to ultracentrifugation, all molecules migrate with the same velocity and one observes only one boundary which moves in course of time and corresponds to the zone where protein concentration passes from 0 (pure solvent) to a certain value C. (see fig. 1-23 B). For a given angular velocity (ω), C is a function of the distance between the molecule and the rotation axis of the rotor (r). In fact, what one follows through the optical system, is the displacement of a peak corresponding to the variations of dc/dr (see fig. 1-23 C). The displacement velocity of this peak, or sedimentation velocity, is r_2-r_1/t_2-t_1, or, at the limit dr/dt; it depends therefore on the gravitation field. This determines a *sedimentation constant*, characteristic of a given protein, $S = (dr/dt)\omega^2 r$, referred to defined conditions (sedimentation at 20°, in water) and expressed in Svedberg Units (S). One Svedberg unit is equal to 10^{-13} cm/s/field unit. For most proteins S is between 1 and 200.

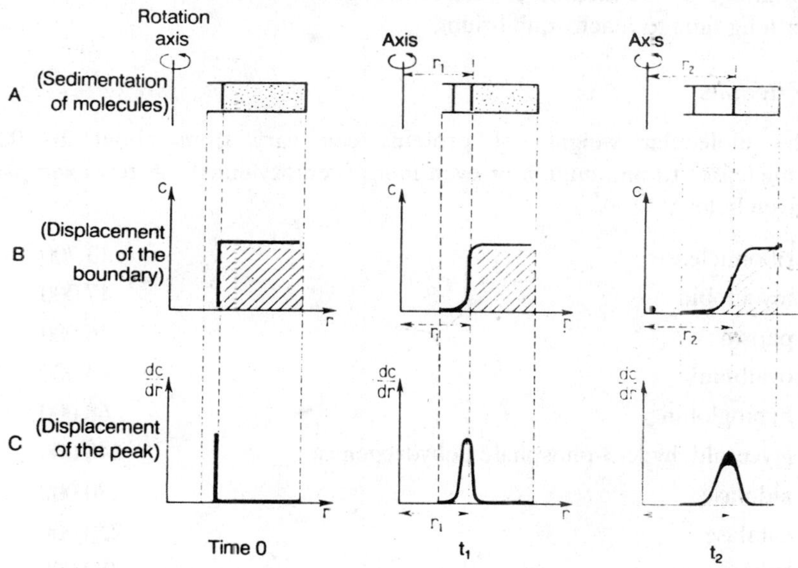

FIG. 1-23. — Ultracentrifugation of a solution containing a pure protein.

From S, we can estimate the molecular weight, with Svedberg's equation:

$M = \dfrac{S \times f \times N}{1-\overline{V}_\rho}$, where N is Avogadro's number, ρ the density of the centrifugation medium, $\overline{V}$ the partial specific volume of the protein and f the friction coefficient of the protein which is a complicated function of its conformation (shape, dimensions, rigidity, degree of preferential solvation).

If we subject to ultracentrifugation a heterogeneous, poly-disperse solution (containing a mixture of proteins, or different oligomers of a same sub-unit) we obtain several peaks corresponding to diverse particles sedimenting at different velocities. We can thus estimate the molecular weights of diverse proteins or diverse oligomers. Ultracentrifugation can also be used as a criterion of purity but the fact that a protein gives only one peak during ultracentrifugation does not mean that it is pure (it may very well be a mixture of two proteins or even more, as revealed by another technique like electrophoresis on polyacrylamide gel for example).

b) **Sedimentation equilibrium.** – A pure protein solution may be subjected to ultracentrifugation until the distribution of the protein in the liquid column reaches a state of stable equilibrium i.e. until displacement of the protein is no longer observed (equilibrium between sedimentation and diffusion phenomena). If we determine protein concentration at different distances from the centre of rotation, we can deduce the molecular weight of the protein. This method does not necessitate the determination of the sedimentation constant; on the other hand, a comparatively small centrifugal force is sufficient, but the disadvantage of the method is that centrifugation must be carried out for a rather long time to reach equilibrium.

D. – Results

The molecular weights of proteins can vary from about 10 000 (ribonuclease) to one million or even more (hemocyanins). A few examples are given below:

ribonuclease	13 700
myoglobin	17 000
pepsin	36 000
ovalbumin	43 000
hemoglobin	68 000
glyceraldehyde-3-phosphate dehydrogenase	140 000
aldolase	180 000
catalase	250 000
urease	480 000
thyreoglobulin	630 000

3. – Amphoteric character

Most amino and carboxylic groups of amino acids are – in proteins – involved in peptide linkages. But we have seen (page 13) that the side chains contain ionizable groups which confer on the proteins an amphoteric character. Depending on the pH of the solution, these groups are more or less ionized according to their pK, so that when pH is raised the protein may exist in the following 3 states:

$$\underset{\text{Cation}}{\text{Protein}^+} \longleftrightarrow H^+ + \underset{\text{Zwitterion}}{{}^-\text{Protein}^+} \longleftrightarrow H^+ + \underset{\text{Anion}}{{}^-\text{Protein}}$$

The *isoionic point* is reached when the number of positively charged groups is equal to the number of those negatively charged; the number of protons combined with the basic groups is then equal to the number of protons liberated by the dissociation of acid groups. As in the case of amino acids (see note, bottom of page 10), this isoionic point (characteristic of a pure protein in solution in water) is near the *isoelectric point* where the total charge of the protein is nil (taking into consideration the other ions in solution, which may fix themselves to the protein molecule). The pH_i can vary widely in different proteins:

	pH_i		pH_i
pepsin	1	hemoglobin	6.9
ovalbumin	4.6	ribonuclease	7.0
fibrinogen	5.5	cytochrome *c*	10.6

Due to the large number of groups capable of reacting reversibly with protons, and in a rather wide range of pH values, proteins have considerable *buffer capacity* in biological media.

From the practical point of view, these properties of proteins lead to – as in the case of amino acids – a number of interesting applications:

– *the solubility of a protein is minimal near its isoelectric point*; it can therefore be precipitated during its separation and purification,

– *proteins can be fractionated by ion exchange chromatography*. Modified celluloses are often used for this purpose; these substances are either anion exchangers like diethylaminoethyl-cellulose (DEAE-cellulose), or cation exchangers like carboxymethyl-cellulose (CM-Cellulose). The principle of this method was explained in connection with the fractionation of amino acids (see page 18),

– *except at the isoelectric point, proteins migrate when placed in an electric field:* if the pH is higher than the isoelectric pH, the protein is charged negatively and migrates towards the anode, but if the pH is lower than the isoelectric pH, the protein is charged positively and migrates towards the cathode. One can therefore *fractionate proteins by electrophoresis*, either in a liquid medium in a Tiselius type apparatus (U-tube), or on a support like cellulose acetate paper, starch gel or polyacrylamide gel[1] (zone electrophoresis). This second technique is being increasingly used because it is more convenient; with the Tiselius apparatus, an optical system must be used to follow the displacement of the boundary between the protein and the buffer by measuring the refractive index; in zone electrophoresis, the position of

---

[1] In the case of starch gel, and more so with polyacrylamide gel, there is additional molecular sieving effect similar to the one described in connection with dextran gels (see pages 35-36). Thus, if at a given pH, two proteins have the same electrophoretic mobility in liquid medium (migration velocity per unit electric field), the one having the lower molecular weight will migrate faster in this type of gel.

different zones containing the proteins which migrated differently is visualised by a specific coloration of proteins (by amido-schwartz for example); this method also permits a quantitative analysis of the different proteins of a mixture, as one can either titrate by photodensitometry the coloration of different zones, or cut out these zones and elute the proteins, or even collect the different fractions at the end of the gel which they reach at different times on account of differences in their electrophoretic mobility. In a given electric field, this mobility mainly depends on the charge and size of the protein molecules. For example, the electrophoresis of serum proteins has very great importance in medicine because it permits the diagnosis of certain diseases which are reflected by anomalies of the electrophoregram like the increase or decrease of a protein fraction. Figure 1-24 shows the electrophoretic profile of proteins of a normal serum.

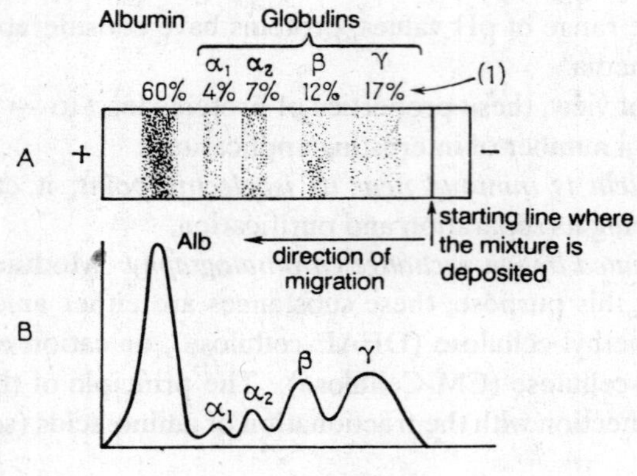

(1) These figures represent the approximate proportions of different protein fractions in a normal serum (total about 75 g/1). The fibrinogen, absent in the serum, migrates between the β and γ globulin during the electrophoresis of blood plasma proteins.

FIG. 1-24. — Fractionation of blood serum proteins by zone electrophoresis (on paper).

A: Paper strip after electrophoresis at pH 8.6 and detection of proteins with a dye.
B: Curve obtained after photodensitometric titration of the coloration of different zones.

Immuno-electrophoresis is an application of electrophoresis as well as of the specific precipitation produced by the mixture of a protein and the corresponding antibody. Proteins have an *antigenic capacity,* i.e. when introduced in a foreign organism they induce the production of specific γ-globulins or *antibodies* which are means of defence of the organism against the protein (or antigen) introduced (for example, vaccination induces the formation of specific antibodies). The protein mixture to be analysed is therefore subjected to a zone electrophoresis and a serum containing a given antibody is allowed to diffuse. If the antibody meets the corresponding protein (i.e. its antigen) we observe an arc-shaped precipitation zone which reveals the presence of this protein (see fig. 1-25).

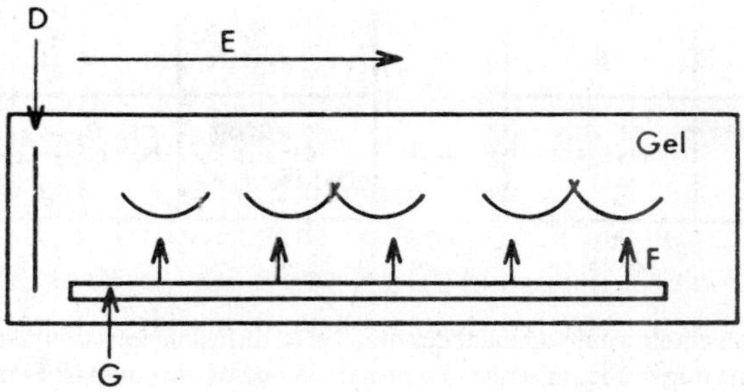

FIG. 1-25. — Immunoelectrophoresis.
D: Deposit of mixture of proteins in the gel acting as support for the electrophoresis (agarose gel for example).
E: Direction of electrophoretic migration.
F: Direction of immune serum diffusion.
G: Slit where the immune serum is deposited after electrophoretic migration. This immune serum contains the antibodies specific to one or several proteins of the mixture and was prepared earlier by injecting an animal (rabbit for example) with the corresponding protein (or proteins).

4. – Osmotic pressure

Let us consider two compartments separated by a membrane permeable to ions but impermeable to proteins. The presence of a protein in one of the compartments causes an unequal distribution of diffusible ions on the two sides of the membrane (whereas in the absence of protein there is equal distribution, at equilibrium).

Figure 1-26 shows that in the initial state, the compartment contains negatively charged protein molecules denoted P^- (anions). There is a disequilibrium in the distribution of Cl^- and the Cl^- ion will therefore diffuse from B to A; but in order to maintain electrical neutrality, an equivalent quantity (x) of Na^+ ions will also move from B to A. A final state is therefore reached where the distribution of Na^+ and Cl^- is unequal on the two sides of the membrane; this is the *DONNAN equilibrium*; its corresponding equation is:

$$(c_1 + x)\,x = (c_2 - x)(c_2 - x)$$

which gives:

$$x = \frac{(c_2)^2}{c_1 + 2c_2}$$

we can write:

$$\frac{[\mathrm{NaCl}]\text{ in B}}{[\mathrm{NaCl}]\text{ in A}} = \frac{[\mathrm{Cl^-}]\text{ in B}}{[\mathrm{Cl^-}]\text{ in A}} = \frac{c_2 - x}{x} = 1 + \frac{c_1}{c_2}$$

(substituting x by its value found above)

In other words, the ratio of NaCl concentrations on the two sides of the membrane is equal to 1+ the ratio of protein concentration in A to the initial NaCl concentration in B. This means that *the higher the protein concentration in A,* the more unequal the final distribution of diffusible ions, and therefore, *the greater the osmotic pressure.*

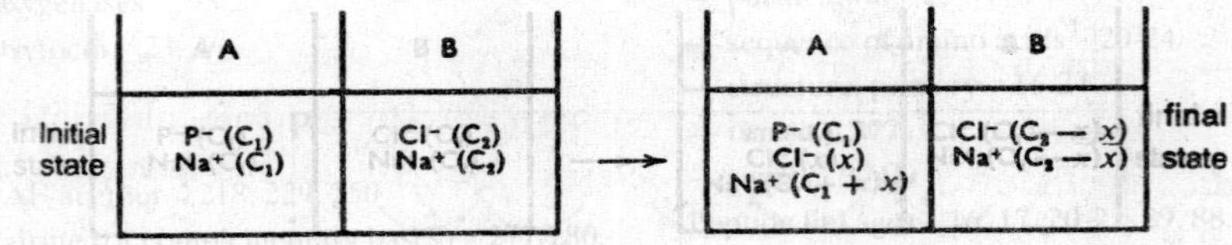

FIG. 1-26. — Donnan equilibrium: unequal distribution of diffusible ions on the two sides of a membrane due to the presence of a protein in one of the compartments.

To the osmotic pressure of protein (proportional to its concentration, but inversely proportional to its molecular weight, because, as seen on page 35, $\Pi = RT\frac{C}{M}$), is therefore added in A, the osmotic pressure due to the excess of diffusible ions: the sum represents the *oncotic pressure.*

It is clear that this excess of ions will be minimum at the isoelectric point of the protein and will increase for values diverging from this pH, because it depends on the degree of ionization of the protein.

In blood plasma, albumin contributes to osmotic pressure to the extent of 75 to 80% (although it represents no more than half the plasma proteins), because its molecular weight is lower and on the other hand, at the blood pH, it is further away from its pH_i value than the globulins. The oncotic pressure due to plasma proteins plays an important role in the passage of extravascular water of the interstitial liquid towards the intravascular compartment. It should be noted that if, in the example of figure 1-26, we replace the diffusible cation Na^+ by H^+, we will have on the two sides of the membrane an unequal distribution in H^+ ions, i.e. a difference of pH, and compartment B will be less acid than compartment A.

On the other hand, if in the initial state, the pH and the isoelectric point of the protein are such that we have cations P^+ (in presence of an equivalent concentration in Cl^- ions), compartment B will be more acid in the final state.

V. — ISOLATION, FRACTIONATION AND PURIFICATION OF PROTEINS

The isolation of a pure protein from a homogenate of cells is generally a long and delicate task, because in most cases we face several hundreds of proteins having somewhat similar chemical properties. Besides, care must be taken not to denature the protein to be isolated. The methods most frequently used were mentioned while studying the properties of proteins; they are mainly:

— dialysis, to eliminate the small molecules,

— precipitation by neutral salts (or salting out); ammonium sulphate is very often used because it is highly soluble, even at the low working temperature adopted to avoid denaturation (see pages 34-35),

– precipitation at the isoelectric point (see page 35),

– precipitation by organic solvents at low temperature (see page 35),

– adsorption chromatography on gel (alumina, calcium phosphate or hydroxyapatite etc.) which allows to retain certain proteins and then elute them by gradually varying the pH or saline concentration,

– ion exchange chromatography (see pages 18 and 39),

– chromatography by filtration on dextran gels (or molecular sieves, see pages 35-36),

– affinity chromatography which allows to purify an enzyme protein for example. The method consists in bonding covalently on a solid support, a steric analogue of the enzyme substrate (competitive inhibitor or modified substrate) to form with the enzyme a complex of the enzyme-substrate type (see pages 64-65). Therefore, during the passage on the column, the enzyme will be delayed with respect to the other proteins which, not recognizing the analogue, will come out of the column more rapidly than the enzyme,

– immunoadsorption is very similar in its principle, to affinity chromatography: the protein is first purified by the previous methods, specific antibodies against this protein are prepared by injecting it to a rabbit. These antibodies are then isolated and chemically fixed covalently on a solid support. The support thus prepared is used to make an immunoabsorbent column which will specifically retain the antigen protein, while all other proteins will pass through such a column without being retained. The antibody-antigen complex thus formed will then be dissociated on the column to recover the purified protein,

– electrophoresis (see pages 13 and 39-40)

– electrofocusing. In this technique, a pH gradient is produced in a liquid column with the help of appropriate electrolytes. After introducing the protein mixture in the column, an electrophoretic migration is effected. A given protein will migrate, depending on its initial charge, towards one of the electrodes, until it reaches the zone corresponding to its isoelectric pH; in that zone its charge becomes zero and its migration stops. By this method, proteins are therefore fractionated according to their pH_i,

– preparatory ultracentrifugation (see pages 36-38),

– crystallization,

Purification may be followed either by *physico-chemical methods* such as electrophoresis or analytical ultracentrifugation, or by biological methods such as *determination of specific activity* in the case of enzymes (catalytic activity per unit of protein mass, which of course, will increase as purification proceeds).

VI. – CLASSIFICATION OF PROTEINS

Different types of classifications have been proposed:

1. – Classification based on molecular shape

As mentioned while studying the three-dimensional conformation of proteins we distinguish:

A. – *Fibrous proteins*

Also called *scleroproteins*, they consist – as indicated by their name – of fibers or fibrils; they are practically insoluble. This group includes silk *fibroin, collagens* found in conjunctive tissues, cartilages, tendons (and which on heating with water in the autoclave give *gelatins*) and *keratins* present in the skin and superficial body growths (hair, fur, nails, horns, feathers etc).

B. – *Globular proteins*

They are also called *spheroproteins* on account of their spherical or ovoid shape; they are generally more easily soluble. This group includes mainly albumins and globulins which will be described in the following.

2. – Classification based on solubility

The following are generally distinguished:

– *albumins*, which are soluble even in distilled water. They precipitate on addition of ammonium sulphate between 70 and 100% saturation. Their isoelectric point is in general, distinctly less than 7 (serumalbumin, 5.3; ovalbumin 4.7); they are therefore acid in character;

– *globulins*, are insoluble in pure water but soluble in dilute saline solutions (e.g., 5% NaCl). They precipitate on addition of ammonium sulphate at 50% saturation. They are often glycoproteins or lipoproteins (see below);

– *protamins and histones* are soluble proteins of comparatively small size (especially the protamins with their molecular weight varying between 2 000 and 5 000; in fact they are polypeptides rather than proteins); they have a very distinct basic character due to the presence of a large proportion of lysine and arginine (up to 90% in certain cases) giving them a high isoelectric point (around 11). On account of this basic character, they can combine with various acid compounds (nucleic acids, acidic proteins) and they are found combined with deoxyribonucleic acids in cell nuclei;

– *globins* differ from the above proteins by a comparatively high histidine content (up to 10%). They constitute the protein part of hemoglobins and myoglobins, chromoproteins whose prosthetic group is heme (see pages 48-53).

– *prolamins and glutelins* are plant proteins, insoluble in water but soluble in acids and dilute bases;

– *scleroproteins*, insoluble in water and dilute saline, acid or alkaline solutions (see above);

– *soluble fibrilar proteins*: these proteins constitute the fibrils in muscle cells (where they are very abundant) or microtubules of the cytoskeleton.

Fibrils of muscle cells contain four types of proteins called contractile proteins:

– myosin, molar weight 460 000, organized in two chains in coiled double helix (130 nm length) ending with a globular head which can link with a second protein, the actin.

– actin, a globular protein of molar weight 42 000, which polymerizes in long filaments containing about 400 monomers; it forms a double helix of 2 μm length. Two other proteins are found along these filaments:

– troponin (molar weight 80 000) a globular protein and tropomyosin (molar weight 70 000) a fibrilar protein.

Within other cells are found similar organizations constituting what is called the cycloskeleton thanks, in particular, to a protein called tubulin which can polymerize into microtubules.

3. – Classification based on composition

There are two broad groups:

– *the holoproteins*, consisting exclusively of amino acids.

– *the heteroproteins*, which contain, on one hand, one or several polypeptide chains and on the other hand, a non-protein part called *prosthetic group* which is bonded covalently.

Heteroproteins form an extremely heterogeneous family, not only by the variety of prosthetic groups, but also the number of groups present per molecule of protein and the nature – more or less labile – of linkages between the prosthetic groups and the polypeptide chains. Heteroproteins include phosphoproteins, glycoproteins and chromoproteins. The proteins can also be associated *in vivo* with very diverse compounds and if these associations are sufficiently stable to withstand the extraction and purification processes, lipoproteins and nucleoproteins may be isolated. There are even ternary associations like the phosphoglycoproteins or phospholipoproteins.

PHOSPHOPROTEINS

These proteins contain phosphoric acid esterifying the alcohol groups of serine and threonine. The best known are milk caseins and vitellin, vitellenin and phosvitin of egg yolk.

Casein is a group of milk proteins, forming stable complexes called micelles in presence of Ca^{++} ions. Casein can be precipitated by heat (thermocoagulation), by neutral salts around pH_i (4.7), and certain enzymes like rennet. It contains about 0.8% phosphorus. It can be fractionated by electrophoresis into α, β and γ caseins; α-casein has a carbohydrate part containing galactose, galactosamine and N-acetylneuraminic acid (the structures of these compounds are given in figs. 4-5, 4-10 and 4-11, respectively), it is therefore a phosphoglycoprotein.

GLYCOPROTEINS

The study of these compounds necessitates a knowledge of the structures of carbohydrates; we will therefore examine glycoproteins in Chapter 4 after the study of carbohydrates structures (see pages 162-167).

LIPOPROTEINS

The first lipoprotein was isolated from serum by Macheboeuf, who gave the name "*cenapses*" to this type of molecular combinations where physical bonds play an important role. It appears, in fact, that the lipids are linked with proteins, partly by Van der Waals type interactions between the hydrophobic chains of lipids and those of certain amino acids, and partly by ionic bonds.

A distinction is made between cellular lipoproteins and transport (circulating) lipoproteins. Some lipoproteins were isolated from blood serum; they were easily separated from other proteins by centrifugation due to their distinctly lower density (decreasing with increasing lipid content). Certain lipoproteins contain up to 80% lipids: triglycerides, phospholipids, cholesterol, sterides (see the structure of these compounds in Chapter 5). The serum β-lipoprotein was studied in detail. It contains two polypeptide chains of molecular weight 250 000, 1 400 molecules of cholesterol esters, 700 cholesterol molecules, 650 molecules of phospholipids and 250 molecules of triglycerides.

Soluble lipoproteins could be isolated from diverse tissues: rhodopsin from the retina, chloroplastin from leaves, mitochondrial lipoproteins from the liver, lipovitellin and lipovitellenin from egg yolk (they are lipophosphoproteins). But there is a large number of cellular lipoproteins, especially in the cellular and intracellular membranes (mitochondria, chloroplasts, microsomes etc.). By virtue of the hydrophilic character of the protein part and hydrophobic character of the lipid part, these cellular lipoproteins can play an important role in cellular metabolism (particularly in allowing the passage of certain compounds from one phase to another, or when there is a change of solubility of the substrates of a reaction).

NUCLEOPROTEINS

It has already been mentioned that protamines and histones, by virtue of their basic character, are combined with deoxyribonucleic acids in the cellular nuclei where they form *deoxyribonucleoproteins*. We also know *ribonucleoproteins*, combination of ribonucleic acids and proteins: the ribosomes; their role in protein biosynthesis will be studied in Chapter 6. Besides, the simplest viruses (particularly the bacteriophages) consist simply of a (ribo- or deoxyribo-) nucleic acid-protein combination and may be considered as nucleoproteins.

CHROMOPROTEINS

As indicated by their name, these are coloured proteins, their prosthetic group often contains a metal element (Fe, Cu, Mg). These compounds are generally classified on the basis of their prosthetic group having, or not having, the tetrapyrrol structure called porphyrin and according to their respiratory or non-respiratory physiological role.

I. – CLASSIFICATION

1. – Porphyrin chromoproteins

A. – Respiratory pigments

They all contain iron.

a) **Hemoglobins** or respiratory pigments of Vertebrates and various Invertebrates; we will study hemoglobins in more details.

b) **Cytochromes**, electron transporters; they will also be studied later (see page 115).

c) **Cruorins**, pigments of Invertebrates

d) **Heme enzymes:**

– the *catalases*, which decompose peroxides:

$$H_2O_2 \longrightarrow H_2O + \frac{1}{2}O_2$$

– the *peroxidases*, which catalyze the oxidation of a substrate at the cost of a peroxide:

$$RH_2 + H_2O_2 \longrightarrow R + 2H_2O$$

B. – Non-respiratory pigments

a) **Chloroplastins**, found in chloroplasts (where photosynthesis takes place); their prosthetic group is a *chlorophyll* with a porphyrin ring somewhat similar to that of hemoglobin but containing magnesium (see fig. 3-10).

b) **Phycobilins**, pigments of blue algae (Phycocyanins) and red algae (Phycoerythrins).

2. – Non porphyrin chromoproteins

A. – Respiratory pigments

a) **Hemerythrins**, pigments of certain Invertebrates; they contain iron.

b) **Hemocyanins**, blue pigments (containing copper) of certain Crustaceans and Molluscs.

B. – Non-respiratory pigments

a) **Carotenoproteins**, containing a carotenoid (see fig. 5-8), found for example in the retina.

b) **Flavoproteins**, enzyme proteins; we will see later their role in oxidation-reduction phenomena (see pages 81-82 and 114-115).

C. – Metalloproteins

1. Containing IRON:

– *Ferritin*, enabling the storage of iron in the liver and spleen,

– *Siderophilin*, enabling the transfer of iron into the plasma.

2. – Containing COPPER:

– *Hemocuprein*, in red blood cells,

– *Ceruleoplasmin*, in the plasma.

II. – HEMOGLOBINS

Animal cells need oxygen. In primitive organisms this need can be met by diffusion of oxygen gas in tissues, but in more complex organisms, particularly man, the transport is achieved by chromoproteins which have the property of binding oxygen to form an easily dissociable combination and which are called respiratory pigments. As mentioned above, there are different types of pigments: hemerythrine, hemocyanins, cruorins and hemoglobins.

There are numerous hemoglobins; their structure varies:

– according to the animal species,

– in the same animal, according to the tissue: for example we distinguish *myoglobin* (earlier called muscle hemoglobin) from *blood hemoglobin.*

– in the same animal, according to age: adult hemoglobin is different from foetal hemoglobin.

– in the same animal species, depending on whether the hemoglobin is normal or pathological.

1. – Structure of hemoglobins

As indicated by their name, hemoglobins consist of a prosthetic group; the *heme* (4%) and a protein part: the *globin* (96%).

A. – Heme

This is the prosthetic group common to various hemoglobins (while globin varies in different hemoglobins). It contains one molecule of protoporphyrin and one iron atom.

a) **The protoporphyrin.** – Four pyrrole rings, linked by methenyl bridges =CH – between their α and α' carbon atoms form the porphin (see fig. 1-27).

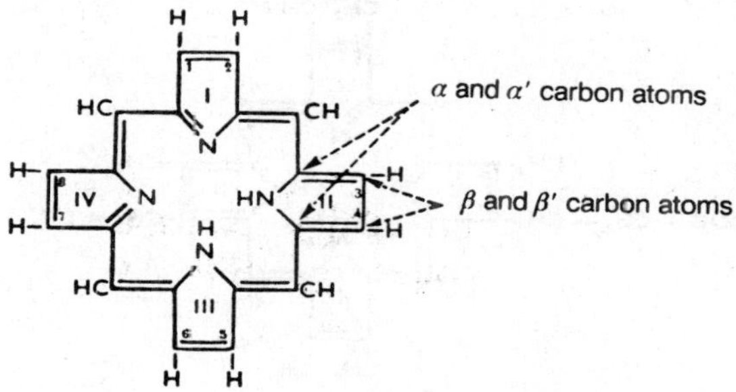

FIG. 1-27. — Structure of porphin
— Roman numerals: numbering of the 4 pyrrole rings.
— Arab numerals: numbering of the 8 β and β' carbon atoms whose H atoms may be substituted.

We must note the alternation of the double bonds which are all conjugated.

Porphyrins are simply derivatives of porphin where the 8 β and β' carbon atoms are carriers of different substituents. It is clear that there are possibilities of isomerism; we will study these, taking the example of *uroporphyrins* where the 8 β and β' carbon atoms carry 4 acetic radicals (A) and 4 propionic radicals (P); figure 1-28 shows that with 2 different substituents there are 4 possible arrangements; among natural pigments, mostly type III and sometimes type I, are found.

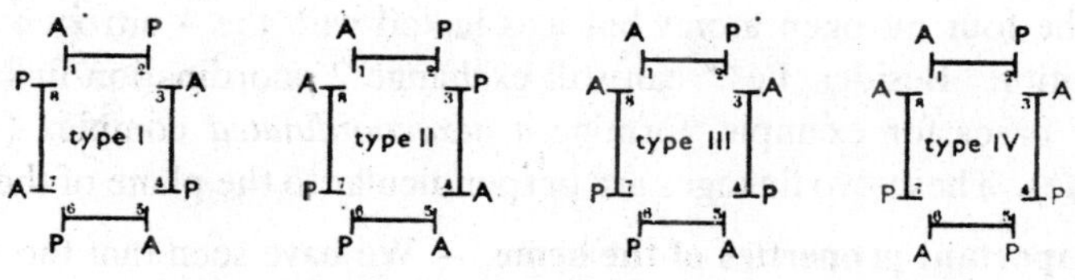

FIG. 1-28. — Diagrammatic representation of 4 types of uroporphyrins.
The figure shows only the β and β' carbon atoms of the porphyrin cycle, carriers of substituents, as the remaining part of the molecule (see fig. 1-27) does not change.
A represents the acetic radical: —CH_2—COOH.
P represents the propionic radical: —CH_2—CH_2—COOH.

Protoporphyrin of hemoglobins is of type III; as may be seen in figure 1-29, it derives from uroporphyrin III by decarboxylation of the 4 acetic radicals (into methyl groups) and decarboxylation and then dehydrogenation of two of the four propionic radicals (which are thus transformed into vinyl groups).[(1)]

This is denoted protoporphyrin IX by certain authors, because with 3 different substituents there are actually 15 possible isomers (and not 4) and the comparison with different porphyrins obtained by synthesis has shown the identity of natural protoporphyrin with the type IX of synthetic isomers. Protoporphyrin therefore contains *4 methyl groups* in 1, 3, 5 and 8; *2 propionic radicals* in 6 and 7; and *2 vinyl groups* in 2 and 4.

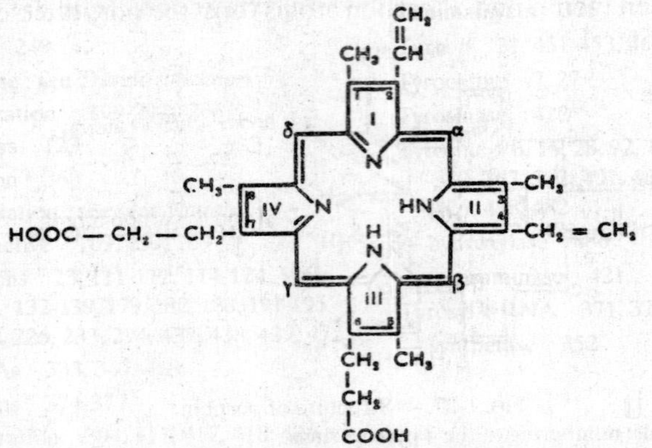

FIG. 1-29. — Structure of protoporphyrin IX (type III).
The 4 pyrrole rings are numbered in Roman numerals. The 4 methenyl bridges are denoted by Greek letters. The whole unit forms a plane structure in space.

b) Linkage of iron and protoporphyrin. – Porphyrins have the property of binding, by their pyrrole nitrogen atoms, metals like Fe, Mn, Ni, Co, Mg; metalloporphyrins containing ferrous iron (Fe^{++}) are called *hemes*; in the particular case of protoporphyrin, the ferroporphyrin is called *protoheme* (the term "heme" is used by abbreviation). Metalloporphyrins containing ferric iron (Fe^{+++}) are called *hematins* (the one corresponding to protoporphyrin is protohematin).

In the protoheme, the iron atom replaces the 2 hydrogen atoms carried by two of the four nitrogen atoms but it is linked with the 4 nitrogen atoms by coordination. Besides, Fe^{++} can still exchange 2 coordination linkages with nitrogen bases for example, forming *a hexacoordinated* complex (as in ferrocyanide). These two linkages are perpendicular to the plane of the heme.

c) Important properties of the heme. – We have seen that the heme is a ferrous compound; under the action of oxidants (like alkaline ferricyanides, for

---

(1) The biosynthesis of protoporphyrin will be studied in connection with the metabolism of glycine (see page 408).

example), it can be oxidised into hematin (where iron is ferric) which can be again reduced into heme (by sodium hydrosulphite, for example).

$$\text{Heme } (Fe^{++}) \underset{\text{reduction}}{\overset{\text{oxidation}}{\rightleftarrows}} \text{Hematin } (Fe^{+++})$$

This reversible reaction forms the basis of the participation of ferroporphyrins, like the cytochromes, in the oxidation–reduction processes (but it does not take place during the conversion hemoglobin ⇄ oxyhemoglobin).

As will be seen below, the heme can link with globin. It can also link with other nitrogen bases (pyridine, nicotins) or with the nitrogen groups of proteins, to form *hemochromogens*.

Similarly, hematin can link with different proteins to form *parahematins*, often having catalase or peroxidase properties.

B. – *Globin*

This is the protein part of the chromoprotein; it is easily obtained by treating hemoglobin with acetone containing 5% HCl (globin precipitates). Globin is the specific part of hemoglobin which varies according to age, species, and in certain diseases.

In *myoglobins* there is a single polypeptide chain (combined with one heme) whose primary, secondary and tertiary structure is known, at least in certain species (see figure 1-19). Molecular weight is about 17 000. Histidine content is comparatively high (6 to 10%).

Blood hemoglobins are tetramers; they contain 4 polypeptide chains (each combined with one heme) and their molecular weight is about 68 000.

The 4 chains are united by linkages which are easy to break; in the case of *adult human hemoglobin* (HbA), 2 types of chains (α and β) are obtained; these are combined in pairs in the original molecule; this is expressed by writing HbA $= \alpha_2^A\ \beta_2^A$ (A means "Adult"). The α chain contains 141 amino acids and the β chain 146; the sequences are fully known.

The *foetal hemoglobin* also contains 4 chains, 2 α chains and 2 γ chains (the latter have about ten amino acids different from those present in β chains). We can write HbF $= \alpha_2^A\ \gamma_2^F$.

The *pathological hemoglobins* differ from HbA, either by anomalies in the *distribution of chains* (there are hemoglobins having 4 β chains or 4 γ chains), or by anomalies in the *sequence* of the α chain or β chain (for example, in sickle-cell anaemia, glutamic acid normally present in position 6 of the β chain is replaced by a valine, and just this difference is sufficient to alter the physiological properties of hemoglobin).

C. – Union of heme and globin

In the myoglobin of the sperm whale, which was thoroughly studied from the structural point of view, the heme is lodged in a fold of the tertiary structure of globin (see fig. 1-19).

In all these ferroporphyrin – protein complexes, the iron atom has an octahedral structure of chemical type d^2sp^3 (double pyramid with square base formed by the 4 pyrrolic nitrogen atoms). The 5th coordination position of the iron is occupied by the imidazole nitrogen N_3 of the histidyl residue in position 93 (F8 or 8th residue of the helix F in the Perutz nomenclature). The 6th coordination position is free in the non-oxygenated or Deoxy state, but it is occupied by oxygen in the oxygenated or Oxy state (or by the carbon of carbon monoxide, competitive inhibitor of oxygen).

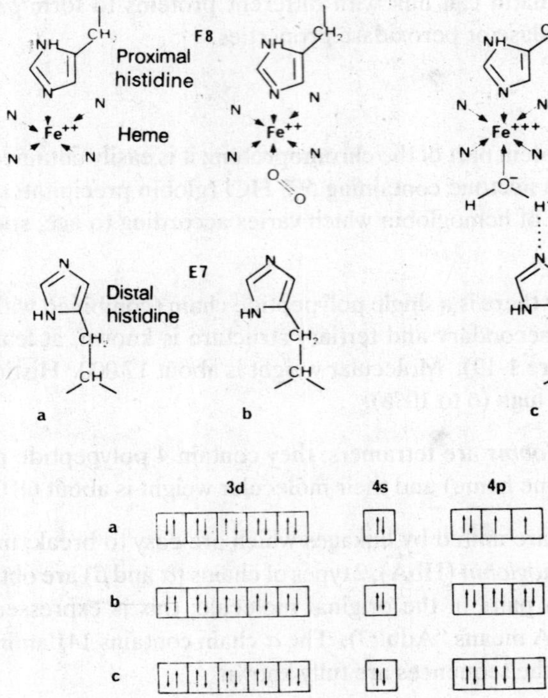

FIG. 1-30. — Linkages between porphyrin and globin through the iron atom (according to THEORELL and PAULING) and external electron layer of iron (sub-layers 3d, 4s and 4p). The 4 N atoms are those of the four pyrrole rings of the porphyrin plane (4th). The Fe—N bonds are perpendicular in pairs to one another. Fe—N HiS F8 and Fe—Oh bonds are perpendicular to this plane.
a) Fe^{2+} Myoglobin or Hemoglobin (α or β chain) Deoxy
b) Fe^{2+} Myoglobin or Hemoglobin (α or β chain) Oxy
c) Fe^{3+} "Met"—Myoglobin or Hemoglobin (α or β chain).
F8 and E7 denote respectively proximal histidine and distal histidine in Perutz nomenclature.

In the case of the α or β chains of hemoglobin, we have exactly the same Deoxy and Oxy structures (fig. 1-30): iron is linked with the nitrogen N_3 of histidine F8 called proximal (position 87 for α and 92 for β).

If the iron atom is oxidized to the ferric state, its 6th coordination position cannot any more be occupied by an oxygen molecule, it is occupied by the oxygen atom of a water molecule, one proton of which is attached by hydrogen bond to the nitrogen N_3 of another histidine called distal and which occupies the position E7 in the Perutz nomenclature (7th residue of the helix E, respective positions 64, 63 and 58 in myoglobin, chain β and chain α). This form called *metmyoglobin* or *methemoglobin* is a stable Deoxy state in which, hexacoordinated ferric iron can no longer play the role of oxygen fixing agent, contrary to the previous case.

It is observed that in the Deoxy state, the external electron layer of the ferrous iron atom contains 16 electrons of which 2 are unpaired. The ferroporphyrin-protein complex is then paramagnetic. In the Oxy state, this external layer is saturated with 18 electrons all paired, and the complex is therefore diamagnetic. In the metmyoglobin or methemoglobin form, the external layer has 17 electrons of which 1 is unpaired, and the complex is paramagnetic. This form can be stabilized by a CN^- cyanide ion which brings 3 electrons to the ferric ion, the external layer of which is saturated with 18 electrons. We then obtain a diamagnetic stable Oxy form where oxygen cannot displace the CN^- ion.

Beside this primary linkage between Fe and N^3 of the proximal histidine, two saline linkages (between 2 propionyl radicals of heme and 2 basic groups of lysine or arginine) and Van der Waals type linkages (between hydrophobic groups of heme and globin) maintain the stability of the structure.

It must be noted that if we have a non-denatured globin, it is possible to effect *in vitro* the combination heme + globin; we obtain a hemoglobin having the properties of the chromoprotein which was used for the isolation of globin (whatever the origin of the heme). Inversely, the combination of a heme with diverse globins produces chromoproteins which differ according to the origin of the globin used.

2. – Properties of hemoglobins

When we centrifuge fresh blood or blood made incoagulable, the red cells sediment. Bringing them back in suspension in a hypotonic medium (for example distilled water) is sufficient to make them burst (hemolysis) and, after elimination of the cellular debris by centrifugation, we obtain a red coloured solution. The solution is dark red for hemoglobin and bright red for oxyhemoglobin (due to differences in the absorption spectra); there is therefore an oxidized form of hemoglobin whose physiological role – as mentioned in the introduction – is to transport oxygen after forming with this gas an easily dissociable combination. We will now examine the most important property of hemoglobins namely their capacity to reversibly bind certain gases particularly oxygen.

A. – Combinations of hemoglobins with gases

a) **Combination with oxygen.** – For all hemoglobins, the maximum quantity of oxygen which can be bound is a function of the quantity of iron: one atom-gram of iron combines with one molecule-gram of oxygen (or one molecule gram of carbon monoxide). Since there are 4 hemes (i.e. 4 iron atoms) per molecule of hemoglobin, we can write:

$$[HB]_4 + 4O_2 \leftrightarrow [HbO_2]_4$$

(but very often, oxyhemoglobin is represented by HbO_2). This is an *oxygenation* and not an oxidation because iron remains in the ferrous state. 17 000 g Hb (corresponding to the monomer which contains 1 iron atom) can therefore combine with 32 g of O_2 and since 1 mole of oxygen occupies 22.4 l, it may be derived that 1 g Hb can bind 1.34 ml O_2. But, *100 ml of blood* contains in average *15 g Hb,* so that it can bind 1.34 × 15, i.e. about *20 ml O_2*; this represents complete saturation in O_2.

The equation of oxygenation reaction shows that *oxygen pressure* is an important factor controlling this equilibrium. In the air that we breathe where mean pressure is 760 mm Hg and oxygen concentration about 20%, the partial pressure of oxygen is 760/5, i.e. about 150 mm. In arterial blood, oxygen pressure is about 80 mm, so that (see fig. 1-31) 19 ml of O_2 can be bound for 100 ml blood (95% saturation). In the capillaries, partial pressure of oxygen falls to 40 mm or even 20 mm, therefore O_2 fixation can be only 77% and 40% of saturation respectively. Figure 1-31 shows that this stage corresponds to the portion of the curve where a comparatively small decrease of oxygen pressure causes a comparatively major liberation of oxygen; when pressure falls from 40 to 20 mm, the quantity of oxygen which can be present in blood falls from 20 × 77/100 i.e. 15.4 ml O_2/100 ml to 20 × 40/100 i.e. 8 ml O_2/100 ml; in other words there is liberation of 7.4 ml O_2/100 ml of blood in favour of the tissues.

It is seen in figure 1-31 that the saturation curve of myoglobin is a branch of hyperbola while that of hemoglobin is a sigmoid (S-shaped). This difference is due to the fact that hemoglobin is a tetramer whose 4 hemes are not independent, because the oxygena-

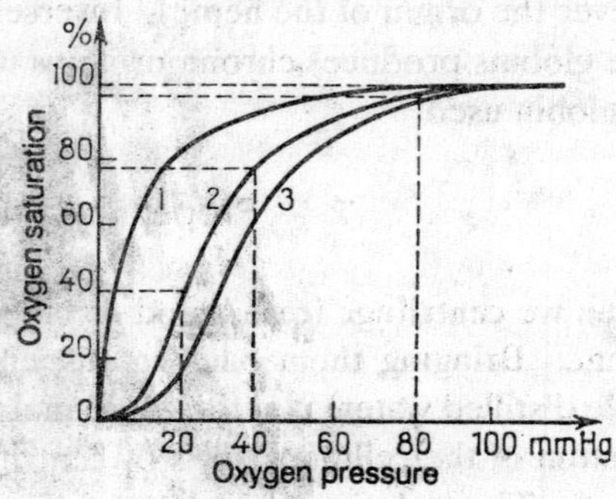

FIG. 1-31. — Oxygen saturation curves of:
— myoglobin (curve 1),
— hemoglobin, for CO_2 pressure p_1, (curve 2),
— hemoglobin, for CO_2 pressure $p_2 > p_1$ (curve 3).

tion of one of them favours that of others (which thus requires less energy). This interaction does not take place directly because the hemes are too distant from one another (about 30 Å); it takes place through the quaternary structure of hemoglobin; to obtain this *cooperative effect* between hemes, the quaternary structure must be intact, and a pathological hemoglobin (by anomaly of the distribution of chains) or a denatured hemoglobin behaves like the myoglobin; furthermore, X-ray studies have actually shown – during the transformation $[Hb]_4 \longrightarrow [HbO_2]_4$ or inversely – a change of the conformation of the molecule, called *allosteric transition,* which takes place only with a hemoglobin whose quaternary structure is native. This is an example of relationship between quaternary structure and biological activity; we will come across other such cases while studying enzymes and we will see that the presence of allosteric enzymes ensures a regulation of cellular metabolism (see pages 74-77 and 472-477).

Lastly, it should be noted that oxyhemoglobin, because of changes of conformation, has a larger number of dissociated acid groups. Therefore, H^+ ion concentration also has an effect on the equilibrium which may be written:

$$[Hb-H]_4 + 4O_2 \longleftrightarrow [HbO_2]_4 + 4H^+$$

In the tissues, there is production of CO_2 which will lead to an increase of H^+ concentration and favour liberation of oxygen. In the capillaries, this liberation is therefore caused, *mostly by the decrease of O_2 pressure, but also by the increase of CO_2 pressure*; figure 1-31 shows that for a given O_2 pressure, the quantity of oxygen which may be combined is smaller when CO_2 pressure is higher (curve no. 3). On the contrary, in the lungs, the O_2 pressure is high and CO_2 pressure falls (because CO_2 is eliminated) and this decreases acidity; these two factors – particularly, the former – favour the formation of oxyhemoglobin.

b) **Combination with carbon monoxide.** – This is very similar to the combination with oxygen; there is binding of one molecule-gram CO per atom-gram iron, but the stability of the combination $(Hb\text{-}CO)_4$ called *carbonyl-hemoglobin* (or carboxyhemoglobin) is about 200 times greater; the dissociation is very difficult; this explains the toxicity of the gas, and the fact that high concentrations of oxygen (oxygenotherapy) are required in the treatment of intoxication by carbon monoxide.

c) **Combination with other gases.** – *Carbon dioxide* can combine with hemoglobin to give carbhemoglobin, but the heme plays no part, because CO_2 binds to the basic groups of globin; it also binds to those of other proteins (but it should be remembered that hemoglobin represents about 75% of the total blood proteins).

About one fourth of the CO_2 carried by blood travels in the form of *carbaminoproteins*:

$$CO_2 + \text{protein}-NH_2 \longleftrightarrow \text{protein}-NH-CO-OH$$

It should be of interest to examine here, how, in general, *CO_2 is transported* from the tissues where it is formed to the lungs where it is eliminated. This transport must prevent, on the one hand, a progressive acidification of blood as it flows through the tissues liberating CO_2, and on the other hand, the formation of CO_2 bubbles, because the solubility of this gas is comparatively limited. These difficulties are overcome by the *buffer capacity of blood proteins* which can bind H^+ ions as mentioned earlier and so, more H_2CO_3 can be ionized and therefore more CO_2 can be dissolved (see fig. 1-32). On the contrary, when blood reaches the lungs, CO_2 is liberated and the reactions of figure 1-32 take place in the opposite direction; the proteins are again ready to bind H^+ ions.

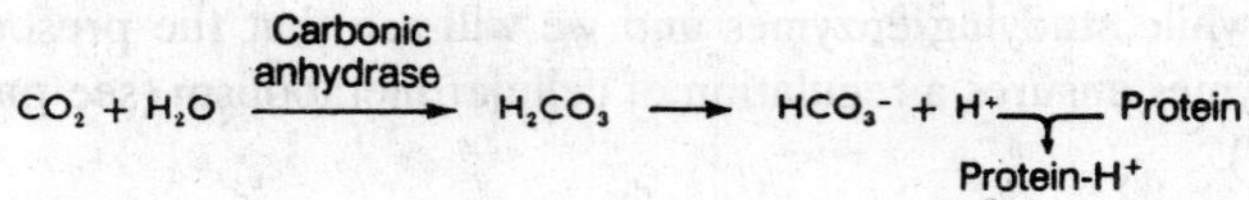

FIG. 1-32. — Series of reactions increasing the solubilization of CO_2 in blood (in the form of carbonic acid and bicarbonate ions).

It must be noted that the system $H_2\,CO_3 \leftrightarrow HCO_3^-$ is a buffer system and that it is largely responsible (together with the system $H_2PO_4^- \leftrightarrow HPO_4^=$) for protecting blood from excesses of acid or base.

B. – *Reversible oxidation of hemoglobin and oxyhemoglobin*

Hemoglobin treated by a mild oxidising agent like potassium ferricyanide is oxidized into *methemoglobin*. Iron changes from Fe^{++} state to Fe^{+++} state, so that the heme is oxidized into hematin; this oxidation is accompanied by a modification of the absorption spectrum and the loss of the property of the prosthetic group to combine with O_2 or CO. If oxyhemoglobin or carboxyhemoglobin is subjected to this oxidation, we also observe the formation of methemoglobin accompanied by the liberation of the gas combined, which incidentally permits the titration of the gas combined with hemoglobin. The

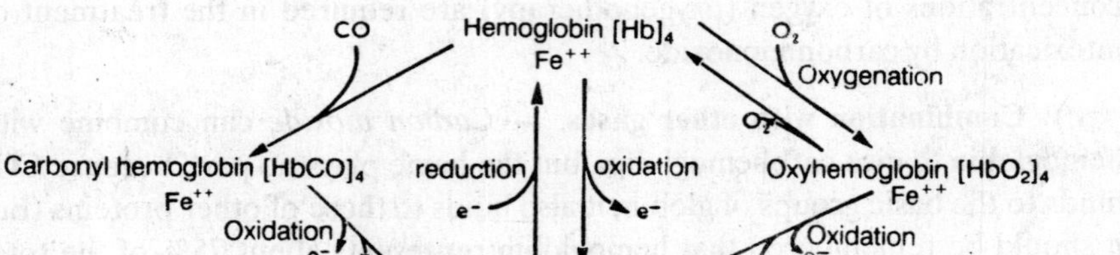

FIG. 1-33. — Diagram representing interconversions between hemoglobin, oxyhemoglobin, carbonylhemoglobin and methemoglobin.

formation of methemoglobin *in vivo*, either as a result of a serious intoxication or after a metabolic disease, has grave consequences because oxygen transport is made impossible. However, methemoglobin can be again reduced into hemoglobin *in vitro* as well as *in vivo* in normal red blood cells, by the action of reducing agents. These interconversions may be summarized in a diagram (see fig. 1-33).

CHAPTER 2[1]

ENZYMES AND ENZYMATIC CATALYSIS

Enzymes are biological catalysts; their presence enables the life-supporting chemical reactions and cell multiplication to proceed faster and with a specificity which excludes the formation of by-products.

CATALYSIS

I. – EQUILIBRIUM CONSTANT OF A REACTION AND FREE ENERGY CHANGE

Any chemical reaction will reach an equilibrium after a sufficient time. A reversible reaction may be written:

$$A + B \underset{k_2}{\overset{k_1}{\rightleftharpoons}} C + D$$

The velocity of the reaction which will produce C and D is written:

$$v_1 = k_1 [A][B]$$

and that of the reaction producing A and B would be:

$$v_2 = k_2 [C][D]$$

k_1 and k_2 are the velocity constants of the reaction.

When equilibrium is reached, velocities v_1 and v_2 are evidently equal

$$k_1 [A]_e[B]_e = k_2 [C]_e [D]_e$$

i.e.
$$\frac{[C]_e[D]_e}{[A]_e[B]_e} = \frac{k_1}{k_2} = k \text{ (equilibrium constant)}$$

The concentrations indicated in this formulation are those reached at equilibrium.

---

(1) Revised by M. LAZDUNSKI and J. BONNET.

The capacity of a molecule to react is characterized by a parameter expressed in kCal/mole and called *free energy*. The free energy change in a reaction of the type A + B $\rightleftarrows$ C + D when one mole of A and one mole of B give one mole of C and one mole of D, while their respective concentrations [A], [B], [C], and [D][(1)] are maintained constant[(2)], is written:

$$\Delta G = \Delta G_0 + RT \ln \frac{[C][D]}{[A][B]} \quad (1)$$

This change therefore depends on two parameters. First, a constant characteristic of the reaction (ΔG_0), the *standard free energy change*, which is the change in free energy in conditions called standard : temperature, 298°K; pressure, one atmosphere; pH = 0; and [A], [B], [C], [D], maintained at 1M. We have then $\Delta G = \Delta G_0$.

ΔG then depends on the concentration of the products and reagents. It is important to note that, theoretically, there are always concentrations of reagents and products such that $\Delta G < 0$. In this case, the reaction is called *exergonic* and is always accompanied by a *decrease in free energy*. It can then yield energy. For example the reaction

$$C_6H_{12}O_6 + 6O_2 \longrightarrow 6CO_2 + 6H_2O$$

has a ΔG of – 686 kcal/mole of glucose, which means that in standard conditions, the oxidation of one mole of glucose will yield 686 kilocalories of free energy. This energy may be dissipated as heat or converted into mechanical energy (muscle contraction), electrical energy (transmission of nerve impulse or formation of ion gradients), or in certain cases, even radiant energy. It can also be conserved as chemical energy in molecules, the most important of which is ATP (see Chapter 3).

Lastly, if the reaction is in equilibrium, $\Delta G = 0$ and one has the following equation:

$$\Delta G_0 = -RT \ln \frac{[C]_e[D]_e}{[A]_e[B]_e}$$

$$= -RT \ln K$$

There is therefore, a direct relationship between the standard free energy change of the reaction and its equilibrium constant.

---

(1) In reality, one should write in this equation, not the concentrations but the *activities* (or effective concentrations), thus taking into consideration the interactions of each molecule with the others and with the solvent; therefore, A should be replaced by a_A (activity of A) which is equal to C_A (concentration) $\times \gamma$ (activity coefficient). But in general, one does not know the activity coefficients for the intracellular concentrations of different substances, and these concentrations are sufficiently small for γ to be near 1 and therefore a_A very near C_A.

(2) This is the most frequent situation in biological systems: the concentrations of most chemical species are maintained constant by the set of biochemical reactions.

II. – ENERGY OF ACTIVATION AND ROLE OF CATALYSTS

An exergonic reaction is a spontaneous reaction; it can therefore yield energy usable for chemical, mechanical or electrical work. However, it should be remembered that *spontaneous* does not mean *rapid*, but only without external energy input. A spontaneous exergonic reaction can be extremely slow. As mentioned earlier, the reaction of glucose with oxygen is exergonic. However, glucose can be left in contact with the oxygen of air for months or years without any appreciable oxidation. The reaction is slow. To accelerate it, one or more catalysts must be used. This is naturally what takes place in living organisms which must, in certain physiological conditions, utilise in few minutes the energy available after the oxidation of glucose.

Figure 2-1 introduces the notion of free energy of activation. To pass from A + B to C + D the reactive path must pass through a "privileged state" of reactivity called activated complex. To transform the molecules A and B into activated complex, free energy must be supplied to the system: this quantity of free energy to be supplied is called free energy of activation. At a given temperature, the greater the concentration of activated complex, the higher the reaction velocity (which is proportional to this concentration). If temperature is raised, more energy is supplied to the system by increasing molecular agitation; the concentration of activated complex increases and the reaction becomes more rapid. Any chemical reaction, with or without catalyst, becomes faster when temperature rises. At a given temperature, the lower the amount of free energy of activation to be supplied, the greater the concentration of activated complex. Catalysts, and among them enzymes which are biological catalysts, act by considerably decreasing the free energy of activation. The smaller the quantity of free energy of activation to be supplied, the more effective the catalyst (see fig. 2-1).

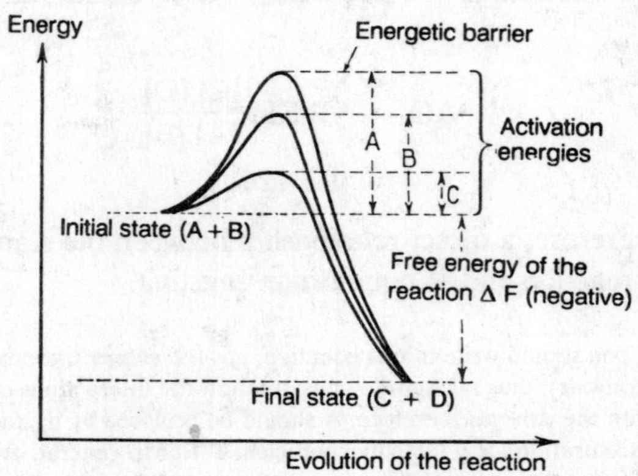

FIG. 2-1. — Activation energies required for triggering an exergonic reaction: in absence of catalyst (A), in presence of a chemical catalyst (B), in presence of the corresponding enzyme (C).

The main properties of a catalyst may be briefly summarised as follows:

– At the end of the reaction the catalyst is found *intact*, although during the reaction it was combined with the molecules which reacted. The catalyst can also accelerate the velocity of numerous successive cycles of a given reaction and that is why it acts in small doses.

– The catalyst, does not in any manner, alter the final equilibrium of a reversible reaction; it simply *allows this equilibrium to be reached more rapidly by accelerating* in the same manner the *velocities of the 2 reactions* proceeding simultaneously in opposite directions.

Enzymes or biochemical catalysts, also possess the above properties; they differ from chemical catalysts by the following properties :

– *They further decrease the energy of activation* (see fig. 2-1). For example, the decomposition of H_2O_2 requires 18 kcal/mole without catalyst and 11.7 kcal/mole in presence of colloidal platinum, but only 2 kcal/mole in presence of the enzyme, catalase. (One molecule of catalase allows the decomposition of 5×10^6 molecules of H_2O_2 in one minute, in optimal conditions of temperature and pH). This decrease is due to the fact that the enzymatic reaction proceeds in a path other than the non-enzymatic pathway comprising in most cases, several successive steps where the energy barriers – which the molecules must cross in order to form an activated complex – are smaller.

– They are all of *protein nature* (exceptions are given in page 341). As will be seen later, this gives them a very high *specificity*. There are hundreds of different enzymes to catalyze the diverse reactions taking place in cells of living organisms, and as will be seen elsewhere in this chapter, even if a particular reaction is considered (dehydrogenation for example), the enzyme which catalyzes it can differ according to the compound (or *substrate*) undergoing the transformation.

– Their activity can be *controlled* which – as indicated later – is of very great importance for cell metabolism (see Chapter 8).

KINETICS OF ENZYMATIC REACTIONS

As mentioned above a catalyst – whether chemical or biological – acts by increasing the velocity of reactions. We will now examine the kinetics of enzymatic reactions. The velocity of an enzymatic reaction is generally followed up by *measuring the quantity of substrate transformed per unit time*; one can also record the quantity of the product (or one of the products) formed by the reaction. Samples of the reaction mixture are therefore collected at different times and titrations made by chemical, chromatographic, colorimetric or manometric methods depending on each case. Continuous analysis techniques may be used sometimes, particularly spectrophotometric methods (for example, one can record the variations of optical density at a given wavelength if the reaction is amenable to it, which is especially the case with NAD or NADP dehydrogenases) or titrimetric methods (for instance, the pH variations occurring during the reaction can be balanced by continuous addition of a base or acid, addition which may be automatically controlled by a pH meter).

Enzymatic activity may be expressed in several ways:

– The *enzymatic unit* is the quantity of enzyme which catalyses the transformation of 1 micromole of substrate in one minute, at 25° in optimal conditions of pH and substrate concentration (but in certain cases one millimole, or one milligram of substrate transformed per minute, are also used).

– The ratio of the number of enzymatic units to the amount (generally to one mg) of protein, gives the *specific activity* (which, obviously, increases as enzyme purification progresses).

– The molar activity is the number of moles of substrate transformed (or product formed) per mole of enzyme per minute; for this mode of expression one must have a highly purified enzyme of known molecular weight.

I. – INITIAL VELOCITY OF THE REACTION

If the quantity of substrate transformed is plotted against time, it is observed (fig. 2-2) that the first portion of the curve is a straight line with constant slope; the curve then bends, i.e. velocity decreases and finally becomes zero when all the substrate which was to be transformed (as a function of the equilibrium of the reaction) has disappeared. It is therefore advantageous to measure the velocity at the beginning of the reaction, when only a small quantity of the substrate has been transformed and when the amounts of reaction products are sufficiently small for the reverse reaction to be negligible (in practice a large excess of substrate is added); the slope is then maximum and the *initial velocity* of the reaction (v_0) can be determined.

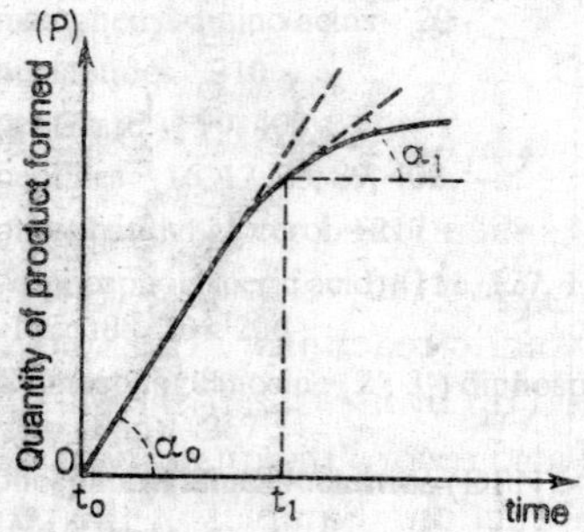

FIG. 2-2. — Determination of the initial velocity of an enzymatic reaction and of the velocity at an instant t_1.

$$v_0 = \tan \alpha_0$$

The reaction velocity (v) at a given time (t) is $d[P]/dt$ (where P is the quantity of product formed in the reaction). It can be determined by measuring the angle of the tangent to the curve at time t. The tangent at the origin gives v_0, the initial velocity of the reaction; this is the velocity referred to in the equations developed in paragraph II.

Some factors can influence the velocity of an enzymatic reaction.

II. – EFFECT OF ENZYME CONCENTRATION ON KINETICS

If several experiments are carried out with increasing quantities of enzymes, it is observed that after a given time (t_1), the quantity of substrate transformed is larger when more enzyme is present, provided one remains in the straight line portion of the curve (i.e. provided initial velocities are measured in each experiment); on the contrary, at time t_2, the proportionality between velocity and enzyme concentration ceases (see fig. 2-3 A).

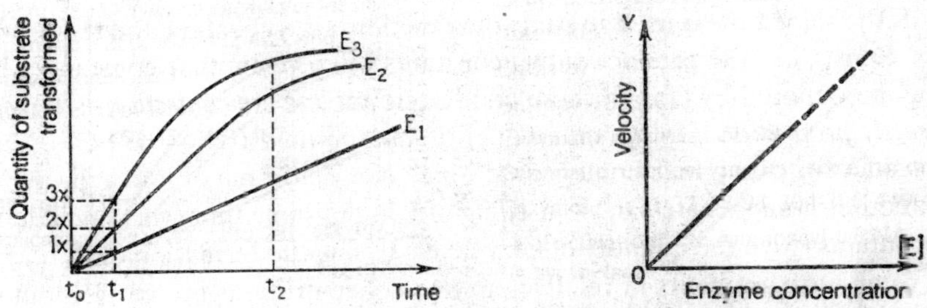

A

Kinetics of an enzymatic reaction in presence of increasing quantities of enzymes ($E_1 = E_2/2 = E_3/3$)

B

Variation of the velocity of an enzymatic reaction as a function of enzyme concentration

FIG. 2-3. — Effect of enzyme concentration on the velocity of enzymatic reactions.

The variation of initial velocity can also be plotted against enzyme concentration (after carrying out several experiments with increasing enzyme concentrations): it is observed that within certain limits, *initial velocity is proportional to enzyme concentration* (see fig 2-3 B); theoretically, for high enzyme concentrations, the velocity curve should bend and velocity should become constant, but in general this phenomenon is not observed because of the limitation imposed by the solubility of protein macromolecules (a 0.01 M solution of a protein of molecular weight 100 000 would require a solution of 1 000 g/liter); therefore, in the curve of figure 2-3 B, the plateau is not reached although – as will be seen later – the enzyme is a molecule which reacts with the substrate and one would expect that it would influence reaction velocity in the same manner as the substrate (see fig. 2-4).

The curve of figure 2-3 B shows that reaction velocity is nil when enzyme concentration is nil; but it was said earlier that enzymes catalyze reactions which could proceed even in their absence and sometimes with appreciable velocities. While studying velocity as a function of enzyme concentration, it is usual to carry out a "control" or "blank" test without the enzyme; the values obtained at different times in this blank test are subtracted from the values observed under various enzyme concentrations. The curve obtained with these corrected values represents exclusively the result of the enzymatic action, (and $v = 0$ for $[E] = 0$ is therefore justified).

The proportionality between reaction velocity and enzyme concentration has important practical applications because it enables the estimation of the relative concentrations of a given enzyme in cell homogenates without the necessity of purifying the enzyme. One has, on the one hand, $v = \text{Const.} \times [E]$, and on the other hand $v = [P]/t$, therefore:

$$\text{Const.} \times [E] = [P]/t, \quad \text{i.e.} \quad [E] = \frac{[P]}{t \times \text{Const.}}$$

If the same time is used to study the reaction, t is a constant and the product $t \times$ Const., i.e., the product of two constants will give another constant (which may be denoted by C). In these conditions *enzyme concentration is therefore directly proportional to the quantity of product formed* ($[E] = [P]/C$). Let us consider for example, a homogenate which can yield x mg of P in a given time; if another homogenate can yield $2x$ mg of P in the same time, we conclude that it contains twice the quantity of enzyme catalysing the formation of P.

Such measurements of relative enzyme concentrations are carried out systematically (on blood for example) in clinical laboratories, because certain variations enable the diagnosis of a pathological condition. In research laboratories, these determinations often represent the initial tests carried out while studying the regulation of the biosynthesis of an enzyme. It must be noted however, that the increase of active enzyme concentration can be due to either an increased synthesis of enzyme or an activation of the pre-existing enzyme molecules.

III. – EFFECT OF SUBSTRATE CONCENTRATION ON KINETICS

If enzyme concentration is maintained constant while substrate concentration [S] is varied, a rapid increase of velocity is observed first; but if [S] continues to increase, the curve bends and for higher values of [S], velocity ceases to increase and tends asymptotically to a *maximum value* (V_{max}), as can be seen in figure 2-4.

1. – Enzyme-substrate complex

For a better understanding of the changes taking place when substrate and enzyme concentrations are varied, one must study in more details the reaction

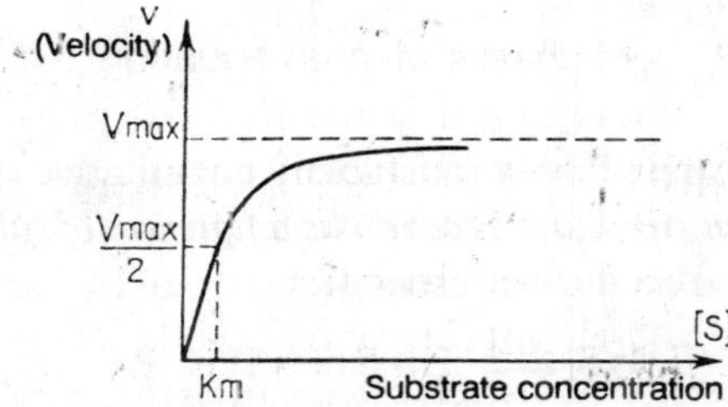

FIG. 2-4. — Variation of the velocity of an enzymatic reaction as a function of substrate concentration (enzyme concentration being constant).

catalyzed by an enzyme. When this reaction is written $S \overset{E}{\rightleftarrows} P$ (admitting of course, that there can be more than one substrate and more than one product), we get the false impression that the enzyme will modify the velocity of this reaction by its mere presence in the medium. But in reality it participates in the reaction by forming transitorily a specific enzyme-substrate complex. The existence of such complexes was suggested early this century, but it is only about thirty years later that their reality could be proved, first in the case of the peroxidase of horseradish (whose porphyrin prosthetic group has a characteristic absorption spectrum which is modified during the formation of the complex with the substrate), then for other enzymes, by various methods, using physicochemical techniques like the study of visible or ultraviolet absorption spectra, fluorescence, nuclear magnetic resonance etc. or by equilibrium dialysis. The modes of interactions between the 2 partners of the complex are now determined very accurately by X-ray diffraction analysis of crystals of enzyme-substrate complexes.

The basic reaction describing the enzymatic catalysis is written:

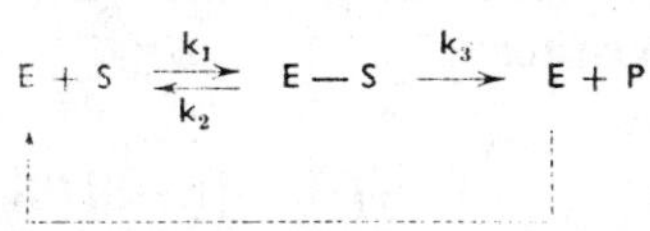

This representation is valid for the *initial period of the reaction when concentration of P is sufficiently low for the reverse reaction to be neglected.* The quantity of P formed will depend directly on the E – S complex concentration so that one is led to study the variations of E – S concentration as a function of the increase of [S]. Posing the problem in this manner, it is clear that in the beginning, if [S] increases, [E – S] will also increase and reaction velocity will therefore increase. But if [S] further increases, it is obvious that [E – S] cannot continue to increase beyond a certain maximum value which depends on the quantity of enzyme available. [E] becomes the limiting factor of the formation of the E – S complex, and an increase of [S] ceases to affect *v*: the plateau is reached (see fig. 2-4).

2. – Michaelis-Menten equation

These two authors carried out a satisfactory quantitative study of the *variations of the velocity of an enzymatic reaction as a function of substrate concentration*. This study is based on the representation

$$E + S \rightleftarrows E{-}S \longrightarrow E + P$$

and on the fact that equilibrium between E, S and E – S is a rapid process compared to the reaction E – S ⟶ E + P (which therefore controls the velocity of the enzymatic reaction). For the study of the velocity of a reaction, the Menten-Michaelis equation offers the advantage that it is not necessary to know either [E] or [E–S], concentrations which are often difficult to determine (and which were even more so in 1913 at the time of these two authors).

As mentioned above, the velocity of the enzymatic reaction is proportional to the E–S complex concentration, therefore $v = k_3$ [E–S]. As for the maximum velocity (V_{max}), it is observed for a substrate concentration such that the entire enzyme is bound to the substrate; the maximum value of [E – S] is equal to the *total enzyme concentration* $[E_T]$: $V_{max} - k_3 [E_T]$. We can write:

$$\frac{v}{V_{max}} = \frac{k[E{-}S]}{k[E_T]} = \frac{[E{-}S]}{[E_T]}$$

The equilibrium constant or Michaelis constant for the dissociation of the complex E – S is [1]:

$$\boxed{K_m = K_s = \frac{k_2}{k_1} = \frac{[E]\,[S]}{[E-S]}}$$

where [E] is the *free enzyme concentration*, equal to $[E_T]$ – [E – S]. Substituting [E] by this value, we can write:

$$K_m = \frac{([E_T] - [E{-}S])\,[S]}{[E{-}S]} = \frac{[E_T]\,[S]}{[E{-}S]} - \frac{[E{-}S]\,[S]}{[E{-}S]} = \frac{[E_T]\,[S]}{[E{-}S]} - [S]$$

This may be written in the form:

$$\frac{[E_T]\,[S]}{[E{-}S]} = K_m + [S] \quad \text{or} \quad [E{-}S] = \frac{[E_T]\,[S]}{K_m + [S]}$$

---

(1) This is valid for the conditions described above; if on the contrary, the *velocity of decomposition* of E—S into E+P becomes greater and ceases to be negligible as compared to the *velocity of dissociation* of E—S into E+S, the Michaelis constant ceases to be equal to the constant of dissociation of E—S into E+S; we no longer have $K_m = K_s = \frac{k_2}{k_1}$, but $K_m = \frac{k_2 + k_3}{k_1}$

or again:

$$\frac{[E-S]}{[E_T]} = \frac{[S]}{K_m + [S]}$$

but $\frac{[E-S]}{[E_T]} = \frac{v}{V_{max}}$ (see above). We can therefore write:

$$\boxed{\frac{v}{V_{max}} = \frac{[S]}{K_m + [S]}}$$

It may be noted that, having determined experimentally the variations of v as a function of [S], one can use the *Michaelis-Menten equation* to obtain the value of V_{max} *and calculate* K_m. The knowledge of enzyme concentration is not necessary; K_m can therefore be determined on incompletely purified preparations. If for a given substrate, an identical K_m is found in two preparations, it may be assumed that they contain the same enzyme.

Taking $v = V_{max}/2$, we have:

$$\frac{1}{2} = \frac{[S]}{K_m + [S]} \quad \text{or } K_m = [S]$$

The *Michaelis constant* (or equilibrium constant of the dissociation of the E−S complex) is therefore equal to the *substrate concentration for which velocity is half the maximum velocity* (see fig. 2-4). In general, the K_m values range between 10^{-2} M and 10^{-8} M. It must be noted that K_m is a measure of the affinity of the enzyme for its substrate. The stronger the E−S interaction, the larger the quantity of enzyme combined with its substrate in the form E−S and the smaller the quantity of free enzyme; therefore [E] will be small and [E−S] large; consequently, K_m will be small. The *affinity* of an enzyme for a substrate is equal to $1/K_m$. If K_m is large (a high substrate concentration is therefore required to obtain a velocity equal to $V_{max}/2$), affinity is low; if K_m is small, affinity is high (a small substrate concentration is sufficient to have a velocity equal to $V_{max}/2$).

Examining the Michaelis-Menten equation written in the form $v = V_{max}\frac{[S]}{K_m + [S]}$, one may note that when [S] is very large compared to K_m, K_m can be neglected, and v tends to V_{max}. On the contrary, when [S] is very small compared to K_m, v is equal to $V_{max}\frac{[S]}{K_m}$, which shows that v is proportional to [S]. This does confirm the experimental results represented by the curve of figure 2-4.

Enzymatic catalysis comprises two steps:

1) *the formation of E−S* which implies the recognition of S by the active site of the enzyme. This first step is characterized by the K_m which, as mentioned earlier, is equal to $\frac{[E][S]}{[E-S]}$;

2) *the chemical operation of the decomposition of E – S into E + P.*

This second step is characterized by V_{max}.

v can be plotted as a function of [S] on the basis of experimental results; a branch of an equilateral hyperbola is obtained (figure 2-4). V_{max} (at least an approximate value) is obtained for large values of [S]. To determine K_m, the point $V_{max}/2$ is plotted on the vertical axis and from the corresponding point of the curve a perpendicular is drawn to the horizontal axis. But the value of V_{max} being only approximate, the determinations of $V_{max}/2$ and thence K_m are rather inaccurate. Furthermore, a large number of experimental points are needed to trace such a curve, particularly at the low substrate concentrations where precision is not very high. But there is another graphical representation which requires only a small number of experimental points obtained at high substrate concentrations where precision in the determination of initial velocity is much greater.

3. – Lineweaver-Burk representation

The Michaelis-Menten equation may be written:

$$v = \frac{V_{max}[S]}{K_m + [S]} \quad \text{or} \quad \frac{1}{v} = \frac{K_m + [S]}{V_{max}[S]}$$

or

$$\frac{1}{v} = \left(\frac{K_m}{V_{max}} \times \frac{1}{[S]}\right) + \frac{1}{V_{max}}$$

The latter form is an equation of the type $y = ax + b$ (where $1/v$ and $1/[S]$ are the variables y and x, while K_m/V_{max} and $1/V_{max}$ are the constants a and b). Then, by plotting the values of $1/v$ (the reciprocals of velocities determined experimentally) as a function of $1/[S]$, we obtain a *straight line* which – as seen in figure 2-5 – intersects the vertical axis ($1/v$) at the point $1/V_{max}$ (since $1/[S] = 0$, we have $1/v = 1/V_{max}$). This straight line intersects the horizontal axis ($1/[S]$) at the point $-1/K_m$; if $1/v = 0$, we have:

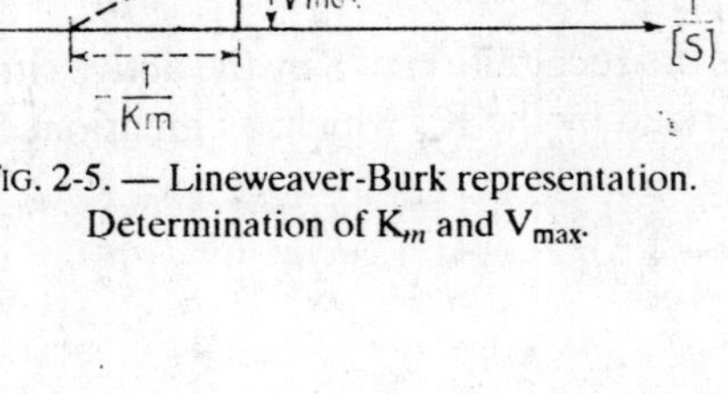

FIG. 2-5. — Lineweaver-Burk representation. Determination of K_m and V_{max}.

$$0 = \left(\frac{K_m}{V_{max}} \times \frac{1}{[S]}\right) + \frac{1}{V_{max}}$$

or

$$\frac{1}{[S]} = -\frac{1}{V_{max}} \times \frac{V_{max}}{K_m}$$

i.e.

$$\frac{1}{[S]} = -\frac{1}{K_m}$$

This graphical representation is more convenient for determining the Michaelis constant (K_m) and the maximum velocity (V_{max}) of the reaction for a given enzyme concentration. Moreover, as will be seen later (see pages 72-74), it permits the distinction between a competitive inhibitor and a non-competitive inhibitor.

IV. – EFFECT OF VARIOUS PHYSICAL AND CHEMICAL AGENTS ON KINETICS

We will first consider the effects of the variations of two physical factors, temperature and pH. We will then examine certain chemical agents (or *effectors*) which can modify the kinetics of enzymatic reactions and are either activators or inhibitors.

1. – Effect of temperature

The study of the initial velocity of an enzymatic reaction as a function of temperature reveals two very distinct phases corresponding to two different phenomena (see fig. 2-6):

– in the zone of the lowest temperatures (in figure 2-6, between 0 and 40° approximately) reaction velocity increases when temperature increases. This increase of velocity with temperature is simply explained as seen earlier, by an increase in activated complex concentration, when more energy in thermal form is supplied to the reacting system.

– then, beyond a certain temperature which varies according to the enzymes (about 45°), one observes the denaturation of the protein (see pages 32-33).

In figure 2-6, the curve in solid line is therefore the resultant of the 2 curves in dotted line: *activation curve* and *denaturation curve*. The *optimal temperature* is the temperature at which the 2 phenomena are in equilibrium; it depends on pH, ionic strength of the medium and also reaction time. The shorter the reaction time, the higher could be the optimal temperature; at 50 or 60°, velocity can be very high but only for a comparatively short time, only one minute for example; (therefore, if one measures the quantity of substrate transformed – or product formed – after an interval of 10 minutes, one would erroneously find a velocity which is only one tenth of the actual velocity of the reaction during the one minute interval when the enzyme functioned before being inactivated).

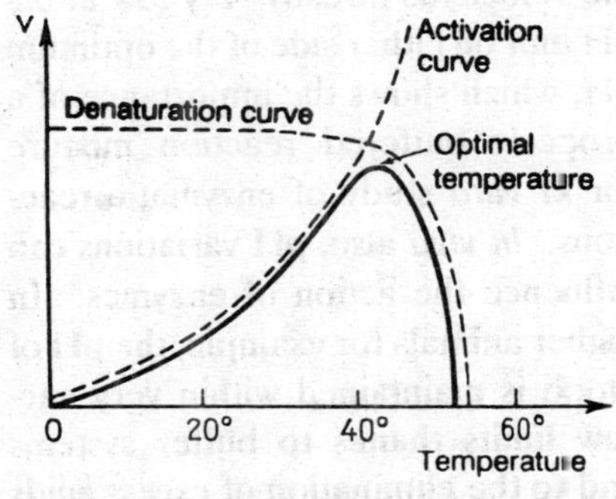

FIG. 2-6. — Effect of temperature on the velocity of an enzymatic reaction.

Some enzymes, especially those of comparatively lower molecular weight and simple structure, stabilized by disulphide bridges less sensitive to temperature rise (myokinase, ribonuclease), are specially heat resistant and can even be heated to the boiling temperature of water. In general, enzymes seem to be less sensitive to heat in presence of the corresponding substrate (there is certainly a stabilization of the three-dimensional structure). Certain thermophile microorganisms live in water at 70 or 80°, their enzymes therefore function at those temperatures. A mutation can impart to an enzyme a thermosensitivity different from that of the enzyme of the wild strain.

2. – Effect of pH

pH variations can act on:

– the *enzyme*, by modifying the degree of ionization of certain functional groups whose charge – positive or negative – is necessary, either for the formation of the enzyme-substrate complex (these are groups belonging to amino acids which constitute the "active site" or "catalytic site" of the enzyme), or for *maintaining the native three-dimensional conformation* of the enzymatic protein (these are groups possibly belonging to amino acids located outside the active site, at various places of the molecule);

– *the substrate*, by changing its degree of ionization, which can lead to, or on the contrary prevent, the formation of the enzyme-substrate complex if the substrate must be in a given ionized form to be able to bind to the active site of the enzyme (in other words, in this case, the *real* substrate of the enzyme is present only in a suitable pH interval).

One can therefore define an *optimum pH* for the enzymatic transformation of a substrate *in a medium of given composition*. As shown by figure 2-7, on either side of this optimum pH, the reaction velocity decreases rapidly and, in general, at two pH units on either side of the optimum pH, velocity is negligible (at least 100 times less). In the case of the pancreatic ribonuclease, the velocity is already very low at 0.5 pH unit on either side of the optimum pH, which shows the importance of a properly buffered reaction mixture for *in vitro* study of enzymatic reactions. *In vivo* also, pH variations can influence the action of enzymes. In higher animals for example, the pH of blood is maintained within very narrow limits thanks to buffer systems and to the elimination of excess acids or bases.

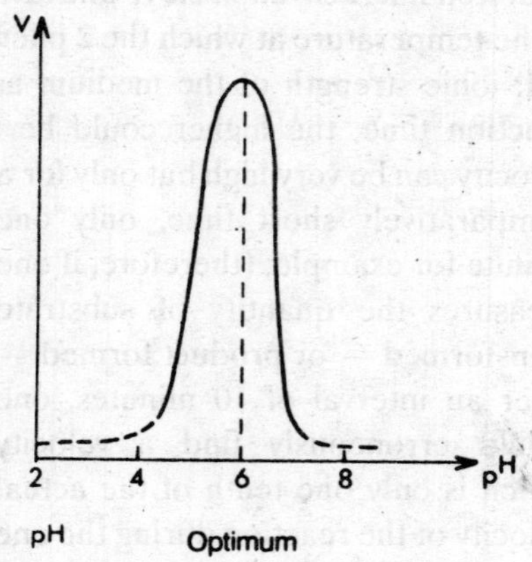

FIG. 2-7. — Variations of the velocity of an enzymatic reaction as a function of pH (example of urease).

The optimum pH varies widely according to the enzymes. It can be highly acidic (between 1.5 and 2 for pepsin) or highly basic (between 9.5 and 10 for arginase), but it is mostly near neutrality (between 6 and 8).

Very often K_m or V_{max} depend on the pH in a manner resembling a titration curve; this allows the determination of the nature of an amino acid participating in the formation of the E–S complex, from the pK of the ionizable group apparently implied in the interaction: a pK of 4-4.5 corresponds to a carboxylic group; a pK of 6.5–7, to an imidazole group (side chain of histidine); a pK of 8.5–9, to a – SH group; a pK of about 10.5, to a phenol group (side chain of tyrosine) or a ε-amino group (side chain of lysine). However, certain residues have rather similar pK values, and besides, these values can be modified when the amino acid is involved in a protein (by the general charge of the protein, the presence of a nearby ionized group, a masking due to steric effects, etc.), so that the nature of the amino acid must then be confirmed by other, more rigorous methods.

3. – Inhibitors

In this study we will consider only specific inhibition, but it may be recalled that certain agents cause the *denaturation* or even *degradation* of proteins. In most cases, this is a sudden, irreversible, *non specific* phenomenon, the study of which gives no information on the mechanisms of enzymatic action.

On the contrary, the use of certain inhibitors enabled, in various cases, the determination of the nature of amino acids belonging to the active site and participating in the formation of the E–S complex. Among substances which can bind to various groups of proteins (hydroxyl, sulphydril, carboxyl, etc.) and cause a loss of catalytic properties, either by modifying the conformation or blocking the active site of the enzyme, one finds ions of heavy metals, or compounds like mono-iodoacetic acid or para-chloromercuribenzoic acid which act on the thiol groups:

Enzyme – SH + ICH_2 – COOH
$\longrightarrow$ HI + Enzyme – S–CH_2 – COOH (irreversible reaction)

Enzyme – SH + ClHg – C_6H_4 – COOH
$\longrightarrow$ HCl + Enzyme–S–Hg–C_6H_4–COOH (reaction reversible by addition of an excess of a reducing substance like cysteine).

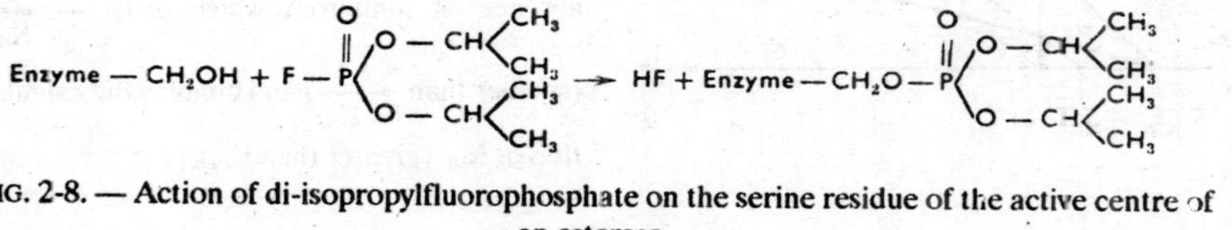

FIG. 2-8. — Action of di-isopropylfluorophosphate on the serine residue of the active centre of an esterase.

One can also cite di-isopropylfluorophosphate which enables the phosphorylation of the serine residue of the active centre of various enzymes (esterases and certain proteolytic enzymes like trypsin) thus causing the irreversible inactivation of the enzyme (see fig. 2-8).

Among the *specific* inhibitors (of an enzyme or group of enzymes) a distinction is generally made between inhibitors having a competitive action and those acting in a non-competitive manner.

A. – Competitive inhibitors

These are compounds which present a *structural analogy* with the substrate of the enzyme and can thus compete with it in order to bind to the active site of the enzyme. The enzyme can therefore combine either with the substrate or with the inhibitor, and the following equilibria may be established:

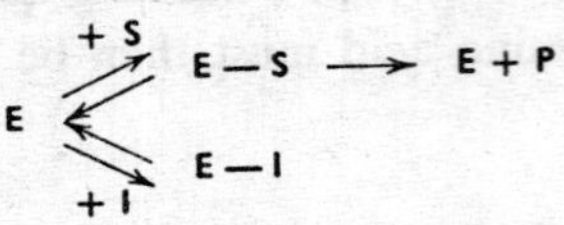

It is clear that the enzyme involved in the E – I complex cannot function as catalyst; only the E – S complex will allow the formation of the reaction products. *Inhibition*, in other words the proportion of enzyme molecules combining with the inhibitor, depends on *substrate concentration*, *inhibitor concentration*, *affinity of the enzyme for the substrate* and *affinity of the enzyme for the inhibitor.*

The addition of a competitive inhibitor favours the dissociation of the E – S complex and *therefore increases the Michaelis constant*, K_m (which means that $1/K_m$, affinity of the enzyme for the substrate, decreases, as shown by figure 2-9). By adding large quantities of the substrate, the inhibitor can be displaced from the active sites it occupies. If $1/v$ is plotted as a function of $1/[S]$ as proposed by Lineweaver and Burk (see pages 68-69) the resulting *straight line intersects the straight line obtained in the absence of inhibitor at an infinitely high*

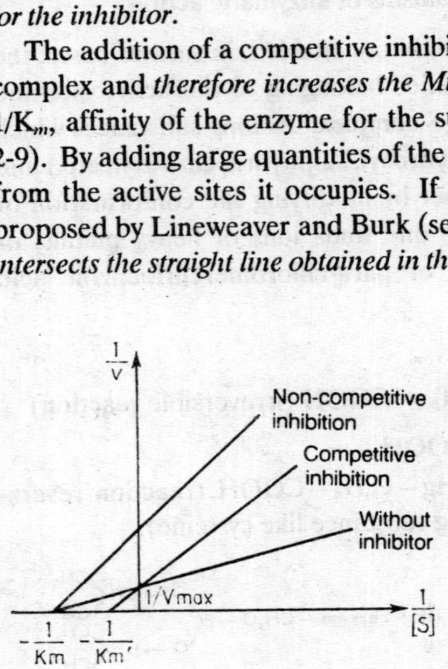

FIG. 2-9. — Lineweaver-Burk representation. Differences between competitive inhibition and non-competitive inhibition.

It may be noted that the straight line corresponding to competitive inhibition intersects the horizontal axis at a point (different from the point of intersection obtained in the absence of inhibitor) which gives $-\frac{1}{K_m'}$, (smaller than $-\frac{1}{K_m}$) and enables the calculation of $K_{m'}$ (greater than K_m).

substrate concentration, i.e. when $1/[S] = 0$, in other words on the *vertical axis* (see fig. 2-9); this is characteristic of a competitive inhibition.

Competitive inhibition exists for all enzymes. A non-metabolizable structural analogue of the substrate is generally a competitive inhibitor. One may cite the example of *succinodehydrogenase* (see fig. 2-10) inhibited by various structural analogues of succinic acid (which differ from it only by the number of $-CH_2-$ between the 2 carboxylic groups).

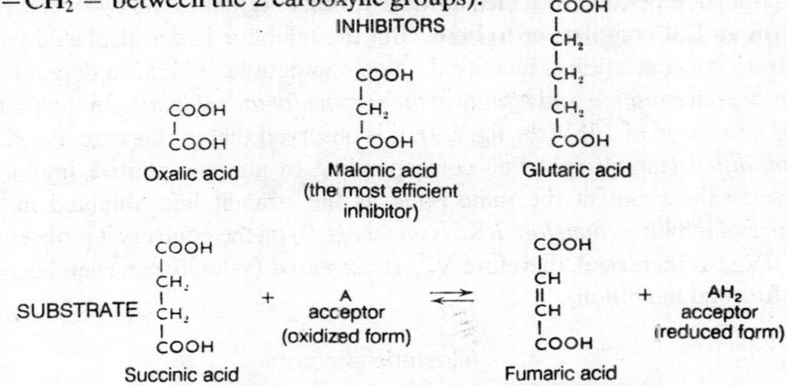

FIG. 2-10. — Competitive inhibitors of succinodehydrogenase.

Numerous practical applications are based on competitive inhibition, particularly in therapeutics: fight against microbes, control of weeds, parasite insects etc. The basic idea is to try to *specifically inhibit an enzymatic reaction which is of capital importance in a given organism* (the one being fought), but is not so in the neighbouring organisms (the host for example), either because the reaction product is dispensable in the host, or because the host can produce it by a different process. A classical example is that of *sulphamides*, analogues of para-amino-benzoic acid (see fig. 2-11), a compound indispensable to many bacteria (but not to man) for the synthesis of folic acid, the coenzyme necessary for the transport of one-carbon units (see fig. 2-19). The sulphamides will compete with para-aminobenzoic acid for the active site of a bacterial enzyme catalyzing the transformation of this derivative into folic acid and this explains their bacteriostatic action.

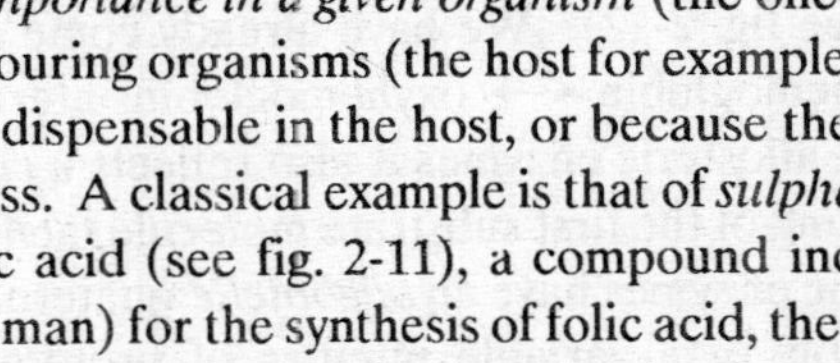

FIG. 2-11. — Structural analogy between para-aminobenzoic acid and a compound of the sulphamide family.

Chemotherapy against cancer is based on numerous *antimetabolites*, which are generally structural analogues of purine or pyrimidine bases (constituents of nucleic acids, as will be seen in Chapter 6), inhibiting the biosynthesis of nucleic acids and therefore blocking cell division (this effect is particularly pronounced in the tumoral tissues which grow much faster, but unfortunately, it manifests itself also in healthy cells).

B. – Non-competitive inhibitors

They bind either to the enzyme (but to a site other than the active site, so that there is no competition with the substrate for this site), or to the E–S complex to form an ESI complex, or to both. But the inhibitor is not displaced when substrate concentration is increased. Non-competitive inhibition depends on *inhibitor concentration* and *affinity of the enzyme for the inhibitor.* In the plot of $1/v$ as a function of 1/[S] (see fig. 2-9), it is observed that in this case, *the* K_m *is not modified* (the straight line corresponding to non-competitive inhibition intersects the x-axis at the same point as the straight line obtained in the absence of inhibitor, *therefore* $1/K_m$ *is unchanged*); on the contrary it is observed that $1/V_{max}$ is increased, therefore V_{max} *is decreased* (velocity can even become zero for total inhibition).

4. – Allosteric effectors

Allosteric enzymes differ from other enzymes in that the plot of v as a function of [S], obtained when studying the kinetics of the reactions catalyzed by allosteric enzymes, does not give a branch of an equilateral hyperbola corresponding to the Michaelis-Menten equation (see fig. 2-4), but a sigmoid curve (S-shaped) (see fig. 2-12). We have already come across such a curve while studying the hemoglobin ←→ oxyhemoglobin interconversion (see fig. 1-31). In the case of allosteric enzymes it also reflects a *cooperative effect*, i.e. the fact that the binding of the first substrate molecule facilitates the binding of the second. Allosteric enzymes have an *oligomeric* quaternary structure resulting from the assembly of a variable number of protomers joined by non-covalent bonds which were studied in the previous chapter (see page 28). These oligomeric structures bear a close resemblance with that of hemoglobin

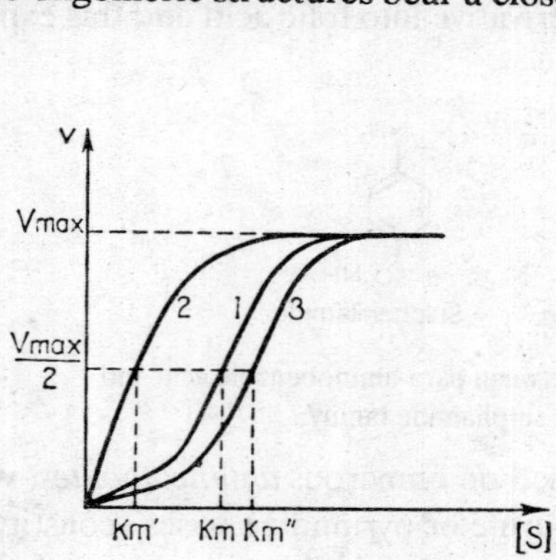

FIG. 2-12. — Kinetics of a reaction catalyzed by an allosteric enzyme.

Curve 1 : in absence of allosteric effector.

Curve 2 : in presence of an allosteric activator.

Curve 3 : in presence of an allosteric inhibitor.

(see pages 54-55); we will study them in more details in the chapter on regulation (see pages 472-477).

In addition to the catalytic site where the substrate binds, these enzymes have one or several *allosteric sites*, which can be located on a different polypeptide chain, and where *allosteric effectors* (activators or inhibitors) having no structural analogy with the substrate, can bind.

A. Allosteric activators

When an allosteric activator binds to the allosteric site, there results a slight modification of the conformation of the enzyme – called *allosteric transition* – (reversible), which causes a change of conformation of the active (catalytic) site. This site acquires a conformation more favourable to the binding of the substrate; *the affinity of the enzyme for the substrate increases* ($K_m' < K_m$). Even the shape of the curve can change from the sigmoid form to the hyperbolic form (see fig. 2-12). As shown by figure 2-13, one can represent schematically, the allosteric transition caused by the binding of the activator and its effects on the active site (this scheme does not claim to represent the real structure of the enzyme; it does not show the various protomers).

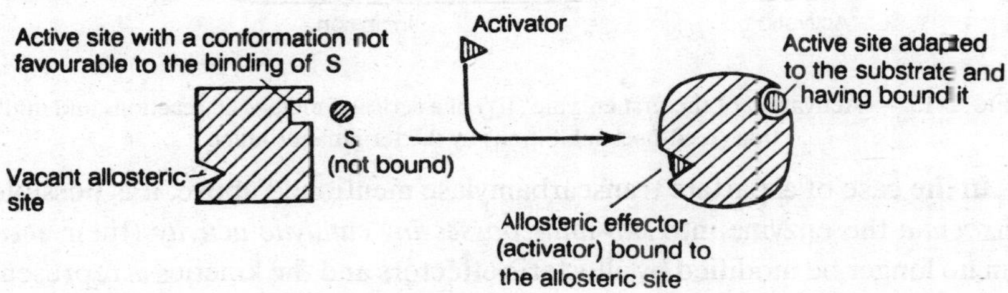

FIG. 2-13. — Allosteric transition caused by an allosteric activator (diagram).

B. – Allosteric inhibitors

When an allosteric inhibitor binds to the allosteric site, an allosteric transition (also reversible) takes place causing a change in the active site which now takes a conformation less favourable to the binding of the substrate (see fig. 2-14). If v is plotted as a function of [S], the curve is again a sigmoid *but a decrease of the affinity of the enzyme for the substrate is observed* ($K_{m''} > K_m$). In fact, this is only one of the various possible mechanisms because there are

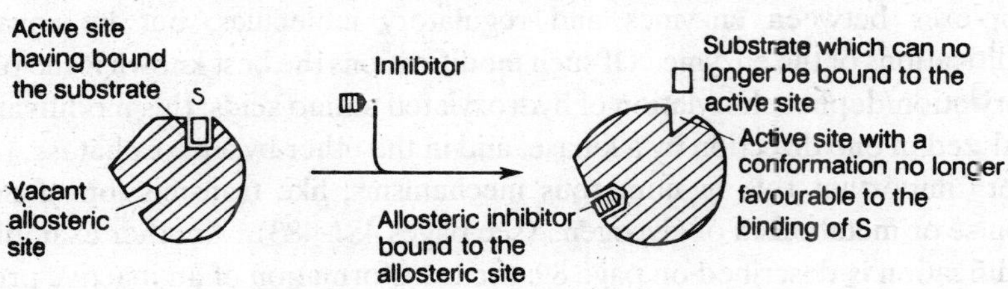

FIG. 2-14. — Allosteric transition caused by an allosteric inhibitor (diagram).

several types of inhibitions; we will study these phenomena in some more details in Chapter 8 (see page 474).

Very often, the enzyme catalyzing the first reaction of a metabolic pathway is an allosteric enzyme whose activity can be controlled by various allosteric effectors (activators and inhibitors). Frequently, one of the terminal products of the metabolic pathway (H in the diagram of figure 2-15) is an allosteric inhibitor of this first enzyme (E_1). This feedback inhibition offers the very great advantage of inhibiting the first enzyme – as soon as there is an excess of terminal product – which represents a significant economy for the cell, especially in the case of a sequence of reactions requiring energy (for example, during the biosynthesis of an amino acid or a nucleotide). Thus, we will see (page 296) that aspartate transcarbamylase, first enzyme of the series of reactions bringing about the synthesis of pyrimidine ribonucleotides, is inhibited by a terminal product of this pathway (CTP) and we will study in Chapter 8, the various aspects of this mode of metabolism regulation by the control of the activity of enzymes (see pages 472-481).

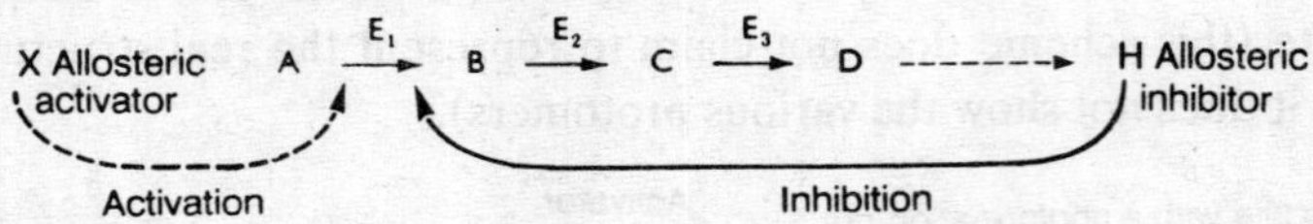

FIG. 2-15. — Activation of the first enzyme (E_1) of a series of metabolic reactions and inhibition (feed-back inhibition) by the terminal product.

In the case of aspartate transcarbamylase mentioned above, it is possible to *dissociate* the enzyme into *sub-units possessing catalytic activity* (their activity can no longer be modified by allosteric effectors and the kinetics is represented by a hyperbola which conforms with the Michaelis equation) and into *regulatory sub-units* capable of binding the allosteric effector (but deprived of catalytic activity), which shows that catalytic and allosteric sites are not only different, but also carried by different polypeptide chains joined in the quaternary structure of the enzyme (see page 475).

5. – Activation or inhibition by covalent modification

The activity of numerous enzymes is controlled, not by the formation of complexes between enzymes and regulatory molecules, but by covalent modifications of the enzyme. Of such modifications the best known is the phosphorylation/dephosphorylation of hydroxylated amino acids; this modification, catalyzed in one direction by a kinase, and in the other by a phosphatase, plays a very important role in numerous mechanisms, like transmission of nerve impulse or metabolism of glycogen. (see pages 481-483). Another example of modification is described on page 89: the transformation of an inactive proenzyme into an active enzyme by proteolytic cleavage (or cleavages).

STRUCTURE OF ENZYMES

Enzymes are *proteins*, having primary, secondary, tertiary and in certain cases, even quaternary structures of the type described in Chapter 1. Sometimes, this protein part (or apoenzyme) is not sufficient for catalytic action which then requires the presence of a co-factor.

I. – POLYPEPTIDE CHAINS

Some enzymes consist of a single polypeptide chain; in most cases these are secreted enzymes like ribonuclease. Other enzymes, in a much larger number, are composed of several chains (or sub-units), identical or different. When enzymes comprise identical sub-units, each chain naturally carries an active centre: a tetrameric enzyme has 4 active centres. This category of oligomeric enzymes includes the allosteric enzymes (representing 10-20% of enzymes with quaternary structure) referred to in the previous paragraph.

In certain cases the dissociation into monomers is relatively easy; in others, stronger agents are needed to break the bonds linking the sub-units. It is only in very rare cases, that sub-units obtained by dissociation of an enzyme of quaternary structure, still possess activity. Whether an oligomeric enzyme is allosteric (sigmoidal velocity curve) or not (Michaelian velocity curve), the interaction of sub-units within the oligomer is generally an absolute condition for the expression of catalytic activity.

The existence of *isoenzymes* i.e. enzymes existing in several molecular forms, should be mentioned: lactate dehydrogenase or lacticodehydrogenase (see fig. 4-30) for example, exists in 5 different forms which can be separated by electrophoresis. These 5 isoenzymes are all tetramers which differ only by the mode of association of the two types of monomers (H and M)[1]; there are 5 possibilities of association : H_4, H_3M_1, H_2M_2, H_1M_3 and M_4. Different isoenzymes of lacticodehydrogenase are found in the same organism depending on the tissue studied and changes in the distribution of these different forms are observed in certain pathological conditions.

Now, in a large number of cases, one knows not only the sequence of amino acids determined by the methods described in Chapter 1, but also details of the three-dimensional conformation thanks to the study of X-ray diffraction patterns of enzyme crystals. One knows the three dimensional structure of hydrolases like lyzozyme, ribonuclease, chymotrypsine, and also that of dehydrogenases like alcohol dehydrogenase or glyceraldehyde-3-phosphate dehydrogenase, or that of kinases like hexokinase, or that of isomerases, etc.

---

(1) These monomers are thus designated because the former predominates in the Heart (H) and the latter in the Muscle (M).

The knowledge of the three-dimensional structures of these enzymes has led to significant advances in the study of the relationship between conformation and catalytic activity.

The *molecular weight* of enzymes is extremely variable, not only because of the presence of chains of varying lengths but also because of the presence of oligomeric enzymes consisting of a varying number of chains. It may vary from about 10 000 (ribonuclease: 13 700) to several hundreds of thousands (β-galactosidase: 520 000).

The enzymes which catalyze a series of metabolic reactions can aggregate and form multienzymatic complexes whose molecular weight can then widely exceed one million. For example, the transformation of pyruvate into acetyl-coenzyme A (see fig. 4-36) is made possible by a multienzyme complex composed of 3 different enzymes.

II. – CO-FACTORS

Some enzymes need a co-factor to exert their catalytic action. There are different types of co-factors.

1. – Metal ions

A large number of enzymes need a metal ion for activity; they are called *metalloenzymes*. The metal cation is supplied in the food to the cell which uses the metalloenzyme : it is an *oligoelement*. One of the ions most frequently found in metalloenzymes is Zn^{2+} : it is necessary for the expression of the activity of carbonic anhydrase, carboxypeptidase, alkaline phosphatases and a variety of other enzymes. This metal is strongly bound to the enzyme (dissociation constants of the order of 10^{-10} M or less). In most cases, the Zn^{2+} ion participates in the recognition (K_m) and the catalytic transformation (V_m) of the substrate. Frequently, the binding of the cation Zn^{2+} to the enzymatic protein also plays a structural role by stabilizing an effective conformation of the active centre. Some enzymes use Zn^{2+}, but Mg^{2+}, Ca^{2+} or even Na^+ and K^+ are also used by a very large number of other enzymes for activity. Metalloenzymes which contain iron or copper in their active centre are most frequently used for the catalysis of the oxidation-reduction processes:

$Fe^{2+} \rightleftharpoons Fe^{3+}$ (cytochromes for example) and $Cu^+ \rightleftharpoons Cu^{2+}$

2. – Prosthetic groups

These are organic co-factors of non-protein nature strongly bound to the enzymatic protein, and whose presence is indispensable for catalytic activity. Thus, in catalases, peroxidases and cytochromes, the porphyrin group is covalently bound with the protein part (or *apoenzyme*), as in the case of heme and globin (see pages 51-53).

3. – Coenzymes

A. – Mode of action of coenzymes

These are also organic molecules, of small size compared to the enzymatic protein. These molecules are called coenzymes for historical reasons: they remained firmly bound to the enzymatic protein throughout the purification procedure.

This designation is extremely misleading. There are two broad categories of coenzymes. The first category which really deserves to be called coenzyme, includes the coenzymes which are really part of the active centre. In other words, the small coenzyme organic molecule assists the protein side chains in the catalysis. The best example of such a coenzyme is pyridoxal phosphate (see page 85 and fig. 7-1).

The second category of molecules called coenzymes does not really deserve its name. It would be preferable to call them substrates (or co-substrates). The prototype of this type of coenzyme is NAD^+ (see fig. 2-16).

If a group X is transferred from a substrate "A" to a substrate "B" in presence of enzyme "E" and coenzyme "co", there are 2 steps:

— the binding of X to the coenzyme:

$$A - X + co \overset{E}{\rightleftharpoons} A + co - X$$

— the transfer of X to the second substrate:

$$co - X + B \overset{E}{\rightleftharpoons} co + B - X$$

The sum is written $A - X + B \overset{E,\, co}{\rightleftharpoons} A + B - X$. It should be stressed that the coenzyme has reacted stoichiometrically and that it is regenerated. There are two cases:

— the above 2 steps are catalyzed by the *same enzyme*. In this case, the compound co-X formed intermediately is generally not released because the coenzyme remains bound to the enzyme. This takes place for example in reactions catalyzed by the transaminases whose coenzyme is pyridoxal phosphate (see fig. 7-1);

— the 2 steps are catalyzed by 2 *different enzymes*. In this case the coenzyme is a transporter of X and it must be able to dissociate easily from its combinations with the two enzymes. These two enzymes therefore have a specific site for the reversible binding of the coenzyme which then behaves as a substrate.

This happens for example in the reactions catalyzed by the dehydrogenases whose coenzyme is NAD^+ (see fig. 2-16).

B. – Relations between vitamins and coenzymes

Vitamins (etymologically the term means: "nitrogenous substances needed for life") are organic substances which cannot be synthesized by some organisms and must therefore be supplied to them regularly but in small quantities – in order to ensure a normal metabolism. Certain vitamins, notably water-soluble vitamins and particularly the B group vitamins, are coenzymes or essential constituents of coenzymes (see below). It must be noted that the same compounds often serve as coenzymes for all living organisms, but are vitamins only for some of them. They may be compared with some amino acids which are found in proteins of all organisms but are "indispensable" or "essential" only for some of them (those which cannot synthesize them). Even if a coenzyme is not a vitamin for a given species, it can be one for an individual of this species having undergone a mutation affecting an enzyme needed for the synthesis of this coenzyme. In animals and man, a prolonged deficiency in a vitamin of this type, called *avitaminosis*, produces a set of pathological symptoms due to the inability of the organism to catalyze some reactions (because the corresponding coenzyme is missing).

C. – Principal coenzymes

Some coenzymes are of nucleotide nature[(1)]. Coenzymes may be classified either according to their structure, or according to the reaction in which they participate (in the latter case distinction may be made between the coenzymes of *oxidation-reduction* reactions and those of *group transfer* reactions).

a) Coenzymes involved in oxidation-reduction reactions. – The following ones may be cited:

NICOTINAMIDE ADENINE DINUCLEOTIDE or *NAD* – Its structure (see fig. 2-16) comprises 2 nucleotides (one consisting of adenine, ribose and phosphate and the other nicotinamide, ribose and phosphate) linked by a pyrophosphate bond. Nicotinamide has the pyridine ring (this is why the coenzyme is also called *DiphosphoPyridine Nucleotide* or *DPN*) with a quaternary ammonium; this explains why the coenzyme is written NAD^+ when it is in oxidised form (in the reduced form nitrogen is ternary).

NAD is the coenzyme of various dehydrogenases; it binds the hydrogen originating from a substrate AH_2 and transforms into $NADH + H^+$ (sometimes written $NADH_2$ for brevity). But in the course of another reaction catalyzed by a second dehydrogenase, it can transfer the hydrogen to another substrate B

(1) A nucleotide consists of a nitrogenous base, a pentose (ribose or deoxyribose) and a molecule of phosphoric acid. The structure of nucleotides, constituents of nucleic acid, is studied in detail in Chapter 6.

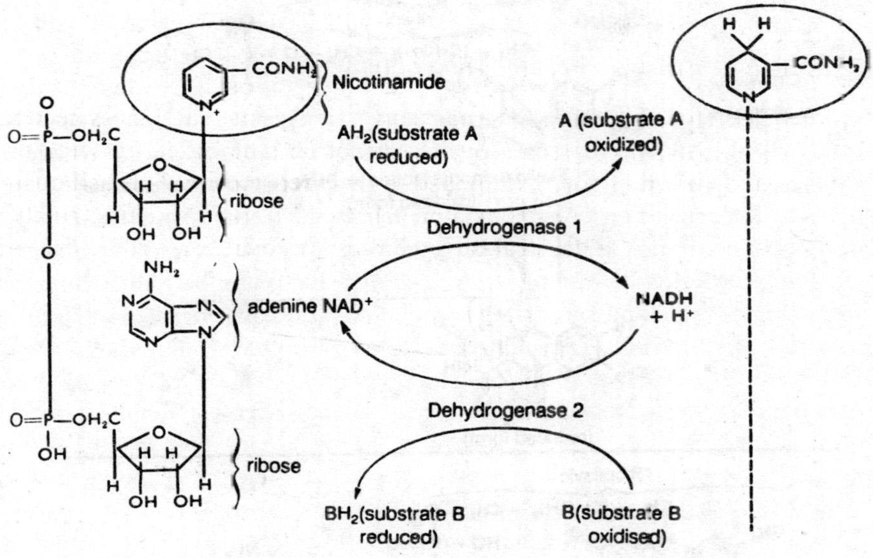

FIG. 2-16. — Structure of NAD and participation of this coenzyme in reactions catalyzed by dehydrogenases.

which will be reduced into BH_2, while the coenzyme is oxidized to the stage NAD^+ (fig. 2-16). During these 2 reactions, the coenzyme really behaves like a substrate (just as A or B).

Just like the nucleotides constituting the nucleic acids, this coenzyme absorbs in U.V. around 260 nm (in oxidized or reduced form), furthermore, the reduced form absorbs around 340 nm, so that the enzymatic reactions involving this coenzyme may be followed up by spectrophotometry.

NICOTINAMIDE ADENINE DINUCLEOTIDE PHOSPHATE or *NADP* – This is also a coenzyme of some dehydrogenases, very similar to the one mentioned above; it differs only by an additional phosphate residue esterified on the hydroxyl in 2′ of the ribose bound to adenine (it was also called *TriphosphoPyridine Nucleotide* or *TPN*). It is a hydrogen transporter. Its mechanism of action is very similar to that of NAD and therefore, it also behaves as a substrate.

The relationship of these 2 coenzymes with *nicotinic acid* or *vitamin PP* ("Pellagra Preventing" i.e. preventing a disease called pellagra) is rather evident.

FLAVINE NUCLEOTIDES – *Flavine Mono Nucleotide (FMN)* and *Flavine Adenine Dinucleotide (FAD)* are also coenzymes of dehydrogenases often called *flavoproteins* or *yellow ferments* (example : Warburg respiratory yellow ferment) because of their coloration due to the chromophore group formed by the conjugated double bond system of the flavine ring. As shown by figure 2-17, these two coenzymes derive from *riboflavin* or *vitamin B_2*. It must

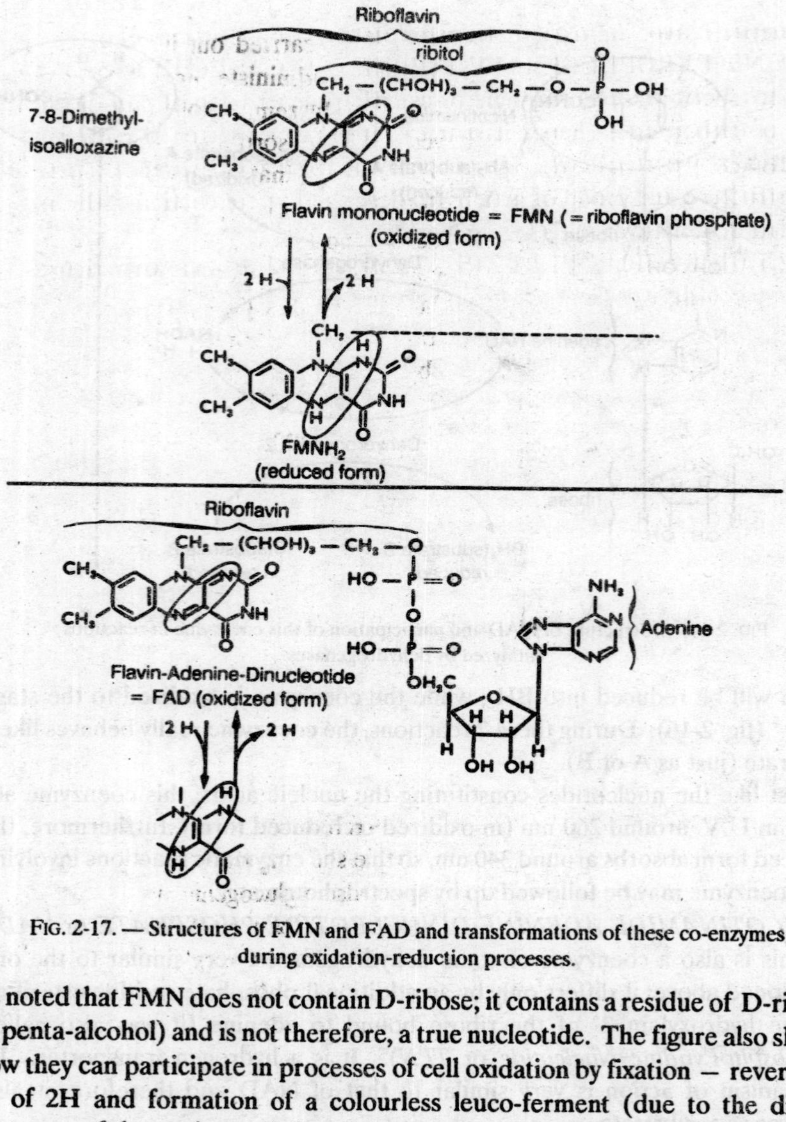

FIG. 2-17. — Structures of FMN and FAD and transformations of these co-enzymes during oxidation-reduction processes.

be noted that FMN does not contain D-ribose; it contains a residue of D-ribitol (a penta-alcohol) and is not therefore, a true nucleotide. The figure also shows how they can participate in processes of cell oxidation by fixation – reversible – of 2H and formation of a colourless leuco-ferment (due to the disappearance of the conjugated double bond system. They behave like substrates. Succinate dehydrogenase, an enzyme of the Krebs cycle, is an example of FAD enzyme (see fig. 4-38).

FERRO-PORPHYRINS – The structure of porphyrins was studied in Chapter 1. These ferroporphyrin coenzymes are associated with *cytochromes* whose role will be studied in Chapter 3 (see page 115). In reduced form, iron is divalent; in oxidized form it is trivalent.

LIPOIC ACID – This coenzyme acts as hydrogen transporter in the *oxidative decarboxylation* reactions, particularly in that of pyruvic acid (structure and mechanism of action shown in figure 4-36).

b) Coenzymes involved in group transfer reactions. – One may cite:

THIAMINE PYROPHOSPHATE (TPP) or CO-CARBOXYLASE – The pyrophosphoric ester of thiamine or *vitamin B_1* is an integral part of the active site of carboxylases (hence its name). It is a transporter of *aldehyde groups*, particularly *acetaldehyde*. It is a coenzyme of *decarboxylases of α-keto acids*. Its structure and mode of action are described in connection with the decarboxylation of pyruvic acid (see fig. 4-36).

COENZYME A (OR ACYLATION COENZYME) – As shown in figure 2-18, coenzyme A derives from a B group vitamin, the *pantothenic acid*, which is combined on the one hand, with a phosphate attached by a pyrophosphate bond to an adenosine 3′-5′ diphosphate molecule and on the other hand, to a thio-ethylamine (or cysteamine) molecule. The *thiol* group of the latter is the active part of the molecule; its H atom can be substituted by various acyl radicals (especially acetyl) to form *acyl-coenzymes A* or activated fatty acids; chemically these are thio-esters, highly reactive compounds (acetyl-coenzyme A, for example). We will study the importance of acyl-coenzymes A particularly in the oxidation of fatty acids (see figs. 5-10 and 5-11). We will also see that *phosphopantetheine* is the active part of a protein (ACP) transporting acyl radicals during the biosynthesis of fatty acids (see fig. 5-16). Coenzyme A behaves as a substrate.

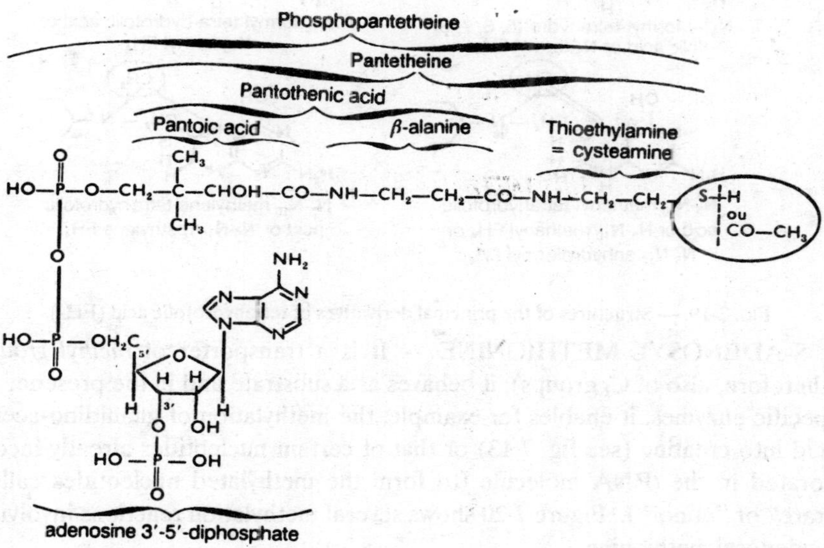

FIG. 2-18. — Structure of coenzyme A (and of acetyl-coenzyme A).

TETRAHYDROFOLIC ACID, FH_4 – It is the active form of a vitamin, *folic acid* or pteroylglutamic acid. It was mentioned earlier (see page 73) that in certain microorganisms, the synthesis of folic acid requires para-amino benzoic acid and that the bacteriostatic action of sulphamides is explained by a structural analogy of sulphamides with this compound, which enables them to bring about a competitive inhibition.

Tetrahydrofolic acid is a coenzyme, *transporter of C_1 units* (or one-carbon radicals), which behaves as a substrate and is involved in a large number of processes: biosynthesis of purine ribonucleotides (see fig. 6-19), formation of thymidylic deoxyribonucleotides (see page 300), glycine ←→ serine interconversion (see fig. 7-9) etc. Various derivatives of FH_4 can participate in these reactions; the structure of the main derivatives is given in figure 2.19.

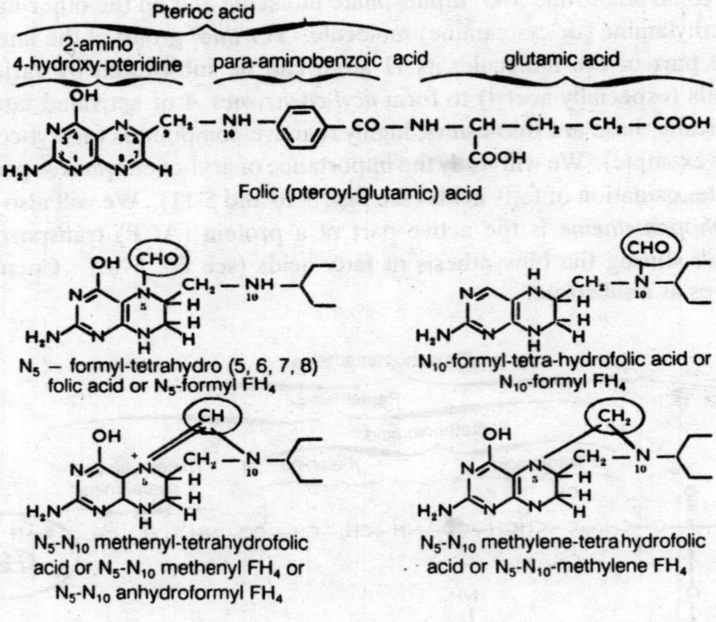

FIG. 2-19. — Structures of the principal derivatives of tetrahydrofolic acid (FH_4).

S-ADENOSYL METHIONINE – It is a transporter of *methyl groups* (therefore, also of C_1 groups); it behaves as a substrate and in the presence of specific enzymes, it enables for example, the methylation of guanidino-acetic acid into creatine (see fig. 7-13) or that of certain nucleotides already incorporated in the *t*RNA molecule (to form the methylated nucleotides called "rare" or "minor"). Figure 7-20 shows several methylation reactions involving S-adenosyl methionine.

BIOTIN – It is the coenzyme of different *carboxylation* reactions. We will particularly study in Chapter 5 the carboxylation of acetyl-coenzyme A into malonyl-coenzyme A and we will see (see fig. 5-15) how biotin binds the CO_2 which it has to transfer and how it is linked by its carboxyl to ε-NH_2 of a lysyl residue of the enzymatic protein.

PYRIDOXAL PHOSPHATE – This coenzyme derives from pyridoxine or vitamin B_6. It is an integral part of the active site of various enzymes involved in the metabolism of amino acids, which we will study in Chapter 7 (see fig. 7-1): transaminases, amino acid decarboxylases, serine dehydratase, cysteine desulphydrase, etc. This shows that the same coenzyme can participate in completely different reactions depending on the enzymatic protein to which it is associated.

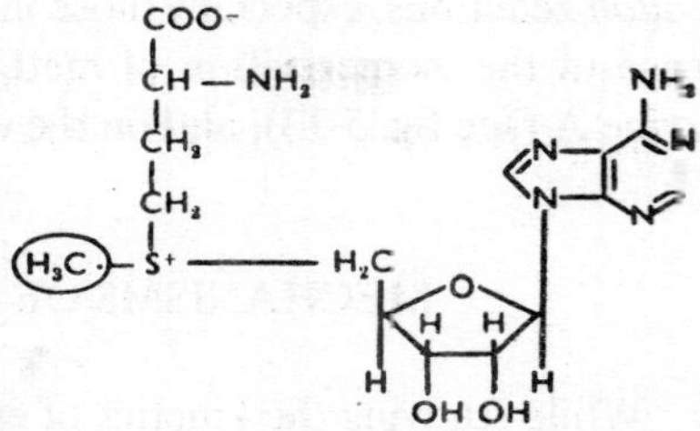

FIG. 2-20. — Structure of S-adenosylmethionine.

COBAMIDE COENZYMES – These coenzymes derive from *cobalamine* or vitamin B_{12}, whose structure is shown (see fig. 2-21) in the form of cyanocobalamine (which is one of the most frequent), with a CN^- anion bound to cobalt. But other anions may be found in place of CN^- (Cl^-, OH^-, NO_2^- etc.). Besides, in place of the 5,6-dimethylbenzimidazole molecule, linked on one hand to cobalt by a coordination bond and on the other hand, to a ribose-3-phosphate, one may sometimes find 5-hydroxybenzimidazole, or adenine. In coenzymes, in place of the anion, a 5′ deoxyadenosine is usually bound to cobalt.

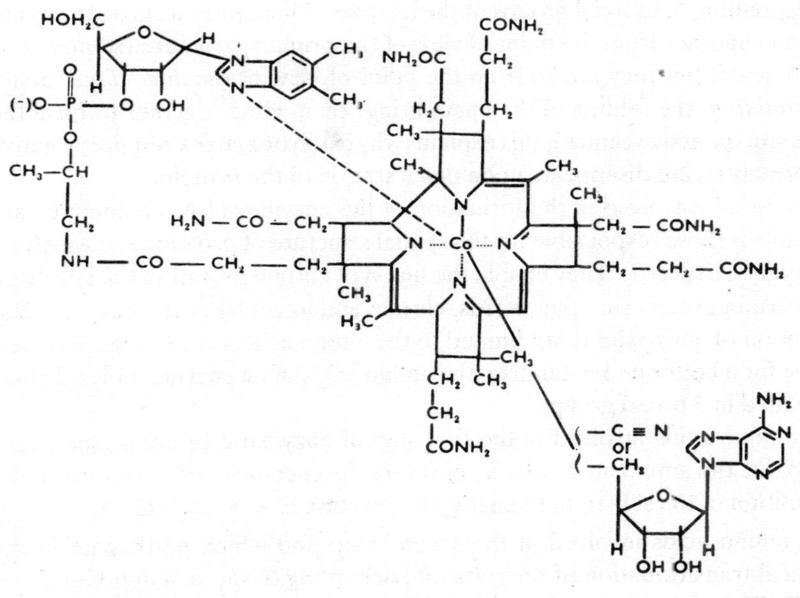

FIG. 2-21. — Structure of cyanocobalamine with the anion (C ≡ N⁻) and of a cobamide-coenzyme (with 5′-deoxyadenosine).

The cobamides-coenzymes participate on the one hand, in various *isomerization* reactions, especially those implying a transfer of *carboxyl group*, as in the case of the isomerization of methylmalonyl-coenzyme A into succinyl-coenzyme A (see fig. 5-13), and on the other hand, in some *methylation* reactions.

MECHANISMS OF ENZYMATIC REACTIONS

While studying the kinetics of enzymatic reactions we noted that Michaelis was led to postulate the formation of an intermediate enzyme-substrate complex right at the beginning of this century. What was then only a hypothesis was later largely confirmed and proved. The great disproportion between the size of the enzyme and that of the substrate obviously suggested that only a limited part of the enzyme participates in this interaction; this part is called *active site* (or active centre), or *catalytic site*. This is where the substrate binds and undergoes the transformation into reaction product (or products).

1. – ACTIVE SITE OF ENZYMES

It will be seen later, that the formation of the enzyme-substrate complex is characterized by a *specificity*, and even a rather narrow *stereospecificity*, due to the fact that the substrate molecule must have several functional groups in a suitable spatial configuration which will allow these groups to react with the corresponding functional groups of the enzyme. These enzyme groups are not near one another from the point of view of the primary structure (sequence of amino acids) but they are so from the point of view of the *three-dimensional conformation*: the folding of the chain brings them close together to form the active site (or active centre); this explains why catalytic activity requires a native conformation and disappears upon denaturation of the protein.

The bonds involved in the formation of the enzyme-substrate complex are the same as those responsible for the spatial structure of proteins and which we studied in Chapter 1. They enable the union of certain groups of the substrate with certain groups (in other words, certain amino acids) of the enzyme. But the notion of active site is not limited to the amino acids involved in these new bonds: for a better understanding, the amino acids of an enzyme molecule may be divided in 3 broad groups:

– amino acids involved in the first step of enzymatic catalysis (see pages 67-68), i.e. the amino acids which constitute the specificity site and ensure the recognition of the substrate to enable the reaction $E + S \rightleftharpoons E - S$;

– amino acids involved in the second step and which participate in the chemical transformation of the substrate according to the reaction $E - S \longrightarrow E + P$. They form the catalytic part of the active centre;

– other amino acids which are necessary for maintaining the adequate conformation of the active centre, maintaining the adequate conformation of the allosteric sites (which control the functioning of the active centre), and adequate positioning of the enzyme within the cell (for example, association of the enzyme with another enzyme to form a multienzyme complex, or association with a membrane).

The active centre comprises amino acids of the first two groups. The study of active sites is rather delicate; in most cases it requires the simultaneous use of several methods. The comparative study of the fine *three-dimensional structure* of the enzyme (with the help of X-rays) when free or involved in the complex, provided valuable information in the case of crystallizable enzymes (lysozyme, for example). We have also already seen that the study of *enzymatic action as a function of pH* enables in certain cases, the determination of the pK of a group implicated in the interaction (see page 71). But *chemical labelling methods* have provided most of the accurate information available on the nature of amino acids which are part of the active sites. After binding the reagent to one (or several) residue(s), the protein is hydrolyzed; this yields the label-carrying amino acid in a peptide, which enables its characterization and the determination of its position in the polypeptide chain. Obviously, one must make sure that the label did bind an amino acid belonging to the active site; this may be verified when the binding of the reagent causes a stoichiometric inactivation of the enzyme or on the contrary, when the binding of the reagent is prevented by a "protective" compound which is known to bind to the active site of the enzyme. Very often, the active site of an enzyme contains an amino acid residue, which if not unique, is at least *highly reactive* and will react preferentially with the label (for example the seryl residue of esterases). In certain cases the label can be the *substrate* itself, provided the E–S complex is formed by sufficiently stable covalent bonds (for example, phosphoglucomutase, which catalyzes the isomerization of glucose-6-phosphate into glucose-1-phosphate, can be labelled by phosphate and phosphoryl-serine can then be isolated), or a *quasi-substrate* capable of forming a stable covalent bond with the enzyme, such as di-isopropylfluorophosphate, which was studied in connection with inhibitors (see fig. 2-8) and which binds to the seryl residues belonging to the active sites of a large number of esterases and proteases. The use of *bi-functional reagents* combining with 2 groups of the active site of the enzyme could confirm that these groups can be part of peptide sequences very remote from one another in the primary structure but sufficiently close together in the spatial conformation, to enable the same reagent molecule to combine with both.

Very recently, with the possibility of isolating the gene of a given enzyme (see pages 383-384) it has become possible to prepare enzyme molecules modified at will, by modification and then expression of this gene. One can thus replace an amino acid of the active site by another or delete a part of the molecule. Based on the study of molecules thus modified one can determine the role of each amino acid of the active site or the function of various regions of the molecule.

The conformation of the active centre is not rigid. It has a great flexibility. The association of the substrate with the active centre induces very regularly a change of conformation which results in an *induced adjustment* or *induced fit* of the structure of the active centre. The side chains which participate in the functioning of the active centre, either at the specificity centre, or in the catalytic part of the active centre, change position during the attachment of the substrate to form the enzyme-substrate complex, but very often also, during the catalytic transformation of the complex to form the reaction products. These position changes of amino acids which constitute the active centre are not necessarily very considerable; they can be only of the order of a fraction of an angstrom (1Å = 1/10 000 000 mm); however, the structural reorganizations taking place at the active centre upon the binding of the substrate may sometimes correspond to position changes of the side chains of the amino acids of the active centre, attaining 5 to 10Å (i.e. 0.5 to 1 nm). It is quite obvious that the mutations which will lead to an enzyme structure, in which the conformational changes taking place in the active centre and which are essential for the expression of catalytic activitv are blocked, will cause an inactivation of the enzyme.

II. – MECHANISMS OF THE CATALYSIS

As an example, figure 2-22 represents the mechanism of action of chymotrypsin, as it is presently admitted. This is a proteolytic enzyme which

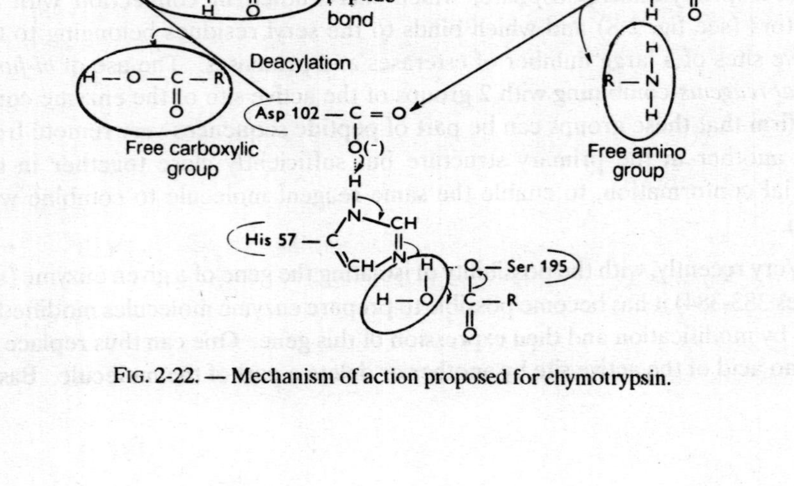

FIG. 2-22. — Mechanism of action proposed for chymotrypsin.

hydrolyzes certain peptide bonds (these will be indicated in figure 2-25) in two steps: first, *acylation* and then *deacylation* with regeneration of the enzyme as it was at the beginning of the reaction. The catalytic part of the active site consists of 3 amino acids which are very distant from one another in the sequence (they occupy respectively the positions 57, 201 and 195) but close together in the three-dimensional conformation. According to the classification given above, histidine, serine and aspartic acid are part of the catalytic site of the enzyme. This enzyme has a specificity for aromatic amino acids. The active site consists of a groove of hydrophobic character containing particularly methionine 192.

Some enzymes – particularly those responsible for the digestion of proteins in our digestive tract (pepsin, trypsin, chymotrypsin) – are synthesized in a catalytically inactive form called *proenzyme* or *zymogen* (pepsinogen, trypsinogen, chymotrypsinogen). Their conversion into active enzyme consists of the cleavage of one or several peptide bonds, reaction which is itself catalyzed by an enzyme: for example, the transformation inactive chymotrypsinogen ⟶ active chymotrypsin is catalyzed by trypsin which breaks the bond arg 15 – ile 16 in chymotrypsinogen. The catalytic site is already present in chymotrypsinogen, but the specificity site does not exist; cleavage by trypsin initiates a reorganization of the spatial structure which reveals the specificity site.

In reactions where there are *2 different substrates*, both must be bound to the enzyme in such a way that their interacting groups are near one another, and the active centre of the enzyme must have a binding site for each substrate.

While studying the general principles of catalysis, we have seen that the enzyme increases reaction velocity. Although it is not possible to explain, for each enzyme, how this is achieved, it may be said that in most cases two factors play a capital role:

– the bonds established between enzyme and substrate cause modifications of the electrons' distribution between the atoms of the substrate and can thus *increase the reactivity of some groups of the substrate*;

– the enzyme, in binding 2 substrates in such a way that their respective interacting functional groups are near one another, *increases the probability of effective collisions* needed for the reaction to take place between the 2 substrates.

III. – ROLE OF CO-FACTORS

We have seen – for example in the case of NAD – that *a coenzyme* is in most cases, one of the substrates of the reaction. But it is bound to the enzyme more strongly than an ordinary substrate, before and after the reaction (hence its name). Moreover, it is *restored to the initial state by a second reaction*, so that the same coenzyme molecule can react successively with a very large number of molecules of the substrate (provided it is brought back to the initial state each time), and this explains why reference is sometimes made to the *cyclic role* of coenzymes. This regeneration of the coenzyme can take place while it remains bound to the same apoenzyme or after its detachment (in the latter case the coenzyme can bind again to a molecule of the same apoenzyme after regeneration).

Sometimes, *metal ions* can act in the same manner, i.e. they may be reduced in a first reaction, then reoxidized in a second reaction, as in the case of cytochromes for example (see fig. 3-6a). But in other cases, they can, by *formation of a complex*, enable the *binding of the substrate to the enzyme*; for example, it is believed that in some enzymatic reactions, Mg^{2+} ions form bridges between the phosphate groups of the substrate and the enzyme molecules and that Co^{2+} or Zn^{2+} ions are necessary for the bond between the dipeptide glycylglycine and the dipeptidase etc.

SPECIFICITY OF ENZYMATIC ACTION

Specificity is one of the main characteristics of enzymes. It prevents the formation of by-products which happens with chemical catalysts. Specificity may be observed on the one hand in the type of reaction catalyzed by the enzyme, and on the other hand in the substrate of the reaction.

I. – REACTION – RELATED SPECIFICITY

A given substrate, an α-amino acid for example, can undergo various transformations (studied in more details, in Chapter 7). Each of the 3 reactions of figure 2-23 is catalyzed by a different enzyme although the substrate is the same.

II. – SUBSTRATE – RELATED SPECIFICITY

Specificity may be very broad (the enzyme acts on a wide variety of compounds) or on the contrary, it may be very narrow (the enzyme acts on one substrate only); there are also numerous intermediate possibilities (the enzyme acts on a family of compounds). We will take the example of *maltase* which, as indicated by its name, hydrolyzes maltose (structure shown in figure 4-15) with formation of two molecules of glucose.

Maltase is a *hydrolase*, catalyzing a reaction of the type $A-B + H_2O \longrightarrow A - OH + B - H$. But a large number of linkages can be hydrolyzed (ester linkages, peptide linkages, oside linkages etc.) and there is one type of specific enzyme for each *type of linkage* (esterases, peptidases, osidases etc.). Maltase is an *osidase* because it breaks the bond joining two monosaccharides. These two monosaccharides are in fact two glucose molecules; maltase is therefore a

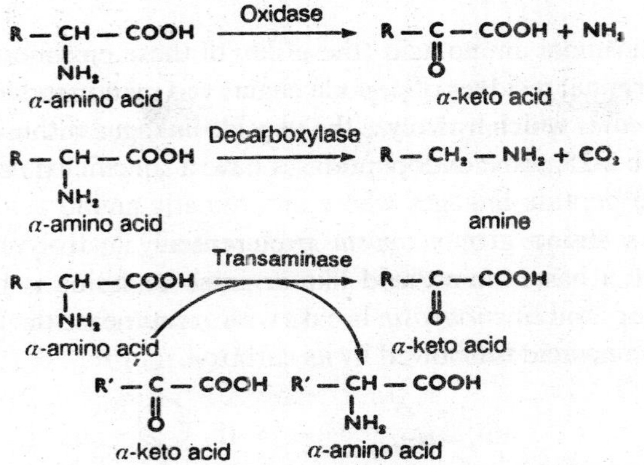

FIG. 2-23. — Various transformations undergone by the same amino acid and catalyzed by different enzymes.

glucosidase. They are coupled by a 1-4 linkage which precisely, maltase can hydrolyze; it is therefore a *1-4 glucosidase*. The linkage is of α-type, and glucose is of the D series, maltase is therefore a *α, D, 1-4 glucosidase*.

1. – Specificity related to the nature of the linkage

As just mentioned in the case of hydrolases, there are osidases, esterases, amidases etc. according to the type of the linkage.

2. – Group specificity

Among *esterases* for example, one may distinguish *carboxy-esterases* – which are specific of ester linkages where organic acids are involved – from *phospho-esterases* which hydrolyze phosphoric esters. As shown by figure 2-24, the latter are again sub-divided in *phosphomonoesterases* or *phosphatases* which split the phosphomonoester linkages (notably in the metabolism of carbohydrates), and *phosphodiesterases* which hydrolyze one of the ester linkages of a diesterified phosphoric acid (example: ribonucleases and deoxyribo-nucleases; see figures 6-10 to 6-12).

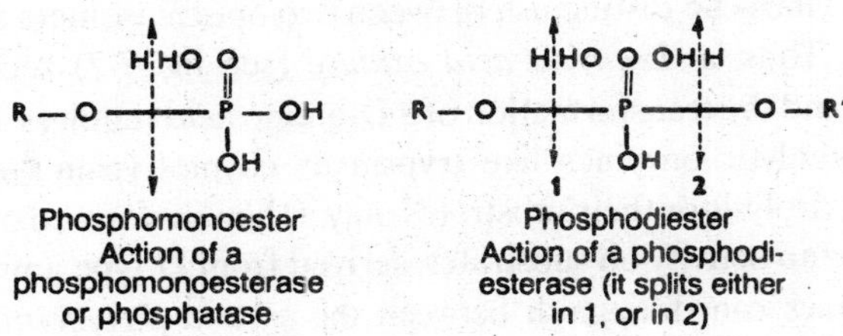

FIG. 2-24. — Action of phosphoesterases.

Among *proteolytic enzymes*, or *proteases*, which hydrolize the peptide linkages; there are *exopeptidases* which act at the end of chains and detach the

N – or C–terminal amino acid (the utility of these enzymes for the determination of terminal residues of peptide chains was mentioned in pages 21-22), and *endopeptidases* which hydrolyze the peptide linkages within the chains. As shown in figure 2-25, some endopeptidases have a specificity: *pepsin* preferentially splits the peptide linkages where an aromatic amino acid (Phe, Tyr) is involved by its amino group; *trypsin* preferentially hydrolyzes the peptide linkages where a basic amino acid like arginine or lysine is involved by its carboxylic group, and *chymotrypsin* breaks with greater ease the linkages where an aromatic amino acid is involved by its carboxyl.

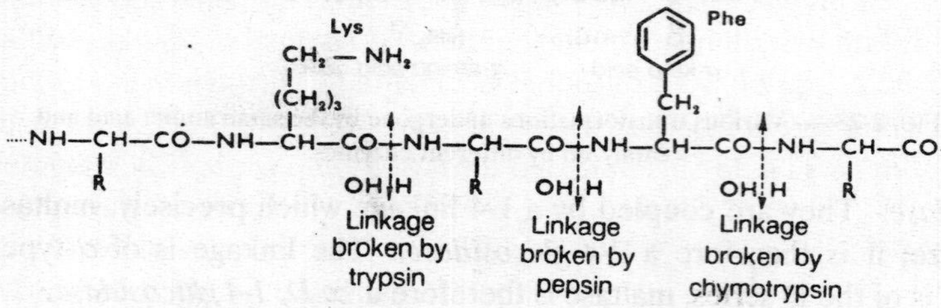

FIG. 2-25. — Specificity of action of some endopeptidases.

3. – Absolute specificity for one substrate only

Some enzymes act on only one substrate : urease acts only on urea; arginase only on L-arginine; fructose 1-6-bisphosphatase only on fructose 1-6 bisphosphate; carbonic anhydrase only on CO_2 etc.

Some oxidoreductases and transferases manifest even a narrower specificity concerning not only the *substrate* or the *group of substrates* they can recognize, but also the *acceptor of the group* whose transfer they catalyze. For example, 3-phosphoglyceraldehyde dehydrogenase (see fig. 4-27) catalyzes the transfer of hydrogen only from 3-phosphoglyceraldehyde (substrate) to NAD^+ (acceptor).

4. – Stereospecificity

Several enzymes can distinguish between two optical isomers and act only on one of them. Thus, a *L-amino acid oxidase* (see fig. 7-2) recognizes only a L-amino acid and the transformation of a D-amino acid requires a D-amino acid oxidase. Proteolytic enzymes like trypsin or chymotrypsin (see above) will catalyze the hydrolysis or their substrates only if they are formed of L-type amino acids and have no activity on substrates derived from D-type amino acids.

Some enzymes can distinguish between the α- *and* β-*conformations* in oside linkages; there are, for example, α- and β-glucosidases, α- and β-galactosidases, etc.

Other enzymes have a stereospecificity vis-a-vis *cis-trans isomers*. Thus, fumarase allows the addition of one molecule of water on the double bond of fumaric acid *(trans)*, but not on that of maleic acid *(cis)*, (see fig. 2-26).

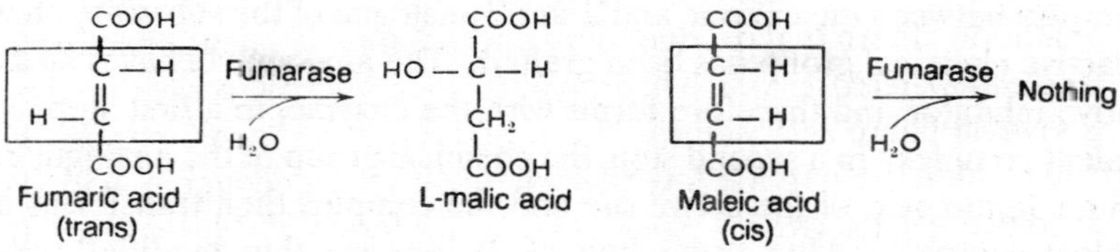

FIG. 2-26. — Stereospecificity of fumarase vis-a-vis cis-trans isomerism.

A stereospecificity which is even more subtle is observed vis-a-vis the substances of the general formula Cx_1x_2yz, like glycerol or citric acid (see fig. 2-27). From the organic chemistry point of view such substances are symmetric (they have a plane of symmetry) but they behave like *asymmetric substances* as substrates of certain enzymatic reactions (in other words x_1 for example, can be attacked in preference to x_2). In fact the free substrate is symmetric but it becomes asymmetric during the formation of the E–S complex and when 3 (at least) functional groups of the substrate must interact with 3 groups of the active site of the enzyme. If the 3 groups of the substrate are all different, the substrate can bind to the enzyme in one way only (model I), and it is clear that x_1 and x_2 are no longer equivalent because one of them is bound and the other not. If on the contrary, x_1 and x_2 are among the 3 functional groups of the substrate involved in the formation of the complex with the enzyme, the asymmetry is less evident but not any the less real (model II); to make sure, one may consider that the left hand represents the enzyme, with the thumb (Z), the index (X_1) and the middle finger (X_2) at the active site, and that the right hand represents the substrate with the 3 same fingers corresponding respectively to z, x_1 and x_2; this shows that only one orientation is possible; therefore, x_1 and x_2, by the fact that they are bound to groups of the enzyme which are not symmetric, cease to be equivalent, and one can understand that only one of them undergoes the transformation catalyzed by the enzyme.

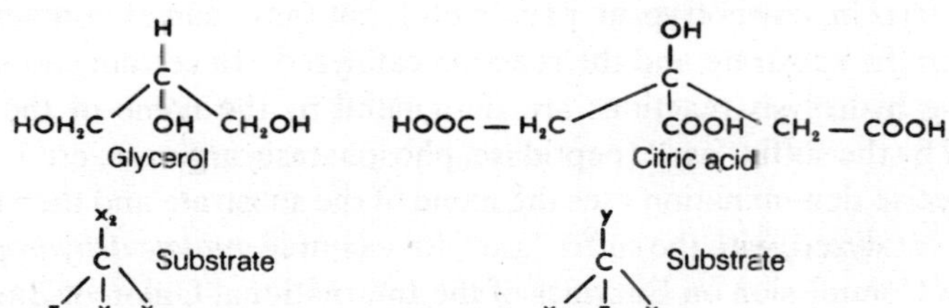

FIG. 2-27. — Asymmetric behaviour of compounds of the type Cx_1, x_2yz in enzymatic reactions.

This notion of stereospecificity has found a direct application called *affinity labeling* for the study of active sites of enzymes. The method consists in forming a complex between the enzyme and a steric analogue of the substrate on which a reactive chemical group has been grafted. The analogue behaves as a competitive inhibitor and therefore forms with the enzyme, in a first step, a non-covalent complex. In a second step, the chemical group of the analogue reacts with an amino acid of the active site and the complex then transforms into a covalent complex. After proteolysis of the enzyme thus modified, one can identify the amino acid which has reacted with the analogue.

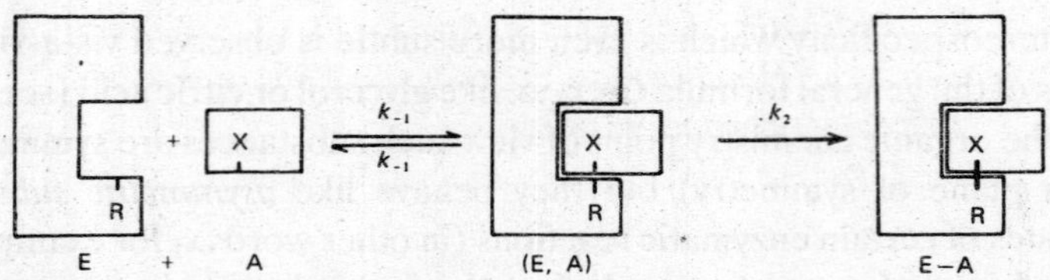

FIG. 2-28. — Affinity labeling of an enzyme by a steric analogue of the substrate. There is first, formation of a reversible complex between the enzyme and the analogue A, then irreversible chemical reaction between the group X of analogue A and an amino acid R of the active site leading to the formation of covalent complex E—A.

With this method it is therefore possible to locate the amino acids of the active site, involved either in the recognition of the substrate, or in the catalytic mechanism, and since it is based on this very important notion of stereospecificity, it is applicable to all types of sites, particularly also to allosteric sites.

Affinity chromatography (see page 43) is also a direct application of specificity.

CLASSIFICATION OF ENZYMES

Some enzymes are often designated by common names based on usage (pepsin, trypsin, chymotrypsin, papain etc); but these names contain no information on the substrate and the reaction catalyzed. In certain cases, enzymes catalyzing hydrolysis reactions are designated by the name of the substrate followed by the suffix "ase" (peptidase, phosphatase, arginase, etc.). A slightly more precise denomination uses the name of the substrate and then that of the reaction catalyzed, with the suffix "ase", for example, *malate dehydrogenase*. In 1961, the Commission on Enzymes of the International Union of Biochemistry established a systematic and much more rigorous classification and nomenclature comprising 6 classes divided into sub-classes; the latter are themselves divided into sub-sub-classes which are numbered. In this new denomination, malate dehydrogenase is called *malate-NAD-oxidoreductase*; this name reflects not only the type of reaction catalyzed, but also the name of the substrate and that of the hydrogen acceptor. The following are a few examples illustrating this classification.

1. – Oxidoreductases

This class comprises the enzymes which were earlier called dehydrogenases, oxidases, peroxidases, hydroxylases, oxigenases, etc.

1.1 Acting on a $>$CH – OH group (of the hydrogen donor):

1.1.1 With NAD^+ or $NADP^+$ as hydrogen acceptor
Ex. L-Malate : NAD-oxidoreductase (1.1.1.37), see fig. 4-38.
L-lactate : NAD-oxidoreductase (1.1.1.27), see fig. 4-30.

1.1.2 With a cytochrome as acceptor
Ex. L-lactate : ferricytochrome c-oxidoreductase (1.1.2.3).

1.1.3 With O_2 as hydrogen acceptor
Ex. glucose oxidase or β-D-glucose : oxygen-oxidoreductase (1.1.3.4).

1.2 Acting on a $>$C = O group (aldehyde or ketone):

1.2.1 With NAD^+ or $NADP^+$ as acceptor
Ex. D-glyceraldehyde-3-phosphate : NAD oxidoreductase (1.2.1.12) see fig. 4-27.

1.2.3 With O_2 as acceptor
Ex. xanthine : oxygen oxidoreductase (1.2.3.2) see fig. 7-32.

1.2.4 With lipoic acid as acceptor.

1.3 Acting on a $>$CH – CH$<$ group :

1.3.1 With NAD^+ or $NADP^+$ as acceptor
Ex. 4.5 dihydrouracil : NAD oxidoreductase (1.3.1.1.1).

1.3.2 With a cytochrome as acceptor.

1.3.3 With O_2 as acceptor
Ex. 4.5-dihydro-orotate : oxygen oxidoreductase (1.3.1.1.1), see fig. 6-22.

1.4 Acting on a $>$CH – NH – group :

1.4.1 With NAD^+ or $NADP^+$ as acceptor
Ex. L-Glutamate : NAD oxidoreductase (1.4.1.2) see fig. 7-3.

etc.

2. – Transferases

2.1 Transferring a monocarbon group (C_1) :
- 2.1.1 Methyl transferases
  Ex. S-adenosyl-methionine : L-homocysteine S-methyl transferase (2.1.1.10).
- 2.1.2 Hydroxymethyl transferases and formyl transferases
  Ex. L-serine : tetrahydrofolate 5,10 hydroxymethyl transferase (2.1.2.1) see fig. 7-9.
- 2.1.3 Carboxyl transferases and carbamoyl transferases
  Ex. carbamylphosphate : L-aspartate carbamyl transferase (2.1.3.2) see fig. 6-22.

2.3 Acyl transferases.

2.4 Glycosyl transferases :
- 2.4.1 Hexosyl transferases
  Ex. UDPG-glucose : D-fructose glucosyl transferase (2.4.1.13).
- 2.4.2 Pentosyl transferases
  Ex. Uridine : Orthophosphate ribosyltransferase (2.4.2.3).

2.5 Alkyl transferases.

2.6 Transferring nitrogen groups :
- 2.6.1 Amino transferases
  Ex. L-aspartate : ketoglutarate amino transferase (2.6.1.1) see pages 392-393.

2.7 Phosphoryl transferases :
- 2.7.1 With an alcohol group as acceptor
  Ex. ATP: D-hexose-6-phosphotransferase (2.7.1.1) see. fig. 4-20.
- 2.7.2 With a carboxylic group as acceptor
  Ex. ATP: 3 Phosphoglycerate 1-phosphotransferase (2.7.2.3) see fig. 4-28
- 2.7.3 With a nitrogen group as acceptor
  Ex. ATP: creatine phosphotransferase (2.7.3.2) see fig. 7-13.

etc.

3. – Hydrolases

3.1 Splitting the ester bonds :
- 3.1.1 Carboxylester-hydrolases
  Ex. Lipase or glycerol-ester hydrolase (3.1.1.3) see page 228.
- 3.1.3 Phosphomonoesterases
  Ex. alkaline phosphatase (3.1.3.1) see fig. 6-13.
- 3.1.4 Phosphodiesterases
  Ex. ribonucleases, deoxyribonucleases, see figs. 6-10 to 6-12 deoxyribonucleate 3′ nucleotido hydrolase (3.1.4.6).

3.2 Splitting oside bonds
  3.2.1 Glucosidases
    Ex. β-glucosidase or β-D-glucoside glucohydrolase (3.2.1.21)
3.4 Splitting peptide bonds
  3.4.1 α-aminopeptido-amino acid hydrolases
    Ex. aminopeptidase or aminoacyl-peptide hydrolase (3.4.1.2) see page 21.
  3.4.2 α-carboxypeptido-amino acid hydrolases
    Ex. carboxypeptidase A or peptidyl-L amino acid hydrolase (4.3.2.1) see pages 21-22.
  3.4.4 Peptido-peptide hydrolases (endopeptidases).
    Ex. trypsin (3.4.4.4) see fig. 2-25.
  etc.

4. – Lyases

Catalyzing the removal of a group by a process other than hydrolysis (often, with formation of a double bond) or on the contrary, catalyzing the addition of a group.

4.1 C-C lyases
  4.1.1 Carboxylases (Carboxylases or Decarboxylases)
    Ex. aspartate decarboxylase or L-aspartate 4 carboxylase (4.1.1.12) see pages 389-391.
  4.1.2 Aldehyde-lyases
    Ex. fructose-bisphosphate aldolase or fructose 1-6 bisphosphate : D-glyceraldehyde-3-phosphate lyase (4.1.2.13) see fig. 4-26.
4.2 C-O lyases
4.3 C-N lyases
  4.3.1 Ammonia lyases
    Ex. L-aspartate-ammonium lyase (4.3.1.1) see fig. 7-5.
  etc.

5. – Isomerases

5.1 Racemases and epimerases
  5.1.1 Acting on amino acids
    Ex. alanine racemase (5.1.1.1) see page 389.
  5.1.3 Acting on oses
    Ex. D-ribulose-5-phosphate-3-epimerase (5.1.3.1) see fig. 4-40.
5.2 *Cis-trans* isomerases
  Ex. 4 maleyl-aceto acetate *cis-trans* isomerase (5.2.1.2) see fig. 7-24.

5.3 Intramolecular oxidoreductases

5.3.1 Catalyzing the interconversion aldose-ketose.

Ex. D-glyceraldehyde 3 phosphate keto-isomerase or triosephosphate isomerase (5.3.1.1) see fig. 4-26.

5.4 Intramolecular transferases

Ex. L-methylmalonyl-coA-coA-carbonyl mutase (5.4.99.2) see fig. 5-13.

etc.

6. – Ligases or synthetases

Enzymes enabling the union of 2 molecules, coupled with the rupture of a high energy potential bond (for example rupture of the pyrophosphate bond of ATP or of another nucleoside-triphosphate).

6.1 Formation of C – O bonds

6.1.1 Aminoacyl-tRNA synthetases (or amino acid tRNA ligases)

Ex. Alanyl-tRNA synthetase (6.1.1.7) see fig. 6-39.

6.2 Formation of C – S bond

Ex. Acetyl-coA synthetase (6.2.1.1).

6.3 Formation of C – N bonds

6.3.1 Acid-ammonia ligases

Ex. L-glutamate : ammonia ligase or glutamine-synthetase (6.3.1.2), see fig. 6-22.

6.3.2 Acid-amino acid ligases

Ex. γ-L-glutamyl-L-cysteine : glycine ligase or glutathione synthetase (6.3.2.3) see page 406.

6.4 Formation of C-C bonds.

etc.

CHAPTER 3[(1)]

BIOENERGETICS

Living beings need energy to produce heat, light, electricity, to permit movements, transports, transmissions etc. In a given system, energy can neither be created nor destroyed, but it can be transferred from one system to another (for example, from one cell to another, or even from one molecule to another). The various forms of energy (thermal, chemical, mechanical, electrical, radiant light energy) allow energy transfers, but within living organisms certain interconversions do not take place. In fact, we will see in this chapter that chemical energy plays a capital role in living cells. The latter draw their energy mainly from two sources:

– certain *light radiations*, particularly solar light, are utilizable by *autotrophic* organisms, i.e. green plants and photosynthetic bacteria containing chlorophyll or other photosensitive molecules, like bacteriorhodopsin. These organisms convert light energy into *chemical energy*, by synthesizing complex organic substances (carbohydrates, protids, lipids) from simple mineral molecules (CO_2, H_2O, NH_3). Light energy is really the principal source of energy on Earth. Photosynthetic organisms can utilise this energy directly and other organisms actually depend on the former (see below);

– *the chemical energy of organic substances* thus synthesized by plants constitutes the source of energy of all other living beings called *heterotrophic*; this source is either direct (e.g. for herbivorous animals) or indirect (for instance, when a carnivorous animal eats a herbivorous animal). The chemical energy of some mineral substances (S, Fe^{2+}, NH_3) can also be utilized, but only by certain microorganisms (see pages 396-397 for example).

Before examining how this chemical energy can be transferred in living beings, one must recall certain notions indispensable for the understanding of the phenomena.

---

(1) Revised by D. GAUTHERON.

I. – NOTIONS OF THERMODYNAMICS

1. – Free energy change

If we consider a system consisting of two compounds A and B which react to give C:

$$A + B \longrightarrow C$$

By reacting, the system can exchange work or energy (heat, light) with the immediate surroundings. For example, it can receive energy and produce an equivalent work.

At the time of mixing, A + B represent the initial state and C, the product formed, represents the final state.

Each of these 2 states is characterized by an internal energy level (U_1 and U_2 respectively) which depends on the chemical nature of the constituents (with their atoms, their bonds), their concentrations and their position in space (absolute temperature T, pressure P, Volume V...). The *internal energy change*, denoted ΔU, resulting from a change of molecular structures (new bonds and different distribution of electrons), depends only on the initial state and final state: $\Delta U = U_2 - U_1$. *It is the maximum quantity of energy which can be made available* to accomplish a work by the transformation A + B into C. For a constant pressure system (as in the case of many biological reactions) there can be volume changes and either a liberation or an absorption of heat.

The work done by pressure is equal to P $(V_2 - V_1)$, i.e. $P\Delta V$. The heat of reaction first defined by ΔQ is equal to the internal energy change plus the work done

$$\Delta Q = \Delta U + P\Delta V$$

If we write H (abbreviation of "heat content") instead of Q (quantity of heat), the above expression becomes

$$\boxed{\Delta H = \Delta U + P\Delta V} \qquad (1)$$

for a given temperature, because we know that PV = nRT (n = number of moles, R = gas constant, T = temperature in degrees Kelvin)

ΔH is the *enthalpy change of the system*. ΔH like ΔQ is mostly expressed in cal. or kcal/mole but ought to be expressed in joules/mole (1 calory = 4.18 joules).

ΔH can be measured in a calorimeter. If $\Delta H < 0$ the reaction is *exothermic* (heat is liberated) and if $\Delta H > 0$ the reaction is *endothermic* (the system absorbs heat). In the past, it was thought that the knowledge of the sign of ΔH is sufficient to predict whether a reaction is spontaneous or not, and it was said that an exothermic reaction is spontaneous while an endothermic reaction is not. It was then found that while these predictions are correct in most cases, there are exceptions, and that some spontaneous reactions are in fact accompanied by absorption of heat by the molecules.

One was therefore led to consider the *entropy* of the system denoted S, a function which measures the disorder of a system, and the variation of which $\Delta S = S_2 - S_1$, depends – like ΔU – on the initial state and the final state. In order to pass from a first stable state, the initial state, to a new stable state, the final state, one must upset the structures of molecules A and B for the reaction to take place. The greater the increase in molecular disorder, the greater the entropy (for example, entropy increases while passing from a given protein to the constituent amino acids in free state). Entropy is related to the complexity of the molecule. The larger the number of atoms, the larger the number of complicated chemical groups, the larger the number of possible structures and the greater the entropy. In a reversible transformation at constant pressure, $\Delta Q = T\Delta S$. One can then write:

$$\Delta H = \Delta G + \Delta Q = \Delta G + T\Delta S$$

or

$$\boxed{\Delta G \text{ (free enthalpy change)} = \Delta H - T\Delta S} \quad (2)$$

ΔG or the free enthalpy change of a constant pressure system is therefore the *maximum energy utilizable to accomplish a work.* All of the ΔH or heat of reaction is not utilizable. The larger the entropy factor, the smaller the ΔG.

When a reaction proceeds from the initial state to the equilibrium state, free energy may be:

– either converted into heat (in this case ΔS can be equal to zero);

– or, used for increasing entropy (in this case ΔH can be equal to zero, which explains that some spontaneous chemical reactions are not accompanied by heat liberation).

Writing $\Delta U + P\Delta V$ for ΔH:

$$\Delta G = \Delta U + P\Delta V - T\Delta S \quad (3)$$

In a reversible transformation at constant volume ($\Delta V = 0$) the *free energy change* is

$$\boxed{\Delta F = \Delta U - T\Delta S} \quad (4)$$

In biochemistry, it may be admitted that in most cases $\Delta V = 0$ because the volume changes for reactions taking place in solution are negligible. In equation (3) giving ΔG, it is therefore admitted that $P\Delta V = 0$, so that ΔG becomes, like ΔF (equation 4), equal to $\Delta U - T\Delta S$ and therefore *ΔF becomes ΔG.* This is particularly true if the reactions involve liquids and solids in solution (and no gas) at physiological temperatures which are low.

Three cases are possible. When $\Delta G < 0$, the reaction is *exergonic* (it can be spontaneous) and free energy is available. On the contrary if $\Delta G > 0$, the reaction is *endergonic*; free energy must be supplied to the system for the reaction to take place. The reaction will take place only coupled with another exergonic reaction whose negative ΔG is greater in absolute value than the positive ΔG of the endergonic reaction. Many biochemical reactions are reversible. $\Delta G = 0$ when dynamic equilibrium is reached. This is very interesting because for very small variations of active masses of the reaction's participants, the reaction can take place in either direction without energy problem as ΔG will vary in either direction. It may be noted that if ΔS is small, as T is comparatively low in biochemistry, the term $T\Delta S$ can be neglected and then $\Delta G = \Delta H$ (see equation 2); in other words, in this case, the terms "exothermic" and "exergonic" on the one hand, and 'endothermic" and "endergonic" on the other, are similar.

2. – Application to chemical reactions

If we consider 2 compounds A and B in solution at 25°C under a pressure of 1 atmosphere, each in a concentration of 1 mole/litre of solvent (1 molal) and which can react completely giving C + D (which will therefore be present in the same concentration after total transformation) the free *energy change* is called "*standard*" and denoted ΔG_0[1]. If $\Delta G_0 < 0$, the reaction is exergonic; it can take place spontaneously, but this is not necessarily the case even if the free energy change is very large. In fact, ΔG_0 expresses the free energy change between an initial state and a final state. However, ΔG_0 gives no information on reaction velocity. A catalyst, an enzyme for example, might be needed to trigger the reaction and adjust its velocity without altering its thermodynamic characteristics. As mentioned in the previous chapter, despite the highly negative value of ΔG_0 (– 686 kcal/mole) of the total oxidation reaction of glucose ($C_6H_{12}O_6 + 6O_2 \longrightarrow 6CO_2 + 6H_2O$), this substance can be left for years in presence of the oxygen of air without any oxidation taking place.

On the one hand, while ΔG_0 indicates a standard state, it does not take into consideration the special circumstances which may prevail, particularly in a biological medium: concentration and activities of compounds, pH etc. Normally reversible reactions proceed up to the equilibrium state and ΔG takes this equilibrium state into account. But the reaction can proceed up to completion if the normal equilibrium state is continuously broken because the reaction products are progressively consumed by a following reaction. This is what happens in metabolic chains where, in fact, a general state of equilibrium is observed.

---

(1) The value of ΔG in conditions other than "standard" was given in the previous chapter (page 59).

On the other hand, reactions which, at the limit, can be reversible *in vitro* can be irreversible *in vivo* because of metabolic conditions.

Actually, *the free enthalpy change ΔG or free energy change ΔF will always depend on metabolic circumstances and will be related to equilibrium states* in vivo.

It was indicated in the previous chapter that:

ΔG_0 depends on the equilibrium constant K of the reaction:

$$\Delta G_0 = -RT \ln K$$

and that in given circumstances, the effective ΔG of a reaction is:

$$\Delta G = \underset{\text{(1st factor)}}{\Delta G_0} + \underset{\text{(2nd factor)}}{RT \ln \frac{[C][D]}{[A][B]}}$$

for a reversible reaction $A + B \rightleftharpoons C + D$ proceeding from left to right. The ΔG of the reverse reaction will have the opposite sign.

Therefore, 2 factors define ΔG, on the one hand, the ΔG_0 which characterises a given reaction, and on the other hand the second factor, characteristic of given circumstances and hence related to concentration, activity, dissociation or protonation of the reaction participants etc. When the ratio of activities of the 2nd factor becomes equal to 1, the 2nd factor becomes zero and $\Delta G = \Delta G_0$.

It must be noted that ΔG and ΔG_0 correspond to pH = 0 and ambient temperature while $\Delta G'$ and $\Delta G_0'$ represent the values in the physiological conditions (pH near 7 and physiological temperature, 30 to 37°) to be specified in each case.

Let us examine the case of a reaction whose ΔG_0 = –3 kcal/mole. It is known that R = 1.987 cal/mole/degree, that at 25°, T = 273 + 25 = 298°K and that ln K = 2.30 log K. We can write:

$$\Delta G_0 = -3\,000 \text{ cal/mole} = -1.987 \times 298 \times 2.30 \log K = -1362 \log K$$

therefore,

$$\log K = \frac{-3\,000}{-1\,362} = 2.2$$

i.e., approximately equal to 2, which corresponds to a value of K equal to about 100.

But it was indicated (page 58) that $K = \frac{[C][D]}{[A][B]}$, therefore [C][D] = 100 [A][B], which means that at equilibrium, transformation of A + B into C + D would have been almost total (99%).

If on the contrary, a reaction has a ΔG_0 = +3 kcal/mole, then K will be approximately equal to 0.01, in other words, the reaction will practically not take place because equilibrium will be reached as soon as only 1% of the reactants will have been transformed.

It follows that the ΔG_0 of a reaction can be calculated if one knows K, i.e., if one can measure the concentrations of the reactants and reaction products (this is possible, except when K is extremely small or extremely large because then the concentrations of the reaction products or reactants are too low to be determined accurately). Figure 3-1 gives the values of ΔG_0 for a series of values of K from 10^{-3} to 10^6.

K	log K	$\Delta G_0 (= -1\,362 \log K)$ in cal/mole
0.001	−3	+4 086
0.01	−2	+2 724
0.1	−1	+1 362
1	0	0
10	+1	−1 362
100	+2	−2 724
1 000	+3	−4 086
10 000	+4	−5 448
100 000	+5	−6 810
1 000 000	+6	−8 172

FIG. 3-1. — Values of ΔG_0 for different values of K.

It is clear that the reactions whose equilibrium constant is >1 are accompanied by a decrease of free energy (this energy becomes available); these are *exergonic* reactions. On the contrary, if K is <1, the reaction is *endergonic*.

3. – Application to reactions in living organisms

The $\Delta G'$ in given conditions will define the direction of the reaction (*and not the* ΔG_0 *or the* ΔH).

Since $\Delta G' = \Delta G_0 + RT \ln \frac{[C][D]}{[A][B]}$ and since ΔG_0 is a constant for a given system which can be found in tables, it is the 2nd *circumstantial* factor which will decide the direction of the reaction. This is why one often indicates an interval of $\Delta G'$ values which a reaction can take in normal physiological conditions. For example, the energy yielded by the cleavage of ATP to ADP + P_i can have a $\Delta G'$ between 7 and 12 kcal/mole in physiological conditions. The reactions close to equilibrium can proceed sometimes in one direction, sometimes in the other.

When products are eliminated from the system as and when they appear (CO_2 for example or by another reaction) equilibrium is displaced in the same direction.

Moreover, even a reaction having a largely positive ΔG_0 can take place in living cells when it is part of a sequence of reactions and when the ΔG of the whole pathway is <0; the products of this reaction will be continuously taken over by the following reaction, so that despite its theoretically unfavourable equilibrium, the reaction with $\Delta G > 0$ can continue to take place, because in fact its ΔG becomes negative as the products are continually consumed.

Lastly, one can consider at this juncture, the problem of the *synthesis of macromolecules* such as polyosides or proteins; the ΔG for the polymerization of oside units into polysaccharides or for the polymerization of amino acids into proteins is very largely positive[1] and in this case there is no following reaction which takes over the products formed, because they are terminal products. Indeed, despite the unfavourable equilibrium, there could still be synthesis of polysaccharides or proteins in small quantities if there were very high concentrations of monosaccharides or amino acids; but precisely the reverse situation prevails in living cells: the concentration of polymers is very high compared to that of monomers (glucose and free amino acids). It must be concluded that the polymers are not formed by simple polymerization of monomers and we will see that indeed, the macromolecules (polysaccharides, proteins, nucleic acids) are synthesized by different processes in which the global free energy change is negative.

The example of polysaccharides or proteins is not unique; it may be said generally, that living cells are far from being systems in equilibrium. The creation and maintenance of such a situation require energy from the environment of the cell; if the external source of energy disappears and consequently energy transfer to the cell is no longer possible, the cellular processes will all tend towards equilibrium: the cell dies.

Once energy has been absorbed by the living cell – in the form of light or chemical substances – *energy transfers* are necessary to permit the synthesis of biochemical compounds like polysaccharides, nucleic acids or proteins. We have already mentioned the fact that in living organisms, thermal energy is not a possible intermediate. The living cell cannot therefore be compared with an internal combustion engine where the chemical energy of molecules is partly transformed, during the combustion (oxidation) of petrol, into kinetic energy (heat) which causes the expansion of gases and thereby the movement of a piston (mechanical energy). Cells do not use this type of energy transfers for their endergonic reactions and it must be noted that an increase of temperature would rapidly cause the denaturation of proteins, particularly enzymes.

Indeed, the cells do call upon the *oxidation* of some substances, but contrary to what happens in an internal combustion engine, cellular oxidations take place without great changes of either temperature or pressure. It is true that a part of the energy is liberated in the form of heat (and this enables the homeotherms to maintain their temperature constant), but this fraction of energy is no longer available for performing the endergonic reactions. In fact, *the major part of the energy remains in the chemical form* and it is transferred directly from one molecule to another.

---

(1) For example, in the reaction: polypeptide + $H_2O \longleftrightarrow$ amino acids, the equilibrium is very clearly on the side of amino acids.

II. – IMPORTANCE OF ATP AND OTHER COMPOUNDS OF HIGH ENERGY POTENTIAL

Living cells must perform some endergonic reactions; these are not only anabolic reactions like those already mentioned (biosynthesis of macromolecules such as polysaccharides, proteins, nucleic acids), but also some reactions of transport, transmission, mechanical work, functioning of the nervous system, catabolic reactions, particularly when the substrates must be activated before they can be degraded (e.g., activation of fatty acids, phosphorylation of sugars). The energy required is supplied by the exergonic reactions thanks to *energetic coupling.* This energetic coupling is one of the important characteristics of living beings. On the one hand, it implies that the exergonic reaction which supplies the energy must have a negative $\Delta G'$ greater in absolute value than the positive $\Delta G'$ of the endergonic reaction to which it is coupled. On the other hand, the two coupled reactions must have a *common intermediate*.

Let us consider two reactions having the common intermediate B:

– endergonic reaction A $\leftrightarrow$ B, $\Delta G_0 = +1\,362$ cal/mole, K = 0.1

– exergonic reaction B $\leftrightarrow$ C, $\Delta G_0 = -4\,086$ cal/mole, K = 1 000

– overall reaction A $\leftrightarrow$ C, $\Delta G_0 = -2\,724$ cal/mole, equilibrium constant $K = 1000 \times 0.1 = 100$[1]

If one starts with 1 011 moles of A, the equilibrium of the first reaction ought to be attained when 10% of the molecules are transformed, i.e. when 101 moles of A are converted to 101 moles of B. But as soon as the product B appears, the second reaction can take place, and here for equilibrium, 99.9% of moles of B must be transformed into C. In the final equilibrium, there will be 10 moles of A, 1 of B and 1 000 of C (which is in conformity with the equilibrium constants of the 2 reactions, 0.1 and 1 000 respectively); in other words 99% (1 000/1 011) of the moles of A present at the start will have been transformed into C. This shows that the coupling of the 2 reactions has enabled the exergonic reaction to "carry along" the endergonic reaction so that there remains only 1% of the molecules of A present initially. The coupling results from the equilibrium constants of the 2 reactions having in common the substance B, and the process can, for example, be started in living cells when the concentration of A increases or when that of C decreases.

It is clear that it would be extremely advantageous for the cells to have a single common intermediate compound, formed in all exergonic reactions and usable by all endergonic reactions; in fact such a compound exists: it is ATP.

---

(1) *In vivo*, the $\Delta G'$ are naturally the determinants.

1. – ATP

Adenosine-triphosphate or ATP, the structure of which is shown in figure 3-2, has 3 O – P bonds;

– one $-CH_2 - O - \overset{\overset{\displaystyle O}{||}}{\underset{|}{P}} -$ linkage, which is an *ester* bond similar to those found in phosphorylated sugars and whose hydrolysis is accompanied by a standard free energy change

$$\Delta G_0 \# -2.5 \text{ kcal/mole}$$

in the following reaction:

$$AMP + H_2O \longrightarrow \text{Adenosine} + P_i$$

– two $- \overset{\overset{\displaystyle O}{||}}{\underset{|}{P}} - O - \overset{\overset{\displaystyle O}{||}}{\underset{|}{P}} -$ linkages, called *phospho-anhydride* or *pyrophosphate* linkages whose hydrolysis – due to rearrangements in electron distribution and a greater stability of the products formed – is accompanied by a *strong liberation of utilizable energy*, i.e. a standard free enthalpy change $\Delta G_0 \#$ –7 kcal/mole in one of the following two reactions:

$$ADP + H_2O \longrightarrow AMP + P_i + H^+$$

$$ATP + H_2O \longrightarrow ADP + P_i + H^+$$

This second type of link is sometimes represented by a squiggle as may be seen in figure 3-2, and called "energy-rich bond", a denomination which is not quite correct because it is the *energy liberated by the hydrolysis which is high* and not the energy of the bond. It is also said sometimes that ATP (as also the other compounds indicated below) is an "energy-rich" compound. Actually, upon complete oxidation into CO_2 + H_2O, glucose can liberate a much larger quantity of energy (as mentioned earlier, oxidation of one molecule of glucose can yield 686 kcal, whereas the transformation of ATP into ADP + P_i yields only 7 kcal), but the superiority of ATP resides in the fact that – contrary to glucose – it can participate directly in energy transfer reactions. It is more correct to state that ATP has *a high energy potential or a high phosphate group transfer potential*, which explains the fact that its terminal phosphate group can be easily transferred to another compound (glucose, for example), thanks to the energy which is made available by the hydrolysis of ATP into ADP + P_i and which is immediately utilizable for the phosphorylation of glucose (an endergonic reaction).

It must be noted that the quantity of energy available varies according to the conditions and particularly as a function of the respective concentrations of the

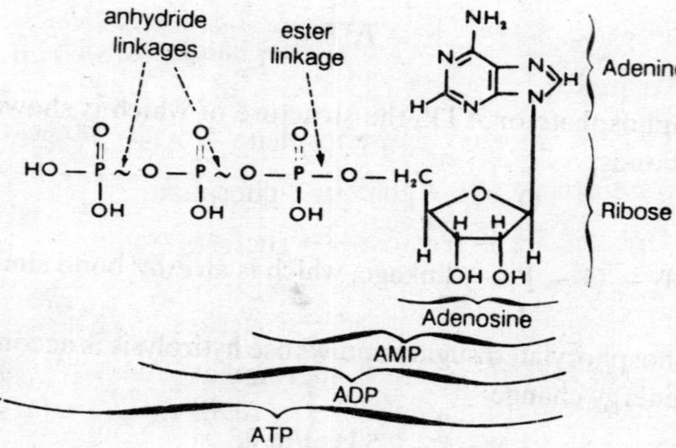

FIG. 3-2. — Structure of ATP.

reactants, which come into play (as mentioned in page 59) whenever the conditions are other than standard (i.e. always in living cells) and therefore, the values of ΔG *will be more or less remote from* ΔG_0. Moreover, all the energy made available is not necessarily transferred entirely inside the system formed by 2 coupled reactions; if only a part of this energy is utilized, *the rest can be transformed into thermal energy.*

ATP occupies a place of prime importance in processes of energy transfer. As stated below, the reactions which supply energy after absorption of light or oxidation of glucose lead to the formation of ATP from ADP + P_i , while the reactions requiring energy are linked either with the hydrolysis of ATP to ADP + P_i (even if the latter appears only subsequently in the form of free orthophosphate), or with the hydrolysis of another compound of high energy potential (see Table page 111). Very often indeed, the formation of this other compound is itself linked with the hydrolysis of ATP as, for example, in the case of the UTP required for the synthesis of polysaccharides (see figs. 4-22 and 4-23), the CTP required for the synthesis of phospholipids (see fig. 5-21), the GTP required for the biosynthesis of proteins (see pages 358-360) or the ribo- and deoxyribonucleosides-triphosphates required for the biosynthesis of the RNAs and DNAs.

To illustrate the manner in which *ATP enables a "chemical work"* i.e. an endergonic reaction to take place, let us take a simple example: the condensation of one molecule of glucose and one molecule of fructose to form one molecule of sucrose (for structures see Chapter 4 and especially figure 4-17):

$$\text{glucose} + \text{fructose} \longrightarrow \text{sucrose} + H_2O \qquad \Delta G_0 \ \# + 5.5 \text{ kcal/mole}$$

The energy required for this reaction is supplied by the hydrolysis of ATP, an exergonic reaction already indicated above:

$$ATP + H_2O \longrightarrow ADP + P_i + H^+ \qquad \Delta G_0 \ \# -7 \text{ kcal/mole}$$

In reality the 2 reactions are not independent as written above; they are coupled as follows[1]:

$$ATP + glucose \longrightarrow glucose\text{-}6\text{-}phosphate + ADP + H^+$$

$$glucose\text{-}6\text{-}phosphate \longleftrightarrow glucose\text{-}1\text{-}phosphate$$

$$glucose\text{-}1\text{-}phosphate + fructose \longrightarrow sucrose + P_i$$

$$\text{Sum}: ATP + glucose + fructose \longrightarrow sucrose + ADP + P_i + H^+$$

One may note that there are 2 intermediates common to the 3 coupled reactions: glucose-6-phosphate for the 1st and 2nd reactions, glucose-1-phosphate for the 2nd and 3rd reactions. In the first reaction a part of the energy of ATP has been transferred to glucose together with the phosphate group; this enables glucose-1-phosphate to react with fructose. The loss of free energy is 1.5 kcal/mole (which may be dissipated in the form of heat) and it can be said that for these 3 coupled reactions the *efficiency of energy storage* is 5.5 × 100/7 # 80%.

Generally, as shown by figure 3-3, the energy – originating from various sources which are not all listed in the diagram – is transferred to ADP + P_i to form ATP. Energy is then transferred to other processes from ATP which splits into ADP + P_i.

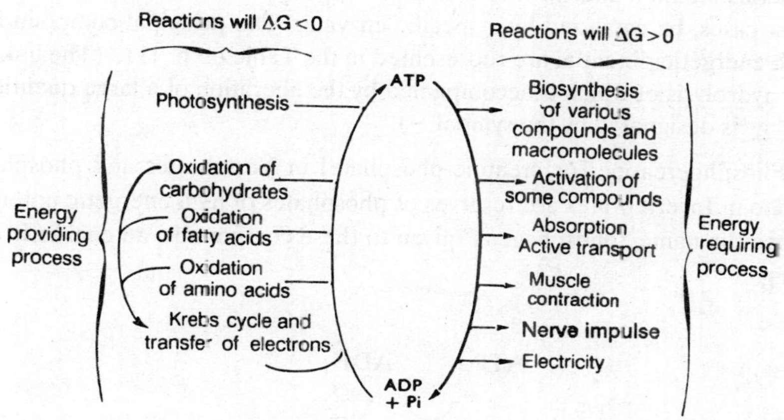

FIG. 3-3. — Cycle of the formation and cleavage of ATP during the processes of energy transfer in living organisms.

(1) The synthesis of sucrose can take place in this manner in certain organisms (for example, Pseudomonas saccharophilia) thanks to a sucrose phosphorylase, but in plants, the mechanism is more complex (it involves UDP-glucose).

The cleavage of ATP into ADP + P_i with liberation of energy is the most frequent reaction, but it is not the only one involving ATP. The latter can transfer:

– *an orthophosphate group* (with release of adenosine-5′-diphosphate or ADP) for example to an alcoholic hydroxyl; these transfers are catalyzed by kinases, and the reaction is reversible only if the energy potential of the compound formed is comparable with that of ATP; otherwise ATP will be regenerated from ADP by another reaction;

– *a pyrophosphate group* (with release of adenosine-5′-monophosphate or AMP) for example during the formation of 5-phosphoribosyl-1-pyrophosphate from ribose-5-phosphate (see fig. 6-18);

– *an AMP group* (with release of pyrophosphate) for example during the activation of fatty acids (see fig. 5-10) or amino acids (see fig. 6-39);

– *an adenosyl group* (with release of orthophosphate and pyrophosphate), for example during the formation of S-adenosyl-methionine, a coenzyme which was studied in Chapter 2 (see fig. 2-20).

2. – Other compounds of high energy potential

Among these compounds, some have a phosphate group (designated by the symbol Ⓟ). Although the ΔG′ corresponding to the hydrolysis is rather strongly negative, the hydrolysis of these compounds is not necessarily rapid in aqueous medium and for it to take place with appreciable velocity it must, in most cases, be catalyzed by a specific enzyme. The principal compounds of high energetic potential are represented in the Table on p. 111. (The linkage, the hydrolysis of which is accompanied by the liberation of a large quantity of energy is designated by the symbol ~).

Phosphocreatine (or creatine-phosphate) in Vertebrates and phosphoarginine in Invertebrates are reserves of phosphates of high energetic potential, hence the name "phosphagens" given to these compounds; an energetic coupling:

ATP ADP

creatine (or arginine) phospho-creatine (or phospho-arginine or other phosphoamidines in Invertebrates)

brings about the storage reaction (ATP + Creatine ⟶ ADP + Phosphocreatine) when ATP is present in excess and inversely, the formation of ATP by the reverse reaction (see fig. 7-13) when the cell needs ATP.

Type of linkage	*Examples of compounds*	*Reaction and* ΔG_0 (kcal/mole)
Phospho-anhydride (pyrophosphate) Ⓟ – O ~ Ⓟ	Nucleosides-triphosphates In addition to ATP: GTP, CTP, UTP, ITP etc.	ZTP ←→ ZDP + P_i −7
	Nucleosides-diphosphates ADP, GDP, CDP, UDP, etc.	ZDP ←→ ZMP + P_i −7
Acyl-phosphate O ‖ R – C – O ~ Ⓟ	1,3-diphosphoglyceric acid (see fig. 4-28) Aminoacyl-AMP (see fig. 6-39)	1,3 diphosphoglyceric acid ←→ 3 phosphoglyceric acid + P_i −10 Amino acyl-AMP ←→ AMP + amino acid −7
Enol-phosphate R – C – O ~ Ⓟ ‖ CH |	Phospho-enol-pyruvate (see fig. 4-29)	Phospho-enol-pyruvate ←→ pyruvate + P_i −12.8
Amidine-phosphate (guanidino-phosphate) R – C – NH ~ Ⓟ ‖ NH	Creatine-phosphate (see fig. 7-13) Arginine-phosphate (see pages 5 and 110)	Creatine-phosphate ←→ creatine + P_i −10.5 Arginine-phosphate ←→ arginine + P_i
Thioester O ‖ R – C ~ S – R′	Acetyl-coenzyme A (see fig. 2-18) Acyl-coenzyme A	Acetyl-coenzyme A ←→ coenzyme A – SH + acetate −8.8

We must now consider the origin of these compounds of high energetic potential; most of them are formed from ATP by a coupling similar to the one we have just seen for the formation of phosphocreatine; the problem is therefore the determination of the origin of ATP.

ATP is formed in living organisms *by phosphorylation of ADP coupled with oxidation reactions* which are accompanied by a large decrease of free energy and which therefore supply the energy needed. We will study the phosphorylations which take place during transfers of electrons *in the respiratory chain* in aerobic cells (where oxygen is the final electron acceptor, i.e. the oxidant of nutritive substances) and *during the light phase of photosynthesis* (which enables the conversion of energy of light into chemical energy); as mentioned later in this chapter, these two processes are somewhat similar and both of them take place in cellular organelles: the mitochondria and the chloroplasts respectively; we will also study the phosphorylation of ADP into ATP in *anaerobic cells* (where the electrons cannot reach oxygen).

III. – FORMATION OF ATP IN THE RESPIRATOR CHAIN

1. – Advantage of an oxidation by steps

Glucose or glucosyl equivalents, are our main source of energy; we will take them as example to illustrate the relations between cellular oxidations and ATP formation. But it must be said that other compounds can be a source of energy for living organisms and we will see particularly in Chapter 5, that triglycerides have (for equal weight) a greater energetic value than carbohydrates.

We have already stated that the complete oxidation of glucose (e.g., by combustion in air) according to the reaction $C_6H_{12}O_6 + 6O_2 \longrightarrow 6CO_2 + 6H_2O$ liberates 686 kcal/mole. If we compare this value to the energy required for the synthesis of ATP by the reaction ADP + $P_i \longrightarrow$ ATP + H_2O (8 to 10 kcal/mole on an average, in the usual intracellular conditions), it appears that theoretically, the oxidation of one mole of glucose can yield sufficient energy for the synthesis of 686/9 # 76 moles of ATP. In fact, as we will see in Chapter 4, the complete oxidation of glucose into $CO_2 + H_2O$ in the usual pathway (glycolysis followed by the Krebs cycle and the respiratory chain) permits the synthesis of 38 ATP per mole of glucose (i.e. half the theoretical figure), and the rest of the energy is liberated as heat; this may be summarized as follows:

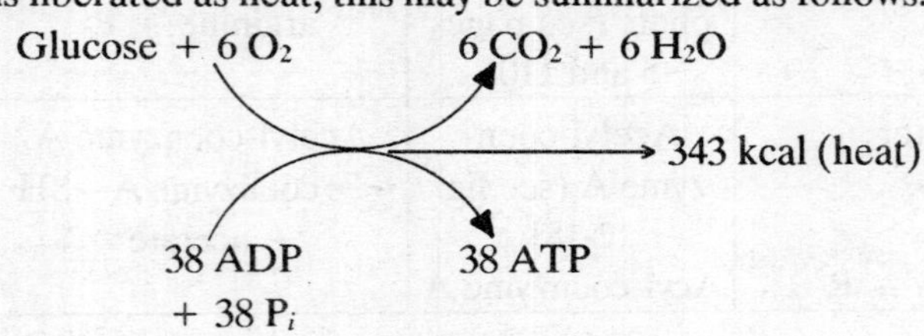

This is actually a diagram which summarizes the whole set of reactions, and not a single reaction because, if there had been only one reaction, only one molecule of ATP would be formed and all the remaining energy would be dissipated in the form of heat. *The oxidation of glucose takes place in successive steps, with liberation of a small fraction of the total energy each time*, thus enabling the formation of several molecules of ATP. Anticipating to some extent on the studies presented in Chapter 4, it may be said that glucose is oxidised in 6 steps: one oxidation step in the glycolysis (reaction 4, fig. 4-35), one in the decarboxylation of pyruvic acid (see fig 4-36), and four in the Krebs cycle (reactions 3, 4, 6 and 8, fig. 4-38).

But one could reasonably wonder how 38 molecules of ATP can be produced during 6 oxidation reactions only, because on examining one of these reactions, for example, the dehydrogenation (or oxidation) of isocitric acid to oxalo-succinic acid (see fig. 3-4), it is observed that isocitric acid loses, to the advantage of oxygen, 2 hydrogen atoms and 2 electrons (since oxygen passes from oxidation state zero to state –2); and although this is the minimum oxidation which isocitric acid can undergo, it would not be very economic from the energetic point of view, if it took place in a single step, because only one molecule of ATP could be formed out of the 5 which could arise thanks to the 50 kcal made available (and some 40 kcal which remain would be dissipated as heat).

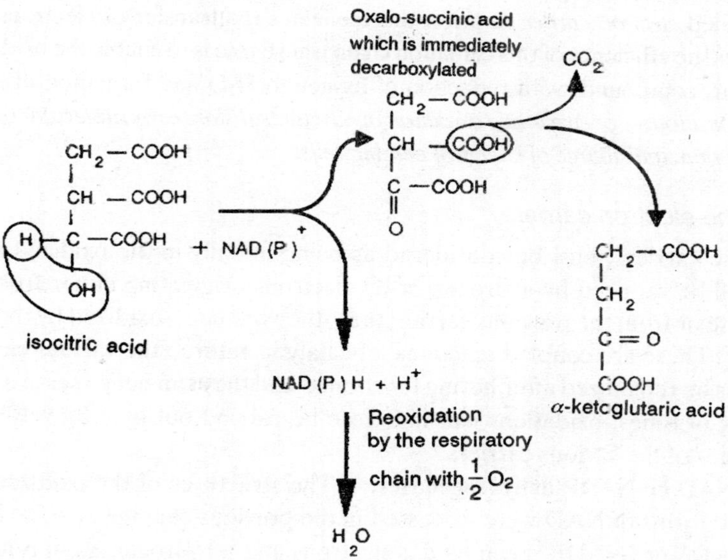

FIG. 3-4. — Oxidation of isocitric acid. (Globally, the reoxidation of the reduced co-enzyme, by the respiratory chain, produces H_2O with a reoxidation ΔG of about 50 kcal.)

In fact, for each molecule of isocitric acid oxidized into oxalo-succinic acid, 3 molecules of ATP (not one) are formed. The reaction – as it is written above – does not reflect the real situation because the electrons do not pass directly from isocitric acid to oxygen. As will be seen in the following, *a series of oxidation-reduction reactions enable a transfer of electrons by steps, and three of them are coupled with the formation of one molecule of ATP.*

2. – Electron carrier system or respiratory chain

In certain cases, the electrons of the substance to be oxidized pass directly to oxygen[1], but in general, electrons are transferred in *successive steps* to a series of carriers, which is much more advantageous from the energetic point of view. One must also note that the sudden oxidation of substances like glucose with liberation of 686 kcal/mole, could not take place in living organisms, because it would be accompanied by an excessive release of heat as only one ATP can form per reaction. Furthermore – and this is another advantage – the electrons originating from the oxidation of very diverse substrates are channelled *towards the same electron carrier system* to finally reach oxygen. This system can vary slightly from one organism to another, but in general it contains: an enzyme (dehydrogenase) having as coenzyme a pyridine nucleotide (NAD or NADP), a flavoprotein, coenzyme Q and several cytochromes. These compounds represent in mitochondria about 1/4 of total proteins; they are grouped in multi-enzymatic aggregates forming complex structures called *electron carrier particles*; this facilitates the transfer of electrons and increases the efficiency of this oxidation chain whose role is to enable the oxidation of various compounds with reduction of oxygen to H_2O and formation of ATP. *This respiration – or enzymatic oxidation of nutritive substances by molecular oxygen – is the principal source of energy of aerobic cells.*

A. – The electron carriers

These carriers must be considered as being initially in the oxidized state; they will be reduced by hydrogen or by electrons originating either from the substrate or from the previous carrier; then, they will be re-oxidized by the next carrier. These are coupled reactions, of catalytic nature : the various carriers can thus be reoxidized after having functioned and they can hence serve a large number of times; oxidations can therefore be carried out by relatively small quantities of the various carriers.

a) **NAD or NADP dehydrogenases** – The structures of the oxidized and reduced forms of NAD were discussed in the previous chapter (see fig. 2-16). The NAD^+ or ($NADP^+$) can bind 2 electrons and a hydrogen atom removed from a substrate (which is thus oxidized), and be reduced to NADH + H^+ (or NADPH + H^+). A wide variety of substances are oxidized in this manner: glyceraldehyde-3-phosphate (see fig. 4-27), isocitric acid, malic acid (see fig. 4-38), glucose-6-phosphate (see fig. 4-39), the hydroxy-acyl-co A during the β-oxidation of fatty acids (fig. 5-11), etc.

b) **Flavoproteins** – We have also already examined (fig. 2-17) the structures of coenzymes derived from riboflavin (FAD, FMN) and the manner in

---

(1) It has been stated in the previous chapter, while studying the classification of enzymes (see page 94) that the oxido-reductases are classified first according to the group on which they act, and then according to the nature of the acceptor which is sometimes oxygen (sub-sub-class 1.1.3, 1.2.3, etc.); hydrogen peroxide (H_2O_2) is then formed and immediately destroyed by catalases or peroxidases.

which they participate in oxidation-reduction processes. In the respiratory chain, the flavoproteins allow the reoxidation of NADH to NAD^+ and upon binding a hydrogen atom of NADH, a H^+ of the solution and 2 electrons of NADH, they are then in the reduced state. We will see (see figs. 3-6 b and 4-38) that in certain cases a flavoprotein can bring about the oxidation of a substrate like succinate (without passing through NAD).

c) **Coenzyme Q** – The reduced flavins can be reoxidized by transfer of 2 hydrogen atoms ($2H^+$ and $2e^-$) to a molecule of coenzyme Q which is then in a reduced state. The structure of this compound is shown in figure 5-9; it is a substituted benzoquinone having an isoprenic side chain, which is related to that found in some lipids (hence its classification in Chapter 5). The number of isoprene units in this chain can vary according to the origin of the compound. The coenzyme Q_{10} (10 isoprene units) also called *ubiquinone* 50 (10 units of 5 carbon atoms = 50 carbon atoms) is the most common form in animal mitochondria. Like quinones in general, the coenzyme Q can undergo a reversible reduction to hydroquinone (see fig. 3-5). The coenzyme Q or ubiquinone is a soluble mobile carrier in the lipidic bi-layer of the mitochondrial inner membrane.

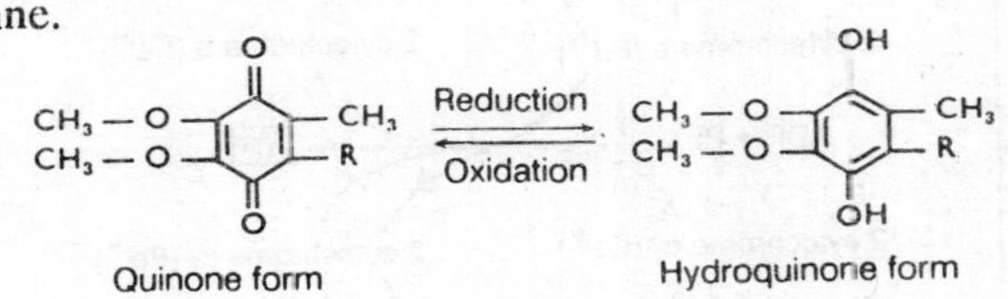

FIG. 3-5. — Oxidized and reduced forms of the coenzyme Q_{10} or ubiquinone 50 (the structure of the side chain is represented in figure 5-9).

Flavoproteins like the coenzymes Q ensure two successive transfers of $1H^+$ and $1e^-$.

d) **Cytochromes** – As indicated by their name, these are coloured substances. Cytochromes are *chromoproteins*, the prosthetic group of which – strongly linked to the protein – is a *porphyrin*, identical or very similar to the heme of hemoglobin (see fig. 1-29). The various cytochromes – 5 of them are known to belong to the mitochondrial system of electron transport – differ either in the protein part, or in the linkages between this protein part and the porphyrin, or in certain cases by the substituents on the β-carbon atoms of the pyrrole rings of porphyrin. As observed below (see Table on page 120), these structure differences result in differences in reactivity and especially in the oxidation-reduction potential. Contrary to the former carriers, cytochromes can transport only one electron at a time and that is why 2 molecules of each cytochrome are involved in the reactions of the respiratory chain in which they participate (see fig. 3-6).

The first to participate in the respiratory chain is cytochrome b which enables the reoxidation of coenzyme Q. It must be noted that at this stage, the 2 hydrogen atoms are not transferred, but 2 protons are liberated in the medium. Since the reduction of one molecule of cytochrome b implies the

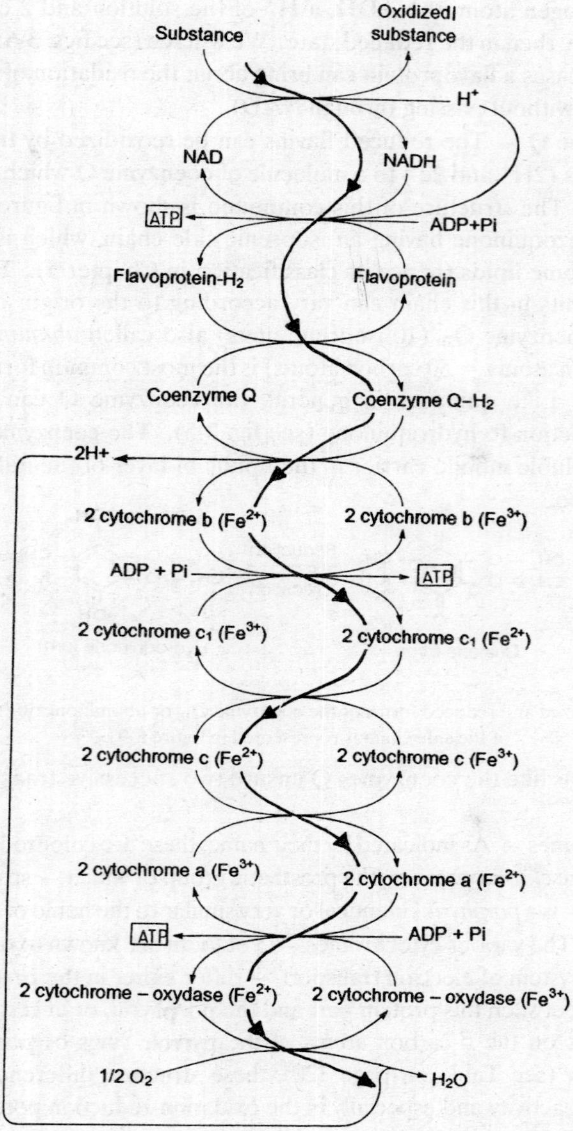

FIG. 3-6 a — The electron carrier system
(the flow of electrons is indicated by the red line).

transfer of only one electron, the participation of 2 molecules of cytochrome b is needed to accept the 2 electrons and allow the reoxidation of the coenzyme Q. It is important to note that *the reduced form of cytochromes contains the ferrous ion* Fe^{2+}, while *the oxidized form has a ferric ion* Fe^{3+} (whereas the

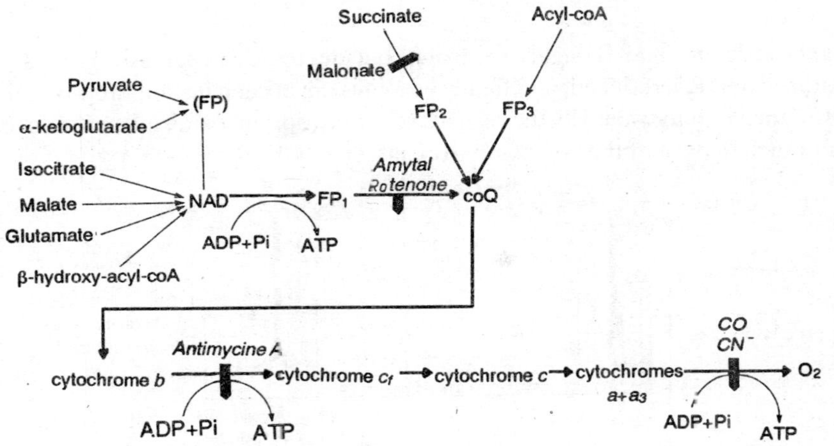

FIG. 3-6 b — The electron carrier system. This diagram shows the source of protons and electrons and their entry points in the respiratory chain. Thick arrows indicate the point of impact of some inhibitors of the respiratory chain. The respiratory chain is an integral part of the mitochondrial inner membrane.

hemoglobin ↔ oxyhemoglobin transformation takes place without any change of the state of oxidation of iron).

The electrons pass from cytochrome b to cytochrome c_1, then to cytochromes *c*, *a*, a_3 by the same process of oxidation and reduction of iron. Finally, the electrons reach the cytochrome a_3 also called *cytochrome-oxidase* which has the peculiarity of being directly oxidizable by molecular oxygen, reaction in which participate more than 9/10th of the oxygen utilized by the cells. This enzyme is inhibited by small concentrations of cyanide (CN^-), so that intoxication by this ion inhibits electron transport and synthesis of ATP by the respiratory chain, which generally results in rapid death. This last reaction of the chain – contrary to the previous reactions – is irreversible, which suffices to drag the electron carrier system in the direction in which it usually functions: from the oxidizable substrate to the oxygen (from top to bottom in the diagram of figure 3-6 a).

B. – Oxidation-reduction potentials

If an electrode of unattackable metal (e.g, Pt) is dipped in a solution of a reducing substance A (i.e. a substance which can yield electrons), the latter will yield electrons to the metal which will thus get negatively charged. If on the contrary, the metal electrode is placed in a solution containing an oxidizing substance B (which can accept electrons), the latter will extract electrons from the metal which will then be positively charged. If the 2 solutions are connected by a conducting bridge and the 2 electrodes by a voltmeter, one can observe on the latter the flow of an electric current through the wire due to the transfer of electrons from the electrode (–) to the electrode (+). This is the principle of the electric cell, where the potential difference is very large at the start if A is highly reducing and B highly oxidizing, but decreases progressively as A is oxidized and B is reduced, and finally becomes nil; at each instant this potential difference is determined by the respective proportions of oxidized and reduced substance in each of the two compartments.

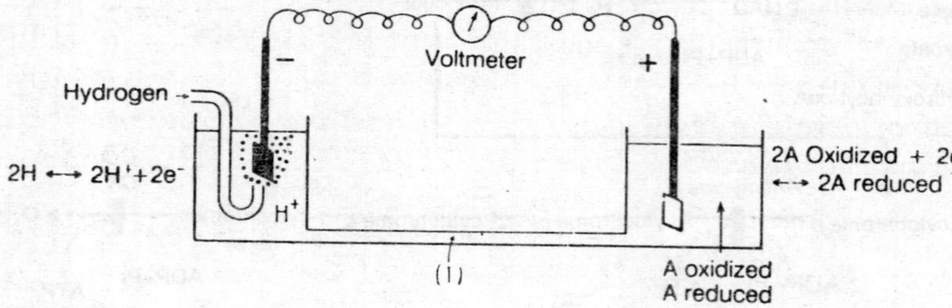

(1) Conducting bridge allowing the passage of ions but not the chemical interactions between the 2 systems.

FIG. 3-7. — Apparatus for measuring the potential difference between a redox system to be studied (A oxidized/A reduced) and a normal hydrogen electrode serving as reference.

Let us now consider one compartment only: depending on whether there is more of A reduced or A oxidized, the electrode will be charged negatively or positively (if the concentrations of the 2 forms are equal, the charge is nil). If we connect this electrode to *a reference electrode*, generally the normal hydrogen electrode (a Pt blade covered with Pt black saturated with gaseous hydrogen, dipping in a solution containing one ion-gram hydrogen per litre), whose potential is, by definition, equal to 0, we will measure a *potential difference* (E), representing the *oxidation-reduction potential* of the system. Such an apparatus is shown in fig. 3-7: on the left we have the reaction $2H \longleftrightarrow 2H^+ + 2e^-$, which will make the Pt electrode negatively charged; on the right, we will have the reaction : 2A (oxidized form) + $2e^-$ (pulled out of the electrode) ←→ 2A (reduced form), so that the electrode will be positively charged. The potential difference is given by the following relation (Nernst formula):

$$E = E_0 + \frac{RT}{n\,\mathcal{F}} \times \ln \frac{[\text{oxidised form}]}{[\text{reduced form}]}$$

or, knowing that $\ln x = 2.30 \log x$, we can write

$$\boxed{E = E_0 + \frac{2.30\,RT}{n\,\mathcal{F}} \log \frac{[\text{oxidised form}]}{[\text{reduced form}]}}$$

where E is the potential difference (in volts or millivolts) between the electrode of the oxidation-reduction system under study and the reference electrode,

E_0 is the normal oxidation-reduction potential, i.e. the potential difference measured between the normal hydrogen electrode and the electrode dipping in the oxidizing-reducing system containing as much oxidized A as reduced A,

R is the gas constant (1.987 cal/mole/degree), T the absolute temperature, n the number of electrons transferred per mole and $\mathcal{F}$ the faraday (96 500 coulombs). As for [oxidized form] and [reduced form], they are the activities of the oxidized and reduced forms which can be – on a first approximation – taken as the molar concentrations (see the note at the bottom of page 59). If these 2 concentrations are equal, their ratio is 1, and since the log of 1 is 0, we have in this case $E = E_0$. It is clear *that variations of the respective concentrations of the oxidized and reduced forms will result in modifications of the value of the oxidation-reduction potential.*

One must note the resemblance between the above relation and the corrected formula giving the free energy change (see page 103). The oxidation-reduction potential represents in fact one way of measuring the free energy of a reduction reaction. We will see in the following, the exact relation between ΔE and ΔG, and how this relation can help in determining the stages in which ATP is formed (see page 120).

It was found that for biochemical oxidation-reductions, the conditions defined for the normal hydrogen electrode, i.e. 25°C, hydrogen pressure = 1 atmosphere, $[H^+] = 1$ (i.e. pH = 0), were not very physiological (particularly the latter); a normal potential E'_0 corresponding to 30°C and pH 7 (which replaces E_0 in Nernst formula) was therefore defined. In these conditions, the potential of the normal hydrogen electrode is not 0, but $E'_0 = -0.42$ volt.

When there are several oxidation-reduction systems (or redox systems) in the same series of reactions, it is of interest to know their oxidation-reduction potentials. For example, when there are 2 systems:

A oxidized + $e^- \longrightarrow$ A reduced (with potential E_A)
B oxidized + $e^- \longrightarrow$ B reduced (with potential E_B)

Let us suppose $E_A > E_B$, i.e. that system A is more oxidizing than system B; this means that if the 2 systems are in contact, the electrons will tend to pass from B to A, in other words B will get oxidized while A will be reduced. *Electrons tend to pass from the carrier having the most negative potential to the carriers having less negative or increasingly positive potentials.*

The oxidation-reduction potentials of a large number of biological oxidation-reduction systems were determined (with respect to a reference system), notably, the potentials of carriers of the respiratory chain which are of special interest to us; the values are given in the table (see p. 120).

This table, showing the values of potential differences of oxidation-reduction at pH 7 and 30°C ($\Delta E'_0$), also contains the values ($\Delta G'_0$) of standard free energy changes at pH 7 and 30°C (see page 103) which enable us to locate the stages of phosphorylation. The $\Delta G'$ values differ depending on the general equilibrium of the respiratory chain, but the effective potential jumps remain the same.

It can be seen that the oxidation-reduction potentials of the various systems increases continuously from pyridine nucleotide enzymes to oxygen; this indicates the theoretical order in which the transfer of electrons must take place (we mentioned above that the transfer of electrons takes place in the direction of a system having a higher potential); this theoretical order is indeed the same as the order in which the electron carriers of the respiratory chain follow one another (see fig. 3-6). It must be noted that the oxidation-reduction potentials of coenzymes (indicated below) can be modified when the coenzyme is linked to the apoenzyme.

Oxidation-reduction system (redox system)	E'_0 (volts) at pH 7 and 30°	$\Delta E'_0$ (volts)	$\Delta G'_0$ in kcal/mole at pH 7 and 30°	*Phosphorylation of ADP to ATP*
Reference: $2H^+/H_2$	−0.42			
$NAD^+/NADH + H^+$	−0.32			
$FAD/FADH_2$	−0.10→	+0.22	−10.1	1
Cytochrome b Fe^{3+}/Fe^{2+}	+0.04→	+0.14	−6.4	
Cytochrome c_1 Fe^{3+}/Fe^{2+}	+0.23→	+0.19	−8.7	1
Cytochrome c Fe^{3+}/Fe^{2+}	+0.26→	+0.03	−1.4	
Cytochrome a Fe^{3+}/Fe^{2+}	+0.29→	+0.03	−1.4	
Cytochrome a_3 Fe^{3+}/Fe^{2+}	+0.55→	+0.26	−12.0	1
$1/2\ O_2/O^{2-}$	+0.81→	+0.26	−12.0	
		+1.13	−52.0	3

C. – Phosphorylations in the respiratory chain

The free energy change between 2 redox systems ($\Delta G'_0$) is related to the difference between the oxidation-reduction potentials of the 2 systems ($\Delta E'_0$) by the following equation:

$$(\Delta G'_0) = -n\,\mathcal{F}\,\Delta E'_0$$

where n is the number of electrons transferred and $\mathcal{F}$ the Faraday; but the use of one Faraday per mole of substrate for a valency modification of one unit, produces an energy of 23.06 kcal/volt; the Faraday is therefore equivalent to 23.06 kcal/volt. If one considers the normal potential difference of oxidation-reduction between the beginning (NAD) and the end (oxygen) of the respiratory chain (1.13 volts), one can – with the help of the above equation – calculate that the oxidation of the H^+ and e^- of $NADH + H^+$ can supply energy amounting to:

$$\Delta G'_0 = 2 \times 23.06 \times 1.13 \ \# \ -52 \text{ kcal}$$

We know that this energy is not liberated in one stroke but in successive steps. The problem is to know which steps will lead to the formation of ATP. Since the formation of ATP from ADP + P_i + H^+ requires about 7 kcal, it may be said that those steps of the respiratory chain in which the energy liberated is greater than 7 kcal, should enable the synthesis of ATP. It is obvious that

$\Delta G'$ for the synthesis or hydrolysis of ATP depends on pH since H^+ is involved. If we write:

$$-7 = -n \mathcal{F} \Delta E'_0$$

we can deduce $\Delta E'_0 = 7/2 \times 23.06 = 0.152$ volt; in other words, *to a normal potential difference greater than 0.152 volt corresponds an energetic variation sufficient for the formation of one molecule of ATP*. The table (page 120) contains the $\Delta E'_0$ and $\Delta G'_0$ values corresponding to each step, and it may be observed that 3 steps should lead to the synthesis of ATP: the passage of electrons from the pyridine coenzyme to the flavin coenzyme, from cytochrome *b* to cytochrome *c*, and from cytochrome *a* to oxygen. In these 3 steps there is possibility of coupling between oxidation-reduction and phosphorylation of ADP to ATP, hence the name *oxidative phosphorylations* or *phosphorylating oxidation-reductions* given to these coupled reactions in the respiratory chain.

The above reasoning is based on *normal* oxidation-reduction potentials (E'_0) for the various electron carriers and on *standard* free energy changes ($\Delta G'_0$), but it is certain that these values must be corrected in order to reflect the intracellular (or more exactly the intra-mitochondrial) conditions. Furthermore, these are theoretical predictions and it remains to be seen whether they have been confirmed experimentally.

If mitochondria are incubated in a suitable medium, in presence of an oxidizable substrate, oxygen, ADP and P_i, one can measure the disappearance of the substrate, ADP or P_i, the consumption of oxygen and the formation of ATP. One can thus define, for the oxidation of each substrate, a *phosphorylation ratio*[(1)] (P/O), i.e. the number of molecules of ATP formed (or molecules of P_i consumed) per atom of oxygen utilized (or per molecule of hydrogen removed from the substrate).

This P/O ratio is equal to 3 when the acceptor is a pyridine coenzyme like NAD^+ [(2)], as for example in the case of the oxidation of isocitric acid or malic acid in the Krebs cycle (see fig. 4-38). But the P/O ratio is only 2 during another oxidation of the Krebs cycle, that of succinic acid (see fig. 3-48), because in this reaction the acceptor is FAD, so that one oxidation-reduction step coupled with a phosphorylation is skipped and only 2 phosphorylating steps remain. The ratio can even be only one when the acceptor is cytochrome *c*, for example,

---

(1) This phosphorylation ratio must not be confused with the respiratory ratio (R.R.) obtained by dividing the volume of CO_2 produced by the volume of oxygen consumed for one gram of food oxidized (R.R. = 1 for carbohydrates, 0.8 for proteins, 0.7 for lipids).

(2) $NADP^+$ can be the hydrogen acceptor in various oxidation reactions, like that of glucose-6-phosphate or 6-phosphogluconic acid (fig. 4-39). We will see that the hydrogen atoms of $NADPH + H^+$ can lead to hydrogenation reactions (for example in the biosynthesis of fatty acids; see fig. 5-18), but they can also be transferred to NAD^+ and then enter the respiratory chain thus enabling the formation of ATP by oxidative phosphorylation.

in the case of the oxidation of ascorbic acid, because 2 steps of phosphorylating oxidation are skipped. On the contrary, the ratio can also be greater than 3; the P/O ratio is 4 for example for the oxidation of α-ketoglutaric acid to succinic acid, because the GTP formed leads to the formation of one additional molecule of ATP (see fig. 4-38).

Experimental results have therefore confirmed the theoretical predictions which suggested that three steps of the respiratory chain are coupled with the formation of ATP. However, this coupling is not absolutely obligatory and there are substances which can *uncouple* the phosphorylating oxidations (thyroxine which is a hormone secreted by thyroid, 2, 4-dinitrophenol, etc.): oxidations continue, but oxidative phosphorylations are inhibited so that the energy, instead of being utilized for the formation of ATP, is dissipated as heat, which brings about a temperature rise in the animal that has been administered an uncoupling substance. However, the phosphorylations bound to the substrate are not affected by the uncoupling agents.

D. – *The shuttles*

Mitochondria possess a folded inner membrane (forming crests), with a very large functional surface (see fig. 3-8). The major oxidation-reduction processes take place in the inner membrane or on the inner face (matrix side) of the inner membrane, and it is in this inner membrane that are found the lipoic acid dehydrogenases of pyruvate and α-ketoglutarate, other dehydrogenases like succinodehydrogenase, enzymes of the Krebs cycle and of β-oxidation of fatty acids, the carriers participating in the respiratory chain and the system of synthesis of ATP coupled with the respiratory chain.

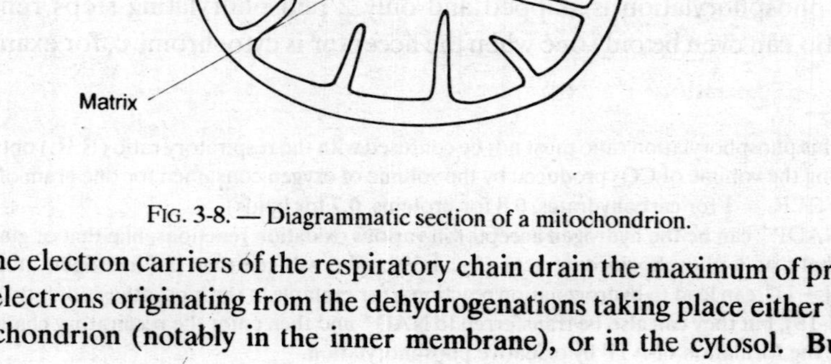

FIG. 3-8. — Diagrammatic section of a mitochondrion.

The electron carriers of the respiratory chain drain the maximum of protons and electrons originating from the dehydrogenations taking place either in the mitochondrion (notably in the inner membrane), or in the cytosol. But the inner membrane has a selective permeability and nucleotides of the type NAD(P) for example, cannot pass through it, so that the electrons and protons from the cytosol are used to temporarily reduce a substrate, like oxaloacetate into malate, which can cross the inner membrane thanks to a carrier and can then be reoxidized (to oxaloacetate) yielding NADH + H^+ which will enter the respiratory chain. This is an example of shuttle, enabling the passage of the reducing equivalents from the cytosol to the mitochondrion (see fig. 3-9).

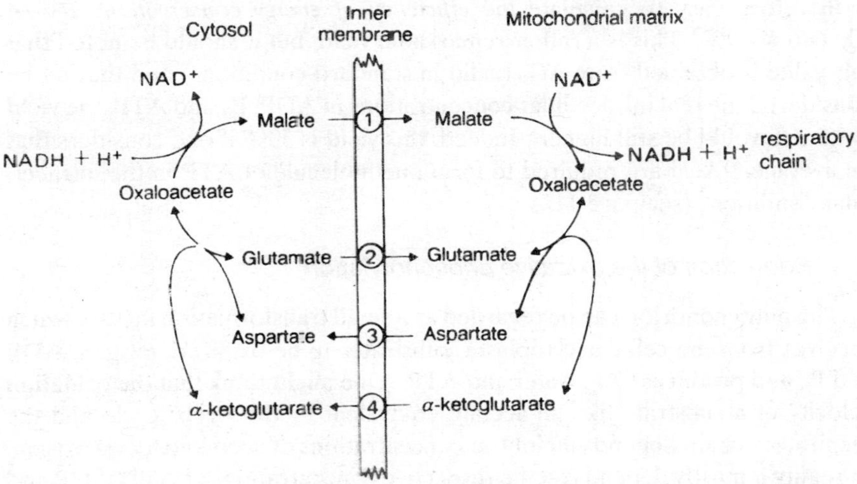

FIG. 3-9. — Diagram showing an example of shuttle allowing the passage of the reducing power from the cytosol to the mitochondrion. This shuttle is based on the presence of malate dehydrogenase and Asp/Glu transaminase in the cytosol and the mitochondrion; it uses carriers of malate (1), glutamate (2), aspartate (3) and α-ketoglutarate (4).

As mentioned above, the substrates originating from the cytoplasm must use *specialized carriers* (or *translocators*) to penetrate into the mitochondrion and be either oxidized or phosphorylated (in the case of ADP and phosphate). One knows for example, carriers of dicarboxylic acids (malate, succinate), tricarboxylic acids (citrate, isocitrate), fatty acids, α-ketoglutarate, glutamate, phosphate, ADP and ATP. These carriers can be selectively inhibited by certain substances, blocking the passage of substrates (and therefore of protons and electrons) through the inner membrane of mitochondria and thereby inhibiting the oxidation-reductions of the respiratory chain.

But cellular oxidation-reductions can also be affected by inhibitors of oxidative decarboxylations (pyruvate, α-ketoglutarate) or of the Krebs cycle, by perturbation of oxygen uptake (especially by substances affecting hemoglobin, like CO), by inhibitors of electron carriers of the respiratory chain (some are indicated in fig. 3-6 b), by uncoupling agents which dissociate the synthesis of ATP from the oxidations of the respiratory chain so that the substrates are oxidized in pure loss with production of heat (by 2, 4-denitrophenol, thyroxine, for example), or by inhibitors of the synthesis of ATP (like oligomycin).

E. — Energy yield of the oxidation of glucose in aerobic cells

At the beginning of this study devoted to the formation of ATP during oxidation processes, glucose was taken as example. It will be seen in Chapter 4 (page 195) that the complete oxidation of one molecule of glucose, through glycolysis followed by the Krebs cycle, allows the formation of 38 molecules of ATP[(1)] which corresponds to about $38 \times 7 = 266$ kcal/mole of glucose. But it is known that the sudden combustion of glucose could yield 686 kcal/mole. It is therefore easy to calculate *the efficiency of energy conservation*: 266 × 100/686 # 40%. This is a rather remarkable yield, but it should be noted that this value is obtained from ΔG_0 (valid in standard conditions) and that if one considered the real intracellular concentrations in ADP, P_i, and ATP, the yield obtained would be still higher. Indeed, this yield is 50% if one considers that on average, 9 kcal are required to form one molecule of ATP in the intracellular conditions (see page 112).

F. — Regulation of the oxidative phosphorylation

The mitochondrion can be regarded as a small transformation factory which receives from the cellular cytoplasm, substrates to be oxidized, oxygen, ADP and P_i, and produces CO_2, water and ATP. One might think that the oxidation velocity of a substrate like an acetate equivalent by the Krebs cycle and the respiratory chain depends mainly on concentrations of acetyl-coA and oxygen. In reality it mostly depends on the respective concentrations of ATP, ADP and P_i and the ATP/ADP ratio has a decisive influence on respiration. For example, in a resting muscle, where the utilization of ATP to give ADP + P_i (reaction required for muscle contraction) is low, the ATP/ADP ratio is comparatively high and the rate of respiration is low, because respiration needs ADP (without ADP, no phosphorylating oxidation); on the contrary, in a muscle at work, the decomposition of ATP into ADP in ADP + P_i is considerable, the ATP/ADP ratio therefore decreases, and the fact that ADP and P_i are available in larger quantities permits an acceleration of respiration (and consequently, a regeneration of the ATP required for muscle work). This regulation by the ATP/ADP ratio obviously assumes that the substrate (here acetyl-coA) and oxygen are in excess in the cell; when these conditions are fulfilled (and they generally are) *the production of ATP by oxidative phosphorylation is therefore controlled by the ATP requirements of the cell*, whether for muscle contraction, to allow endergonic reactions, or for an altogether different purpose (see fig. 3-3). The mechanisms of phosphorylating oxidations are discussed in paragraph VI.

---

(1) Out of these 38 ATP, 8 result from glycolysis and 30 from the oxidation of pyruvic acid by the Krebs cycle and the respiratory chain.

IV. – FORMATION OF ATP IN ANAEROBIC CELLS

We studied first the formation of ATP in the respiratory chain because it is the most common process in living organisms. The *strict anaerobes*, i.e. species which can live only in the absence of oxygen (in the ground or mud, for example) are comparatively rare. On the contrary, there are some *facultative anaerobes*, organisms which preferentially utilize oxygen to oxidize their nutritive substances when oxygen is present but which can also live in absence of oxygen, *then drawing the energy they require from fermentation processes*. This capacity of conserving the energy of food in anaerobiosis (production of ATP during fermentation) is found not only in some microorganisms but also in numerous more advanced organisms, and the formation of ATP during the fermentation of glucose is a process of energy conservation which has reached man; as we will see below, the anaerobic degradation of glucose precedes its complete oxidation by the Krebs cycle and the respiratory chain.

Various substances (sugars, notably glucose, fatty acids, amino acids) can be utilized by bacteria in anaerobiosis, which also helps in identifying or classifying these bacteria. Furthermore, these bacteria – even when they utilize glucose for example – can also be distinguished according to the mode of fermentation; some degrade glucose into ethanol, others into acetone, butanol, butyric acid, propionic acid or lactic acid. But in a large number of cases, glucose is split into 2 fragments, of which one is oxidized by the other, and this oxidation-reduction process liberates energy, a part of which is stored in the form of ATP. We will study here only *the fermentation of glucose* to lactic acid, a mechanism found not only in certain bacteria (e.g., of the genus *Lactobacillus*), but also in higher animals (anaerobic glycolysis).

1. – Energy yield of lactic fermentation

The transformation of glucose into lactic acid corresponds to the following overall equation :

$$C_6H_{12}O_6 \longrightarrow 2CH_3-CHOH-COOH \qquad \Delta G_0 = -52 \text{ kcal/mole}$$

In reality, in living cells, this cleavage of glucose implies a dozen of reactions and first the utilization of 2 ATP for the formation of glucose-6-phosphate and fructose-1,6-bisphosphate, then the formation of 4 ATP, which gives the following overall equation:

$$C_6H_{12}O_6 + 2ADP + 2P_i \longrightarrow 2CH_3-CHOH-COOH + 2\,ATP + 2H_2O$$

$$\Delta G_0 = -38 \text{ kcal/mole}$$

On the basis of ΔG_0 values and admitting that the formation of one ATP requires at least 7 kcal/mole, one can calculate *a minimum energy conservation efficiency of about 27%* (2 × 7 × 100/52) with reference to the energy liberated by the cleavage of glucose into lactic acid; but if this yield is expressed with reference to the 686 kcal which can be liberated by the sudden oxidation of glucose, then it would be 14 × 100/686 # 2%, i.e. a very low yield compared to the 40% obtained in aerobiosis when glycolysis is followed by the Krebs cycles and by electron transfer in the respiratory chain (see above, page 124)[1]. It is quite clear that in anaerobiosis, only a very small part of the total energy of the glucose molecule can be "recovered" and consequently, to accomplish an identical work, *anaerobic cells must consume much more* (up to 20 times more) *glucose per unit time than the aerobic cells.*

Glycolysis therefore appears to be a more primitive, less elaborate means of storing energy in the form of ATP; this is further confirmed by the fact that the enzymes (about ten) catalysing the various steps are in free state in the soluble part of the cellular cytoplasm, and not grouped in pluri-enzymatic systems of complex structure like the enzymes catalyzing the reactions of respiration and photosynthesis in mitochondria and chloroplasts.

2. – Mechanisms of energy conservation. Substrate-related phosphorylation

Referring to Chapter 4 and particularly figure 4-35 which summarizes the various steps of glycolysis, one observes that paradoxically the first steps of this process (which must lead to the formation of ATP) consume ATP: these steps are the phosphorylation of glucose and that of fructose-6-phosphate. But then, for each C_3 fragment there are 2 steps of phosphorylation of ADP into ATP and therefore, 4 ATP are formed as one molecule of glucose gives 2 C_3 fragments. Finally, 2 ATP are consumed and 4 ATP formed, which is conform to the gain of 2 ATP indicated by the overall reaction given in the previous paragraph. We will now see the mechanisms by which these ATP molecules are formed during the glycolysis.

In addition to phosphorylations which take place along the respiratory chain (see page 120) and to photosynthetic phosphorylations (page 134), there are substrate-related phosphorylations; two examples are given below.

A. Formation of ATP during the oxidation of glyceraldehyde 3-phosphate into 3-phosphoglyceric acid

The oxidation of an aldehyde to acid:

$$R-CHO + H_2O \longrightarrow R-COOH + 2H$$

is an exergonic reaction (ΔG_0 # –7 kcal/mole). In the cells, this energy is not necessarily dissipated as heat; it can be utilized for the endergonic reaction of ATP formation:

---

(1) Estimations were made on the basis of ΔG_0 values but it is quite obvious that the real yield is determined from $\Delta G'$ in physiological conditions.

$$ADP + P_i \longrightarrow ATP + H_2O \ (\Delta G_0 \# +7 \text{ kcal/mole})$$

The energy produced by the oxidation of aldehyde is then stored in the ATP formed and one can write an overall reaction, the free energy change of which is practically nil:

$$R-CHO + ADP + P_i \longrightarrow R-COOH + 2H + ATP \ (\Delta G_0 \# 0)$$

As will be seen in Chapter 4 (see figs. 4-27 and 4-28), this is a process involving 2 enzymes – a dehydrogenase, then a kinase – with intermediate formation of an acyl-phosphate, a high energy potential compound (see table on page 111) which can liberate energy of the order of 10 kcal/mole during its hydrolysis. This common intermediate is, on the one hand, the oxidation product of 3-phosphoglyceraldehyde (first reaction), and on the other hand, the donor of the phosphate group to ADP to form ATP (second reaction), as shown by figure 3-10. The free energy of hydrolysis of this compound is considerable because of the high electron density of the anhydride bond between the phosphate and the carboxyl, and of the proximity of the 2 phosphate groups with their double negative charge.

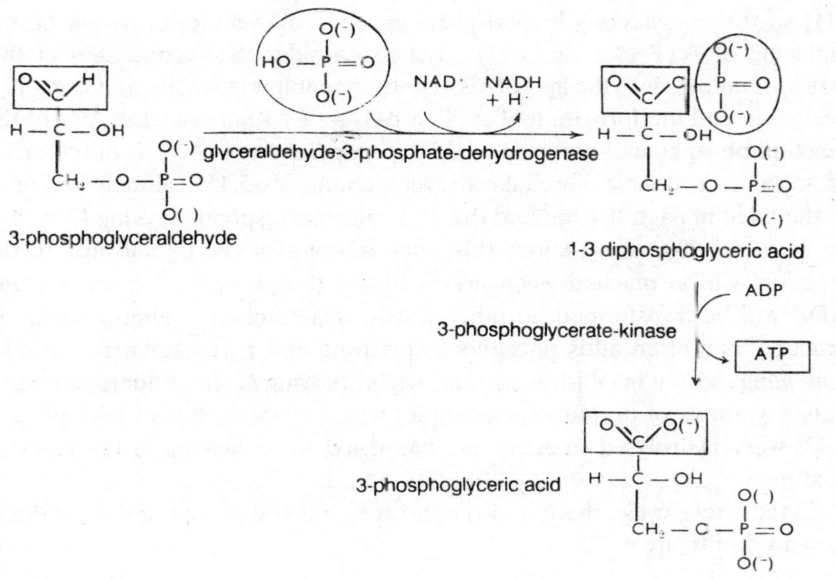

FIG. 3-10. — Oxidation of 3-phosphoglyceraldehyde to 3-phosphoglyceric acid coupled with the formation of ATP.

One can see that during the dehydrogenation of 3-phosphoglyceraldehyde there is formation of NADH + H⁺. When the respiratory chain functions, the transfer of electrons from NADH + H⁺ to oxygen enables – via the shuttles of the reducing power (see fig. 3-9) – the formation of 3 ATP, and since one glucose molecule gives 2 C_3 fragments, 6 additional ATP can be formed when glycolysis is followed by the aerobic phase, but not during the strict anaerobic glycolysis.

B. – Formation of ATP during the conversion of 2-phosphoglyceric acid into pyruvic acid

It was mentioned above that – in glycolysis as well as in the respiratory chain – energy is stored in the form of ATP during the oxidation-reduction reactions. But, on examining 2-phosphoglyceric acid and pyruvic acid, it is not evident that the passage from one of these compounds to the other implies such a process. In fact, the study of the detailed mechanism of this transformation (fig. 4-29) shows a displacement of electrons within the molecule which can be regarded as an *intramolecular oxidation-reduction*. There is here also, formation of an intermediate compound – *phosphoenol-pyruvic acid* – having a high energetic potential due to high electron concentration around the carbon 2, partly due to the formation of a double bond (2 pairs of electrons) and also to a decrease of distances and bond angles around this carbon. This compound has a free energy of hydrolysis of the order of 12.8 kcal/mole (see table on page 111) so that, on yielding its phosphate group, it allows the formation of one molecule of ATP (see fig. 3-11). Let us consider the second step of this transformation: since the hydrolysis of phosphoenolpyruvate has a ΔG_0 # –12.8 kcal/mole and the formation of ATP, a ΔG_0 # +7 kcal/mole, the ΔG_0 of the reaction phosphoenol-pyruvate + ADP ⟶ pyruvate + ATP is of the order of –5 to – 6 kcal/mole; if we take an average value of –5.45 kcal/mole and refer to the table in page 104, we find that this value corresponds to a log K = +4, i.e. K = 10 000; as we know, this value means that the equilibrium of the reaction will be reached when practically all the phosphoenolpyruvate and ADP will be transformed, in other words, this reaction is almost total. In practice, one often adds phosphoenolpyruvate and pyruvate-kinase as *ATP generating mixture* in cell-free systems while studying *in vitro* endergonic reactions (synthesis of proteins for example) which could no longer take place if ATP were hydrolysed to ADP and happened to be lacking in the reaction mixture.

In the Krebs cycle, there is also a substrate-related phosphorylation; this is step no. 5 of figure 4-38.

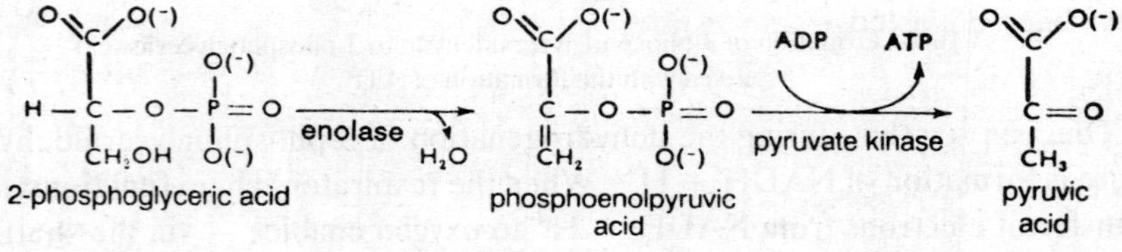

FIG. 3-11. — Conversion of 2-phosphoglyceric acid into phosphoenolpyruvic acid transposed spontaneously into pyruvic acid coupled with the formation of ATP.

V. – FORMATION OF ATP BY PHOTOPHOSPHORYLATION[1]

1. – Photosynthesis: light phase and dark phase

Heterotrophic organisms collect from the external medium, comparatively simple organic molecules which can serve as carbon-containing skeletons for their syntheses or as energy source (ATP) when they are oxidized during mitochondrial reactions of respiration according to the general reaction :

$$C_6H_{12}O_6 + 6O_2 \longrightarrow 6CO_2 + 6H_2O + \text{energy (ATP)}$$

These organic molecules are ultimately synthesized in their very large majority by photosynthetic *autotrophic* organisms. *Higher plants* and most algae do have structures, chloroplasts and an enzymatic equipment which enables them to utilize light energy to perform the synthesis of simple sugars from CO_2 and H_2O according to the general photosynthetic reaction :

$$6CO_2 + 6H_2O + \text{light energy} \longrightarrow C_6H_{12}O_6 + 6O_2$$

Experiments with water containing the isotope ^{18}O showed that the oxygen released originates from the cleavage of water. It therefore appears that, in higher plants, photosynthesis reactions utilize light energy to extract hydrogen atoms from water while releasing oxygen and transfer them to CO_2 in order to form reduced carbon-containing products.[2] This shows the difference with respiration where on the contrary, various reduced substrates are oxidized into CO_2, yielding hydrogen atoms which combine with oxygen to form water while energy (ATP) is liberated.

The photosynthetic reduction of CO_2 involves a complex system of reactions which can be grouped in two principal phases:

– *the light phase* in which light energy absorbed by chlorophyll is converted into chemical energy usable by the cell and during which ATP and NADPH are formed;

– *the dark phase* in which, thanks to ATP and NADPH thus formed, CO_2 is reduced into simple sugars during a cycle of enzymatic reactions called Calvin cycle. For a proper understanding of these reactions one would need a good knowledge of the metabolism of carbohydrates, which is the reason why we are studying them in detail at the end of Chapter 4 (see fig. 4-45).

After being synthesized in the chloroplast, these simple sugars are exported to the cytoplasm where they will play a role equivalent to that of organic molecules collected from the external medium by heterotrophic organisms. They will enter the cellular metabolism and will be transformed through anabolic reaction chains into diverse molecules. Those molecules which will be degraded through catabolic reaction chains, will be oxidized into CO_2 and H_2O

---

(1) Revised by G. DUBERTRET.

(2) Some photosynthetic bacteria utilize not water as donor of hydrogen but other molecules like H_2S which they oxidize while accumulating sulphur.

in the mitochondria by the mechanisms of respiration, with formation of the ATP required for the various cellular activities.

In plant cells, *chloroplasts and mitochondria therefore co-exist* and bring about simultaneously photosynthetic reactions which consume light energy and respiratory reactions of degradation which liberate energy usable by the cell (ATP), in varying proportions depending on external conditions (light, temperature, humidity etc.).

2. – Chloroplasts and chlorophyll

All the photosynthesis reactions in higher plants and eucaryotic algae take place in *chloroplasts*, organelles which present several structural and functional similarities with the mitochondria, both fulfilling the energetic functions of the cell. The number of chloroplasts per cell varies from one (in some algae) to about ten (in higher plants) and their size is around 10 micrometers. Like the mitochondria, they are surrounded by an *envelope* consisting of two membranes: *the outer membrane*, very permeable to ions and small molecules and *the inner membrane*, impermeable and in which numerous transport proteins perform the exchanges with the cytoplasm. The chloroplastic compartment thus defined contains the *stroma*, which is similar to the matrix of mitochondria. It consists of a non-structured mixture of soluble molecules including the enzymes of the Calvin cycle, several copies of the chloroplastic chromosome as well as the systems of replication, transcription and translation of this genetic information. The chloroplastic space also contains an important system of membranes, similar to the crests of mitochondria. These membranes, called *thylakoids*, very rich in galactolipids form flattened, interconnected vesicles arranged in a stack to form the *grana*. The primary steps of

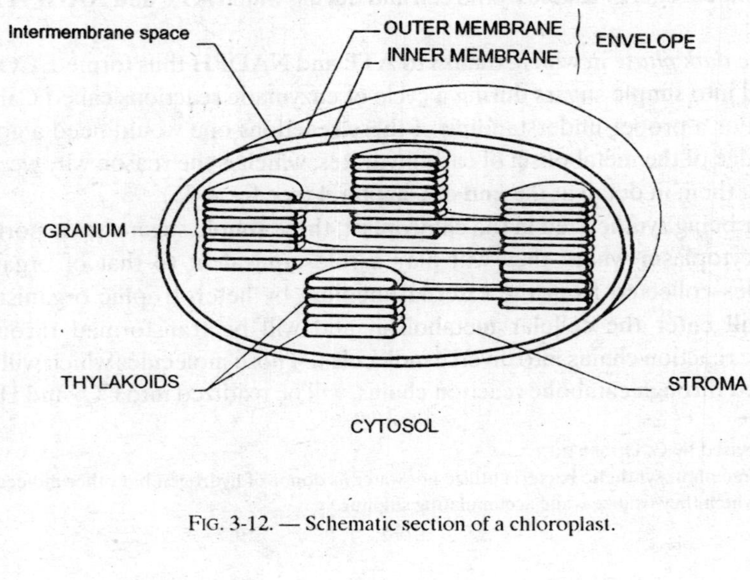

FIG. 3-12. — Schematic section of a chloroplast.

photosynthesis corresponding to light reactions take place in thylakoids. They contain numerous multimolecular assemblies consisting of proteins, chlorophyll, cytochromes, quinones... (see fig. 3-15) which bring about the reactions of light energy absorption, photochemical conversion and electron transport with formation of ATP and NADPH.

Chlorophylls are the pigments characteristic of all photosynthetic organisms. *Chlorophyll a* and *chlorophyll b*, present in all chloroplasts of higher plants and numerous algae are very similar in structure (see fig. 3-13). Both have a *tetrapyrrole ring* similar to that of the heme (see fig. 1-29) but with a magnesium atom. A long hydrocarbon *phytyl* side chain confers on these molecules a liposoluble character which enables them to establish hydrophobic bonds with some of the intrinsic proteins of thylakoids to form *chlorophylls-proteins complexes*. However, chlorophyll *b* has a formyl group on carbon β of one of the pyrrole rings in place of a methyl group present on chlorophyll *a*.

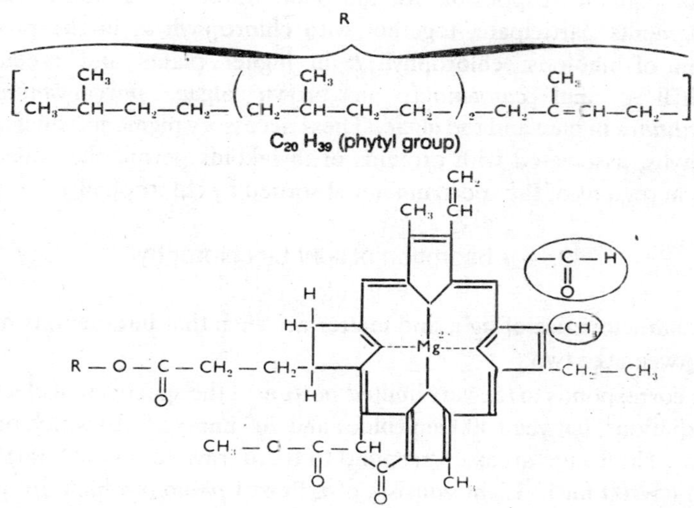

FIG. 3-13. — Structure of chlorophyll *a*.

The colour of chlorophylls is due to their structure which enables them to absorb light at some wavelengths as per the mechanisms described in the following paragraph. The two peaks near 430 nm and 680 nm visible on the *absorption spectrum* (solid line in figure 3-14) show that chlorophyll *a* absorbs light in the violet-blue and red regions respectively. Its green colour therefore corresponds to the portion of the spectrum which has not been absorbed and is therefore transmitted or reflected.

The absorption spectrum of chlorophyll *a* coincides fairly well with the *action spectrum* of photosynthesis, i.e. with the curve which represents the efficiency of photosynthesis as a function of wavelengths (dotted line in figure

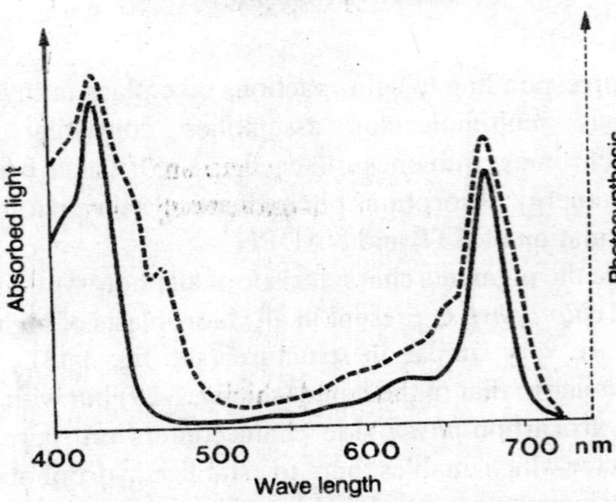

FIG. 3-14. — Comparison of the absorption spectrum of chlorophyll *a* (solid line) with the capacity of light to permit photosynthesis (dotted line) or action spectrum of photosynthesis.

3-14). Chlorophyll *a*, therefore appears, at least in higher plants and algae, as the major pigment responsible for the photosynthetic reactions. However, other pigments participate, together with chlorophyll *a*, in the process of collection of photons: chlorophyll *b* in higher plants and green algae, chlorophyll *c* and *carotenoids* in brown algae, *phycocyanines* and *phycoerythrines* in blue and red algae. These accessory pigments which are, like chlorophylls, associated with proteins of thylakoids permit the collection of photons in regions of the spectrum not absorbed by chlorophyll *a*.

3. – Absorption of light by chlorophyll

The characteristics of light and matter are such that interactions may take place between the two.

Light corresponds to the very limited portion of the spectrum of electromagnetic radiations, between 400 nm (blue) and 700 nm (red), to which our eye is sensitive. The fringe areas correspond to the ultraviolet (<400 nm) and the infrared (>700 nm). Light consists of a flux of *photons* which are particles charged with a defined quantity of energy called *energy quantum*. The energy of photons is inversely proportional to their wavelength and therefore decreases from the violet end to the red end of the spectrum.

Matter consists of nuclei surrounded by electrons placed on orbitals, of energy determined by their distance to the nucleus. These electrons can be shifted by the energy of an absorbed photon. Thus, when a light beam strikes a selenium surface, the energy of photons can extract from the latter a number of electrons which increases as the intensity of light increases (principle of the photoelectric cell). In the case of coloured molecules and chlorophyll in particular, the energy of an absorbed photon can shift electron from its initial orbital up to an orbital of higher energy, more distant from the nucleus. The energy of the absorbed photon is therefore transformed into *energy of electronic excitation*. However, only those photons will be absorbed whose energy is equal to the difference of energy between the two orbitals; the others, corresponding to the non absorbed wavelengths will be reflected or transmitted, conferring their characteristic colour on the molecules thus exposed to white light.

In the case of selenium atoms, the energy of photons is sufficient to pull out of the attraction of the nucleus, electrons which can then create an electric current. In the case of coloured molecules, the excited state of an electron which has changed its orbital under the impulse of a photon is very unstable. In most cases, the displaced electron falls back on its original orbital (return to the fundamental state) *liberating* the stored energy in the form of *heat* or *fluorescence* (emission of photons). In some cases, this excited state can be *transferred* by resonance to a neighbouring identical molecule which is then excited in its turn. Sometimes, the excited electron can be *shifted* to another molecule of different nature; this reaction which thus produces an oxidized molecule (loss of electron) and a reduced molecule (gain of electron) is a *photochemical reaction*.

4. – Reaction centers and antenna complexes

The excited state of a chlorophyll molecule, created by the absorption of a photon, is therefore very unstable. *In vitro*, the displaced electron falls back on its starting orbital, liberating the potential energy in the form of heat or fluorescence. *In vivo*, this electron can on the contrary, be transferred to a neighbouring molecule and produce in it a reducing power utilizable by the cell. This photochemical reaction takes place in the reaction centers where a primary electron donor chlorophyll molecule *a* is closely associated with a primary electron acceptor molecule. A series of oxidation-reduction reactions involving chains of electron carriers very similar to those involved in mitochondrial respiratory chains then permit the flow of this electron towards a terminal acceptor which is then in a reduced form (see paragraph 5).

But the rate of absorption of the photon by a chlorophyll molecule, even in bright sunshine, is too small to produce an appreciable flow of electrons. On the other hand, detailed studies have shown that a very small proportion of chlorophylls (about 1/200) is involved in the photochemical conversion. The great majority of chlorophylls are organized in antenna complexes comprising each about 200 molecules of chlorophyll and each associated with a reaction center. The unique function of these antenna complexes is to absorb light energy; transfers by resonance between the chlorophyll molecules then channel the excitation energy towards the reaction centers whose photochemical activity is multiplied by about 200. These energy transfer and photochemical conversion reactions are very rapid (a few nanoseconds) and require a very accurate structural organisation. This essential condition is fulfilled by the association of the molecules involved (chlorophylls, primary acceptors...) with various proteins which are themselves organized in stable multimolecular structures (see fig. 3-16).

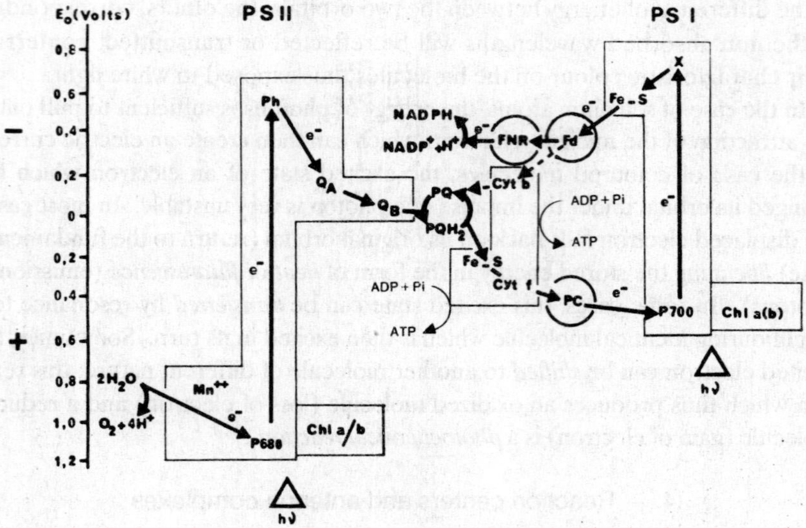

FIG. 3-15. — Diagram of the photosynthetic transport of electrons. Each of the principal intermediates known is placed at the stage corresponding to its oxidation-reduction potential (E'0); to make a comparison with the carriers of the respiratory chain, one can refer to the table in page 120.
PS II, PS I : photosystems II, I; Chl: chlorophyll; P680, P700: chlorophylls of the reactive centers PS II, PS I; Ph: pheophytin. QA, QB : quinones A, B; PQ, PQ H_2 : oxidised, reduced plastoquinones; Cyt *f*, Cyt b_6 : cytochromes *f*, b_6; Fe—S : iron-sulphur protein; PC: plastocyanin; X: primary acceptor of the photosystem I; Fd: ferredoxin; FNR: ferredoxin-NADP reductase.

5. – Electron transport and photosynthetic phosphorylation

We have seen in the foregoing, that in the plant cell, the "excited" electrons can leave the chlorophyll molecules activated by light to be taken over by an electron carrier system. In fact, it appears that the electrons removed from chlorophyll *a* of the reactive center can take two different pathways; one leading to the synthesis of ATP and the other to the formation of ATP and NADPH. We will study these two possibilities and see how – in each case – the problem of the regeneration of chlorophyll *a* can be resolved. All the mechanisms of photosynthesis are not yet known accurately; we are presenting here a simplified diagram taking into consideration the processes which are apparently the most probable at the present time.

We know that in green plants, chlorophyll is involved in two photosensitive systems playing a role in two different photochemical reactions : photosystem I (PS I) and photosystem II (PS II).

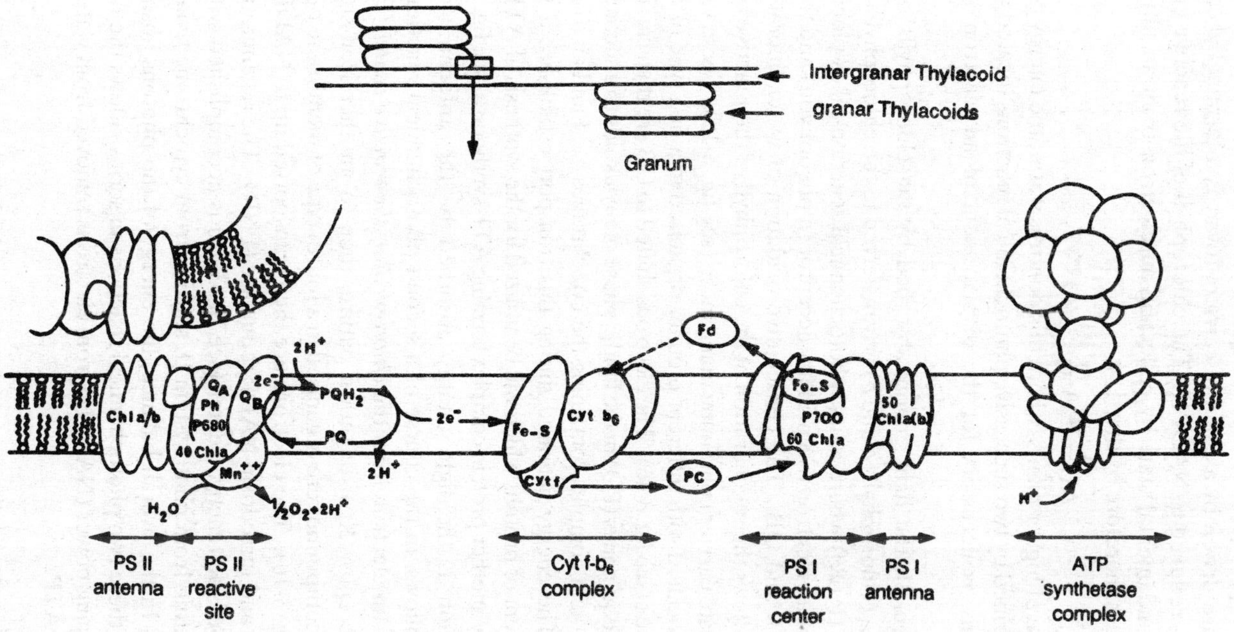

FIG. 3-16. — Physical organisation of the components of the photosynthetic chain of electron carriers in the photosynthetic membranes of higher plants. Abbreviations as in figure 3-15.

A. – Non-cyclic photophosphorylations with formation of ATP and NADPH

In this principal pathway, the electrons removed from the chlorophyll of the reactive center are driven by a chain of carriers towards a terminal acceptor $NADP^+$ which is reduced to NADPH. The chlorophyll of the reactive center returns to the fundamental state by collecting an electron from an ultimate donor AH which is therefore oxidized

$$AH \longrightarrow A + e^- + H^+$$

But in the case of higher plants, where this donor is water, the energy of two photons, transformed by two successive photochemical reactions located in the two photosystems, is necessary for the displacement of an electron up to $NADP^+$.

In photosystem II, the light energy absorbed by the chlorophylls of the collector antenna (chlorophylls *a* and *b*) is transferred to the chlorophyll of the reactive centers (P 680) where a first photochemical reaction takes place and allows the shift of an electron to a primary acceptor (a pheophytin) and then to the quinones QA and QB. The consecutive oxidation of water liberates an electron which allows the regeneration of the chlorophyll of the reactive center via an intermediate carrier containing manganese (see fig. 3-15). These various elements are associated with different proteins organised in a first type of multimolecular complex (see fig. 3-16). The electrons shifted to QB are then carried by a set of free plastoquinones (PQ) in the lipidic phase of thylakoids to a second type of membrane protein complex containing the cytochromes b_6, f and an iron-sulphur protein. The energy liberated during this transport establishes a transmembrane gradient of protons which will be utilized for the synthesis of ATP (see paragraph VI) by another protein complex forming ATP synthetase (see fig. 3-16).

In photosystem I, the light energy absorbed by the antenna (mostly chlorophyll *a*) allows, in the course of a second photochemical reaction, the displacement of an electron from the *chlorophyll of the reactive center* (P 700) to a primary acceptor X of unknown nature, then to another iron-sulphur protein. These compounds are organised in a fourth type of membrane protein complex (see fig. 3-16). The electrons are then transported up to $NADP^+$ via ferredoxin (Fd) and ferredoxin NADP reductase (FNR). The regeneration of the chlorophyll of the reactive center (see fig. 3-15) is accomplished with the electron originating from photosystem II, starting from cytochrome *f* via plastocyanin (PC). In the overall balance, the energy of the photons absorbed, transformed by these two photochemical reactions in series, allows the formation of the reducing power (NADPH) from electrons removed from water and the synthesis of ATP.

B. – Cyclic phosphorylations and formation of ATP only

The importance of these reactions is not yet fully understood; they involve only photosystem I. Their juxtaposition to the non-cyclic phosphorylations, leading to a formation of ATP alone, probably adjusts the ATP/NADPH ratio to the requirements of the chloroplast. The electrons displaced by the photochemical reactions are, as in the previous case, carried up to ferredoxin. But instead of being directed towards $NADP^+$, they fall back on cytochrome b_6 of the complex cytochrome f, b_6, iron-sulphur protein. With the participation of plastoquinones, they establish a transmembrane gradient of protons utilised for the synthesis of ATP. The electrons then join, via the plastocyanins, the chlorophyll P 700 of the reactive center which is thus regenerated. In this case, the energy of the photons absorbed leads only to a cyclic flow of electrons allowing only the synthesis of ATP.

In conclusion, respiration as well as photosynthesis, involve movements of electrons in specialised membranes. In mitochondria these electrons originate from various organic substrates subjected to the action of the corresponding dehydrogenases whereas in chloroplasts, they are removed from the chlorophyll by the energy of light. In mitochondria, the electrons finally reach the oxygen, the reduction of which forms water; in chloroplasts, the electrons can either allow the reduction of $NADP^+$, or return to the chlorophyll and allow the restoration of its fundamental state.

As in mitochondria, the flow of electrons along the chain of carriers makes it possible to store a part of the energy in the form of a transmembrane gradient of protons utilized for the formation of ATP. Only the source of this energy differs; in mitochondria, it is the chemical energy of various organic substrates, while in the chloroplasts, it is the light energy of the absorbed photons.

Finally, it must be indicated that photosynthetic organisms (green plants, microorganisms etc.) possess relay systems of respiratory chains to form ATP through oxidative phosphorylations in the absence of light. Plants thus have mitochondria in addition to chloroplasts.

C. – *Bacteriorhodopsin and photophosphorylation*

Chlorophylls are not the only photosensitive molecules which can collect the energy of photons and utilize it to form reductants and ATP for the purpose of syntheses. Some bacteria carry out their photosynthesis thanks to the presence of bacteriorhodopsin, a protein linked to retinal (see pages 138-139, the mechanism of purple membranes of *Halobacterium halobium*).

VI. – ENERGY TRANSDUCING MEMBRANES

It is known that electron carrier chains are implanted in specialized membranes. Whenever an energy recovery and a synthesis of ATP are coupled with the reoxidation of a chain of e^- carriers, the membrane is called "energy transducer"; it plays an essential role in the synthesis of ATP, in mitochondria (inner membrane) as well as in chloroplasts (thylakoids) and bacteria (plasmic membrane, respiration or photosynthesis).

In order to explain the phosphorylating oxidations coupled with the respiratory chain, it was tried for years to find a mechanism of the type of glyceraldehyde-3-phosphate dehydrogenase coupled with 3-phosphoglycerate kinase (see page 126-127). This hypothetical mechanism which did not take the membrane into consideration was refuted after years of research.

Around 1961, the role of protons in the mechanism of phosphorylating oxidations was suggested independently by P. Mitchell and R.J.P. Williams. Since then, Peter Mitchell in collaboration with Jennifer Moyle developed the elegant and convincing *chemosmotic theory*. Numerous experimental arguments have been advanced in its favour. This theory proposes a general mechanism for the synthesis of ATP linked with the transducer membranes (mitochondria, chloroplasts, bacteria). The membrane has a functional asymmetry. The chain of e^- carriers implanted in the membrane has an anisotropic functioning and owing to the orientation of its sites, it causes the translocation of protons and their ejection on a specific face of the membrane. Since the membrane is impermeable to ions except at the level of specific and controlled translocators, the functioning of the chain creates a gradient of protons on either side of the membrane and a protomotive force which are utilized for the synthesis of ATP thanks to the recapture of the ejected protons by a transmembrane ATPase-ATP synthetase. On the face of the membrane where they have been ejected, the protons penetrate into the proton canal of the enzymatic complex and thus proceed towards the sites of ATP-synthesis, oriented towards the other face of the membrane. ATP is thus synthesised at the cost of the gradient of protons (see fig. 3-17).

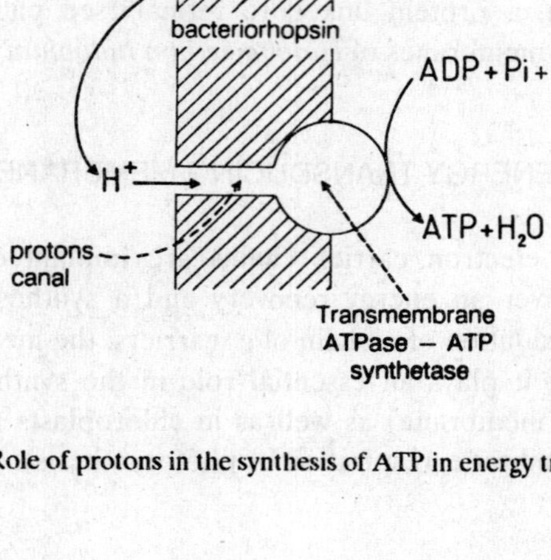

FIG. 3-17. — Role of protons in the synthesis of ATP in energy transducing membranes.

This mechanism is valid even for the simplest transducing membranes of photosynthesizing bacteria (*Halobacterium halobium*). The plasmic membrane of these bacteria is strewn with purple membranes consisting of a single protein, a photosensitive bacteriorhodopsin which, under the effect of light, catalyzes the ejection of protons outside, thus creating a gradient of protons and a protomotive force. The pumping of protons through canals can permit the synthesis of ATP thanks to a transmembrane ATPase-ATP synthetase, or the protons gradient can be utilized directly for the active transports of molecules or ions across the membrane. A complete chain of electron carriers is therefore not indispensable. However, at night, in the absence of light, the respiratory chain implanted in the plasmic membrane (outside the purple membranes) also catalyzes the outward ejection of protons which will be pumped again and utilized for the synthesis of ATP.

The theory of localized protons advanced by R.J.P. Williams (1978) is based on the same concept though it presents some different aspects on the location of ejected protons and their migration.

Lastly, let us cite the conformational hypothesis suggested by P.D. Boyer around 1964. According to this hypothesis, the functioning of the chain of electron carriers, by "energizing" the membrane, produces conformational changes which are utilized, at least partly, for the synthesis of ATP.

CHAPTER 4[1]

STRUCTURE AND METABOLISM OF CARBOHYDRATES

STRUCTURE OF CARBOHYDRATES

Carbohydrates form a group of very important compounds. Some represent a source of energy for living organisms, usable either immediately (glucose), or as reserve (starch, glycogen); other carbohydrates have a structural role (cellulose, chitin, hyaluronic acid); lastly, others have an important biological role like signal recognition (glycans of glycoproteins and glycolipids). The term "*sugar*" is still found in certain works; the term "carbohydrates" derives from the fact that a large number of these compounds correspond to the formula $Cn(H_2O)_n$, but there are several exceptions; for example, in the general formula ($C_5H_{10}O_4$) of deoxyribose, a constituent of deoxyribonucleic acids (see fig. 6-1), hydrogen and oxygen are not in the ratio 2/1; moreover, some compounds containing nitrogen or sulphur are also classified among carbohydrates.

From the chemical point of view, carbohydrates may be defined as *polyhydroxyaldehydes* or *polyhydroxyketones* (or compounds derived from these substances), or polymers which can liberate these compounds by hydrolysis; monosaccharides or simple sugars are thus distinguished from oligo- and polysaccharides which yield several monosaccharides by hydrolysis.

THE MONOSACCHARIDES

The monosaccharides or simple sugars are molecules containing several *alcohol* groups as well as a reducing group, either *aldehyde* or *ketone*. Monosaccharides are classified according to the number of carbon atoms of their molecules (trioses, tetroses, pentoses, hexoses, heptoses, etc.) on the one hand, and the nature of the reducing function (aldoses and ketoses), on the other.

---

(1) Revised by J. MONTREUIL.

The simplest monosaccharides have three carbon atoms; they include an aldotriose – glyceraldehyde – and a ketotriose – dihydroxyacetone (see fig. 4-1).

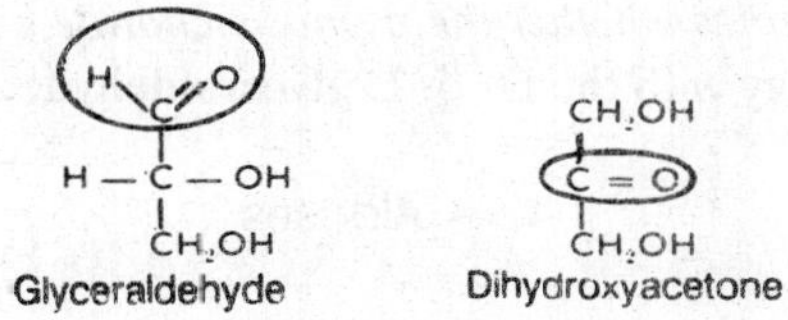

FIG. 4-1. — Formulae of glyceraldehyde (aldotriose) and dihydroxyacetone (ketotriose).

I. – ISOMERISM OF MONOSACCHARIDES

If we consider the projection of the three-dimensional structure of glyceraldehyde (see fig. 4-2) we find that there are 2 possibilities: in one case, the hydroxyl carried by the carbon next to the primary alcohol group (this carbon atom is called asymmetric because it carries 4 different substitutes, and it is indicated by an asterisk) is situated to the right of the plane formed by the carbon chain; this is the D-configuration; in the other case, the hydroxyl is situated to the left of this plane; this is the L-configuration[1]. These 2 forms are mirror images of each other; they represent a couple of optical isomers (or enantiomers), the physical and chemical properties of which are all practically identical.

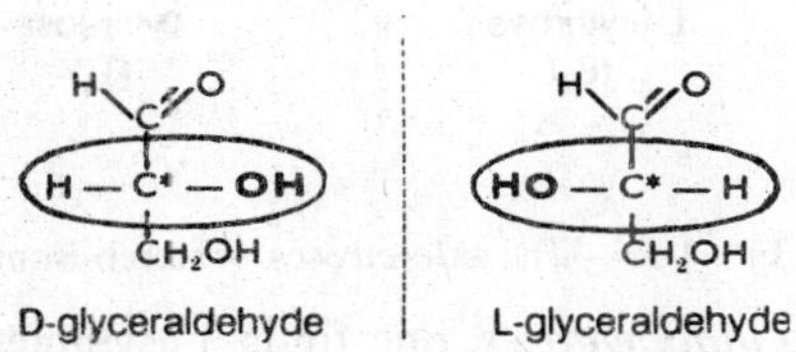

FIG. 4-2. — Projections of the structures of D and L configurations of glyceraldehyde (representation proposed by Fischer).

The spatial structure of other monosaccharides having a larger number of carbon atoms derives from that of glyceraldehyde, *and we will have aldoses of the D- or L-series depending on whether the hydroxyl carried by the asymmetric carbon next to the primary alcohol group is in a configuration identical to that of the D- or L-glyceraldehyde* (i.e. depending on whether it is projected on the right or on the left).

---

(1) Here again, as in the case of amino acids (see page 8), the denominations D and L have no significance relative to the direction of deviation of polarized light by the substance. The direction of the rotatory power may be indicated by the signs (+) and (–). Examples: D(+) glyceraldehyde, D(+)-glucose, D(–)-fructose.

In the case of ketoses which possess a primary alcohol group at each end, this definition may lead to confusion. For the definition to be more general, it may be stated that it is *the configuration of the secondary alcoholic hydroxyl carried by the asymmetric carbon farthest from the reducing group* (aldehyde or ketone) *which determines whether the monosaccharide belongs to the D- or L-series* (again by analogy with the D- or L-glyceraldehyde).

1. – Aldoses

Let us examine the case of *aldotetroses* (aldoses with 4 carbon atoms). We can apply the rule we have just enunciated, but the situation is somewhat complicated by the fact that there are two asymmetric carbon atoms. We will have the 2 configurations D and L of erythrose (E_1 and E_2) forming a couple of enantiomers because they are mirror images of each other, and we will also have the 2 configurations D and L of threose forming a couple of enantiomers (T_1 and T_2). On the contrary, E_1 and T_1, E_1 and T_2, E_2 and T_1, E_2 and T_2 do not form couples of enantiomers; they are *diastereo-isomers*; their physical and chemical properties are distinctly different, so that it is easier to separate 2 diastereo-isomers than 2 enantiomers (see fig. 4-3).

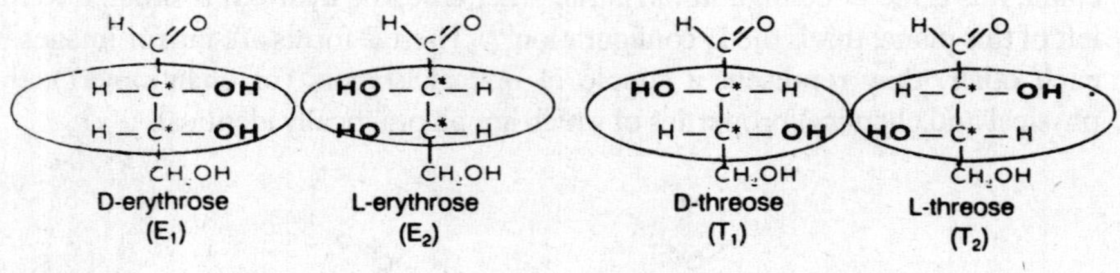

FIG. 4-3. — The aldotetroses : 4 stereo-isomers.

Now considering *aldopentoses*, one finds 3 asymmetric carbon atoms, and therefore 8 possible stereo-isomers(1). Only the 4 D-forms are shown in figure 4-4 (to each one of them corresponds a L-form, i.e., its mirror image).

As regards *aldohexoses*, which have 4 asymmetric carbon atoms, there are 16 possible stereo-isomers (i.e. 8 couples of enantiomers). Figure 4-5 shows only the 3 isomers most frequently found in living organisms.

It is observed that these 3 isomers are all of the D-series. In fact, natural monosaccharides, whether aldoses or ketoses, are *mostly of the D-series* (the only notable exceptions are L-arabinose and L-fucose).

---

(1) The number of possible stereo-isomers (x) is related to the number of asymmetric carbon atoms of the monosaccharide (n) by the expression: $x = 2^n$. We have, $2^2 = 4$ isomers for tetroses, $2^3 = 8$ isomers for pentoses, $2^4 = 16$ isomers for hexoses, etc.

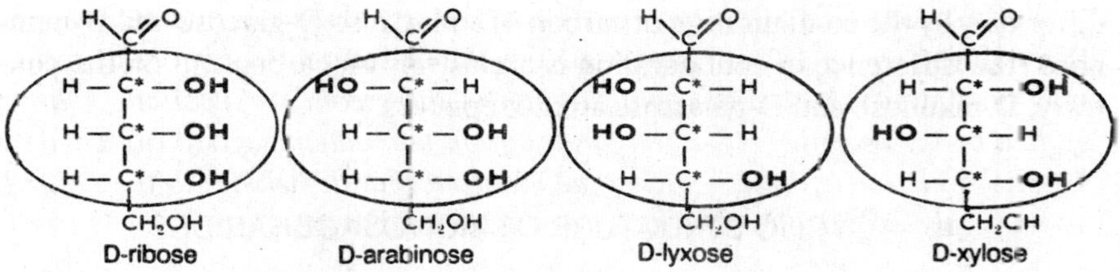

FIG. 4-4. — The aldopentoses : 8 stereo-isomers (only the D forms are represented).

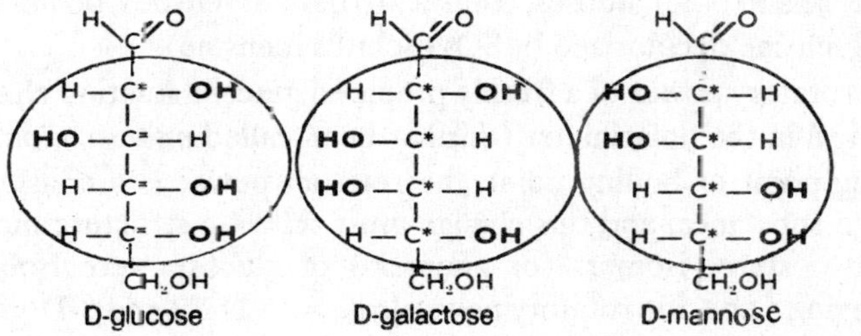

FIG. 4-5. — The aldohexoses : the 3 most important stereo-isomers.

2. – Ketoses

Regarding the D- or L-configuration of ketoses, the same rule – given earlier – applies. Figure 4-6 shows the structure of some ketoses which play an important role in carbohydrate metabolism. It will be observed that they all have the ketone group in position 2 and – excepting fructose – their names are appended with the suffix *-ulose* which is characteristic of ketoses.

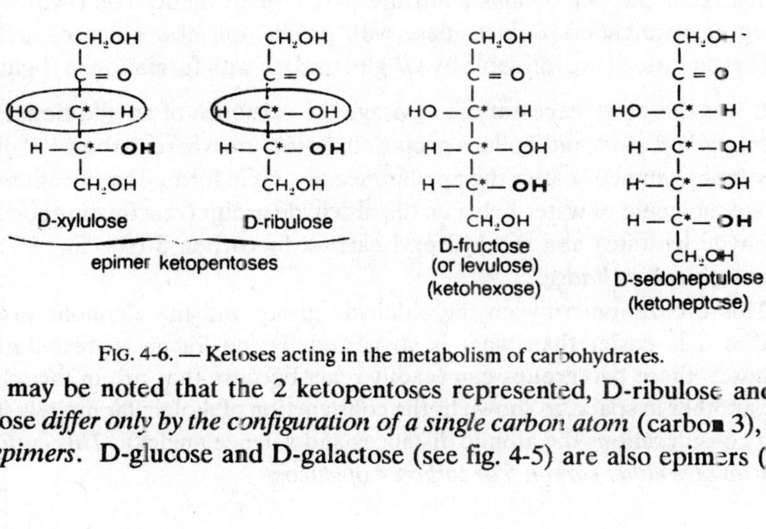

FIG. 4-6. — Ketoses acting in the metabolism of carbohydrates.

It may be noted that the 2 ketopentoses represented, D-ribulose and D-xylulose *differ only by the configuration of a single carbon atom* (carbon 3), they are *epimers*. D-glucose and D-galactose (see fig. 4-5) are also epimers (they differ only by the configuration of carbon 4) and so also D-glucose and D-mannose (the difference in configuration is exclusively on carbon 2); on the contrary, D-mannose and D-galactose are not epimers.

II. – CYCLIC STRUCTURE OF MONOSACCHARIDES

We have, so far, written the formula of monosaccharides in the form of a linear chain. But this representation is not satisfactory because it does not explain some observations.

1. Glucose and other aldoses, contrary to most aldehydes, do not restore the colour of fuchsine decolorized by SO_2 (Schiff's fuchsine).

2. The rotatory power of a freshly prepared glucose solution changes when it is observed in the polarimeter (this has been called *mutarotation*). But like the melting point or boiling point, the rotatory power is a constant characteristic of a substance, and this change must reflect a structure modification. Besides, two stereo-isomers (or anomers) of glucose were isolated: α-D-glucose having a specific rotatory power $[\alpha]_D$ = +112°2 and β-D-glucose with a rotatory power = +18°7; but, once in solution, each of these 2 forms undergoes a conversion and the mixture of anomers α and β finally obtained at equilibrium has a rotatory power of +52°7 at 20°C.

3. When an aldehyde is treated by methanol in acid medium, 2 molecules of methanol react with one molecule of aldehyde to form an acetal.

$$R-CHO + 2CH_3-OH \rightarrow R-CH(OCH_3)_2 + H_2O$$

But with D-glucose, the monosaccharide combines with only one molecule of methanol and one obtains a mixture of α-D-methylglucoside (hydrolyzable by an enzyme called α-glucosidase with production of α-glucose) and β-D-methylglucoside (hydrolyzable by a β-glucosidase with formation of β-glucose).

It was therefore necessary to envisage the existence of an additional *asymmetric carbon* atom and Tollens proposed a structure where carbon 1 of glucose becomes asymmetric after the appearance of a cycle formed by the elimination of one molecule of water between the aldehyde group (reacting in the form of aldehyde hydrate) and the hydroxyl carried by carbon 5 (see fig. 4-7), thus forming an *oxide bridge*.

This cyclization between the aldehyde group and the alcoholic group of carbon 5 is easier than what is suggested by the linear representation of glucose; these two groups can readily react because they are in fact close to one another in space, as shown by the construction of molecular models (taking into considerations the atomic distances and valence angles). *This cyclization may involve either carbon 5 or carbon 4 of glucose.*

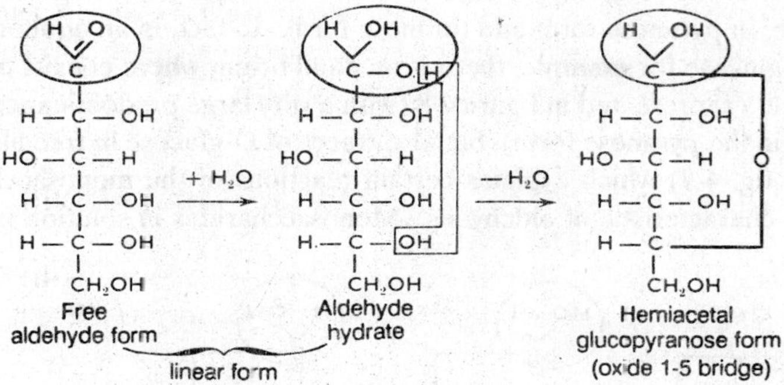

FIG. 4-7. — Passage of D-glucose from the linear form to the glucopyranose form.

In the case of the oxide bridge 1-5, one has a hexagonal cycle containing 5 carbon atoms and one oxygen atom; it is a *pyranose* ring[1]. In the case of the oxide bridge 1-4, the pentagonal cycle contains 4 carbon atoms and one oxygen atom; it is a *furanose* ring[1]. In both cases the reducing group is in the *hemiacetal* (or *pseudo-aldehyde*) form.

The same cyclic forms (pyranose and furanose rings) exist for ketoses, particularly for fructose; in this case, one speaks of a *pseudo-ketone* group for the hemiacetal form. Carbon 2 (i.e. the ketone group) is involved here in the cyclization and is linked by an oxide bridge either to carbon 6 (pyranose ring), or to carbon 5 (furanose ring).

We therefore have a new asymmetric carbon (carbon 1 in the case of aldoses, carbon 2 in the case of ketoses) and depending on the position of the hydroxyl carried by this carbon we will have the α or β isomer. Using Tollens' formula, the α isomer is represented by placing this hydroxyl on the same side as the hydroxyl which determines whether the monosaccharide is of the D- or L-series, i.e., on the same side as the oxide bridge. For simplicity, since monosaccharides are generally of the D-series, it may be said that in this series, α isomers are those with the hydroxyl carried by carbon 1 (aldoses) or 2 (ketoses) situated to the right, while isomers β are represented with this hydroxyl to the left.

---

(1)

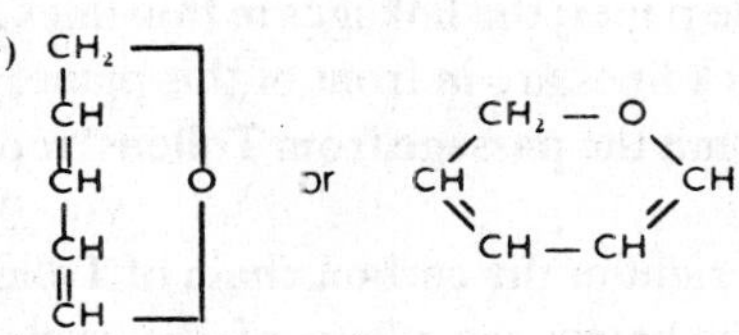

Structure of pyran

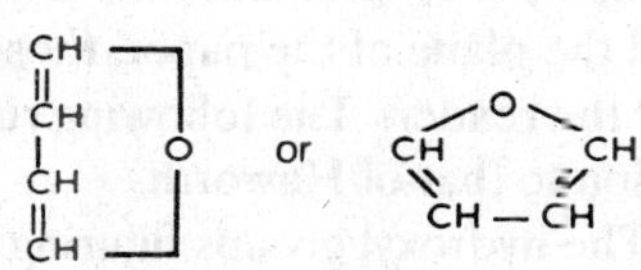

Structure of furan

Figure 4-8 shows the structures of α and β isomers of D-glucose and D-fructose, in pyranose form and furanose form. In fact, in an aqueous solution of D-glucose for example, there is an equilibrium where coexist not only the 4 isomers represented in figure 4-8 (with a very large predominance of the 2 isomers in the pyranose form), but also traces of D-glucose in free aldehyde form (see fig. 4-7) which explains certain reactions of the monosaccharides which are characteristic of aldehydes. Monosaccharides in solution in water

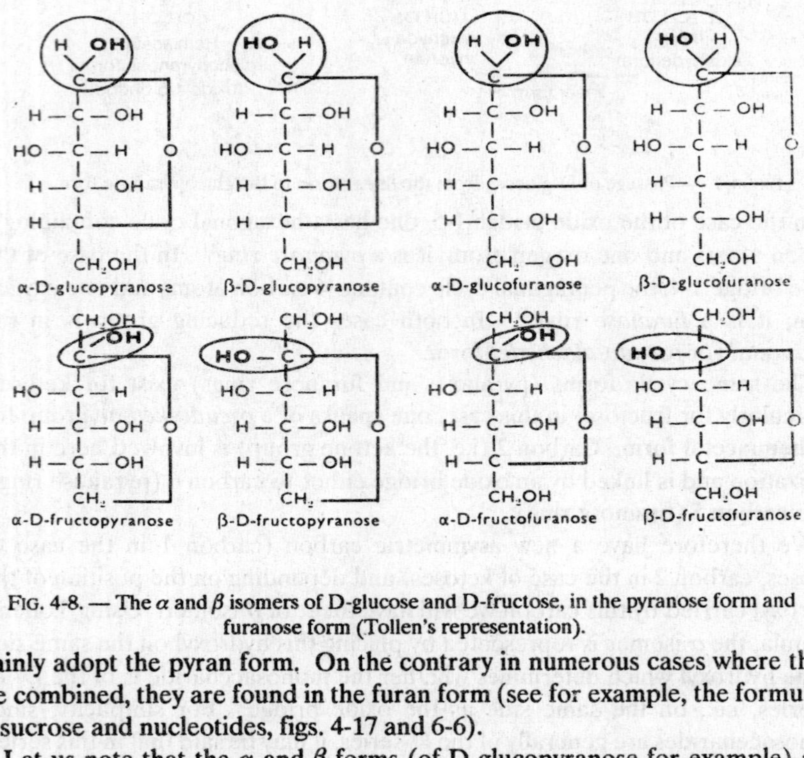

FIG. 4-8. — The α and β isomers of D-glucose and D-fructose, in the pyranose form and furanose form (Tollen's representation).

mainly adopt the pyran form. On the contrary in numerous cases where they are combined, they are found in the furan form (see for example, the formulae of sucrose and nucleotides, figs. 4-17 and 6-6).

Let us note that the α and β forms (of D-glucopyranose for example) are optical isomers, but not enantiomers, because they are not mirror images of each other. They have been named *anomers*, because they differ only by the configuration of substituents on the hemiacetal asymmetric carbon atom. The α anomer is defined as the most dextrorotatory compound.

The representation proposed by Haworth is increasingly used at present: the ring is perpendicular to the plane of the paper; the linkages in thin lines are behind the plane of the paper, those in thick lines are in front of this plane, i.e. nearer the reader. The following rules permit the passage from Tollens' representation to that of Haworth.

1. The hydroxyl groups figuring on the right of the carbon chain of Tollens' formula, are represented downwards, i.e. below the plane of the cycle of Haworth's formula. Inversely, those represented on the left of the chain of

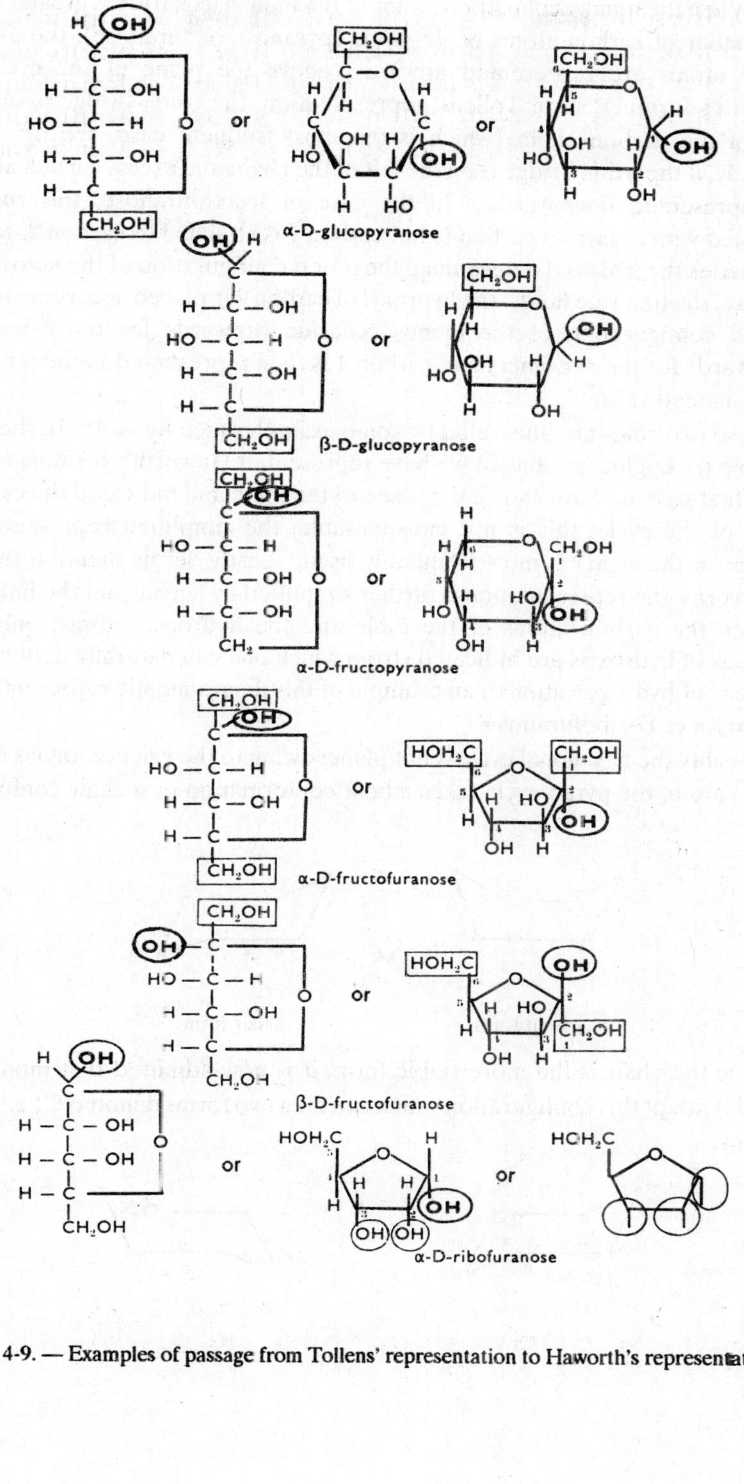

FIG. 4-9. — Examples of passage from Tollens' representation to Haworth's representation.

Tollens' formula are shown upwards, i.e. above the plane of the cycle in Haworth's representation.

2. When the number of carbon atoms of the monosaccharide is greater than the number of carbon atoms of the cycle (pyranose or furanose), the excess carbon atoms are represented upwards (above the plane of the cycle in Haworth's formula) if, in Tollens' representation, the oxide bridge is on the right of the carbon chain (which is the most frequent case, see fig. 4-8). Inversely, if the oxide bridge is on the left of the chain, the excess carbon atoms are represented downwards. In the case of fructofuranose, this rule is respected with regard to carbon 6; but carbon 1 is attached to carbon 2, which also carries the hydroxyl determining the α or β configuration of the ketose; in this case, the first rule holds: the hydroxyl of carbon 2 is placed according to the α or β configuration of the monosaccharide (upwards for the β-isomer, downwards for the α-isomer) and carbon 1 is then represented in the position that remained vacant.

These two rules are illustrated by some examples (see fig. 4-9). In the first example (α-D-glucopyranose) we have represented Haworth's formula twice: in the first case we have shown the plane of the cycle and indicated the carbon atoms of the cycle; this is not indispensable; the simplified representation (shown on the right) is most frequently used. Lastly, let us mention that in some works, the representation is further simplified by leaving out the linkages between the carbon atoms of the cycle and the hydrogen atoms; only the positions of hydroxyls are indicated (from which one can naturally deduce the positions of hydrogen atoms); an example of this diagrammatic representation is given for α-D-ribofuranose.

In reality the hexagonal cycle is not plane: owing to the valence angles of the carbon atom, the pyran cycle takes a boat conformation or a chair conformation:

Boat form

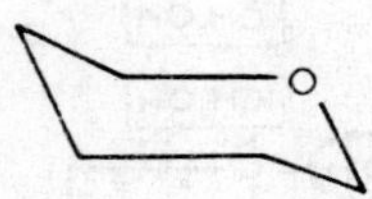

Chair form

Since the chair is the more stable form, it is now admitted that monosaccharides adopt this configuration which can take two forms denoted C1 and 1C, by Reeves.

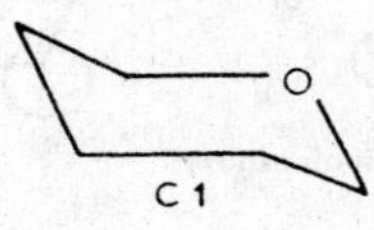

C1

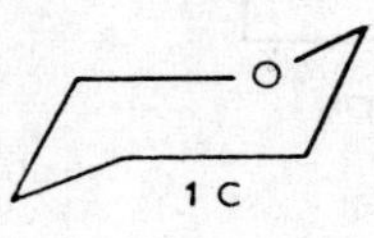

1C

We are giving below as examples, the axial or equatorial positions taken by the groups of substituents in the case of α and β glucoses of configuration C1.

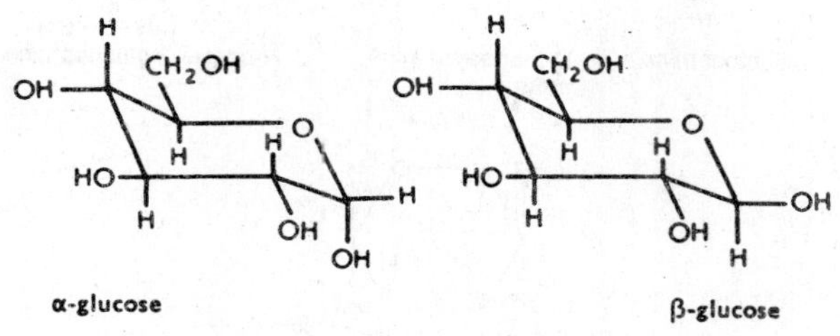

III. – THE DIFFERENT TYPES OF MONOSACCHARIDES

Monosaccharides are divided into four categories: "neutral" monosaccharides, osamines, uronic acids and sialic acids (or neuraminic acids).

1. – "Neutral" monosaccharides

The "neutral" monosaccharides include carbohydrates which possess only alcoholic groups in addition to the aldehyde or ketone group: D-glucose, D-galactose, D-mannose, D-xylose, etc. may be cited as examples.

This group also comprises the *deoxyoses*, which are monosaccharides having lost 1 or 2 oxygen atoms. Among them, can be mentioned the 6-deoxyhexoses which may be considered, either as aldohexoses, the terminal $-CH_2OH$ of which is replaced by a $-CH_3$, or as aldopentoses in which one hydrogen of carbon 5 is replaced by a methyl (examples: *L-rhammose* or 6-deoxy-L-mannose, *L-fucose* or 6-deoxy-L-galactose). The most important is a pentose, the *2-deoxy-D-ribose*, which is the monosaccharide found in deoxyribonucleic acids and derives from D-ribose (see fig. 4-9) by simple replacement of the hydroxyl of carbon 2, by a hydrogen (its structure is indicated in figure 6-1).

2. – Osamines

They derive from the neutral monosaccharides by replacement of a hydroxyl (generally the one carried by carbon 2) by an amine group. They are often found in *polysaccharides* which are described in the following. The amino group is frequently acetylated. Figure 4-10 shows the structure of the 4 most common hexosamines (the structures of glucosamine and muramic acid are represented in the N-acetylated form).

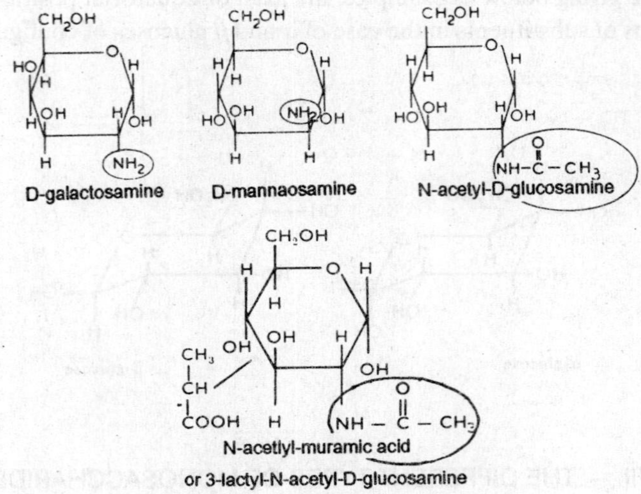

FIG. 4-10. — Structure of some hexosamines.

3. – Uronic acids

Uronic acids derive from aldoses by oxidation of the primary alcohol group into a carboxylic group (and therefore maintain the aldehyde group)(1). D-glycuronic (also called "glucuronic") acid is one of the most widely distributed uronic acids; it is found in various *polysaccharides*; furthermore, it participates in *detoxification* processes: some compounds are eliminated by the higher organisms in the form of heterosides called glucuronides (or glucorono-conjugates). Its structure is shown in figure 4-11.

4. – Sialic acids

Sialic (or neuraminic) acids are derivatives (generally acetylated) of *neuraminic acid* which itself consists of a molecule of pyruvic acid (carbons 1, 2, 3) condensed with a molecule of D-mannosamine (carbons 4 to 9). N-acetyl-neuraminic acid (see fig. 4-11) is obtained if the amino group of this osamine is acetylated. The other acetylations, leading to various sialic acids, occur on hydroxyls (especially in positions 4 and 7). Sialic acids are constituents of various glycoproteins and glycolipids. There are also N-glycolyl-neuraminic acids (the OH of which can also be acetylated).

(1) Acids resulting from the oxidation of the aldehyde group (example: *D-gluconic acid*; see figs. 4-11 and 4-39), and the diacids formed by the oxidation of both the primary alcohol group and the aldehyde group are also known, but they are less important.

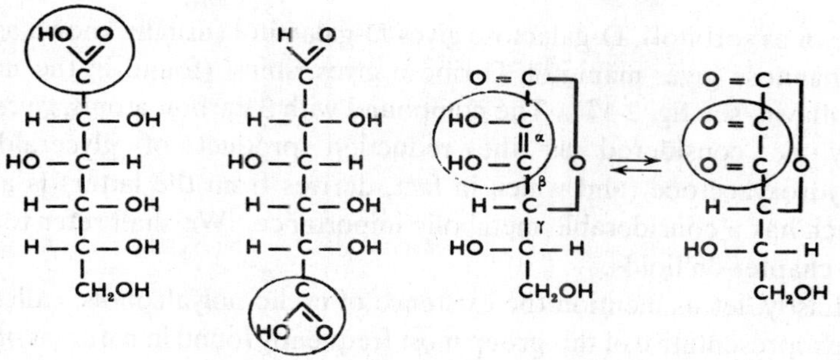

D-gluconic acid D-glucuronic acid L-ascorbic acid dehydro-L-ascorbic acid

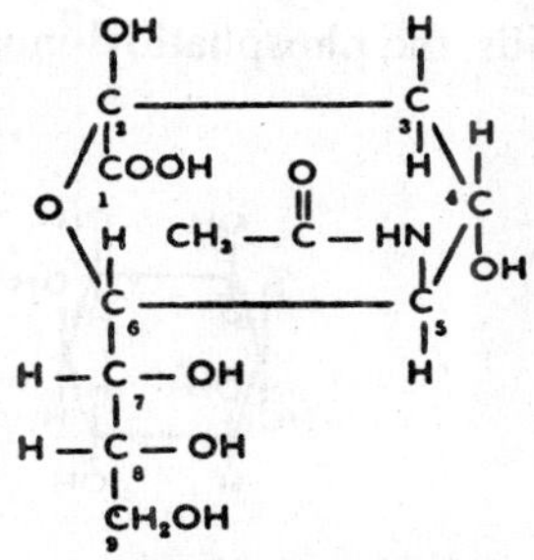

N-acetylneuraminic acid

FIG. 4-11. — Structures of some acid monosaccharides and some acids derived from monosaccharides.

IV. – COMPOUNDS DERIVED FROM MONOSACCHARIDES

1. – L-ascorbic acid (vitamin C)

A study of its structure (see fig. 4-11) reveals that it is the γ-lactone[1] of an hexonic acid (which itself derives from an aldohexose by oxidation of the aldehyde group to acid). It is further characterised by a double bond between 2 carbon atoms, each carrying a hydroxyl (ene-diol). This substance readily oxidises to dehydroascorbic acid (reversible reaction), which enables its participation in cellular oxidation-reduction processes.

2. – Polyalcohols (or polyols)

Polyalcohols are obtained by reduction of the aldehyde or ketone group to alcohol group. For example, by reduction, D-glucose gives D-glucitol (usually

---

[1] A lactone is an internal ester formed by elimination of a molecule of water between a carboxyl and an alcoholic hydroxyl of the same molecule (here, the hydroxyl carried by the γ carbon).

known as sorbitol), D-galactose gives D-galactitol (usually known as dulcitol), D-mannose gives mannitol, D-ribose gives ribitol (found in the molecule of riboflavin, see fig. 2-17). The compound with 3 carbon atoms, *glycerol*, which may be considered as the reduction product of glyceraldehyde or dihydroxyacetone (and which in fact, derives from the latter) is a trialcohol which has a considerable metabolic importance. We shall refer to it again in the chapter on lipids.

Lastly, let us mention the existence of cyclic polyalcohols, called cyclitols. The representative of this group most frequently found in nature is *myo-inositol* (see fig. 4-12); it is present either in the free state, or hexa-esterified by phosphoric acid (*phytic acid*, which exists in the form of a complex salt of Ca^{2+} and Mg^{2+} – called phytin – in the envelope of cereal grains), or as constituent of certain phospholipids, the phosphatidyl-inositols (see fig. 5-3).

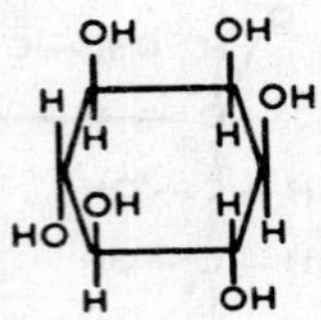

FIG. 4-12. — Structure of myo-inositol, one of the isomers of inositol (hexa-alcohol derivative of cyclo-hexan).

V. – NOMENCLATURE OF MONOSACCHARIDES

The nomenclature of monosaccharides is at present regulated by international rules. Generally, the common name of the carbohydrate is used together with indications relative to the nature of the anomerism, form of the cycle, D-series or L-series and the rotatory power: for example, α-D(+)-glucopyranose. Sometimes, the terminology is complicated as in the case of D-glucosamine which becomes 2-amino-2-deoxy-D(+)-glucopyranose.

Lastly, initials have been adopted in order to facilitate the writing; the following are the most common: Glc: glucose; Gal: galactose; Man: mannose; Fuc: fucose; Xyl: xylose; Ara: arabinose; Rha: rhamnose; GlcN: glucosamine; GlcNAc: N-acetyl-glucosamine; Mur NAc: N-acetylmuramic acid; Glc AU: glucuronic acid; NeuAc: N-acetylneuraminic acid. The nature of the cycle and of the D- or L-series are also indicated: for example, D-glucopyranose becomes D-Glcp.

VI. – CHEMICAL PROPERTIES OF MONOSACCHARIDES

As mentioned above, monosaccharides are polyhydroxyaldehydes or polyhydroxyketones: they will generally give reactions characteristic of alcoholic hydroxyls and of carbonyl groups. Of these numerous possibilities we will consider only some important reactions either from the view point of carbohydrate metabolism, or because they enable the characterisation of monosaccharides.

1. – Formation of esters

As will be seen in the following, phosphoric esters of monosaccharides are of very great importance in the metabolism of carbohydrates. In most cases *the primary alcohol group* is esterified, giving for example, glucose-6-phosphate, fructose-1-phosphate, fructose-6-phosphate, fructose-1,6-bisphosphate, ribose-5-phosphate, ribulose-5-phosphate, glyceraldehyde-3-phosphate, dihydroxyacetone-phosphate, etc.

The hemiacetal group can also be esterified (example: glucose-1-phosphate) and in some cases the monosaccharide is esterified both on its hemiacetal group and its primary alcohol group (example: 5-phosphoribosyl-1-pyrophosphate, see fig. 6-18).

2. – Alkylation

In presence of alkaline agents (sodium hydroxide, silver oxide, dimethyl-formamide) and methyl sulphate or iodide, the free hydroxyls of monosaccharides are replaced by *methoxy* groups (OCH_3). Alkylation does not take place on the blocked OH, like the one involved in the oxide bridge or those engaged in a glycoside linkage, in oligo- or polysaccharide molecules. This property is utilized in the determination of the modes of linkage between the molecules of monosaccharides in polysaccharides (see page 155).

3. – Oxidation of monosaccharides

The aldehyde or ketone group of monosaccharides can be oxidized; aldoses and ketoses will therefore behave like *reductants* and in particular, they will be able to reduce metal salts (in alkaline solution) up to the metal stage or up to a lower degree of oxidation. One of the reagents most frequently used for detecting the presence of reducing sugars for titrating them is based on a cupric salt (Cu^{2+}) maintained in solution by a double tartrate of sodium and potassium which forms a complex (Fehling's liquor); the principle of the reaction is as follows:

$$\text{reducing sugar} + \underset{\text{(blue colour)}}{2\,Cu^{2+}} \longrightarrow \text{oxidized sugar} + 2Cu^{+}$$

Since the medium is alkaline, we obtain:

$$2Cu^{+} + 2OH^{-} \longrightarrow 2CuOH \longrightarrow Cu_2O + H_2O$$

which is less soluble and gives a brick-red precipitate.

This reaction will be produced by all molecules of aldoses and ketoses which have, if not a free aldehyde or ketone group, at least a pseudo-aldehyde or pseudo-ketone group capable of yielding a free group as the equilibrium existing in the solution is displaced by the reaction. On the contrary, the molecules whose pseudo-aldehyde or pseudo ketone group is involved in an oside linkage will not have any reducing character (except if this linkage is hydrolyzed). Examples will be given in the following while referring to heterosaccharides, disaccharides and polysaccharides.

4. – Action of concentrated acids

Under the effect of a concentrated acid and in heat, aldoses and ketoses give *furfural* (in the case of pentoses), a hydroxymethylated derivative of furfural (in the case of hexoses), or furoic acid (in the case of uronic acids). In the case of glucose the reaction is as follows (see fig. 4-13).

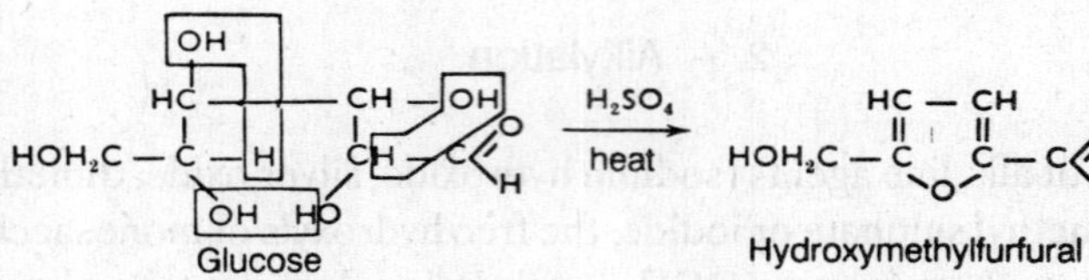

FIG. 4-13. — Formation of a derivative of furfural by action of a concentrated acid on glucose.

Furfural and its derivatives can condense with various phenols (naphtol, orcinol, etc.), and give colorations enabling either the characterization or the titration of monosaccharides, excepting osamines and sialic acids.

5. – Action of phenylhydrazine

In a first step, a molecule of phenylhydrazine reacts with a molecule of aldose or ketose to form a *hydrazone*. Then, an *osazone* is obtained in presence of an excess of phenylhydrazine. Osazones are crystallized products whose characteristics (form of crystals, melting point) can allow the identification of monosaccharides.

However, it should be borne in mind that in all cases, carbons 1 and 2 participate in the formation of osazones, so that the same osazone will be given by:

– two epimer aldoses which differ only by the configuration of the hydroxyl carried by carbon 2; this is the case of glucose and mannose (see fig. 4-5).

– an aldose and a ketose isomers having the same configuration of the carbon atoms carrying secondary alcohol groups; this is the case of glucose and fructose (see fig. 4-14).

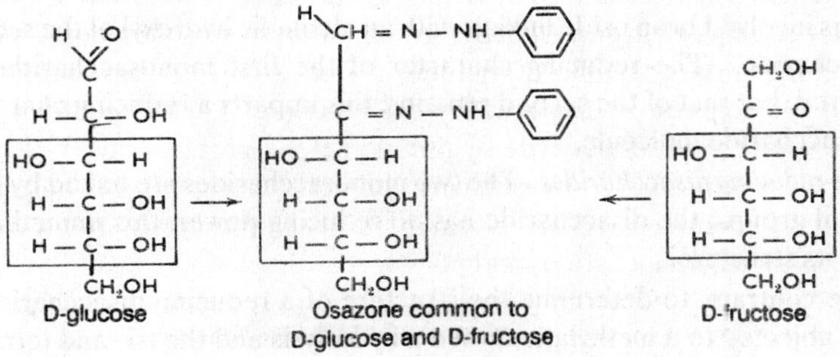

FIG. 4-14. — Osazone formed from D-glucose or D-fructose.

6. – Action of alcohols

As mentioned above, monosaccharides mostly exist in a hemiacetal cyclic form. The pseudo-aldehyde or pseudo-ketone hydroxyl of monosaccharide can also react with an alcoholic hydroxyl to form an *acetal*. Depending on whether this alcoholic hydroxyl does or does not belong to another molecule of a monosaccharide, a disaccharide or a heterosaccharide will be obtained; this leads us to the study of oligo-saccharides.

THE OLIGO- AND POLY-SACCHARIDES

We defined oligo- and poly-saccharides as substances which liberate one or several monosaccharides on hydrolysis. Among these substances there are:—

– *holosides*, which liberate only monosaccharides on hydrolysis,

– *heterosides* which, on hydrolysis, liberate non-carbohydrate substances or aglucones, in addition to monosaccharides.

I. – THE OLIGO- AND POLYSACCHARIDES

Holosides therefore consist exclusively of monosaccharides. Depending on the number of molecules of monosaccharides liberated during hydrolysis there are disaccharides, trisaccharides, etc. We will study particularly two groups of compounds: the disaccharides and the polysaccharides.

1. – Disaccharides

There are 2 types of natural disaccharides depending on the mode of linkage of the 2 molecules of monosaccharides:

– *reducing disaccharides.* The hemiacetal group of one of the monosaccharides is involved in an oside linkage with an alcoholic hydroxyl of the second monosaccharide. The reducing character of the first monosaccharide has disappeared, but that of the second remains; this imparts a reducing character to the disaccharide molecule.

– *non-reducing disaccharides.* The two monosaccharides are linked by their hemiacetal groups; the disaccharide has no reducing power; this immediately indicates its structure.

On the contrary, to determine the structure of a reducing disaccharide, it must be subjected to a methylation, then a hydrolysis and the tri- and tetra-O-methyl-monosaccharides obtained must be characterized (only the free hydroxyls will have been methylated). For example in the case of maltose, the hydrolysis of permethylmaltose will give 2, 3, 4, 6-tetra-O-methyl-glucose and 2, 3, 6-tri-O-methyl glucose. On the other hand, an oxidation or a reduction followed by hydrolysis, will determine the monosaccharide whose reducing group has remained free. Glucose, for example, will give gluconic acid or sorbitol. Lastly, the use of specific enzymes (α- and β-glucosidases, α- and β-galactosidases, etc.) will indicate the configuration of the monosaccharide involved by its hemiacetal group.

As in the case of monosaccharides, there are nomenclature rules giving a precise description of the structure of oligo-saccharides. The use of initials for designating monosaccharides (see page 152) considerably simplifies the expression of the structure of oligosaccharides (see below for example, for maltose or lactose).

A. – *Reducing disaccharides*

a) **Maltose.** – It is an intermediate product of the hydrolysis – acid or enzymatic – of polysaccharides like starch and glycogen. It is composed of the union of 2 molecules of D-glucose by a α-1,4-glucosidic linkage; it is a 4-D-glucopyranosyl-α-D-glucopyranose (see fig. 4-15); for simplicity, it may be written, Glc (α 1-4) Glc, or α D-Glcp-(1→4)-D-Glcp.

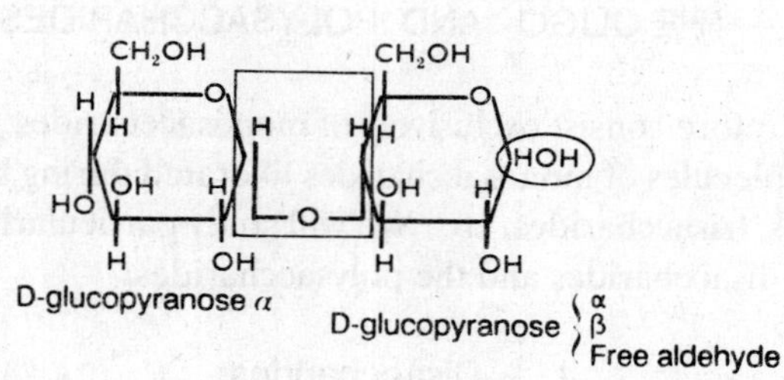

FIG. 4-15. — Structure of maltose or 4-D-glucopyranosyl-α-D-glucopyranose or α-D-glucopyranosyl-(1→4)-D-glucopyranose. The free hemiacetal group can take any one of the 2 possible configurations, that is why the form is not indicated either in the name, or in the formula.

The hemiacetal group remaining free (on the right in the figure) can, in solution, exist under 2 configurations which are in equilibrium: the α-form (α-maltose) and the β-form (β-maltose). The other hemiacetal group, on the contrary, is fixed in the α-1,4-glucosidic linkage. This hemiacetal group which has remained free is responsible for the reducing power of maltose.

b) **Lactose.** – It is found in milk of mammals. It is formed by the union of a molecule of D-galactose (involved by its hemiacetal group) and one of D-glucose (involved by its hydroxyl in position 4); it is a β-1,4-galactosidic linkage (which will be hydrolyzable by a β-galactosidase). Lactose is a 4-D-glucopyranosyl-β-D-galactopyranose (see fig. 4-16). For convenience, it may be written Gal (β1-4) Glc or β-D-Galp-(1⟶ 4)-D-Glcp. As in the case of maltose, there will be, in solution, equilibrium between the 2 possible forms, depending on the configuration taken by the hemiacetal group which has remained free (that of glucose, on the right in the figure). This hemiacetal group is responsible for the reducing power of lactose.

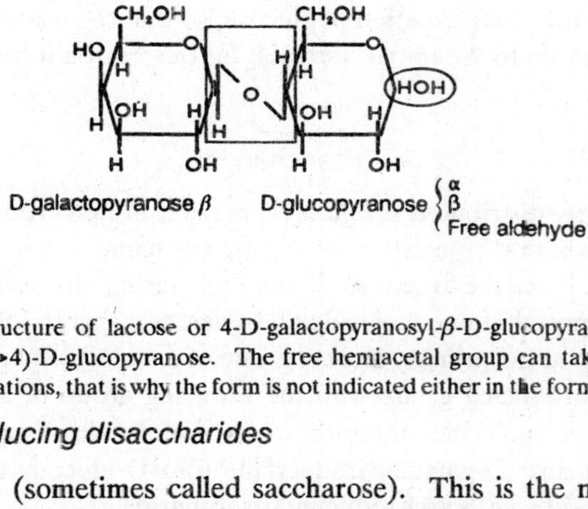

FIG. 4-16. — Structure of lactose or 4-D-galactopyranosyl-β-D-glucopyranose or β-D-galactopyranosyl-(1⟶4)-D-glucopyranose. The free hemiacetal group can take any one of the 2 possible configurations, that is why the form is not indicated either in the formula, or in the name.

B. – Non-reducing disaccharides

Sucrose – (sometimes called saccharose). This is the most abundantly distributed non-reducing disaccharide and it is the only compound examined here. Found in numerous plants, it is particularly abundant in sugar-beet and sugarcane. It is formed by the union of a molecule of D-glucose and a molecule of D-fructose, both involved by their hemiacetal group in the oligosaccharide linkage. There is, therefore, only one possible configuration and no reducing power. Sucrose is the 1-α-D-glucopyranosyl-2-β-D-fructofuranoside (see fig. 4-17). Let us note that fructose is here in the furanose form, while the pyranose form predominates in free fructose.

Sucrose is dextrorotatory ($[\alpha]_D$ = +66°7). When hydrolyzed, either by acids or intestinal invertase, it gives a mixture of equimolecular quantities of D(+) glucose (in mutarotation equilibrium of the α and β-pyranoses forms, $[\alpha]_D$ = +52° approximately) and D(–) fructose (in mutarotation

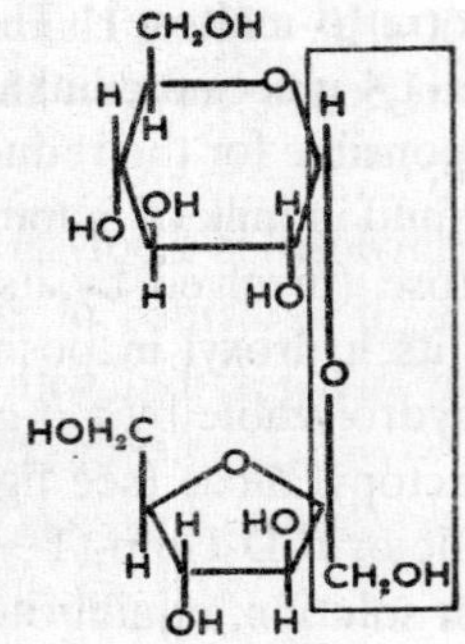

FIG. 4-17. — Structure of sucrose or 1-α-D-glucopyranosyl-2-β-D-fructofuranoside or α-D-glucopyranosyl-(1⟶2)-β-D-fructofuranoside.

equilibrium, $[\alpha]_D = -92°$ approximately); the levorotation of fructose is therefore greater than the dextrorotation of glucose, so that the *mixture obtained* is levorotatory (contrary to the initial sucrose); for this reason it has been called *invert sugar*.

2. – Trisaccharides

The most widely distributed is *raffinose*, present in numerous plants, particularly in beet where it co-exists with sucrose (its name also indicates that it can be obtained from the fractions eliminated during the refining of beet sugar). It is composed of one molecule of α-galactose (pyran form) linked by its reducing group to the primary alcohol group of a molecule of glucose which is involved by its reducing group with the reducing group of a molecule of fructose (these two molecules therefore form a molecule of sucrose). Raffinose is therefore an α-D-galactopyranosyl (1→6)-α-D-glucopyranosyl (1→2)-β-D-fructofuranoside, i.e. a non-reducing trisaccharide.

3. – Polysaccharides

Polyholosides, or *polyosides*, or *polysaccharides* are formed by the condensation of a large number of molecules of monosaccharides, either all identical (homopolysaccharides), or of different types (heteropolysaccharides).

A. – Homopolysaccharides

Glucosans (composed exclusively of glucose units) must be mentioned first.

a) **Starch.** – It is the *carbohydrate reserve form in plants*; it is generally found in the form of starch grains whose morphology varies according to the plant species. In most cases it is composed of 2 constituents:

– *amylose*, a polysaccharide with linear chains, is composed of D-glucose units joined by α-1,4 glucosidic linkages (as in maltose); its molecular weight varies from 150 000 to 600 000. These chains are of varying lengths and can combine through hydrogen bonds between the hydroxyls, thus forming rather compact structures.

– *amylopectin* (or isoamylose) is a polysaccharide whose molecular weight can reach several millions; it is formed of main chains identical to those of amylose (α-1,4 glucosidic linkages) but with side chains joining them – by α-1,6 glucosidic linkages – which have the same structure as the main chains (see fig. 4-18) and which are 20 to 25 glucose residues long.

Some starches contain only amylopectin. In those containing amylose and amylopectin, the respective proportions of these two constituents and the length of chains vary according to the type of starch.

b) **Glycogen.** – It is the *storage form of glucose in animals* (it occurs especially in the liver and muscles). The structure of glycogen is the same as that of amylopectin (see fig. 4-18); however, glycogen is often branched to a greater degree and therefore contains more α-1,6-glucosidic linkages. Besides, the average length of chains varies from 10 to 15 glucose residues. Its molecular weight can attain several tens of millions. It has been recently proved that glycogen is conjugated to proteins through the OH group of tyrosine residues.

c) **Cellulose.** – This is the substance mainly *responsible for the structure* of cell walls of plants. It is not hydrolyzable by enzymes present in the digestive tract of man and has therefore no food value unlike starch (however, the ruminants and various xylophagous insects can utilize cellulose because they have in their digestive tract, microorganisms whose enzymes hydrolyze this polysaccharide). Cellulose consists of long chains composed of D-glucose units joined by β-1,4-glucosidic linkages (and not α-1,4 as in amylose). These chains (see their structure in figure 4-19) are closely attached to one another by hydrogen or Van der Waals type bonds thus forming compact and insoluble fibrous structures which permit the industrial use of cellulose (paper fibers, fabrics, etc.)

d) **Dextrans.** – These polysaccharides consist of α-glucose units linked in 1,6. Shorter chains are grafted (by 1,4 linkages) on this main chain. Dextrans are found in various bacteria. The lengthening of chains is catalyzed by *dextran-sucrase*, an enzyme which can split a sucrose molecule, liberating fructose and adding glucose at the end of a dextran chain which is thus lengthened by one unit (see page 207). Dextrans are used for chromatography by gel filtration (see page 35); some dextrans (of molecular weight between 50 000 and 100 000) are used in therapeutics as substitute for blood plasma.

e) **Chitin.** – It is a polymer of N-acetylated glucosamine; its molecules are joined in linear chains by β-1,4 linkages. In combination with mineral salts and proteins, it forms the exoskeleton of Arthropoda.

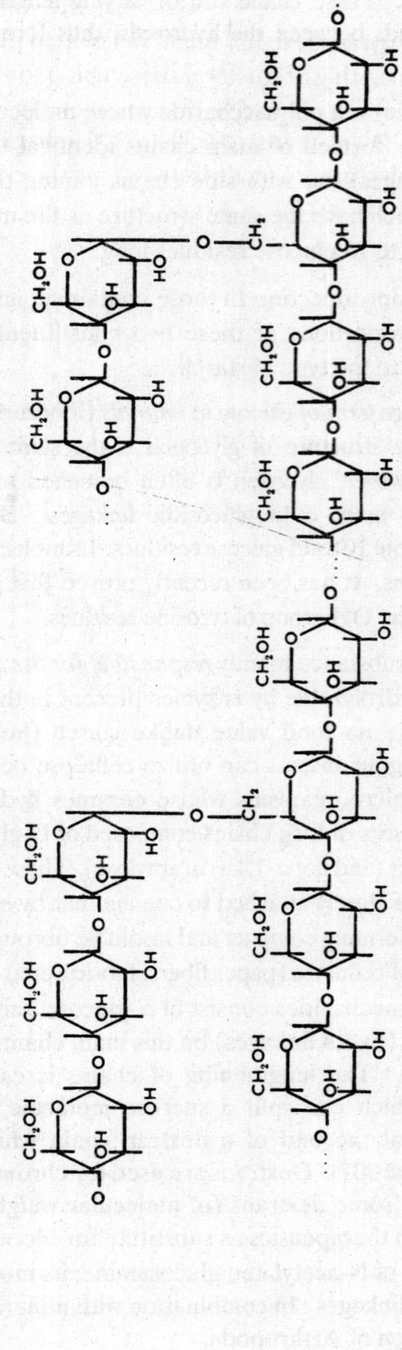

FIG. 4-18. — Structure of amylopectin.

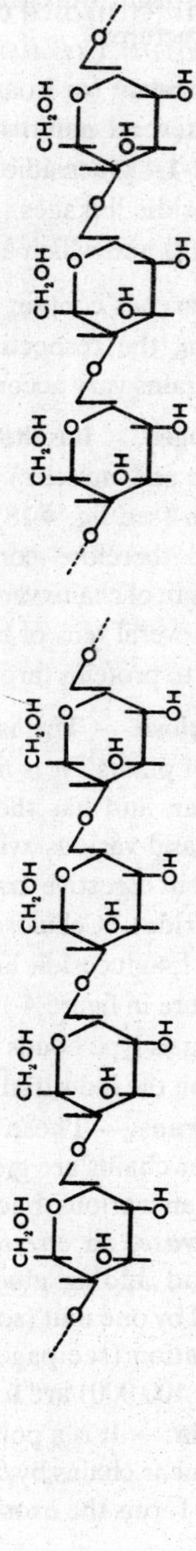

FIG. 4-19. — Structure of cellulose.

In addition to glucosans, the homopolysaccharides also include:

– *arabans* (consisting of α-arabofuranose units generally linked in 1,5) found in pectic matters or pectins, present in fruits and responsible for the phenomenon of jelly formation.

– *xylans* (composed of xylopyranose units) found in lignified plant tissues;

– *fructosans*, like *inulin* (consisting of fructofuranose units) found in tubers of some plants;

– *pectins* are polymers of D-galacturonic acid (α-1,4 linkages); some of their carboxylic groups are in the form of methylesters.

B. – *Heteropolysaccharides*

Their hydrolysis can liberate not only various neutral monosaccharides but also uronic acids, osamines and sialic acids.

In this group are found plant polysaccharides the structure of which is often not yet fully understood, for example, *gums* (branched structures containing D-galactose, L-arabinose, L-rhamnose and D-glucuronic acid), *hemicelluloses* (composed of D-glucose, D-xylose, L-arabinose and D-glucuronic acid), *mucilages* (containing D-mannose, D-galactose, sometimes esterified by sulphuric acid).

To this group belong *the polysaccharides constituting the capsule of bacteria* like pneumococci or streptococci; each type of pneumococcus for example has its own polysaccharide which confers on it an *antigenic specificity*.

As an indication, the structure of the polysaccharide of pneumococcus III is given below:

$$\xrightarrow{\beta\text{-}1,3}\left[\text{D-GlcAU}\xrightarrow{\beta\text{-}1,4}\text{D-Glc}\xrightarrow{\beta\text{-}1,3}\right]_n$$

Lastly, this group sometimes includes oligo- or polysaccharides which, in the native state, are conjugated either with proteins or with lipids. In this respect they rather belong to the group of heterosides. We shall examine some of them in our study of glycoproteins: acid mucopolysaccharides of the connective tissue, like chondroitin sulphuric acids (see page 163), murein (see page 164), etc.

II. – HETEROPOLYSACCHARIDES

The hemiacetal group of a monosaccharide can react with a compound which is not of carbohydrate nature. A heteropolysaccharide is then obtained. The non-saccharide part is called *aglucone*. Depending on the nature of the saccharide, there are *glucosides*, *galactosides*, etc.

Let us first recall a simple example of heteropolysaccharide, mentioned in connection with arguments supporting the cyclic structure of saccharides (see page 144), i.e. the combination of the pseudo-aldehyde group of glucose with the alcohol group of methanol, giving either α-D-methylglucoside, or β-D-methylglucoside. Having studied the cyclic structures of monosaccharides, we may now state that these 2 compounds can exist either in the furanose form or in the pyranose form.

Heteropolysaccharides are very widely distributed in plants. A large number have pharmacodynamic properties and are used in therapeutics. We shall just mention the heterosides having cardiotonic properties (especially in the digital), the aglucones of which are derivative of sterols.

Let us recall to memory, the glucurono-conjugates which are heteropolysaccharides, the monosaccharide of which is glucuronic acid (see page 150).

It may be observed that in most cases, one has O-heterosides where an alcoholic or phenolic hydroxyl of the aglucone is combined with the reducing group of a monosaccharide. But nucleosides and nucleotides (studied in Chapter 6), where the reducing group of ribose or deoxyribose is linked with the nitrogen of a purine or pyrimidine base, can be considered as N-heterosides.

It must be pointed out that glycoproteins and glycolipids are also heterosides, the carbohydrate part of which is called glycan (see below).

GLYCOPROTEINS

Glycoproteins result from the association, by covalent linkage, of a protein with a carbohydrate moiety of variable molecular weight, called *glycan*. They belong to the group of glycoproteins with glyco-amino acids (association of a glycan with an amino acid) and glycopeptides (glycan associated with a peptide). Together with the glycolipids they form the family of *glycoconjugates*.

I. – NATURE OF THE CARBOHYDRATE PART

1. – Monosaccharide constituents

The monosaccharides constituting the glycans belong to the 4 categories described above (see page 149). The most common are:

– *neutral monosaccharides*: D-galactose, D-mannose, L-fucose, L-rhamnose, D-xylose;

– *osamines* (in N-acetylated form): D-glucosamine, D-galactosamine , D-muramic acid;

– *uronic acids*: D-glucuronic acid, L-iduronic acid;

– *sialic acids*: N-acetylneuraminic acid, N-glycolylneuraminic acid, N.O-acetylneuraminic acids.

2. – Structure of glycans

There are linear glycans and branched glycans. As examples, we will indicate the structures of some glycans belonging to each of these two categories and briefly describe the glycoproteins themselves.

A. – *Sub-maxillary mucins*

One of the best known is the sub-maxillary mucin of sheep which contains about 40% carbohydrates, mainly consisting of N-acetylneuraminic acid and N-acetylgalactosamine. These two sugars are conjugated in disaccharide units, and about 800 of these units are linked to the peptide chain in the following manner:

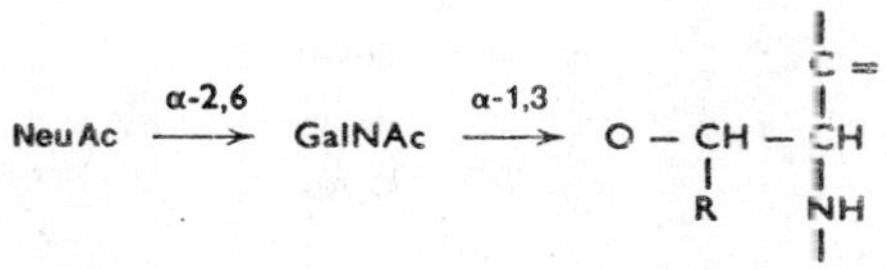

R = H (serine) or CH_3 (threonine). The meaning of initials is given in pages 151-152.

Mucins give highly viscous solutions, but this viscosity disappears under the action of a neuraminidase (or sialidase).

B. – *Proteoglycans of the fundamental substance*

They result from the association – by O-glycosidic linkages with serine residues – of proteins with glycans having an acid character and called acid *mucopolysaccharides*. These glycans are composed of linear chains, of molecular weight varying from 50 000 (chondroitin sulphuric acid A) to several millions (hyaluronic acid), resulting from the polymerization of disaccharide units. These units always contain a N-acetylhexosamine associated either with a neutral monosaccharide (galactose in keratan sulphate), or with a uronic acid. Esterification by sulphuric acid confers on these molecules a highly acid character which reinforces that of uronic acid.

As an indication, one may represent the structure of the disaccharide repetition unit (called hyalobiuronic acid) of *hyaluronic acid*:

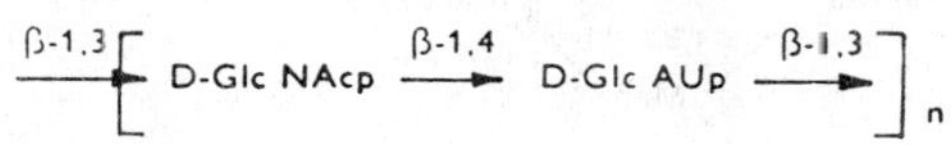

and the general structure diagram of the proteoglycan of *chondroitin sulphuric acid A* (or chondroitin-4-sulphate):

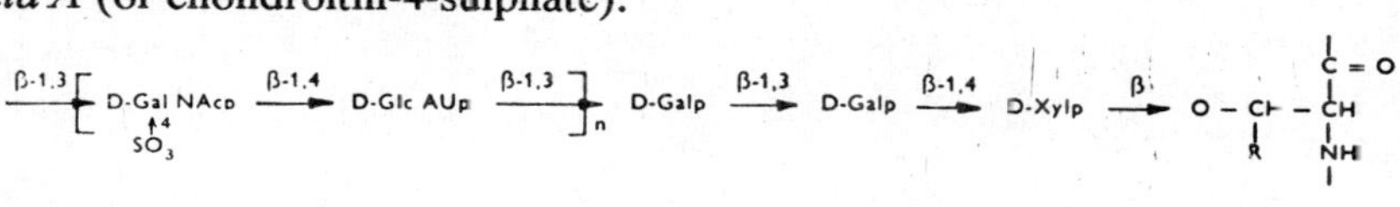

C. – Muropeptides of murein

Murein is one of the constituents of bacterial cell walls. It is composed of muropeptides joined by their peptide chains. These muropeptides result from the linkage of short peptides with glycan chains consisting of disaccharide repetition units; these units are composed, for example, of N-acetyl-D-glucosamine and N-acetyl-D-muramic acid in the case of the muropeptide of *Staphylococcus aureus*.

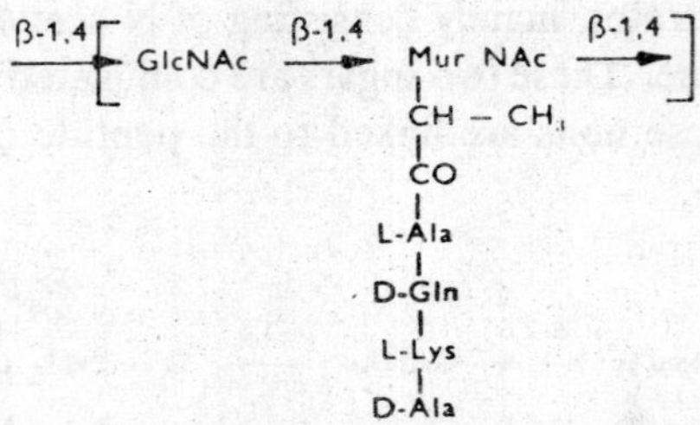

D. – Complex glycans

Glycans of very complex structures are found in numerous animal glycoproteins like ovalbumin, human serum transferrin, thyroglobulin, etc. In transferrin there are two identical glycans having the following structure:

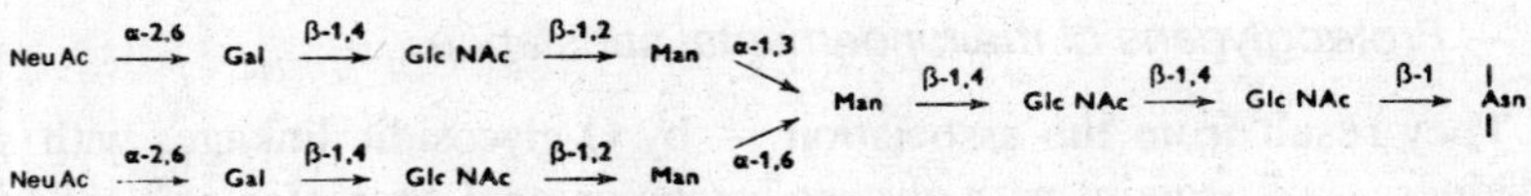

Such a structure is called "bi-antennate" and of the N-acetyl-lactosaminic type, due to the presence of the N-acetyl-lactosamine pattern: β-Gal-(1 → 4)-GlcNac. In other glycoproteins, there are glycans which contain only N-acetyl-glucosamine and mannose, the residues of which are grouped in an oligomannosidic structure. Such glycans, like the one of unit A of thyroglobulin illustrated in the diagram below, are said to be of the oligomannosidic type.

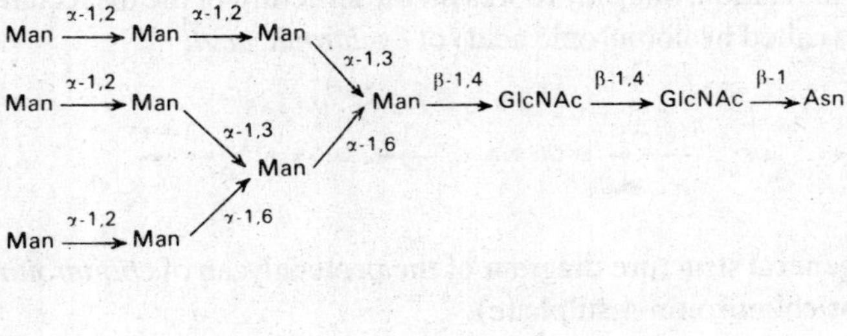

II. – NATURE OF THE GLYCAN-PROTEIN LINKAGE

The above examples show that there are mainly two types of linkages.

1) The amide linkage is present:

– on the one hand, in numerous animal glycoproteins, where it results from the conjugation of the reducing group of N-acetylglucosamine with the amide group of asparagine:

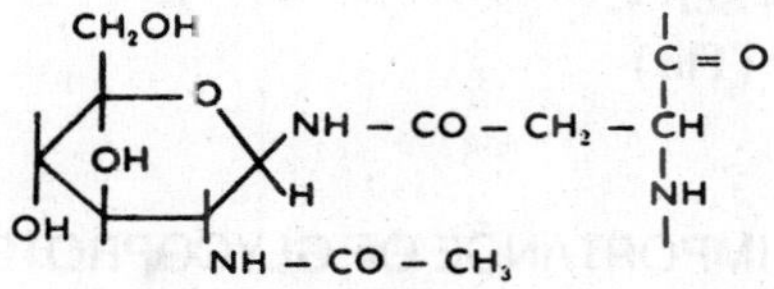

This type of linkage, called "N-glycosylaminic" linkage, defines the class of N-glycosylproteins;

– on the other hand, in the murein of bacterial cell walls, where it results from the condensation of the carbonyl group of N-acetyl-muramic acid with the amino group of alanine:

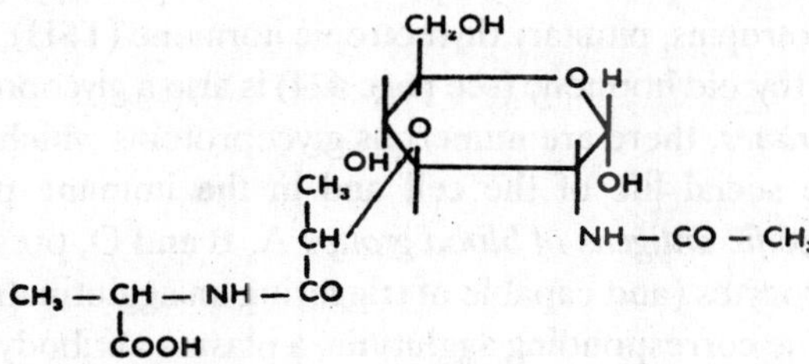

2) The O-glycosidic linkage, characteristic of O-glycosylproteins where the terminal reducing group of a glycan (generally, that of N-acetyl-D-galactosamine) is involved with the hydroxyl of a serine (R = H) or threonine (R = CH_3), as mentioned in the case of sub-maxillary mucins, for example:

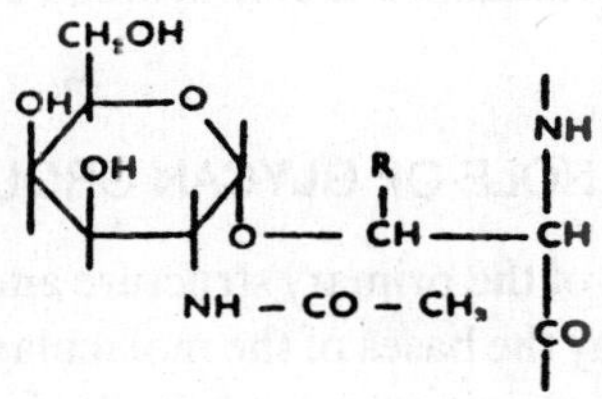

III. – BIOSYNTHESIS OF GLYCOPROTEINS

The biosynthesis of glycan groups involves glycosyltransferases and nucleotides-monosaccharides, in conditions which will be examined later in

this chapter (see pages 205-211) and which are to a great extent common to the synthesis of free and conjugated polysaccharides.

There are signals on the peptide chain which are recognized by the glycosyltransferase fixing N-acetylglucosamine on the protein. For example, in the case of an "asparaginyl-N-acetylglucosamine" linkage (see page 165), the monosaccharide is fixed on Asn only if a Ser or Thr residue is in position β towards the C-terminal end. It may be said that the recognition sequence is therefore: $-$Asn$-$X$-$Ser$-$.
(Thr)

IV. – IMPORTANCE OF GLYCOPROTEINS

Glycoproteins are very widely distributed in animal and plant tissues, as well as in micro-organisms and viruses.

The *fundamental substance* which, in the connective tissue, fills the mesh of glycoprotein fibers of elastin, reticulin and collagen, contains a large proportion of acid mucopolysaccharides linked to proteins (for example, chondroitin-sulphuric acids).

Numerous *hormones* are of glycoprotein nature: pituitary (LH, FSH) and chorionic gonadotropins, pituitary thyreotropic hormone (TSH). Thyroglobulin, precursor of the thyroid hormone (see page 424) is also a glycoprotein.

In *cell membranes*, there are numerous glycoproteins which play an important role in the social life of the cell and in the immune phenomena: for example, the *specific antigens of blood groups* A, B and O, present in the walls of human erythrocytes (and capable of triggering an agglutination if they come in contact with the corresponding agglutinin, a plasma antibody which does not exist in the same individual but may be present in the plasma of an individual belonging to another blood group) are glycoproteins or glycolipids.

Biological fluids are generally very rich in glycoproteins (saliva, urine, bile, milk, tears, blood). For example, in blood, practically all proteins – except serumalbumin – are glycoproteins: antibodies, orosomucoid, transferrin (form of transport of iron), ceruleoplasmin, haptoglobins, prothrombin and fibrinogen (the latter two substances play a fundamental role in blood coagulation).

V. – ROLE OF GLYCAN GROUPS

Our present knowledge of the primary structure and spatial conformation of glycans has enabled us to lay the bases of the molecular biology of glycoproteins and show that glycans play an important role in the following mechanisms:

1) maintenance of the structure of the protein in a biologically active conformation, through glycan-glycan or glycan-protein secondary linkages (hydrogen bonds, ionic bonds and hydrophobic interactions);

2) protection of the peptide chain against attack by proteolytic enzymes;

3) decrease in antigenic power of proteins by the masking of epitopes;

4) control of the imbibition of membranes by-water and, consequently, of the diamembrane passage of mineral ions and small organic molecules. In this connection, the observation that the glycan fractions of membrane glycoconjugates are profoundly modified in cancerous cell (increase in molecular weight of glycans of the glycoproteins and in their rate of sialylation) could explain the modifications noted in the permeability of the membranes of these cells;

5) action in the cellular phenomena of recognition: recognition and association of cells involving proteids which specifically recognize oligosaccharide structures and are called *lectins*; recognition of target-cells by hormones, toxins, bacteria and viruses and also by other glycoproteins (the desialylation of circulating glycoproteins causes their instantaneous capture by a lectin present in the membranes of hepatocytes and which specifically recognizes the residues of galactose in non-reducing terminal position, this capture being followed by the internalization of asialoglycoproteins and their destruction by the lysosomial enzymes). The above mentioned fact that glycans of the glycoconjugates of cancerous cell membranes undergo profound modifications on the one hand, and the observation that glycans play a role in the contact inhibition phenomenon (arrest of mitoses when normal cells come in contact with each other), on the other hand, could explain a) the perturbations manifest in the social life of cancerous cells: loss of contact inhibition and metastatic diffusion and b) the appearance of surface neo-antigens capable of inducing the immune reaction of the organism.

METABOLISM OF CARBOHYDRATES

Monosaccharides, and particularly glucose, owe their importance to the fact that *their oxidation provides the living organisms with a major part of the energy they require*. Furthermore, the carbon atoms of glucose can be found in a large number of compounds (amino acids, fatty acids, sterols, glycerol, etc.). In the study of the metabolism of carbohydrates we will therefore come across substances belonging to proteins, lipids or nucleic acids; this shows that any division of cellular metabolism would be arbitrary, because the latter forms a set of well integrated reactions and it should therefore be borne in mind that divisions are made only for facilitating the presentation.

DIGESTION AND ABSORPTION OF CARBOHYDRATES

Only monosaccharides pass easily through the cell membranes of living organisms. But the carbohydrates present in the food of animals or microorganisms are often disaccharides like lactose and sucrose, and polysaccharides like starch and glycogen. They will have to be hydrolyzed before they are absorbed and metabolized.

I. – Digestion of carbohydrates

The hydrolysis of polysaccharides is catalyzed by *exoenzymes*, i.e. enzymes liberated outside the cells in which they were synthesized. We will study here only the hydrolysis of the polymers of glucose – starch and glycogen – as it takes place in the intestine of man; the other polysaccharides are generally hydrolyzed only by microorganisms (example: hydrolysis of cellulose by bacterial cellulases). Starch and glycogen are hydrolyzed by *amylases*, enzymes catalyzing the scission of α-1,4-glucosidic linkages; two of them are known:

– *β-amylase*, is mainly found in plants, especially in starch-storing grains; it attacks starch from the non-reducing end of chains by detaching maltose units, which, although in the α-form in the polysaccharide, are liberated in the form of β-maltose: α-Glc-(1⟶4)-β-Glc;

– *α-amylase* is formed during the germination of cereals. Besides, it is present in saliva and pancreatic juice. It also breaks the α-1,4 glucosidic linkages, but at any place inside the polysaccharide molecule; in the case of linear chains (amylose), total hydrolysis into maltose units is finally obtained. But in the case of amylopectin and glycogen, the hydrolysis of α-1,4-glucosidic linkages is not complete near the branch points and besides, the α-1,6 glucosidic linkages which constitute these branch points are not attacked. A mixture of α-maltose: α-Glc-(1⟶4)-α-Glc and *residual dextrins* composed of branched oligo- and polysaccharides is obtained.

Amylo-α-1,6 glucosidase, secreted by the cells of the intestinal mucosa, and also called "debranching enzyme", enables the hydrolysis of these residual dextrins by catalyzing the hydrolysis of the α-1,6-glucosidic linkage of the branch points.

Lastly, the hydrolysis of polysaccharides is completed by a α-glucosidase, present in the membrane of the brush-shaped border of enterocytes, called *maltase*, which splits the maltose units (resulting from the action of α-amylase and oligo-α-1,6 glucosidase) giving 2 molecules of glucose.

Furthermore, two other enzymes are found in the intestine; they are also membranous, and attack the disaccharides:

– *invertase* (β-fructosidase) catalyzing the hydrolysis of sucrose into glucose + fructose.

– *lactase* (β-galactosidase) catalyzing the hydrolysis of lactose into galactose + glucose.

Finally, the digestion of carbohydrates included in diet mainly leads to glucose (from starch, glycogen, lactose, sucrose), galactose (from lactose) and fructose (from sucrose); the problem to be examined is therefore the absorption of these monosaccharides.

II. – ABSORPTION OF MONOSACCHARIDES

The absorption of monosaccharides through the intestinal mucosa does not take place by simple diffusion. Various hexoses which diffuse at comparable rates are absorbed at different rates; some pentoses are absorbed more slowly than glucose or fructose although their smaller molecular weight enables them to diffuse more rapidly; lastly, the absorption of monosaccharides proceeds from a compartment where concentration is low, towards a compartment of higher concentration. These observations can also be generalized, because they apply to the absorption of substances other than monosaccharides, and to absorption through cell membranes in general. Membranes have complex structures which confer on them a *selective permeability*; they allow certain substances to pass through them and oppose the passage of other compounds. The concentration of a product on one of the sides of the membrane is a process which requires energy; for this reason it is spoken of as *active transport*.

The mechanisms of cellular absorption of carbohydrates is not yet fully understood. It is however known that glucose and galactose cross the membrane of the intestinal epithelial cell with the help of a Na^+ dependent carrier. On the other hand, in some cases, especially among microorganisms, it was possible to show the participation of an enzymatic type of protein called *permease*, specific of a sugar and this is possibly a general phenomenon. In any case, specificity is an important characteristic of absorption. This is confirmed by observations made with insulin, a hormone secreted by the pancreas and which – through a phenomenon which is not sufficiently known – promotes the absorption of glucose; as a matter of fact, a lack of insulin affects the absorption of glucose but not that of fructose.

Once absorbed, monosaccharides will proceed towards the liver through the portal vein.

METABOLISM OF MONOSACCHARIDES

In the following discussion we will mainly refer to glucose. As mentioned below, the other monosaccharides (especially fructose and galactose) are readily converted into derivatives of glucose and their metabolism links up with that of glucose.

I. – PHOSPHORYLATION OF GLUCOSE

As just mentioned, the phosphorylation of monosaccharides, the first step of their metabolism, takes place very rapidly after their absorption, and we will see that the metabolism of monosaccharides mainly calls upon the phosphorylated monosaccharides.

The phosphorylation of glucose in the cell cannot take place by direct reaction between the monosaccharide and the phosphate ion because the equilibrium is too unfavourable. A phosphate group is given by ATP; there is breakdown of a phosphoanhydride linkage and formation of a phosphoester linkage (see fig. 4-20). The equilibrium of this reaction is largely in favour of the formation of the ester; it is practically irreversible, because it is highly exergonic ($\Delta G_0 \# -5$ kcal/mole). This is an example where the energy of ATP is utilized for the formation of a compound which would be very difficult to synthesize otherwise; we will see many other reactions of this kind.

This reaction consists of a transfer of the phosphate group, it is a transphosphorylation, catalyzed by a transphosphorylase or *kinase*, here a *glucokinase*, catalyzing specifically the phosphorylation on the alcoholic hydroxyl carried by the carbon 6 of glucose. Some organisms possess less specific enzymes; for example, yeast has a hexokinase which catalyzes the phosphorylation of glucose, fructose and mannose into the corresponding hexose-6-phosphate. Besides, some monosaccharides like galactose are phosphorylated on their hemiacetal group. There are therefore C_1- and C_6- hexokinases. It will be noted that Mg^{2+} is necessary for the reaction; this ion is very often the co-factor in the reactions involving phosphorylated compounds. In most cases, the action of the ion consists in maintaining the ATP molecule in a compact conformation which is generally the physiologically active form of this compound.

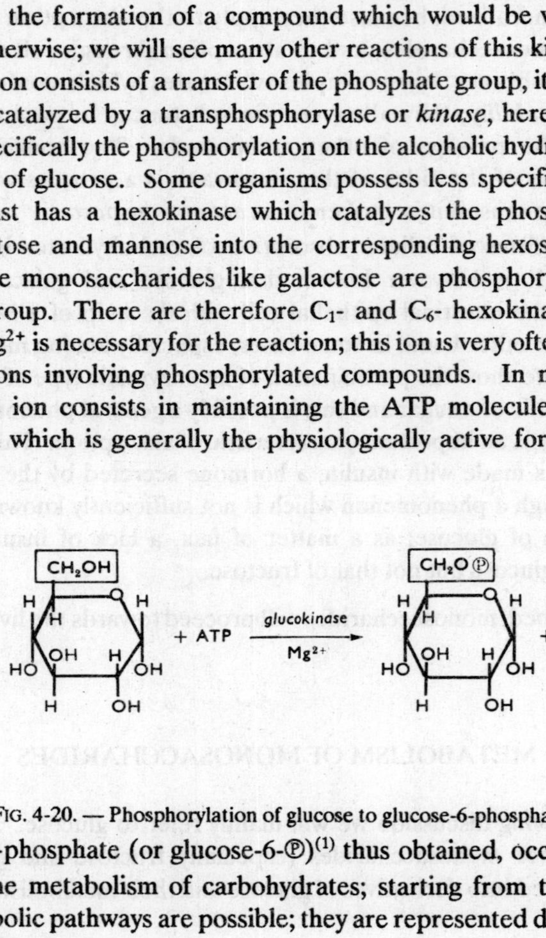

FIG. 4-20. — Phosphorylation of glucose to glucose-6-phosphate.

Glucose-6-phosphate (or glucose-6-Ⓟ)[1] thus obtained, occupies a central position in the metabolism of carbohydrates; starting from this compound, various metabolic pathways are possible; they are represented diagrammatically in figure 4-21.

---

(1) — Ⓟ represents —PO_3H_2 or $-\overset{\overset{\displaystyle OH}{/}}{\underset{\underset{\displaystyle O}{\|}}{P}}-OH$; in the following we will adopt this representation to simplify the writing of phosphoric esters.

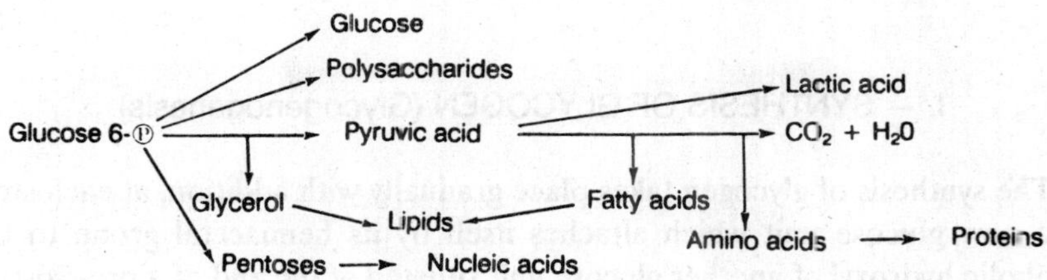

FIG. 4-21. — Diagram of various metabolic pathways starting from glucose-6-Ⓟ.

Before taking up the study of these diverse metabolic pathways, one should try to answer the following question: why would the molecules of glucose-6-Ⓟ take up one of these paths rather than another, in other words, what decides the fate of the glucose-6-Ⓟ molecules? This is a problem of *regulation* which we will consider again at a later stage, but it may be already stated that the proportion of molecules of glucose-6-Ⓟ taking one of these metabolic pathways is variable in a given organism or even in a given cell, depending on the physiological state of this organism or cell. Thus, when large quantities of ATP are necessary, a major proportion of molecules of glucose-6-Ⓟ will undergo oxidation (generating ATP, as will be seen in the following); furthermore, there will be hydrolysis of the reserve polysaccharides and formation, from lipids, of larger quantities of compounds which may be oxidized. On the contrary, when ATP requirements are diminished, a large proportion of molecules of glucose-6-Ⓟ will be converted into glycogen; besides, a major fraction of compounds formed intermediately during the first steps of oxidation of glucose-6-Ⓟ will be directed, not towards the subsequent steps of oxidation (which are the most efficient for the production of ATP), but towards the synthesis of lipids.

II. – FORMATION OF GLUCOSE FROM GLUCOSE-6-Ⓟ

Phosphorylated monosaccharides cannot pass through the cell membrane and enter the external medium; they must first lose their phosphate group. We know (see previous paragraph) this cannot take place by the reverse reaction of phosphorylation. The hydrolysis reaction of the ester linkage is catalyzed by an enzyme called *glucose-6-phosphatase*, absent in many cells (which cannot therefore liberate glucose and have no reason for doing so), but present in hepatic cells which must liberate glucose into the blood for its distribution to other tissues.

$$\text{Glucose-6-Ⓟ} + H_2O \xrightarrow{\text{glucose-6-phosphatase}} \text{Glucose} + P_i^{(1)}$$

(1) For simplicity, we will write P_i (inorganic phosphate) for the phosphate ion. It should be remembered that at neutral pH, this group is highly ionized.

SYNTHESIS AND DEGRADATION OF POLYSACCHARIDES

I. – SYNTHESIS OF GLYCOGEN (Glycogenogenesis)

The synthesis of glycogen takes place gradually with addition, at each step, of a new glucose unit which attaches itself by its hemiacetal group to the alcoholic hydroxyl of another glucose unit situated at the end of a pre-existing glycogen chain. But the direct reaction between a hemiacetal group and an alcoholic OH does not take place spontaneously (the equilibrium is not favourable); for this reason; the hemiacetal group must first be *activated*, i.e. made more reactive in view of the formation of the glucosidic linkage. This activation proceeds in 2 steps:

1) transformation of glucose-6-Ⓟ into glucose-1-Ⓟ, catalyzed by *phosphoglucomutase*;

2) formation of uridine-diphosphate-glucose (UDPG) by action of uridine-triphosphate (UTP) on glucose-1-Ⓟ. The energy required for the synthesis of the glucosidic linkage is supplied by UTP, while in general, mostly ATP yields energy for the biosynthesis reactions. But ATP acts indirectly to reconstitute UTP from UDP formed in the reaction (see diagram; fig. 4-23).

Then UDPG transfers the glucose portion of its molecule to a pre-existing glycogen chain, lengthening the latter by one unit (reaction catalyzed by glycogen synthetase).

These 3 reactions are represented diagrammatically in figure 4-22.

It must be noted that glycogen synthetase catalyzes only the formation of 1,4 glucosidic linkages. The branches are due to a *branching enzyme* capable of transferring an oligosaccharide fragment (4 glucose units for example), collected from a longer linear chain, to the carbon 6 of a glucose of this linear chain.

The synthesis of starch and cellulose in plants takes place by similar mechanisms, with activation of glucose in the form of ADP-glucose in the case of starch and GDP-glucose in the case of cellulose[1].

II. – DEGRADATION OF GLYCOGEN (Glycogenolysis)

It does not take place in the reverse direction of synthesis, but by *phosphorolysis* (a reaction where the phosphate ion plays a role similar to that of water in hydrolysis). The reaction is:

$$[\text{glucose}]_n + P_i \xrightleftharpoons{\text{phosphorylase } a} [\text{glucose}]_{n-1} + \text{glucose-1-Ⓟ}$$

---

[1] The structures of adenosine-diphosphate (ADP) and guanosine-diphosphate (GDP) are studied in the chapter on nucleic acids (see. fig. 6-7).

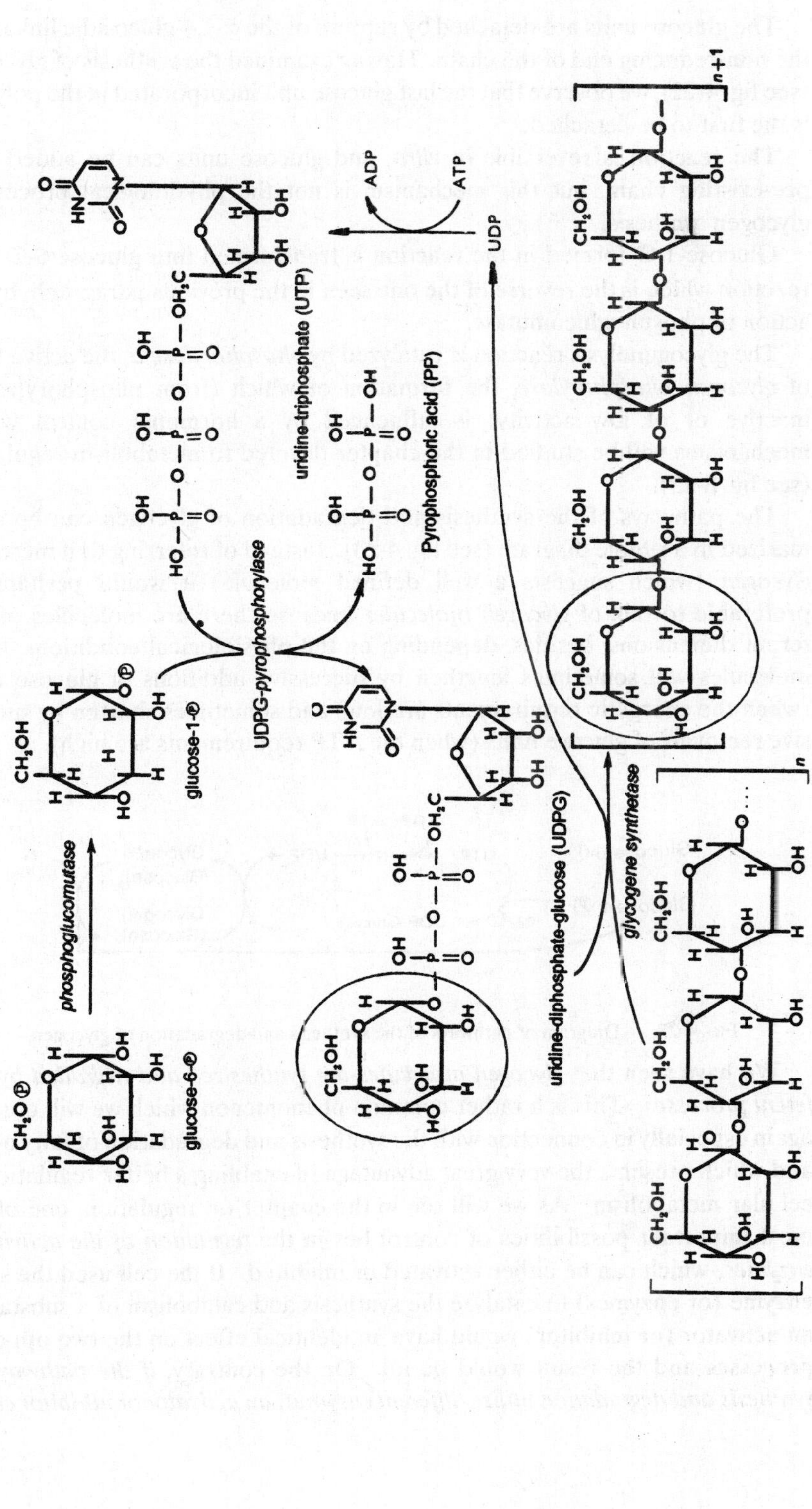

FIG. 4-22. — Synthesis of α-1,4-glucosidic linkages of glycogen.

The glucose units are detached by rupture of the α-1,4-glucosidic linkage at the non-reducing end of the chain. Having examined the synthesis of glycogen (see fig. 4-22), we observe that the last glucose unit incorporated in the polymer is the first to be detached.

The reaction is reversible *in vitro*, and glucose units can be added to a pre-existing chain, but this mechanism is not the physiological process of glycogen synthesis.

Glucose-1-Ⓟ formed in the reaction is transformed into glucose-6-Ⓟ by a reaction which is the reverse of the one seen in the previous paragraph, by the action of phosphoglucomutase.

The glycogenolysis reaction is catalyzed by *phosphorylase a*, the active form of *glycogen phosphorylase*, the formation of which (from phosphorylase *b*, inactive or of low activity) is influenced by a hormonal control whose mechanisms will be studied in the chapter devoted to metabolism regulation (see fig. 8-15).

The pathways of the synthesis and degradation of glycogen can be summarized in a simple diagram (see fig. 4-23). Instead of referring to it merely as *glycogen* (which suggests a well defined molecule) it would perhaps be preferable to talk of *glycogen molecules* because there are molecules of different dimensions; besides, depending on the physiological conditions, these molecules will sometimes lengthen by successive additions of glucose units (when the energetic requirements are low) and sometimes shorten by successive removals of glucose units (when the ATP requirements are high).

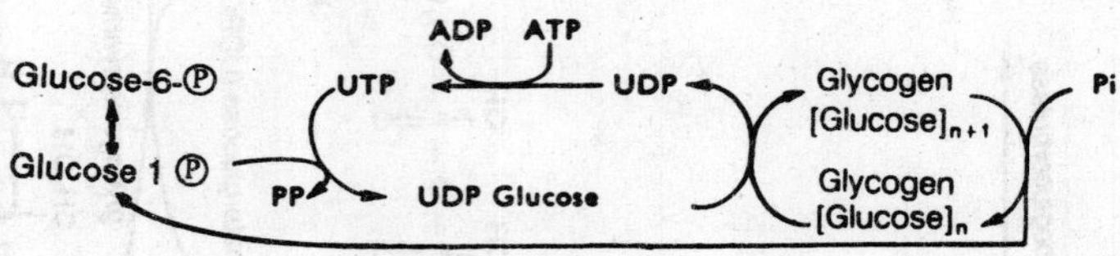

FIG. 4-23. — Diagram of pathways of the synthesis and degradation of glycogen.

We have seen that *glycogen molecules are synthesized and degraded by different processes*. This is a rather common phenomenon which we will observe again especially in connection with the synthesis and degradation of fatty acids, and which presents the very great advantage of enabling a better regulation of cellular metabolism. As we will see in the chapter on regulation, one of the most important possibilities of control lies in the *regulation of the activity of enzymes*, which can be either activated or inhibited. If the cell used the same enzyme (or enzymes) to catalyze the synthesis and catabolism of a substance, an activator (or inhibitor) would have an identical effect on the two opposite processes and the result would be nil. On the contrary, *if the pathways of synthesis and degradation utilize different enzymes, an activator or inhibitor could affect one of them without affecting the other.* For example, the degradation of glycogen can be more pronounced when there is an increase in the quantity of phosphorylase *a*, the active form of the enzyme, which happens (as mentioned above) by the effect of hormones like adrenaline and glucagon (see page 481); this leads to an increased formation of glucose-6-Ⓟ; this action on phosphorylase affects only the glycogenolysis and has obviously no effect on the synthesis of glycogen because the phosphorylase has no part in it.

GLYCOLYSIS

It is usually considered that oxidation of glucose consists of 2 series of steps: the first is extra-mitochondrial and leads to the production of 2 molecules of pyruvic acid; the second is intramitochondrial and permits the complete oxidation of pyruvic acid to CO_2 and H_2O. The first series of reactions is called *glycolysis*, or *Embden-Meyerhof glycolytic pathway*, or *anaerobic phase of the oxidation of glucose* (because these reactions can take place in the absence of oxygen). As will be seen in the following discussion, this series of reactions can also, in certain cases, lead to the formation of lactic acid or ethyl alcohol (the terms lactic fermentation or alcoholic fermentation are then used sometimes). We will now consider these glycolytic reactions and indicate the main characteristics of some of them.

I. – THE GLYCOLYTIC REACTIONS

1. – Phosphohexose isomerase

This enzyme isomerizes glucose-6-Ⓟ (resulting, via glucose-1-Ⓟ, from phosphorolysis of the reserve glycogen or from the phosphorylation of glucose having entered the cell) to fructose-6-Ⓟ (see fig. 4-24).

This reaction is catalyzed by phosphoglucose isomerase (or glucose-6-phosphate isomerase), and there exists a very similar enzyme, phosphomannose isomerase (or mannose-6-phosphate isomerase), which transforms mannose-6-Ⓟ (an epimer of glucose-6-Ⓟ, which differs from it only by the configuration of the hydroxyl on carbon 2: see fig. 4-5) also into fructose-6-Ⓟ.

2. – Phosphofructokinase

Fructose-6-Ⓟ, formed by isomerization of glucose-6-Ⓟ, or by phosphorylation of fructose (resulting from the hydrolysis of diet sucrose), undergoes a second phosphorylation and is converted into fructose-1, 6-bisphosphate (see fig. 4-25). As in the case of the reaction catalyzed by glucokinase or

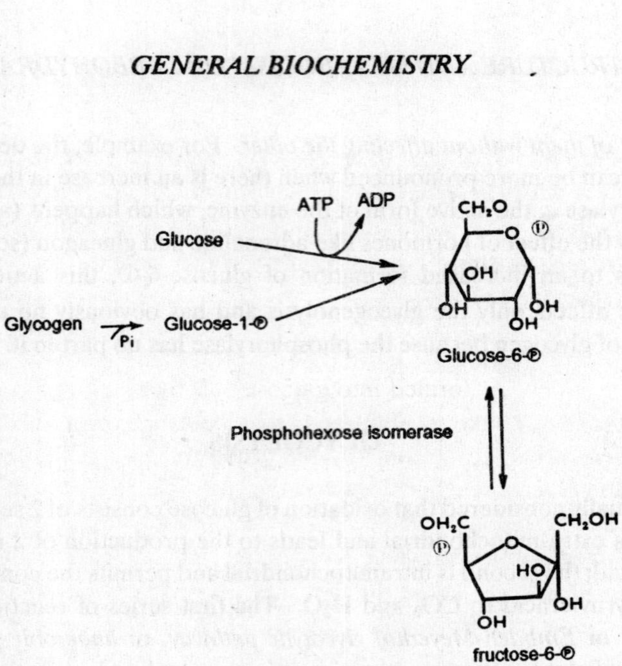

FIG. 4-24. — Isomerization of glucose-6-Ⓟ to fructose-6-Ⓟ.

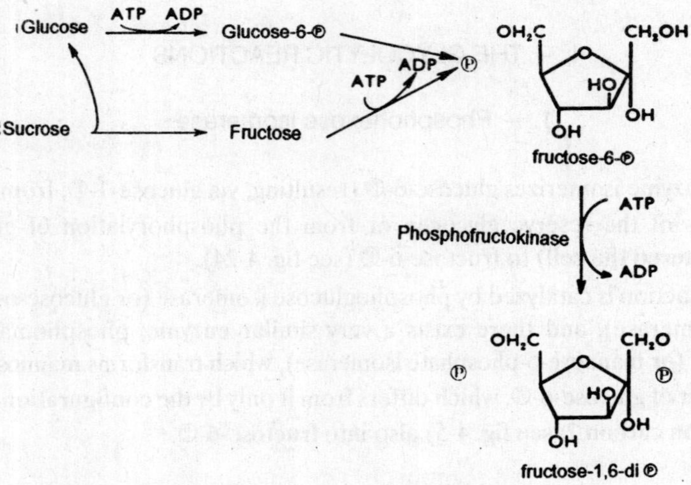

FIG. 4-25. — Transformation of fructose-6-Ⓟ into fructose-1,6-bis-Ⓟ.

hexokinase, this is a transphosphorylation taking place at the cost of ATP. The break of a phosphoanhydride bond is accompanied by the formation of a phosphoester linkage, which results in a considerable decrease in free energy

(ΔG_0 # – 4 kcal/mole), so that the reaction is also not reversible (here also it is a phosphatase which will permit the reconversion from 1, 6 bis-Ⓟ to fructose 6-Ⓟ). We will see in the following that the activity of phosphofructokinase is subjected to an allosteric type of control which enables the regulation of glycolysis (see page 187).

3. – Aldolase or fructose-bisphosphate aldolase

This enzyme owes its name to the fact that it catalyzes an aldol condensation (the addition of an activated methylene group on the double bond of a carbonyl) or the reverse reaction (which is the case here in glycolysis). Fructose, 1,6-bis-Ⓟ is thus split into 2 molecules of triosephosphates, glyceraldehyde-3-Ⓟ and dihydroxyacetone-Ⓟ, which are isomers and interconvertible (like the hexose-phosphates; see above the first reaction of glycolysis) under the action of a triosephosphate-isomerase (see fig. 4-26). The equilibrium of this isomerization reaction is very largely in favour of dihydroxyacetone-Ⓟ (96.5%), but gradually the equilibrium moves towards the formation of glyceraldehyde-3-Ⓟ which alone is implied in the following glycolytic reactions. Thus, dihydroxyacetone-Ⓟ is transformed into glyceraldehyde-3-Ⓟ, and while drawing the balance of glycolysis, it must be remembered that 2 molecules of the latter are in fact formed from each molecule of fructose 1, 6 bis-Ⓟ, i.e. from each molecule of glucose.

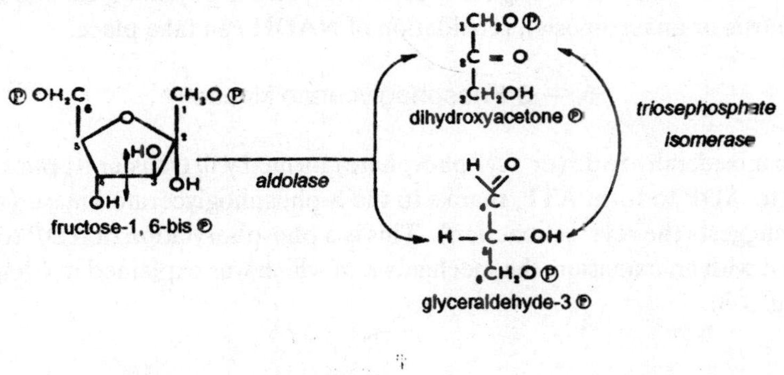

FIG. 4-26. — Scission of fructose-1,6-bis-Ⓟ into 2 molecules of triose-Ⓟ.

4. – Glyceraldehyde-3-Ⓟ dehydrogenase

As indicated by the name of this enzyme, this step will comprise an *oxidation*, the first in this sequence of glycolytic reactions; it will affect the aldehyde group of glyceraldehyde-3-Ⓟ and one will finally obtain a mixed anhydride between the carbonyl thus formed and phosphoric acid. Apparently it seems that the following 3 reactions take place (illustrated in figure 4-27).

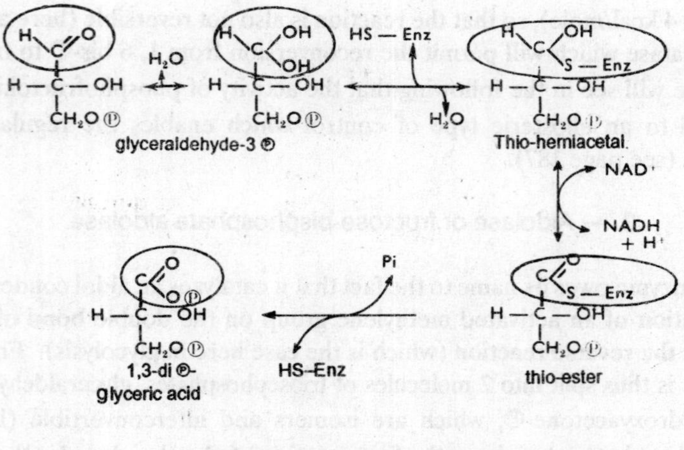

FIG. 4-27. — Transformation of glyceraldehyde-3-Ⓟ into 1,3-di-Ⓟ-glyceric acid by glyceraldehyde-3-Ⓟ dehydrogenase.

a) Reaction of the aldehyde group with a sulphydryl group of the enzyme (which must be in the reduced form to exert its catalytic activity) and formation of a thio-hemiacetal.

b) Oxidation (dehydrogenation) of this thio-hemiacetal to thioester by NAD^+ linked to the enzyme.

c) Phosphorolysis of this thio-ester with formation of the mixed anhydride (1,3-di-Ⓟ-glyceric acid) which has a "high potential energy" bond.

We will see in the following how, depending on the physiological conditions (aerobiosis or anaerobiosis), reoxidation of NADH can take place.

5. – 3-Phosphoglycerate kinase

The mixed anhydride (or acyl-phosphate) formed will transfer its phosphate group to ADP to form ATP, thanks to the 3-phosphoglycerate kinase (whose name suggests the reverse reaction). This is a phosphorylation of ADP to ATP coupled with an oxidation, the mechanism of which was explained in Chapter 3 (see fig. 3-8).

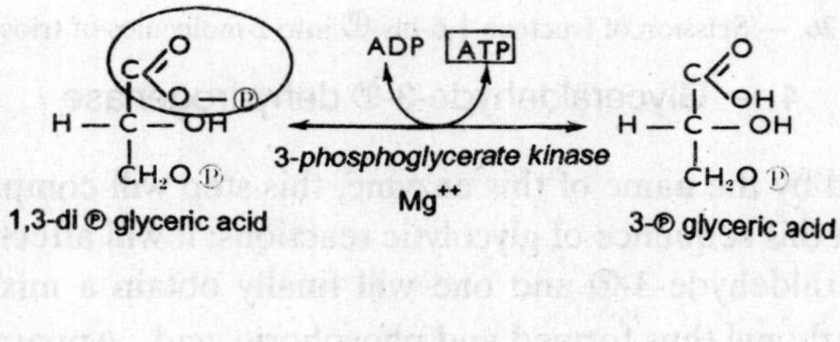

FIG. 4-28. — Transformation of 1,3-diphosphoglyceric acid into 3-phosphoglyceric acid.

6. – Phosphoglycerate mutase

This enzyme catalyzes the migration of the phosphate group and thus permits the conversion of 3-phosphoglyceric acid into 2-phosphoglyceric acid (see fig. 4-29).

7. – Enolase

It catalyzes the dehydration of 2-phosphoglyceric acid which is transformed into phosphoenolpyruvic acid, a compound of "high potential energy" (ΔG_e of hydrolysis $\neq -12.8$ kcal/mole).

8. – Pyruvate kinase

The phosphoenolpyruvic acid can transfer a phosphate group to ADP to form *ATP* (see fig. 3-9) and thus give pyruvic acid (it is probable that the enol form appears first and then undergoes a rearrangement to give the keto tautomeric form). The latter three reactions are illustrated in figure 4-29.

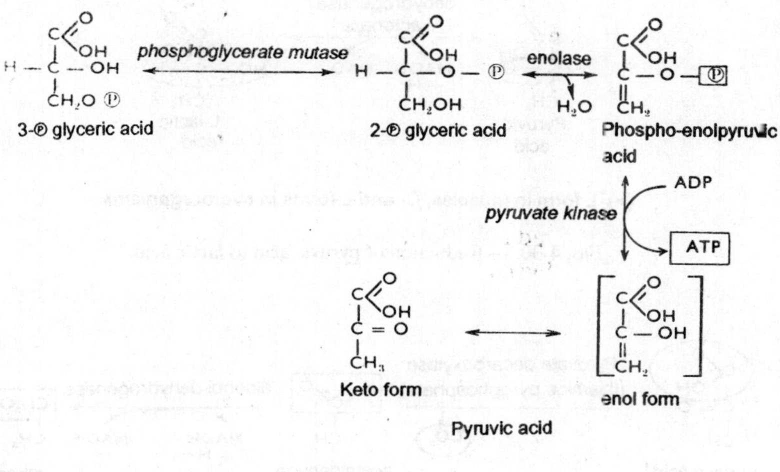

FIG. 4-29. — The last three steps of glycolysis leading to the formation of pyruvic acid.

Glycolytic reactions can take place in aerobiosis or anaerobiosis; but the NAD^+ content of cells is very low, so that glycolysis may be blocked at the stage of the oxidation of glyceraldehyde-3-Ⓟ, a step which requires NAD^+ (see fig. 4-27), if there is no process enabling the reoxidation of NADH to NAD^+. We have already seen (see fig. 3-6) how this reoxidation takes place in aerobiosis, thanks to the electron carrier system and it should be remembered that for each molecule of NADH oxidized, 3 molecules of ATP are formed. But we must now examine what happens in anaerobiosis.

II. – POSSIBILITIES OF TRANSFORMATION OF PYRUVIC ACID AND REOXIDATION OF NADH IN ANAEROBIOSIS

When glycolysis proceeds in absence of oxygen and the electron carrier system does not therefore function, the cell must find other reactions to reoxidize NADH. We will study three possibilities (the first two are more frequently used than the third):

1. – Production of lactic acid

In animals, in muscle cells (where anaerobiosis conditions often prevail when oxygen utilization is more rapid than its input in cells) or in lactic bacteria (which live in anaerobiosis), pyruvic acid[1] is reduced to lactic acid while NADH is oxidized to NAD^+ (see fig. 4-30). This reaction is catalyzed reversibly by *lactate dehydrogenase*.

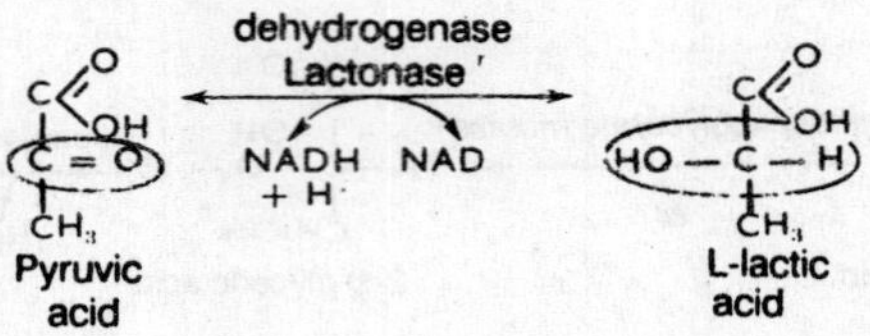

(*) L form in muscles, D- and L-forms in microorganisms.

FIG. 4-30. — Reduction of pyruvic acid to lactic acid.

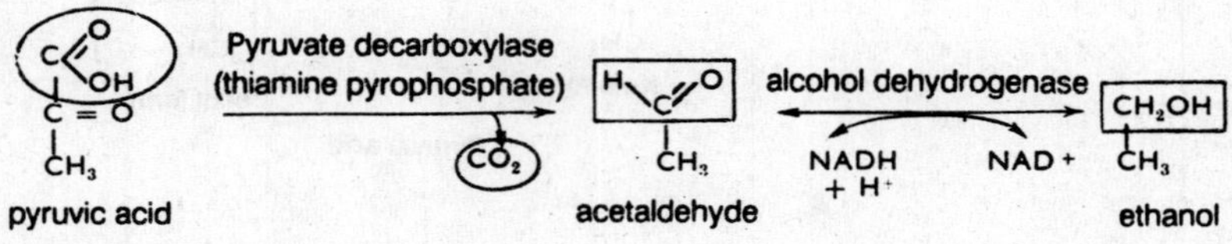

FIG. 4-31. — Transformation of pyruvic acid into ethanol.

---

(1) We are referring to pyruvic acid, lactic acid and, in connection with the tricarboxylic cycle we will refer to citric acid, etc., but it should be remembered that at the pH of the cells these acids exist mostly in the dissociated form (for this reason, some authors use the terms pyruvate, lactate, citrate, etc.).

2. — Alcoholic fermentation

In other microorganisms, especially yeast, pyruvic acid is, in a first step, decarboxylated to CO_2 + acetaldehyde, thanks to a *pyruvate decarboxylase* which has *thiamine pyrophosphate* (or vitamin B_1) as coenzyme and contains zinc (Zn^{2+}). Then in a second step, this acetaldehyde is reduced to ethanol while NADH is oxidized to NAD^+, the reaction being catalyzed by an *alcohol dehydrogenase* (see fig. 4-31).

3. — Formation of α-glycerophosphate and glycerol

The fermentation of glucose by yeast is always accompanied by the formation of small quantities of glycerol. As a matter of fact, yeast has a α-glycerophosphate dehydrogenase which catalyzes the reduction of dihydroxyacetone phosphate to L-α-glycerophosphoric acid[(1)], while NADH is oxidized to NAD^+ (fig. 4-32). Then a specific phosphatase can split the ester linkage and give rise to glycerol. This possibility seems to be used by yeast when there is not sufficient acetaldehyde enabling the reoxidation of NADH; in fact, this possibility has been utilized to produce glycerol by fermentation, by blocking acetaldehyde in a combination with disulphite and thus forcing yeast to reduce dihydroxyacetone phosphate for reoxidizing its NADH.

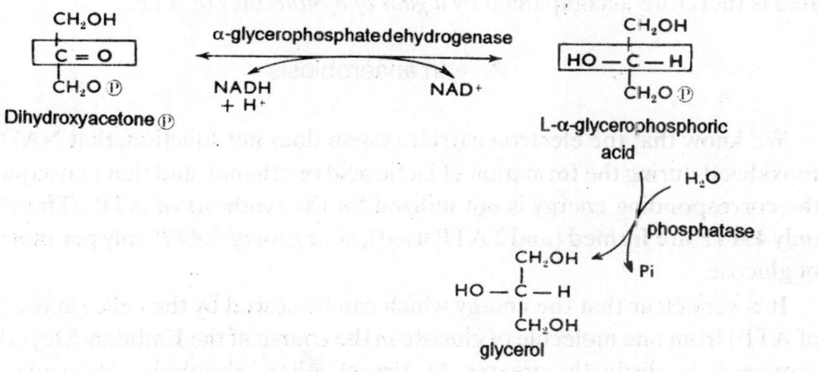

FIG. 4-32. — Formation of α-glycerophosphoric acid and glycerol from dihydroxyacetone Ⓟ.

(1) α-L-glycerophosphoric acid will be again referred to in Chapter 5, because it is the starting point of the synthesis of various lipids (see fig. 5-20).

III. – ENERGY BALANCE OF GLYCOLYSIS

While drawing this balance, one should bear in mind that one molecule of fructose-1, 6-di-Ⓟ gives 2 molecules of triose-Ⓟ and that thanks to triosephosphate-isomerase, these 2 molecules can continue to follow the pathway of glycolysis.

1. – In aerobiosis

	ATP used	ATP formed
Glucose ⟶ Glucose-6-Ⓟ	1	
Fructose-6-Ⓟ ⟶ Fructose 1, 6-bis-Ⓟ	1	
2 × (1,3-di-Ⓟ-glyceric acid ⟶ 3-Ⓟ glyceric acid)		2
2 × (phosphoenolpyruvic acid ⟶ pyruvic acid)		2
2 × (reoxidation of NADH by the electron carrier system)		6
Total	2	10

The transformation of one molecule of glucose into 2 molecules of pyruvic acid is therefore accompanied by *a gain of 8 molecules of ATP*.

2. – In anaerobiosis

We know that the electron carrier system does not function, that NADH is reoxidized during the formation of lactic acid or ethanol, and that consequently the corresponding energy is not utilized for the synthesis of ATP. Therefore, only 4 ATP are formed (and 2 ATP used), i.e. *a gain of 2 ATP* only per molecule of glucose.

It is very clear that the energy which can be stored by the cells (in the form of ATP) from one molecule of glucose in the course of the Embden-Meyerhof's sequence is distinctly greater (4 times) when glycolysis takes place in aerobiosis. We will see in the following that, when oxygen is available, oxidation can continue (through Krebs cycle) and that it is then accompanied by a considerable gain of ATP (see page 195).

Only some microorganisms can manage with anaerobic glycolysis to meet their energy requirements. In animals (and man) this process functions only for limited times, till oxygen is again available; the terminal products of anaerobic glycolysis – lactic acid and ethanol – are also comparatively toxic, and if their concentration exceeds a particular threshold value, they arrest the metabolism in the cells concerned. At the very beginning of this century,

Hopkins had observed that the muscle can contract (which requires energy) in anaerobiosis and that – in such conditions – lactic acid is produced in much larger quantities than if the muscle worked in presence of oxygen and is accumulated till the exhaustion of the muscle; he had further observed that if oxygen is supplied to this exhausted muscle, the lactic acid disappeared and the capacity of the muscle to contract was restored. We will see in the following how, in aerobiosis, lactic acid can be reconverted into glucose or glycogen.

IV. – METABOLISM OF OTHER MONOSACCHARIDES IN RELATION WITH GLYCOLYSIS

We will see briefly how other monosaccharides can join the glycolytic pathway and also how they can be formed from glucose by mechanisms called "conversion of monosaccharides".

We have already stated (page 170) that some organisms, especially yeast, possess a *hexokinase* capable of phosphorylating glucose, fructose and mannose to the corresponding hexose-6-Ⓟ. Fructose-6-Ⓟ is an intermediate of glycolysis. As for mannose-6-Ⓟ, we have seen (in the study of phosphohexose isomerase, (page 175) that it could, just as its epimer glucose-6-Ⓟ, be isomerized to fructose-6-Ⓟ.

Lastly, we will make a quick study of the principal reactions of the metabolism of *galactose*, because on the one hand, it represents a comparatively frequent source of carbohydrates in diet (it is produced by the hydrolysis of lactose or milk sugar), and on the other hand, it will provide an example of metabolic disorder.

We will recall that D-galactose is an epimer of D-glucose (it differs from it only by the configuration of the hydroxyl on carbon 4, see fig. 4-5); this will enable us to study the transformations of this monosaccharide without the necessity of writing the structures of all the compounds.

1) Galactose is first phosphorylated by ATP in presence of a *kinase*:

$$\alpha\text{-D-galactose} + \text{ATP} \xrightarrow{\text{galactokinase}} \alpha\text{-D-galactose-1-Ⓟ} + \text{ADP}$$

2) Then, *UDP-glucose: α-D-galactose-1-phosphate uridyl-transferase* (also called *phosphogalactose-uridyl-transferase*) catalyzes the following reaction:

$$\text{galactose -1-Ⓟ} + \text{UDP - glucose} \longleftrightarrow \text{UDP - galactose} + \text{glucose-1-Ⓟ}$$

As may be seen in figure 4-33, this reaction consists of an exchange between galactose-1-Ⓟ and the glucose-1-Ⓟ involved in UDP-glucose, the mode of formation of which was already described (see fig. 4-22).

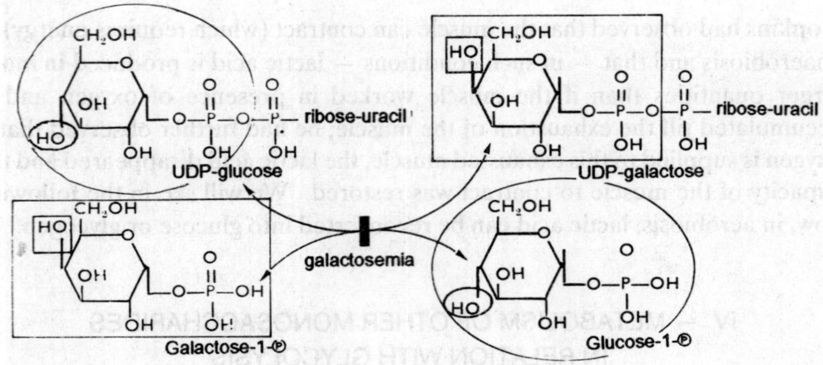

FIG. 4-33. — Reaction catalyzed by UDP-glucose: α-D-galactose-1-phosphate uridyl-transferase (absent in galactosemia).

This UDP-galactose is either incorporated in glycoproteins (see page 163) and various sphingoglycolipids like the cerebrosides (galactolipids) and the gangliosides (see page 219) or converted by a *4-epimerase* into UDP-glucose by a simple change of configuration at carbon 4 (and it then follows the metabolism of glucose).

$$\text{3) UDP-galactose} \xleftrightarrow{\text{UDP-galactose-4-epimerase}} \text{UDP-glucose}$$

But in infants suffering from a genetic disease called *galactosemia, UDP-glucose: α-D-galactose-1-phosphate uridyl-transferase is absent* and the new born is therefore incapable of metabolizing galactose-1-Ⓟ because reaction 2 cannot take place (we have denoted this by a bar across the arrow). Grave disorders appear (intestinal disorders, icterus, cataract, etc.) which can even lead to death if milk lactose is not immediately replaced by another sugar. This hereditary disease is due to an alteration of the gene of the transferase, which is therefore absent in these individuals.

The reversible conversion reaction of galactose into glucose with monosaccharides in conjugated form with a nucleotide (reaction no. 3 above) is a particular example of a very general mechanism. It is in the conjugated form that monosaccharides are transported, converted into one another, or incorporated into molecules of polysaccharides (see for example, the biosynthesis of glycogen, fig. 4-22).

The following are a few examples of these conjugated forms of monosaccharides (initials of monosaccharides have been explained in page 152) and the structure of nucleotides can be found in Chapter 6:

UDP-Glc (or UDPG), UDP-Gal, UDP-Glc NAc, UDP-Glc AU, GDP-Man, GDP-Fuc, CMP-NeuAc, ADP-Glc (or ADPG).

Some reactions of conversion of conjugated monosaccharides are described in the following diagram:

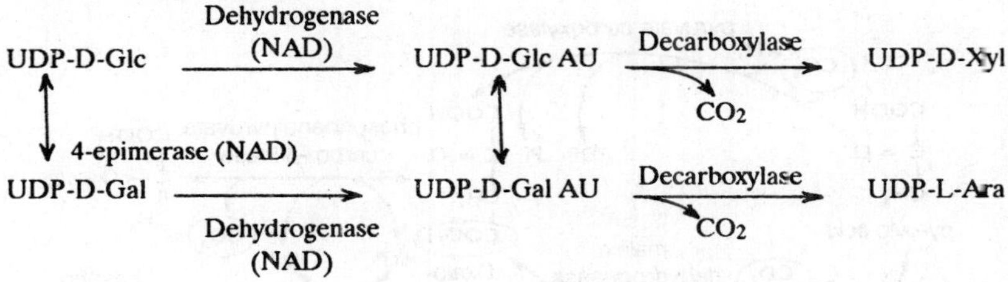

V. – REVERSIBILITY OF GLYCOLYSIS: NEOGLUCOGENESIS

It was observed that pyruvic acid and lactic acid formed in the muscle during glycolysis can be reconverted into glucose (or glycogen) when energy is available (in aerobiosis) and one could wonder whether this *neoglucogenesis* (or *gluconeogenesis*, or *neoglycogenesis*) follows – in the reverse direction – the glycolytic path. In fact, among the reactions leading from glucose to pyruvic acid (or lactic acid) there are 3 which, owing to the great variation of free energy taking place, are not easily reversible in physiological conditions; these 3 reactions are catalyzed by kinases:

– transformation of phospho-enol-pyruvic acid into pyruvic acid (pyruvate kinase);

– phosphorylation of fructose-6-Ⓟ to fructose 1, 6-bis-Ⓟ (phosphofructokinase);

– phosphorylation of glucose to glucose-6-Ⓟ (glucokinase or hexokinase)

Neoglucogenesis can therefore take place by reverse glycolytic reactions except at these 3 steps where different reactions (catalyzed by different enzymes or enzymatic systems) will have to be used.

1. – Conversion of pyruvic acid into phospho-enol-pyruvic acid

The phosphorylation of pyruvic acid by ATP in presence of pyruvate-kinase has been reported, but its contribution is meagre. Phospho-enol-pyruvic acid is generally formed in two steps: first, the carboxylation of pyruvic acid to oxaloacetic acid and then the phosphorylating decarboxylation of the latter to phospho-enol-pyruvic acid.

A. – Carboxylation of pyruvic acid into oxaloacetic acid

Two pathways are possible (fig. 4-34):

a) a direct path, catalyzed by *pyruvate-carboxylase* (whose coenzyme is biotin) by a reaction mechanism similar to that of acetyl-coenzyme A-carboxylase (described in figure 5-16);

b) an indirect path, implying the intermediate formation of malic acid followed by oxidation to oxaloacetic acid.

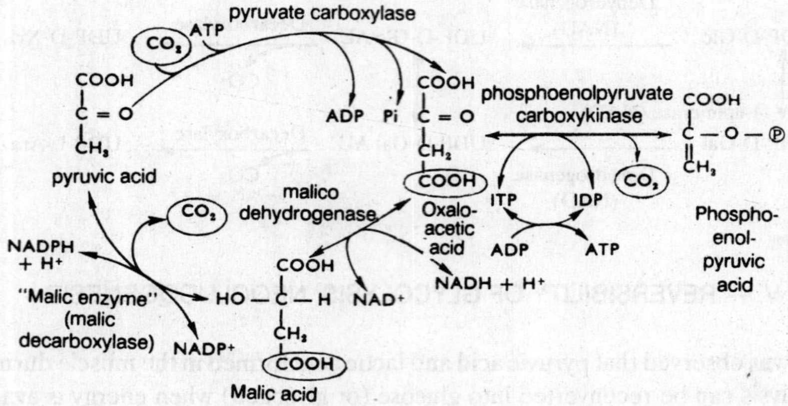

FIG. 4-34. — Formation of phospho-enol-pyruvic acid from pyruvic acid.

B. — Phosphorylating decarboxylation of oxaloacetic acid into phospho-enol-pyruvic acid

It is brought about by *phospho-enol-pyruvate-carboxykinase*, in presence of inosine-triphosphate (ITP) or guanosine triphosphate (GTP)(1). The IDP (or GDP) formed will be rephosphorylated to ITP (or GTP) by ATP (fig. 4-34).

The sum of these reactions can be written as follows (neglecting the trans-hydrogenation which takes place while passing through malic acid):

$$\text{pyruvic acid} + 2\text{ATP} \longrightarrow \text{phospho-enol-pyruvic acid} + 2\text{ADP} + P_i$$

It must be stated right now that compounds like the fatty acids (and also about ten amino acids precisely called glucoformers) which can be transformed into pyruvic acid or oxaloacetic acid (through the Krebs cycle, for example) can contribute to the phenomenon of neoglucogenesis.

2. — Conversion of fructose-1, 6-bis-℗ into fructose-6-℗

This hydrolysis is catalyzed by fructose-1, 6-bis-phosphatase. Regulation phenomena are observed at this stage; they will be referred to in the following paragraph.

3. — Conversion of glucose-6-℗ into glucose

It has already been stated (see page 171) that this hydrolysis is catalyzed by *glucose-6-phosphatase*.

---

(1) The structures of IDP, ITP, GDP and GTP will be studied in the chapter on nucleic acids (see fig. 6-7).

The fact that the pathway of glycolysis and that of neoglucogenesis differ at three stages offers the advantage (already explained in connection with the synthesis and degradation of glycogen) of a different regulation of the two processes; we will now consider an aspect of this regulation; in this study we may refer to figure 4-35 and consult the recapitulative diagram representing the reactions of anaerobic glycolysis and some reactions directly connected with it.

VI. INHIBITION OF GLYCOLYSIS BY OXYGEN – PASTEUR EFFECT

Some controls, particularly hormonal, are carried out on glycolysis. We will examine here only the effect of *oxygen* known as the Pasteur effect because Pasteur was the first scientist to observe that in yeast, glucose was less rapidly consumed in aerobiosis than in anaerobiosis. In other words, the introduction of oxygen brings about a decrease of the rate of glycolysis. The advantage of this phenomenon is obvious when one examines the energy balances as was just done above, and when one knows the high energy balance of aerobic oxidation to CO_2 + H_2O: since in aerobiosis a much greater number of molecules of ATP are formed per molecule of glucose, the cells need use less glucose to meet their energy requirements. While the advantages of the Pasteur effect are obvious, the mechanisms involved are far from clear. We will now review some of the existing hypotheses.

1) First of all, it is possible – in some cases at least – that the phenomenon is more apparent than real, because in presence of oxygen, a very large proportion of lactic acid formed can be reconverted into glucose (as mentioned above), which decreases consumption of glucose.

2) An *allosteric* regulation takes place at the level of *phosphofructokinase*. This enzyme is *activated by AMP and P_i* (available in anaerobiosis) and *inhibited by ATP* (formed in aerobiosis at the cost of ADP and P_i). If one considers that the ultimate objective of glycolysis followed by the Krebs cycle is the production of ATP, and that phosphofructokinase is the first enzyme of this metabolic pathway (which is justified because the previous enzyme, phosphohexo-isomerase does not act solely in glycolysis), it may be said that inhibition of phosphofructokinase by ATP is a case of feedback inhibition, where the first enzyme of the metabolic pathway is inhibited by the final product. In reality, the phenomena are even more complex, because one must take into consideration the inhibiting effect – on phosphofructokinase – of citric acid which can also be considered as a final product of glycolysis, because – as will be seen in the following – it is formed at the beginning of the Krebs cycle.

Furthermore, fructose-1,6-bisphosphatase (enzyme acting like phosphofructokinase, but in the direction of neoglucogenesis) is on the contrary inhibited by AMP. Thus, in anaerobiosis, the increase in AMP concentration has an activating effect on glycolysis and an inhibiting effect on neoglucogenesis whereas in aerobiosis the increase in ATP concentration inhibits glycolysis and favours neoglucogenesis.

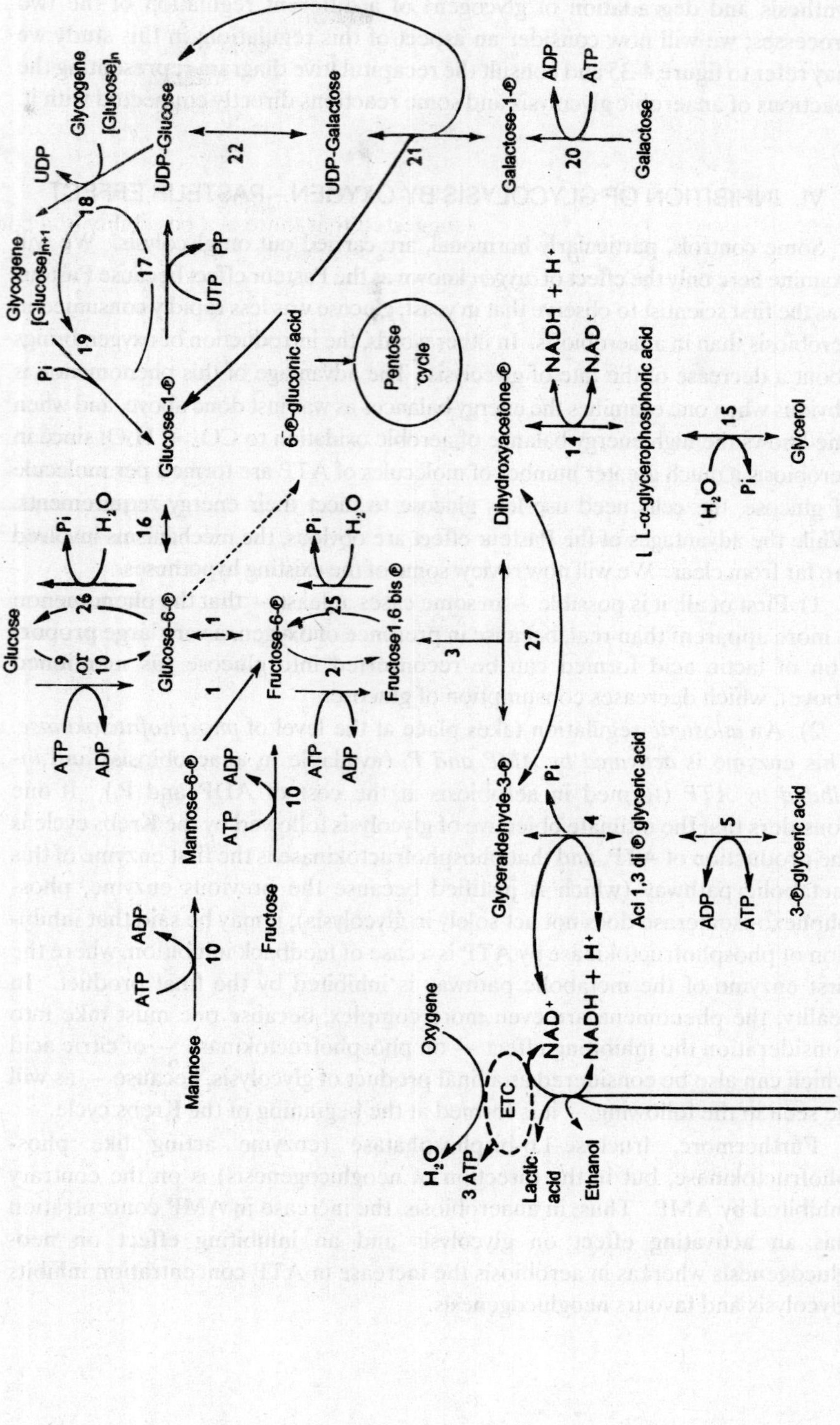
Glycogene
[Glucose]n+1
Glycogene
[Glucose]n
UDP
UDP-Glucose
UDP-Galactose
Galactose-1-℗
Galactose
ADP
ATP
Glucose
Glucose-6-℗
Glucose1-℗
6-℗-gluconic acid
Pentose cycle
UTP
PP
Pi
H2O
Mannose
Mannose-6-℗
Fructose
Fructose-6-℗
Fructose1,6 bis ℗
Dihydroxyacetone-℗
Glyceraldehyde-3-℗
L-α-glycerophosphoric acid
Glycerol
Acl 1,3 di ℗ glyceric acid
3-℗-glyceric acid
NADH + H+
NAD+
Oxygene
ETC
H2O
3 ATP
Lactic acid
Ethanol

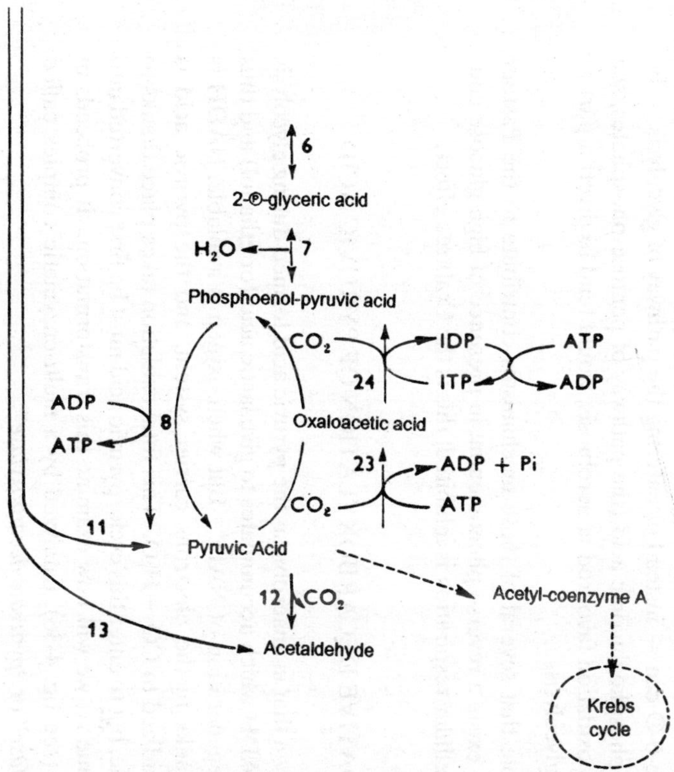

ENZYMES

1. Phosphohexose isomerase.
2. Phosphofructokinase
3. Aldolase.
4. Glyceraldehyde 3-Ⓟ-dehydrogenase.
5. Phosphoglycerate kinase.
6. Phosphoglycerate mutase.
7. Enolase.
8. Pyruvate kinase.
9. Glucokinase.
10. Hexokinase.
11. Lacticodehydrogenase .
12. Pyruvate decarboxylase.
13. Alcoholdehydrogenase.
14. α-glycerophosphate dehydrogenase
15. Phosphatase.
16. Phosphogulcomuatase.
17. UDP-glucose-pyrophosphorylase.
18. Glycogen-synthetase.
19. Phosphorylase a.
20. Galactokinase.
21. Phosphogalactose-uridyl-transferase.
22. UDP-galactose-4-epimerase.
23. Pyruvate-carboxylase.
24. Phospho-enol-pyruvate-carboxykinase.
25. Fructose-1,6-bisphosphatase.
26. Glucose-6-phosphatase.
27. Triosephosphate-isomerase.

FIG. 4-35. — Recapitulative diagram of the reactions of anaerobic glycolysis and some directly associated reactions. The dotted lines (ETC-electron transport chain, oxidative decarboxylation to acetyl-coenzyme A, Krebs cycle) indicate that these reactions take place only in aerobiosis.

It appears that the regulation of enzymatic activity, especially in the case of phosphofructokinase, is the most important factor of the control of glycolysis. However, in order to show the complexity of this problem, we will mention 3 more hypotheses which have been advanced.

3) In aerobiosis, ADP and P_i are utilized to form ATP during the oxidation of pyruvate to CO_2 + H_2O. But, these three reactions of glycolysis – those catalyzed by glyceraldehyde-3-Ⓟ-dehydrogenase, phosphoglycerate kinase and pyruvate kinase – have precisely P_i or ADP as substrate. It was therefore suggested that there is a possibility of a competition, for ADP and P_i between respiration (i.e. the mitochondrial oxidations-reductions) and glycolysis, although these 2 processes take place in different cellular compartments.

4) Some enzymes of glycolysis, particularly glyceraldehyde-3-Ⓟ- dehydrogenase, the acting mechanism of which has been explained in some more details, contain –SH groups required for catalytic activity. An oxidation of these groups in aerobiosis would cause an inhibition of glycolysis.

5) Glucose-6-Ⓟ can – instead of entering the pathway of glycolysis – be oxidized to 6-phosphogluconic acid (the pathway of pentose-phosphates, see fig. 4-39); this oxidation favoured in aerobiosis, would tend to divert a part of glucose from glycolysis.

It is possible that several of these mechanisms contribute to the Pasteur effect. There exists a reverse phenomenon: in presence of high glucose concentrations, cellular respiration is inhibited: this is the Crabtree effect.

OXIDATIVE DECARBOXYLATION OF PYRUVIC ACID

We have seen that in anaerobiosis the pyruvic acid formed during glycolysis reacts with NADH which accumulates to give lactic acid (or ethanol) and thus permits the regeneration of NAD^+. But when oxygen is available, NADH is reoxidized thanks to the electrons carrier system, and the pyruvic acid is completely oxidized to $CO_2 + H_2O$. This total oxidation takes place thanks to the Krebs cycle, but to enter this cycle, pyruvic acid must be first converted into acetyl-coenzyme A; we will now examine this transformation. It proceeds in several steps (see fig. 4-36), catalyzed by a multi-enzymatic complex called "*pyruvate oxidase*" or "*pyruvate decarboxylase*".

I. – DECARBOXYLATION REACTION

The coenzyme implied in this first step is *thiamine pyrophosphate* on which pyruvic acid attaches itself to be decarboxylated to acetaldehyde.

$$CH_3-CO-COOH + E_1-TPP \longrightarrow CO_2 + E_1-TPP-CH_3CHO$$

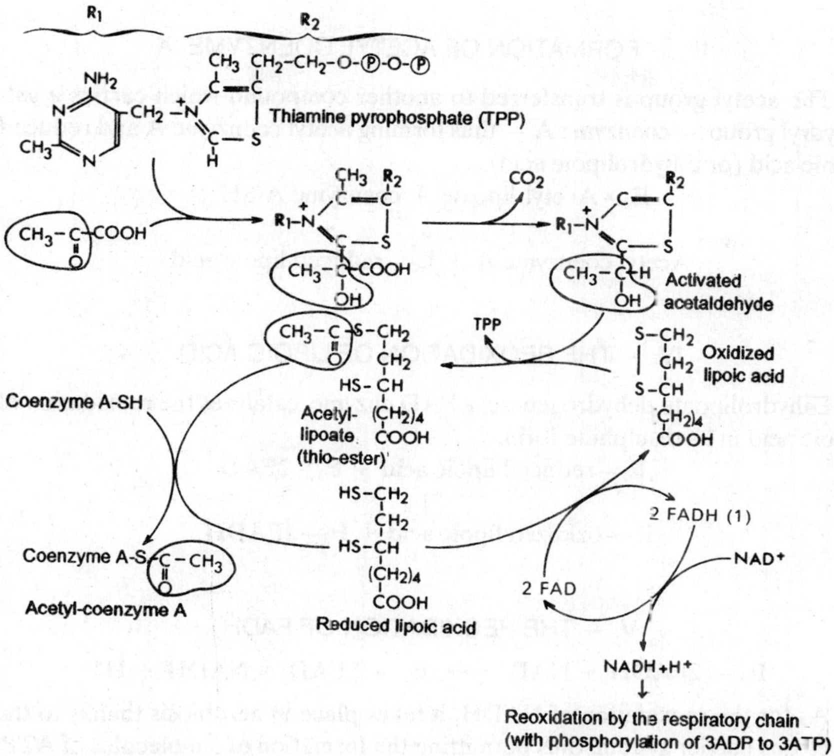

FIG. 4-36. — Oxidative decarboxylation of pyruvic acid, catalyzed by a 3-unit multi-enzymatic system: a decarboxylase linked with thiamine pyrophosphate, an aldehyde dehydrogenase linked with lipoic acid (or thioctic acid) and a dihydrolipoate dehydrogenase which is a FAD enzyme enabling the reoxidation of dihydrolipoic acid to lipoic acid.
It is observed that 5 coenzymes are necessary for this process: TPP, lipoic acid, coenzyme A, FAD and NAD.

II. — OXIDATION REACTION

The acetaldehyde thus activated reacts with lipoic acid, coenzyme in oxidized form (or disulphide): acetaldehyde is oxidized and linked in the form of a thio-ester to lipoic acid which is reduced (as a matter of fact, the thio-ester formed is the derivative of a di-thiol)

$$E_1-TPP-CH_3CHO + E_2-\text{Oxidized lipoic acid}$$
$$\downarrow$$
$$E_1-TPP + E_2-\text{Acetyl lipoate}$$

(1) Two molecules of FAD (linked to the enzyme) are partially reduced to FADH, with a E'_0 value of about – 0.36 volt, which permits the transfer of electrons to the system $NAD^+/NADH$ (see page 119).

III. – FORMATION OF ACETYL COENZYME A

The acetyl group is transferred to another compound which carries a sulphydryl group – coenzyme A – thus forming acetyl coenzyme A and reduced lipoic acid (or dihydrolipoic acid).

$$E_2\text{–Acetyl-lipoate} + \text{coenzyme A-SH}$$
$$\downarrow$$
$$\text{Acetyl coenzyme A} + E_2\text{–reduced lipoic acid}$$

IV. – THE REOXIDATION OF LIPOIC ACID

Dihydrolipoate dehydrogenase, a FAD enzyme, catalyzes the reoxidation of lipoic acid in its disulphide form.

$$E_2\text{–reduced lipoic acid} + E_3\text{–2FAD}$$
$$\downarrow$$
$$E_2\text{–oxidized lipoic acid} + E_3\text{–2FADH}$$

V. – THE REOXIDATION OF FADH

$$E_3 - 2\text{FADH} + \text{NAD}^+ \longrightarrow E_3 - 2\,\text{FAD} + \text{NADH} + \text{H}^+$$

As for the reoxidation of NADH, it takes place in aerobiosis thanks to the electrons carrier system, thus permitting the formation of 3 molecules of ATP.

The overall reaction of the oxidative decarboxylation of pyruvic acid can be written as in figure 4-37.

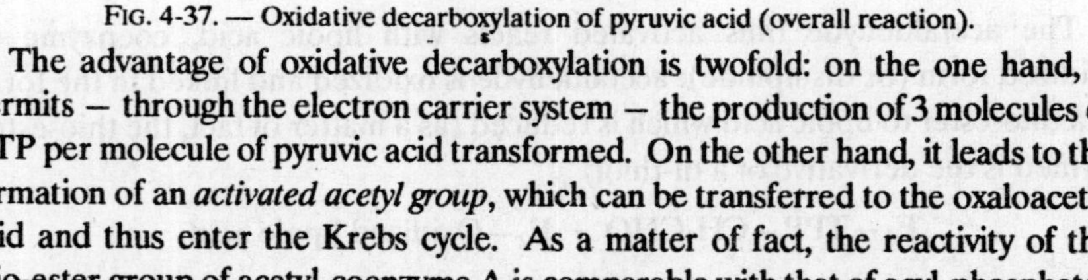

FIG. 4-37. — Oxidative decarboxylation of pyruvic acid (overall reaction).

The advantage of oxidative decarboxylation is twofold: on the one hand, it permits – through the electron carrier system – the production of 3 molecules of ATP per molecule of pyruvic acid transformed. On the other hand, it leads to the formation of an *activated acetyl group*, which can be transferred to the oxaloacetic acid and thus enter the Krebs cycle. As a matter of fact, the reactivity of the thio-ester group of acetyl-coenzyme A is comparable with that of acyl-phosphates (ΔG_0 of the hydrolysis of acetyl-coenzyme A $\neq$ –9 kcal/mole see table on page 111). The sulphydryl group of the molecule of coenzyme A (its complete structure is represented in figure 2-18) constitutes the reactive part of the molecule and is implied in the "acetyl" transfer reactions; for this reason, one generally writes: "coenzyme A-*SH*" or "C_oA-*SH*". The letter "A" indicates that it is a coenzyme acting in the transfer reactions of the Acetyl group. Having thus designated this coenzyme, it was also found that it acts not only in the metabolism of acetic acid but also in that of fatty acids in general (see Chapter 5).

THE KREBS CYCLE

Krebs made an outstanding contribution to the discovery of the operating mechanisms of this cycle which is also called *tricarboxylic cycle* or *citric cycle*, owing to the participation of various tricarboxylic acids, including citric acid.

Before studying this cycle – which will take up acetyl coenzyme A, whose formation was explained above – it would be useful to place it correctly in the context of the entire cellular metabolism. Since, acetyl coenzyme A is formed by oxidative decarboxylation of pyruvic acid which itself results from glucose by a series of transformations (glycolysis), the Krebs cycle – which permits the complete oxidation of glucose – can logically be included in the study of the metabolism of carbohydrates. But one should bear in mind that the *Krebs cycle is situated at the crossroads of metabolisms of carbohydrates, lipids and proteins* and that it will achieve the complete oxidation to CO_2 + H_2O, of not only *glucose*, but also *fatty acids* whose catabolism leads to the formation of acetyl coenzyme A, and numerous amino acids whose metabolism leads either to the formation of pyruvic acid (and therefore of acetyl coenzyme A), or directly to the formation of one of the intermediates of the Krebs cycle (α-ketoglutaric, succinic, fumaric or oxaloacetic acid).

The steps of Krebs cycle are represented in fig. 4-38; we will now give a few details on the reactions which take place in this cycle.

I. – REACTIONS OF THE KREBS CYCLE

In a first step, catalyzed by a *condensing enzyme*, or *citrate-synthetase*, the acetyl group is transferred from the acetyl coenzyme A to oxaloacetic acid to give citric acid, and coenzyme A is liberated. Then, under the effect of one and the same enzyme, *aconitase*, citric acid is dehydrated to *cis*-aconitic acid (which is perhaps not liberated and would remain in that case linked to the enzyme), and rehydrated to isocitric acid; these three compounds coexist in equilibrium.

Then proceeds the oxidation of isocitric acid (with probable intermediate formation of oxalosuccinic acid), catalyzed by *isocitrate dehydrogenase* whose coenzyme is either NAD^+ or $NADP^+$, depending on the organism studied. The same enzyme catalyzes the decarboxylation of oxalosuccinic acid to α-ketoglutaric acid.

Now, α-ketoglutaric acid undergoes an *oxidative decarboxylation* under the effect of an enzymatic complex (called α-keto-glutarate dehydrogenase), having as coenzymes, thiamine pyrophosphate, lipoic acid, coenzyme A, FAD and NAD^+; the mechanisms are similar to those we studied in connection with the oxidative decarboxylation of pyruvic acid to acetyl coenzyme A (see fig. 4-36).

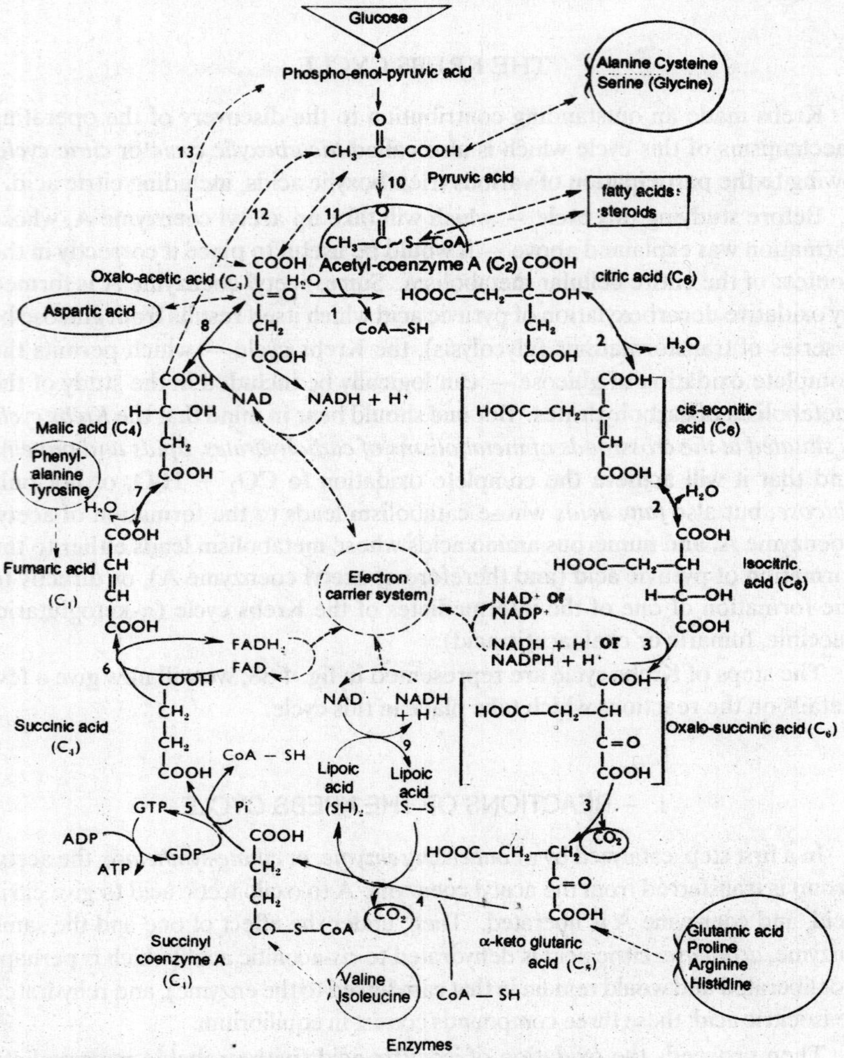

Enzymes

1. Condensing enzyme = citrate synthetase
2. Aconitase.
3. isocitrico dehydrogenase.
4. « α-ketoglutarate oxydase ».
5. Succinate thiokinase.
6. Succinate dehydrogenase
7. Fumarase.
8. Malico-dehydrogenase
9. Lipoic acid dehydrogenase
10. « Pyruvate oxydase. »
11. Pyruvate carboxylase.
12. "malic enzyme"
13. Phosphoenolpyruvate-carboxykinase.

FIG. 4-38. — Reactions of the Krebs cycle and (in dotted line) some associated reactions. Relations with carbohydrates ∇, lipids □ and proteins O.

Succinyl coenzyme A is therefore obtained; it is very quickly split into succinic acid and coenzyme A, while the energy contained in the thio ester molecule is used for the synthesis of a molecule of ATP under the effect of *succinate thiokinase*:

$$\text{Succinyl-CoA} + P_i + \text{GDP}^{(1)} \rightleftarrows \text{Succinic acid} + \text{GTP}^{(1)} + \text{CoA-SH}$$

$$\text{GTP} + \text{ADP} \rightleftarrows \text{GDP} + \text{ATP}$$

Succinic acid is oxidized to fumaric acid, thanks to *succinate dehydrogenase*, the coenzyme of which is FAD. $FADH_2$, will pass on its electrons to the electrons transport system but, as seen in Chapter 3, the entry of electrons at this stage takes place beyond the first site of formation of ATP (which is, so to say, "skipped"), so that contrary to what happens with NAD^+ or $NADP^+$ dehydrogenases only 2 molecules (and not 3) of ATP will be formed per molecule of $FADH_2$ reoxidized.

Fumaric acid is hydrated, thanks to *fumarase*, to malic acid and the latter is oxidized by the effect of *malate dehydrogenase*, a NAD^+ enzyme, to oxaloacetic acid, thus closing the cycle. The oxaloacetic acid formed can again bind a new acetyl group which, in its turn, will be oxidized to $CO_2 + H_2O$ after one turn of the cycle and so on.

The CO_2 formed in the Krebs cycle can be used in various reactions (carboxylations, synthesis of urea, biosynthesis of nucleotides) or eliminated by the lungs after transport by the blood stream (see pages 55-56).

II. – ENERGY BALANCE

To calculate the number of molecules of ATP formed per molecule of glucose, we must envisage, not one turn of the cycle (acetyl – CoA $\longrightarrow$ $2CO_2$), but two turns, because 1 molecule of glucose gives 2 molecules of pyruvic acid (and therefore 2 molecules of acetyl-CoA).

	ATP formed
3 oxidations by NAD^+ or $NADP^+$ enzymes, which thanks to the electron transport system, will each permit the formation of 3 ATP (isocitrate-dehydrogenase, α-ketoglutorate-dehydrogenase, malate-dehydrogenase	9
1 oxidation (succinate-dehydrogenase) by a FAD enzyme, which, thanks to the electron transport system, will permit the formation of 2 ATP	2
Cleavage of succinyl-CoA	1
Total for one turn of the cycle	12

We can now draw the balance for all the reactions of glucose oxidation in aerobiosis:

1 glucose $\longrightarrow$ 2 pyruvic acid (glycolysis) 10—2 =	8
2 pyruvic acid $\longrightarrow$ 2 acetyl—CoA + 2 CO_2 (oxidative decarboxylation)	6
2 acetyl—CoA $\longrightarrow$ 4 CO_2 (2 turns of Krebs cycle)	24
Total for 1 glucose $\longrightarrow$ 6 CO_2	38

(1) The structures of GDP and GTP will be studied in the chapter on nucleic acids (see fig. 6-7).

This total of 38 ATP is considerable compared to the mere 2 ATP, which can be obtained in anaerobiosis (see page 182). It must be indicated that the enzymes in Krebs cycle are mainly located in the mitochondria just as the enzymes involved in electron transport and oxidative phosphorylation, which most probably enhances the efficiency of coupling between the oxidation of acetyl coenzyme A and production of ATP.

However, whatever its importance, the energetic aspect of glucose oxidation must not overshadow the fact that glucose is not only a source of energy but also a *source of carbon* (glucose is often used as the sole source of carbon for growing bacteria). In other words, from the products of glucose catabolism, the cells can synthesize a very wide variety of compounds: for example (as will be seen in the chapter on lipids), fatty acids and steroids are synthesized from acetyl coenzyme A, and some amino acids (like alanine, glutamic acid, aspartic acid) are formed from the corresponding keto-acids (pyruvic, α-ketoglutaric and oxaloacetic acids respectively).

III. – FORMATION AND DECARBOXYLATION OF OXALOACETIC ACID

If some intermediates of the Krebs cycle leave the latter because they are involved in other metabolic pathways, the quantity of oxaloacetic acid will decrease and the capacity of the cycle to accept acetyl coenzyme A will therefore decrease. To restore the *oxidation capacity* of the cycle, *new molecules of oxaloacetic acid must be synthesized*. This can take place in several ways:

1) *by deamination or transamination* of either aspartic acid (which gives directly oxaloacetic acid) or glutamic acid (giving α-ketoglutaric acid which will be converted to oxaloacetic acid by the reactions of the cycle). This is a possible solution, provided that the same reactions in the reverse direction (in other words the necessity for the cells to form these amino acids) are not precisely the cause of the decrease in the quantity of oxaloacetic acid available for the oxidation of the acetyl coenzyme A.

2) *by carboxylation reactions* of various acids having 3 carbon atoms. We have already studied these CO_2 fixation reactions leading to the formation of oxaloacetic acid, while discussing neoglucogenesis (see fig. 3-34); we will only briefly mention them here:

– carboxylation of pyruvic acid to oxaloacetic acid catalyzed by pyruvate carboxylase

– carboxylation of pyruvic acid to malic acid, catalyzed by "malic enzyme", and oxidation of malic acid to oxaloacetic acid, catalyzed by malate dehydrogenase

– carboxylation of phosphoenolpyruvic acid to oxaloacetic acid by phosphoenolpyruvate carboxykinase.

The latter reaction, considered in the direction oxaloacetic acid → phosphoenolpyruvic acid, which was stated to allow the "reversibility" of glycolysis after the pyruvic acid is, in a first step, converted to oxaloacetic acid (see figs. 3-34 and 4-35), *therefore also allows the intermediates of the Krebs cycle to be converted into glucose*. The cycle cannot indeed turn in the opposite direction although most of its reactions are reversible (because the oxidative decarboxylation of α-ketoglutaric acid is not reversible), so that the main pathway through which the intermediates of the cycle (or compounds which can give these intermediates, like some glucoformer amino acids) can "return" to glucose, precisely passes through this decarboxylation reaction of oxaloacetic acid into phosphoenolpyruvic acid.

PENTOSE PHOSPHATES CYCLE

This series of reactions enables the cells to metabolize glucose-6-Ⓟ without using the glycolytic pathways; it is therefore also called "*hexosemonophosphate shunt*", or "*phosphogluconate pathway*" (after the name of one of the intermediates of the cycle), or "Dickens-Horecker pathway" (in view of the contribution of these two authors to the study of this cycle).

This pathway distinguishes itself from glycolysis by two main characteristics: the *oxidation does not require any ATP*, and it is *mainly an aerobic process* because there is no specific mechanism (other than the electron transport chain) permitting the reoxidation of the coenzymes reduced during the oxidation reactions; however some reactions (especially during the biosynthesis of fatty acids or cholesterol, see figs. 5-19 and 5-22) use NADPH and therefore generate $NADP^+$.

It may be said that the advantage of this pathway is fourfold:

— it provides pentose phosphates which are indispensable for the synthesis of nucleotides found in various coenzymes and in nucleic acids;

— it permits the formation of NADPH required for various reactions particularly for the synthesis of fatty acids and steroids;

— it can perform the oxidation of glucose into CO_2, with formation of ATP in appreciable quantity, if transhydrogenation (from NADPH to NAD^+) can take place and provided the NADPH is not used to carry out reactions requiring this reduced coenzyme;

— in photosynthetic organisms, it permits the synthesis of carbohydrates (and therefore of other organic compounds) from CO_2.

I. — OXIDATION OF GLUCOSE-6-Ⓟ INTO RIBULOSE-5-Ⓟ

The reactions permitting this transformation are represented in figure 4-39. *Glucose-6-Ⓟ-dehydrogenase*, a $NADP^+$ enzyme (discovered by Warburg) catalyzes the oxidation of glucose-6-Ⓟ to 6-Ⓟ-gluconic acid. It appears that there is intermediate formation of 6-phosphoglucono-δ-lactone, which hydrolyzes itself either spontaneously or by the effect of a lactonase.

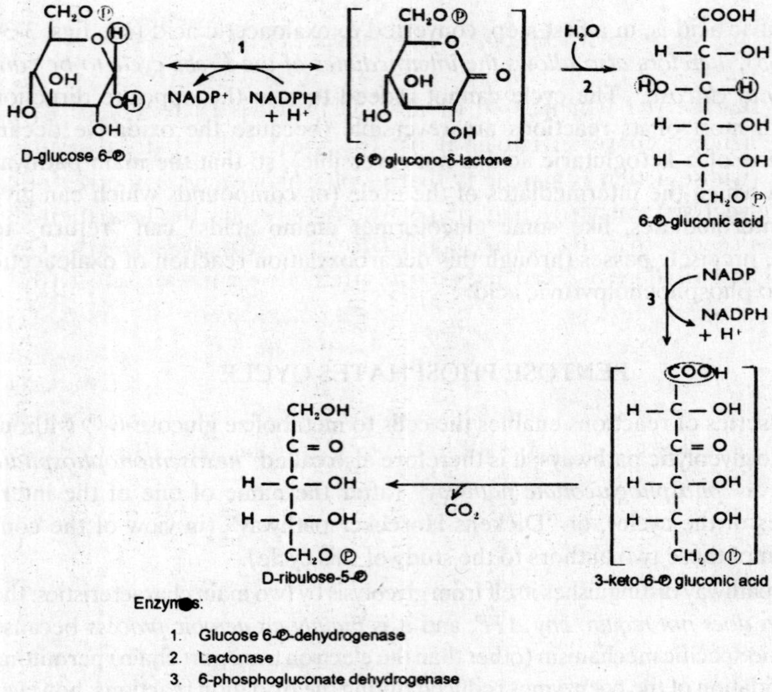

FIG. 4-39. — Oxidation of glucose-6-Ⓟ to ribulose-5-Ⓟ.

Then, a second oxidation takes place under the effect of *6-Ⓟ-gluconate dehydrogenase*, also a $NADP^+$ enzyme, with possible formation of an intermediate (3-keto-6-Ⓟ-gluconic acid) which is decarboxylated to D-ribulose-5-Ⓟ.

For each carbon atom leaving in the form of CO_2 there are therefore two oxidations both accompanied by a formation of NADPH, the electrons of which can be transferred to NAD^+, and from there to the electron transport chain; 3 ATP will be formed per pair of electrons, i.e. *6 ATP*. Since the cycle starts from glucose-6-Ⓟ, 1 ATP (required for the phosphorylation of glucose) must be deducted in order to draw the balance starting from glucose; the oxidation to ribulose-5-Ⓟ is then represented by the departure of a CO_2 and a *net gain of 5 ATP*.

Besides this energetic advantage of NADPH, it should be stressed that this *reduced coenzyme is necessary for some important reactions*, particularly for the biosynthesis of fatty acids (see fig. 5-19), the synthesis of sterols (fig. 5-22), the reduction of dihydrofolic acid to tetrahydrofolic acid, the hydroxylation of a wide variety of compounds (for example, the hydroxylation of tyrosine and various sterols), the carboxylation of pyruvic acid to malic acid (fig. 4-34), etc. The two oxidations of the cycle of pentose-phosphates which generate NADPH, therefore provide the reduced coenzyme indispensable to other important reactions of cellular metabolism.

II. – ISOMERIZATION OF RIBULOSE-5-℗

As may be observed in figure 4-40, ribulose-5-℗ which is a ketopentose, can, on the one hand, be converted into an aldose, the ribose-5-℗, by *phosphoribose isomerase* (whose action is similar to that of phosphohexose isomerase observed in glycolysis) thus providing the pentose required for the formation of nucleotides needed to *synthesize nucleotide coenzymes and nucleic acids*. On the other hand, ribulose-5-℗ can also be converted into an epimer ketose (the change in configuration is relative to carbon 3), *xylulose-5-℗*, by *phosphoketopentose epimerase*. These two reactions are important because they represent the junction point between the first part of the cycle (oxidation of glucose-6-℗ to ribulose-5-℗) and the second part (reversible interconversion of pentose-℗ into hexose-℗ by transaldolization and transketolization) as may be observed in the diagram of figure 4-43.

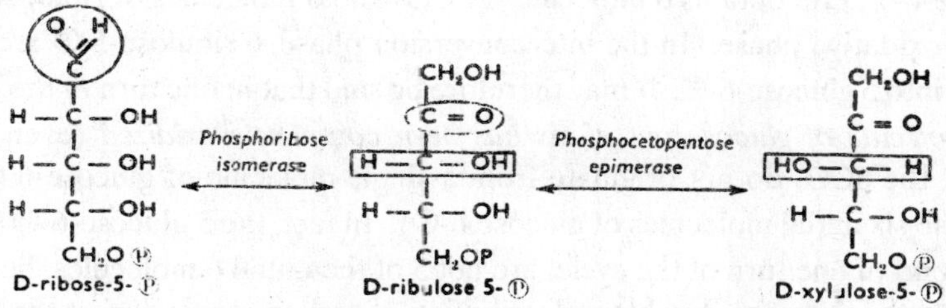

FIG. 4-40. — Isomerization of ribulose-5-℗ to ribose-5-℗ or xylulose-5-℗.

III. – INTERCONVERSIONS OF PENTOSE-℗ AND HEXOSE-℗ BY TRANSALDOLIZATION AND TRANSKETOLIZATION

Transaldolization is a transfer of a group of 3 carbon atoms (dihydroxyacetone) from a donor which is a ketose phosphate (fructose-6-℗, sedoheptulose-7-℗) to an acceptor which is an aldose phosphate (glyceraldehyde-3-℗, erythrose-4-℗). This transfer takes place thanks to the formation of an intermediate combination (enzyme-substrate) between the εNH_2 group of a molecule of lysine of the enzyme and the C=O of the dihydroxyacetone group. The principle of transaldolization together with an example, are represented in figure 4-41.

Transketolization consists of a transfer of a group of 2 carbon atoms (ketol or glycolaldehyde) from a donor which is a ketose phosphate (xylulose-5-℗, fructose-6-℗, sedoheptulose-7-℗) to an acceptor which is an aldose phosphate (glyceraldehyde-3-℗, erythrose-4-℗, ribose-5-℗). Transketolase has the coenzyme *thiamine pyrophosphate*, on which is fixed the glycoaldehyde group, thus forming an intermediate combination similar to the one described in our study of oxidative decarboxylation of pyruvic acid (see fig. 4-36). The principle of transketolization, together with two examples, are represented in figure 4-42.

As shown by figure 4-43, the cycle of pentose-phosphates proceeds by a set of reactions including two transketolizations, a transaldolization (these three reactions are represented in detail in figures 4-41 and 4-42) and an aldolization followed by the action of fructose-1, 6-bisphosphatase (two reactions which we studied in connection with glycolysis).

IV. – ENERGY BALANCE

We have seen that in the oxidative part of the cycle, one molecule of glucose-6-Ⓟ yielded one molecule of CO_2. Starting from 6 molecules of glucose-6-Ⓟ, one obtains 6 molecules of CO_2 and 6 molecules of ribulose-5-Ⓟ in this oxidative phase. In the interconversion phase, 6 ribulose-5-Ⓟ are converted into 5 glucose-6-Ⓟ. It may therefore be said that in one turn of the cycle, *one molecule of glucose out of six has been completely oxidized* (even if, in reality, the $6CO_2$ do not originate from a single molecule of glucose-6-Ⓟ but from the six initial molecules of glucose-6-Ⓟ). In fact, the 5 glucose-6-Ⓟ found at the end of one turn of the cycle, are not 5 of the initial 6 molecules, because there has been a considerable redistribution of carbon atoms during the transaldolization and transketolization reactions. But on the basis of the set of reactions (represented in figure 4-43), the balance of one turn of the cycle may be written as follows:

$$6 \text{ glucose-6-Ⓟ} + 12\,NADP^+ \longrightarrow 5 \text{ glucose-6-Ⓟ} + 6\,CO_2 + P_i + 12\,NADPH + 12\,H^+$$

or simplifying,

$$1 \text{ glucose-6-Ⓟ} + 12\,NADP^+ \longrightarrow 6\,CO_2 + P_i + 12\,(NADPH + H^+)$$

It is observed that the pentose-phosphates cycle yields, for each molecule of glucose oxidized, 12 molecules of NADPH which can be either used in reactions requiring the reduced form of this coenzyme, or enter (if transhydrogenation from NADPH to NAD^+ takes place) in the electron transport chain and then enable the formation of $3 \times 12 = 36$ molecules of ATP. To refer this energy balance to glucose, one must deduct 1 ATP required for the phosphorylation of glucose to glucose-6-Ⓟ, which gives a gain of 35 ATP (a figure that can be compared with the gain of 38 ATP during the complete oxidation of glucose by the glycolytic pathway followed by Krebs cycle, and the gain of 2 ATP only during anaerobic glycolysis).

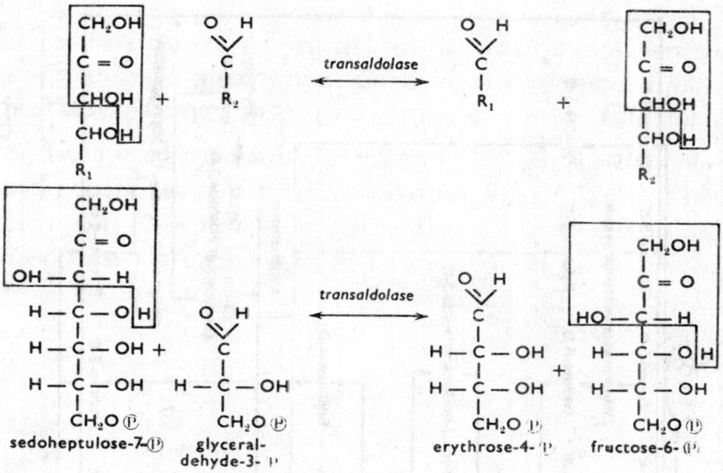

FIG. 4-41. — **Principle and example of transaldolization.**

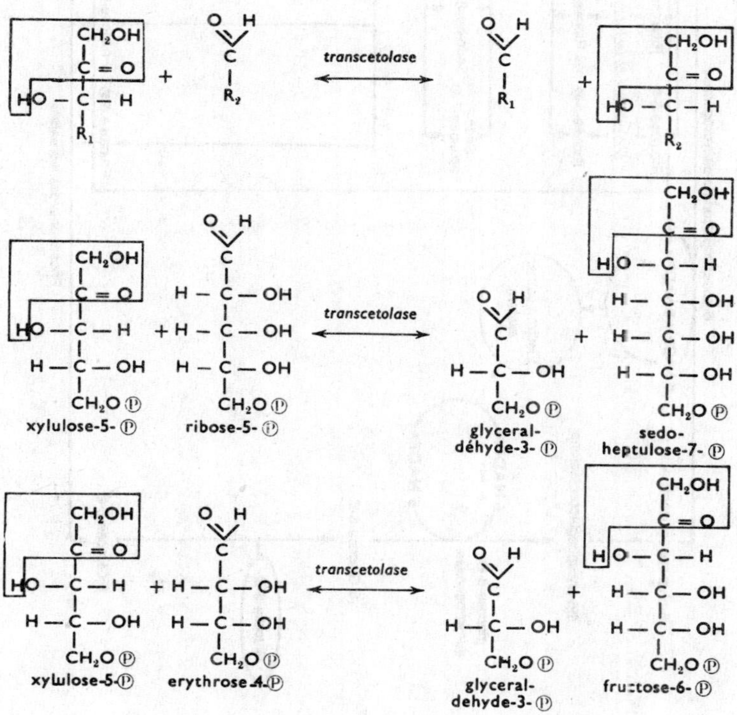

FIG. 4-42. — **Principle and examples of transketolization.**

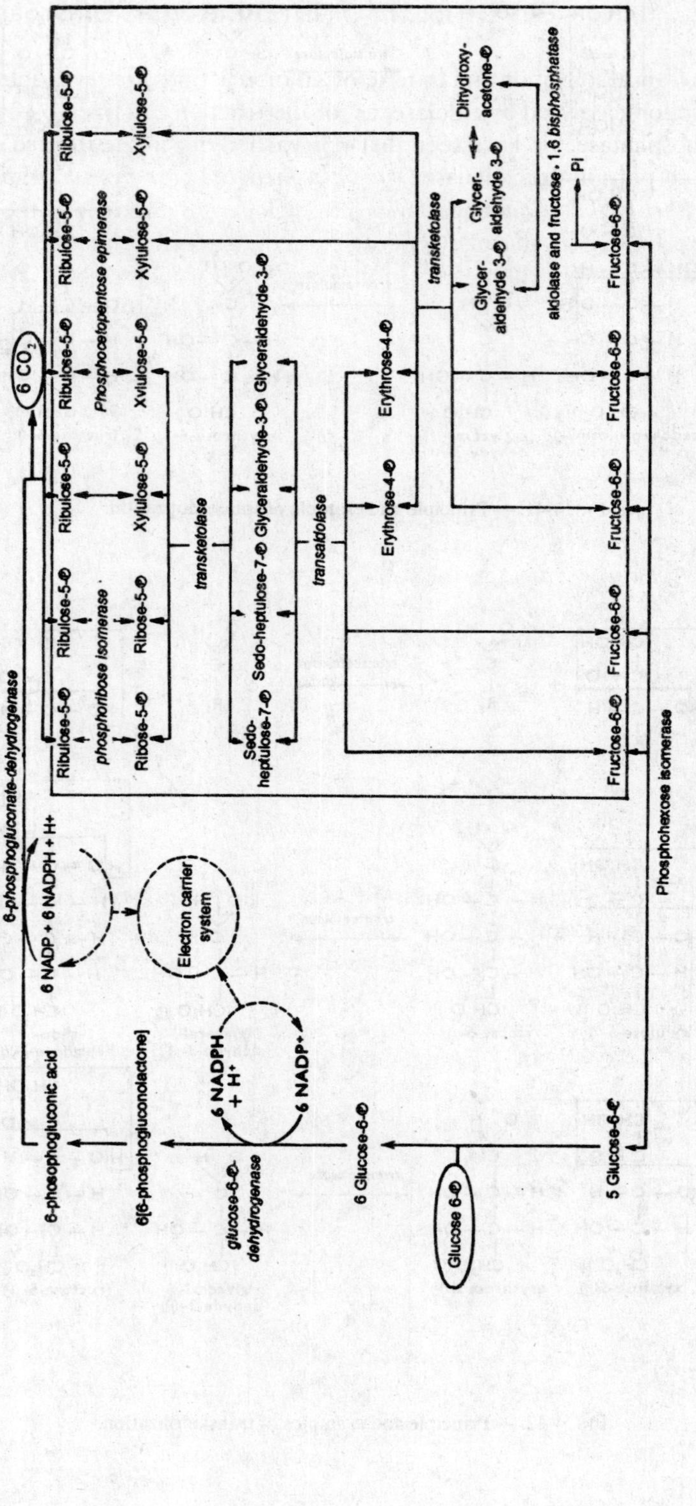

FIG. 4-43. — Recapitulative diagram of reactions of pentose phosphates cycle (boxes show the interconversions which can be summarized as: 6 Ribulose-5-℗ ⟷ 5 Fructose-6-℗).

V. – REVERSIBILITY OF INTERCONVERSIONS

The transaldolization and transketolization reactions are reversible; so also, the reaction catalyzed by aldolase; as for the reaction catalyzed by fructose-1, 6-bisphosphatase, we have seen that it is carried out in the reverse direction thanks to phosphofructokinase. *One can therefore pass from hexoses to pentoses, either by the oxidative pathway* (i.e. in clockwise direction in the diagram of figure 4-43), *or by interconversions* (i.e. in the opposite direction). The cells use both these possibilities to form ribose-5-Ⓟ which is necessary for the synthesis of nucleic acids; it must be noted that the oxidative pathway requires $NADP^+$, and it is only if NADPH can be reoxidized that the cells can continue to utilize this pathway for the synthesis of ribose-5-Ⓟ. But the $NADP^+$/NADPH ratio in cytosol of the rat liver cell is about 0.01 (while the $NADP^+$/NADH ratio is about 700) and the reactions permitting the reoxidation of NADPH into $NADP^+$ are limited (see for example the biosynthesis of fatty acids in fig. 5-19, and that of cholesterol in fig. 5-22). The cells make greater use of the pathway of interconversions than the oxidative pathway on the one hand because the former is more economic than the latter (by the pathway of interconversions, 5 hexoses are sufficient to form 6 pentoses whereas 6 hexoses are required by the oxidative pathway, as may be seen in fig. 4-43), and on the other hand, because the $NADP^+$ required for the oxidative pathway is present in limiting quantities in the cells.

PHOTOSYNTHETIC CYCLE OF REDUCTION OF CO_2 INTO CARBOHYDRATES

We have seen in the chapter on bioenergetics that the transformation of light energy into chemical energy results in the production of ATP and NADPH (see fig. 3-12). Our present knowledge of the metabolism of carbohydrates will enable us to study what is sometimes called the *dark phase* of photosynthesis, i.e. the series of reactions permitting the *reduction of CO_2 to carbohydrates* thanks to ATP and NADPH formed during the "light phase".

It was observed that under illumination, starch accumulated in plant leaves. As it was easy to envisage, on the one hand, the reduction of CO_2 to formaldehyde and, on the other hand, the polymerization of formaldehyde (CH_2O) to glucose ($C_6H_{12}O_6$), it was originally thought that glucose was the product directly formed during the dark phase of photosynthesis. It is now known that this is not true and that in reality, CO_2 is first incorporated in *3-phosphoglyceric acid*. Calvin, using green algae (such as *Chlorella*) which he exposed to light and radioactive CO_2 (^{14}C-labeled) for very short periods (of the order of a few seconds), and then dipped the cells in boiling alcohol to denature the enzymes and block all reactions, showed that the first stable compound formed from radioactive CO_2 is 3-phosphoglyceric-acid. After contact times with $^{14}CO_2$ of the order of 30 seconds, a large number of compounds (trioses-phosphates, pentoses-phosphates, hexoses-phosphates) could be characterized by a radioactive spot following two-dimensional paper chromatography, but after 5 seconds, the only radioactive spot was that of 3-phosphoglyceric acid whose carboxylic group contained the major part of radioactivity. It was then showed that CO_2 binds, in presence of *carboxydismutase (or ribulose-1, 5-bisphosphate carboxylase)*, to ribulose-1, 5-bisphosphate thus forming a ketoacid of 6 carbon atoms, very unstable, which is hydrolysed to 2 molecules of 3-phosphoglyceric acid (see fig. 4-44). Besides this incorporation of CO_2 in compounds in C_3, the fixation of CO_2 in compounds in C_4 could be observed in some plants (like maize) and it was shown that CO_2 binds to phosphoenol pyruvic acid to give oxaloacetic acid thanks to the action of *phosphoenol pyruvate carboxylase*, an enzyme catalyzing a reaction very similar to the one catalyzed by *phosphoenol pyruvate carboxykinase* (see fig. 4-34).

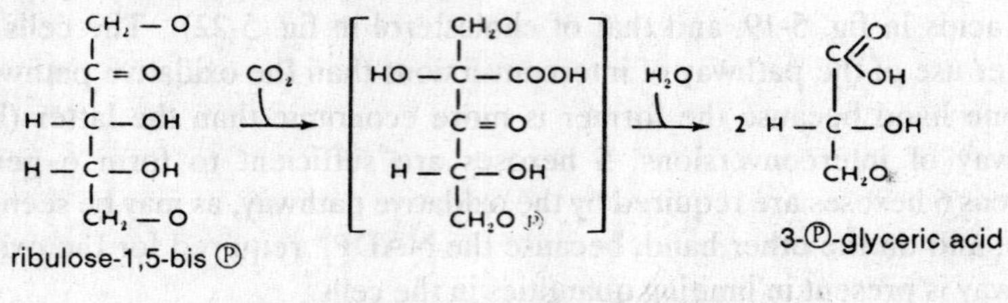

FIG. 4-44. — Fixation of CO_2 on ribulose-1,5-bis-Ⓟ under the effect of carboxydismutase (or ribulose-bisphosphate carboxylase) with formation of two molecules of 3-Ⓟ-glyceric acid.

As may be observed in the diagram of figure 4-45, 3-phosphoglyceric acid is transformed into 1,3-diphosphoglyceric acid by *phosphoglycerate kinase* in presence of ATP and then converted into glyceraldehyde-3-Ⓟ by *glyceraldehyde-3-Ⓟ dehydrogenase* in presence of NADPH; these two reactions studied during glycolysis operate here in a direction opposite to that of carbohydrate catabolism. Furthermore, one clearly sees the purpose served by ATP and NADPH, indispensable compounds formed – as mentioned previously – during the light phase of photosynthesis (see fig. 3-12).

Two molecules of glyceraldehyde-3-Ⓟ (in equilibrium with dihydroxyacetone Ⓟ) can give, thanks to *aldolase*, one molecule of fructose-1, 6-bis-Ⓟ which will be converted by *fructose-1, 6-bisphosphatase* into fructose-6-Ⓟ (these two reactions were also observed during glycolysis). Or, 12 molecules of triose-Ⓟ will give 6 molecules of fructose-6Ⓟ. From 5 of these fructose-6-Ⓟ molecules, 6 molecules of ribulose-5-Ⓟ will be formed (by the series of interconversions of the pentose-phosphates cycle described above and summarized in the diagram of figure 4-43, but here, the interconversions proceed in the reverse direction) and there is *one excess molecule of fructose-6-Ⓟ*. The 6 molecules of ribulose-5-Ⓟ formed will be phosphorylated by a kinase to ribulose-1, 5-bis-Ⓟ and will then again bind CO_2 and thus enter a new turn of the cycle. One turn of the cycle may be summarized as follows:

$$6\,CO_2 + 6\text{ ribulose-1, 5-bis-Ⓟ} + 18\text{ ATP} + 12\,(\text{NADPH} + H^+) \longrightarrow$$
$$6\text{ ribulose-1, 5-bis-Ⓟ} + \text{fructose-6-Ⓟ} + 18\text{ ADP} + 18\,P_i + 12\,\text{NADP}^+$$

In short, it is observed that 6 CO_2 were reduced to one molecule of fructose-6-Ⓟ which can be isomerized to glucose-6-Ⓟ and then stored in the form of starch after the transformations we already studied (see fig. 4-22).

In fact, a certain proportion of various intermediates can leave the cycle and enter other important metabolic pathways. For example, ribulose-5-Ⓟ can isomerize to ribose-5-Ⓟ required for the synthesis of nucleic acids. On the other hand, triose-Ⓟ can give α-glycerophosphoric acid (see fig. 4-32) which will be used for the synthesis of lipids (see fig. 5-21). Even the first compound formed, 3-phosphoglyceric acid, can take another direction: it can be oxidized by the glycolytic pathway followed by the Krebs cycle. Therefore, the output in glucose units cannot be expected to be the theoretical output described above.

It must be noted that photosynthesis has a determining effect on the direction of the glycolytic reactions and the cycle of pentose-phosphates: under illumination, thanks to the production of ATP and NADPH in high concentrations, the photosynthetic cycle will permit the reduction of CO_2 to hexose-Ⓟ; as just mentioned, this cycle uses the series of inter-conversions of the pentose-phosphates cycle in the direction hexoses-Ⓟ ⟶ pentoses-Ⓟ, i.e. in the direction opposite to that enabling the pentose-phosphates cycle to oxidize glucose with production of CO_2 (see fig. 4-43); moreover, the photosynthetic cycle follows some glycolytic reactions, but in the reverse direction. On the contrary, in the dark, ATP and NADPH concentrations will fall (there is no light energy for their synthesis and yet their utilization continues in order to bring about the metabolic reactions requiring energy) so that the plant cells will catabolize glucose to CO_2 just like animal cells, i.e. by glycolysis followed by Krebs cycle, or by the pentose-phosphates cycle. This is why, in the dark, contrary to what happens under illumination, plants are found to *breathe* like animals, absorbing oxygen and eliminating carbon dioxide.

BIOSYNTHESIS OF OLIGO AND POLYSACCHARIDES

I. – FROM "FREE" CARBOHYDRATES

The biosynthesis of polysaccharides rarely takes place from free carbohydrates. The following examples may however be given:

1) Transfer of short chains (about 7 residues of glucose conjugated by α-1,4 linkages) to growing glycogen molecules catalyzed by a *branching enzyme* which grafts by α-1, 6 linkages.

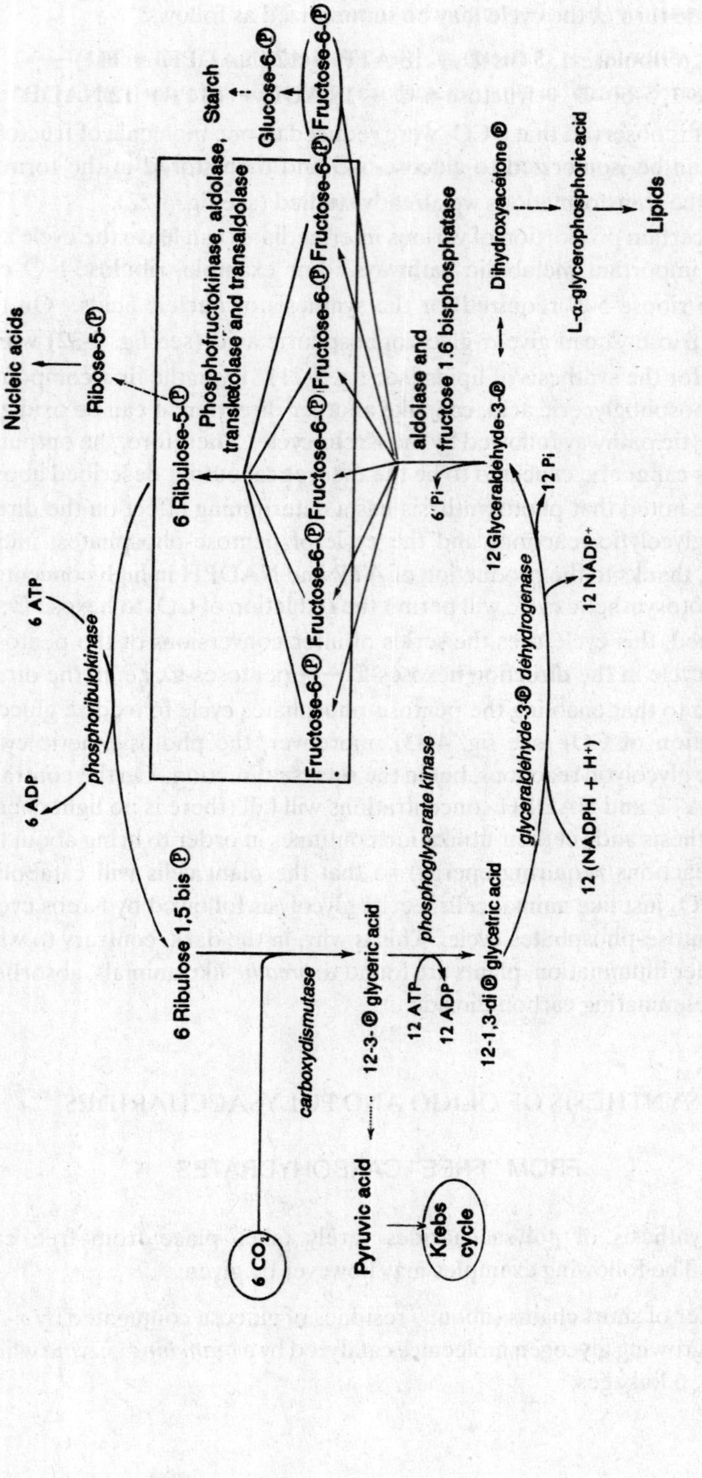

FIG. 4-45. — Diagram showing the reactions of the photosynthetic cycle of reduction of CO_2 to carbohydrates and (in dotted lines) some associated reactions. (The rectangular box shows the interconversions reactions detailed in figure 4-43 and which can be summarized as: 5 Fructose-6-Ⓟ ⟷ 6 Ribulose-5-Ⓟ).

2) Action of *levan saccharase* (or levan sucrase) which synthesizes levans (or fructosans) by the following mechanism:

$$\textbf{Sucrose} + (\textbf{Fructose})_n \longrightarrow (\textbf{Fructose})_{n+1} + \textbf{Glucose}$$

or, *dextran saccharase* (or dextran sucrase) which leads to dextrans by a similar mechanism:

$$\textbf{Sucrose} + (\textbf{Glucose})_n \longrightarrow (\textbf{Glucose})_{n+1} + \textbf{Fructose}$$

II. – FROM MONOSACCHARIDE 1-PHOSPHATES

Phosphorylases can catalyze *in vitro*, the formation (or lengthening) of polysaccharide chains from glucose-1-Ⓟ and thus enable the synthesis of glucosans in which the glucose residues are joined by α-1, 4 linkages. However (as mentioned in page 174) this mechanism does not seem to occur in physiological conditions.

But the following synthesis of sucrose has been described, catalyzed by *sucrose-phosphorylase*

$$\textbf{Glucose-1-Ⓟ} + \text{Fructose} \longleftrightarrow \textbf{Sucrose} + \textbf{P}_i$$

III. – FROM NUCLEOTIDES - MONOSACCHARIDES

The pathway of the synthesis of oligo- and polysaccharides most commonly followed by the cells is the one which uses the nucleotides-monosaccharides, a partial list of which was given above (page 184). The transfer of the monosaccharide is performed by transferases which bear the name of the carbohydrate they conjugate: galactosyl-, mannosyl-, N-acetylglucosaminyl-, sialyl- (or neuraminyl-), glucuronosidyl-transferases, for example. The glycosyl-transferases involved in the biosynthesis of glycans of glycoproteins are enzymes

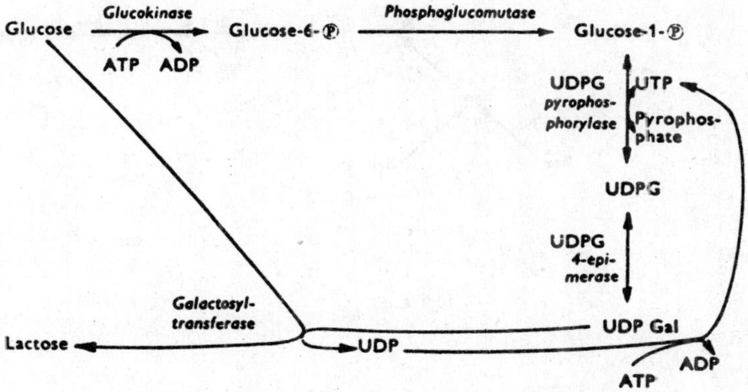

FIG. 4-46. — Synthesis of lactose in the mammary gland.

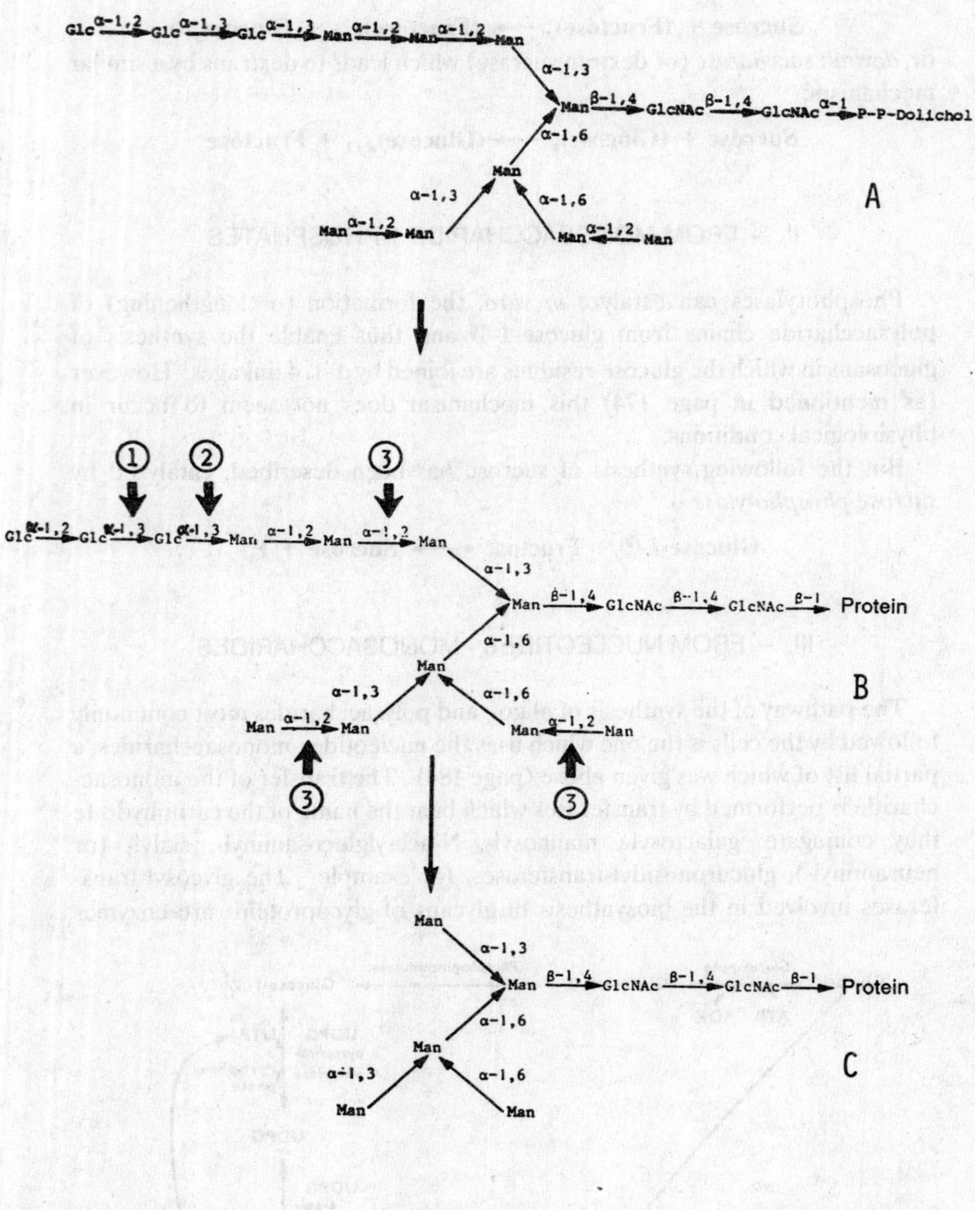
Glc α-1,2 Glc α-1,3 Glc α-1,3 Man α-1,2 Man α-1,2 Man
α-1,3
Man β-1,4 GlcNAc β-1,4 GlcNAc α-1 P-P-Dolichol
α-1,6
Man
α-1,3
α-1,6
Man α-1,2 Man
Man α-1,2 Man
A
1
2
3
Glc α-1,2 Glc α-1,3 Glc α-1,3 Man α-1,2 Man α-1,2 Man
Man β-1,4 GlcNAc β-1,4 GlcNAc β-1 Protein
B
C
Protein

FIG. 4-47

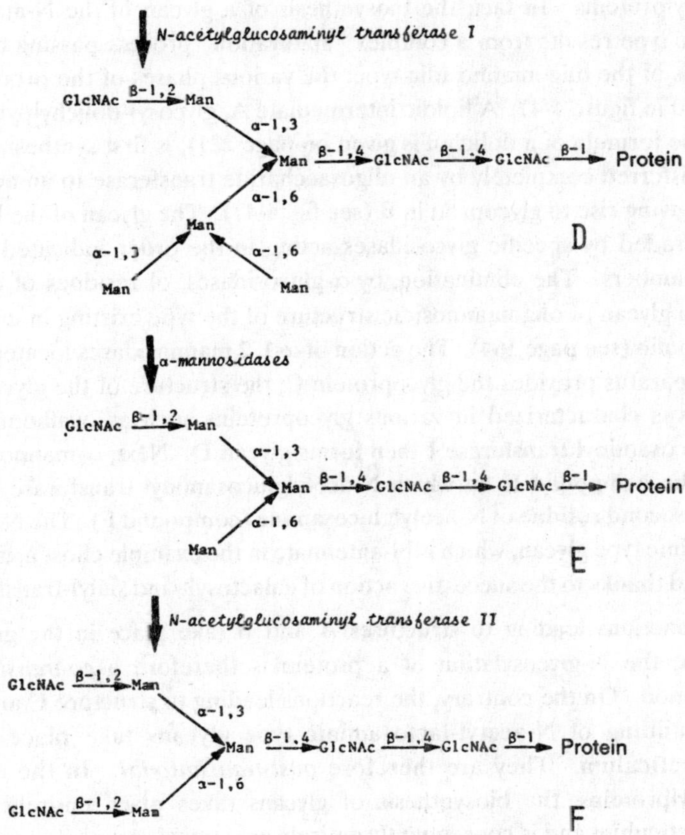

FIG. 4-47. — Synthesis of N-glycosylproteins. The biosynthesis of glycans A and B takes place in the granular fraction of the endoplasmic reticulum whereas that of glycans C to F takes place in the smooth reticulum, particularly in Golgi apparatus.

integrated in the luminal face of membranes of the endoplasmic reticulum. For information, we will restrict ourselves to the description of the biosynthesis of a disaccharide, the lactose (see fig. 4-46) on the one hand, and on the other hand, of the more complex process (see fig. 4-47) of the biosynthesis of the biantennate glycan of the N-acetyl-lactosaminic type described on page 164.

1) *Synthesis of lactose.* The synthesis of lactose in the mammary gland is performed by the transfer of galactose from UDP-Gal to glucose, catalyzed by a *galactosyl transferase* and necessitating the presence of *α-lactalbumin*, the milk protein. The association of these two proteins constitutes the *lactose synthetase* system.

2) *Synthesis of a glycan of the N-acetyl-lactosaminic type.* It was long believed that the residues of monosaccharides were conjugated one after the other by direct transfer from glycosylnucleotides as in the case of

O-glycosylproteins. In fact, the biosynthesis of a glycan of the N-acetyllactosaminic type results from a complex "maturation" process passing through structures of the oligomannosidic type; the various phases of the process are illustrated in figure 4-47. A lipidic intermediate A, glycosyl-dolichylpyrophosphate (the formula of a dolichol is given on page 221), is first synthesized and then transferred completely by an oligosaccharide transferase to an acceptor protein, giving rise to glycoprotein B (see fig. 4-47). The glycan of the latter is then degraded by specific glycosidases acting in the order indicated by the circled numbers. The elimination, by α-glucosidases, of residues of glucose leads to a glycan of oligomannosidic structure of the type existing in unit A of thyroglobulin (see page 164). The action of α-1, 2 mannosidases located in the Golgi apparatus provides the glycoprotein C; the structure of the glycan thus formed was characterized in various glycoproteins such as ovalbumin. N-acetylglucosaminyl transferase I then forms glycan D. Next, α-mannosidases lead to the compound E on which N-acetylglucosaminyl transferase II conjugates a second residue of N-acetylglucosamine (compound F). The N-acetyllactosaminic type glycan, which is bi-antennate in the example chosen, is finally completed thanks to the successive action of galactosyl- and sialyl-transferases.

The reactions leading to structures A and B take place in the granular reticulum; the N-glycosylation of a protein is therefore a *co-translational* phenomenon. On the contrary, the reactions leading to structure C and then to the building of N-acetyl-lactosaminic type glycans take place in the smooth reticulum. They are therefore *post-translational*. In the case of O-glycosylproteins, the biosynthesis of glycans takes place entirely in the smooth reticulum and is consequently entirely *post-translational*.

As for the glycoproteins thus synthesized, either their peptide chain will be liberated from the ergatoplasmic membrane and they will be secreted out of the cell, or they will remain integrated in it and externalized by the membrane flow, becoming the constituents of the plasmic membrane.

Lastly, the glycans of oligomannosidic structure can be phosphorylated in the position 6 of some mannose residues and, according to a mechanism proposed by Sly, the residues of mannose-6-phosphate of glycans carried by the lysosomial hydrolases then represent the signal which permits the segregation of theses enzymes from the "externalized" glycoproteins of the cell. As a matter of fact, only the glycoproteins which carry the mannose-6-phosphate signal will be firmly attached to a receptor of the Golgi membrane and their maturation towards a N-acetyl-lactosaminic structure will be blocked. They will be still present on the internal face of membranes of vesicles which are then formed and which will become primary lysosomes. On the contrary, the non-phosphorylated glycoproteins evolve towards the N-acetyl-lactosaminic type by the cascade of transformations described in figure 4-47.

The set of mechanism of intracellular "distribution" of glycoproteins are diagrammatically represented in figure 4-48. They demonstrate the fundamental importance of glycan structures as recognition signals.

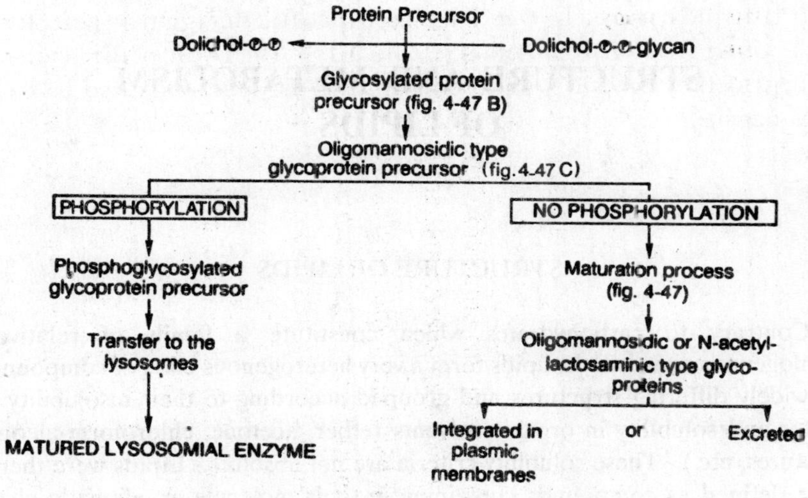

FIG. 4-48. — Role of phosphorylation of oligomannosidic type glycans in the evolution of glycoproteins: lysosomal glycoenzymes or membrane and excreted glycoproteins.

CHAPTER 5[1]

STRUCTURE AND METABOLISM OF LIPIDS

STRUCTURE OF LIPIDS

Contrary to carbohydrates which constitute a family of relatively homogenous compounds, lipids form a very heterogenous class of compounds of widely differing structures and grouped according to their insolubility in water and solubility in organic solvents (ether, acetone, chloroform-alcohol mixtures, etc.). These solubility criteria are not absolute. Lipids were therefore defined as compounds containing in their molecule an aliphatic chain (chain consisting of $-CH_2-$) of at least 8 carbon atoms. Some short-chain fatty acids (like butyric acid, in C_4) are the only exceptions to this rule.

The term *fats* and *oils* denote mixtures of lipids respectively solid (lard) or liquid (olive oil) at ordinary temperature; one must avoid using these terms to designate esters of glycerol (only the industry still uses them).

In earlier manuals one finds the distinction between simple lipids (yielding an alcohol and one or several fatty acids on hydrolysis) and complex lipids (the hydrolysis of which liberates not only an alcohol and fatty acids but also phosphoric acid, carbohydrates, etc.). As structures of the so-called complex lipids were progressively elucidated a more precise and more rational classification appeared possible. We are following a classification based on structural characteristics.

1. – FATTY ACIDS

They are found in small quantities in free state, but in large quantities involved in *ester* (or sometimes *amide*) linkages. As a general rule, these are monocarboxylic, straight unbranched chain acids containing an even number of carbon atoms (between 4 and 36). They may be saturated or unsaturated and sometimes hydroxylated or branched.

---

(1) Revised by G. Rebel

1. – Saturated fatty acids

Their general formula is: $CH_3-(CH_2)_n-COOH$. The most frequent are palmitic acid (C_{16}) and stearic acid (C_{18}). In lower concentration are found the fatty acids with 14 or 20 carbon atoms. Longer fatty acids (up to 36 carbon atoms) are present in numerous cells (bacteria, unicellular eucaryotes, plants vertebrates). They are generally present in some types of lipids. Milk on the contrary, is rich in short-chain fatty acids. Besides the even-carbon fatty acids are generally found small quantities of fatty acids having 15, 17 or 19 carbon atoms.

2. – Unsaturated fatty acids

Fatty acids are numbered from the terminal carboxyl (carbon 1) to the CH_3 group (carbon n). The double bond is indicated by the sign Δ, accompanied by the number corresponding to the first carbon atom participating in the double bond. The sign : is being increasingly used; it is followed by the number of double bonds, the position of the latter being indicated within brackets. There is also a biochemical nomenclature. In this case carbon 1 is the terminal methyl. The place of the last double bond is indicated by ω followed by the number of atoms of the carbon existing up to this double bond. In practically all biological unsaturated fatty acids, the double bond, has a *cis* isomerism. The principal unsaturated fatty acids are:

A. – Monounsaturated fatty acids (1 double bond).

Oleic acid (C_{18}), double bond between carbon atoms C_9 and C_{10}, abbreviated as: (C_{18}, Δ^9 or 18 : 1(9) or 18 ω 9).

$$CH_3-(CH_2)_7-CH = CH-(CH_2)_7-COOH$$

B. – Polyunsaturated fatty acids (several double bonds)

In the most common of such acids, the non-conjugated double bonds are separated by a methylene group. Plants can however contain fatty acids with conjugated double bonds, for example, eleostearic acid.

Linoleic acid (C_{18}, $\Delta^{9,12}$ or 18:2 (9, 12) or 18 ω 6)

$$CH_3-(CH_2)_4-CH = CH-CH_2-CH = CH-(CH_2)_7-COOH$$

Linolenic acid (C_{18}, $\Delta^{9,12,15}$ or 18 : 3 (9,12,15) or 18 ω 3)

$$CH_3—CH_2—CH = CH—CH_2—CH = CH—CH_2—CH = CH—(CH_2)_7—COOH$$

Arachidonic acid (C_{20}, $\Delta^{5,8,11,14}$).
Docosahexaenoic acid (C_{22} $\Delta^{4,7,10,13,16,19}$).
Eleostearic acid (C_{18}, $\Delta^{9,11,13}$).

In mammals, polyunsaturated fatty acids can have up to 22 carbon atoms and 6 double bonds, but in plants these acids do not exceed 18 carbon atoms and 4 double bonds.

An important physical property of fatty acids is their melting point; it decreases with increasing number of double bonds. For example, the melting point of stearic acid (18 : 0) is 70°C, whereas that of oleic acid (18 : 1) is 13°C, that of linoleic acid (18 : 2), –5.8°C and that of arachidonic acid (20 : 4), – 49.5°C.

3. – Hydroxylated fatty acids

Plants can synthesize a series of hydroxylated fatty acids like ricinoleic acid for example:

$$CH_3-(CH_2)_5-\underset{}{\overset{\displaystyle OH}{\overset{|}{C}H}}-CH_2-CH=CH-(CH_2)_7-COOH$$

Some of these hydroxylated fatty acids lead to the formation of cutin.

Other types of hydroxylated fatty acids are found in mammals. Some glycolipids contain large quantities of α-hydroxylated acids (OH on carbon 2) with 22, 23, 24 and 25 carbon atoms. Moreover, cells of the epiderm have lipids containing very long-chain ω hydroxylated acids which play a role in the structure of this particular tissue.

4. – Branched fatty acids

Example : 15 methylhexadecaenoic acid

$$(CH_3)_2-CH-(CH_2)_{13}-COOH$$

The above type of fatty acid is particularly abundant in Gram$^+$ bacteria.

5. – Prostaglandines, leukotriens and peroxides

Prostaglandines and leukotriens are derived from polyunsaturated fatty acids with 20 carbon atoms ω 6 and ω 3 (hence their general name, eicosanoids) and especially from arachidonic acid, under the action of cyclooxygenase (prostaglandines) and lipoxygenase (leukotriens).

In mammals, these are compounds having hormonal action with various biological effects. Prostaglandines E are powerful activators of adenylate cyclase. Prostaglandines F and leukotriens B,C,D, activate the contraction of various smooth muscles. These compounds have a short half-life because they are metabolized by the tissues into biologically inactive derivatives. In vertebrates, eicosanoids are synthesized by numerous tissues. Insects form prostaglandines from the polyunsaturated fatty acids present in food. Plants have lipoxygenases which metabolize linoleic acid into compounds analogous to leukotriens.

Besides these compounds, numerous organisms can "oxidize" fatty oxides to lipidic peroxides. The hydroxylated fatty acids of plants are formed by such a mechanism.

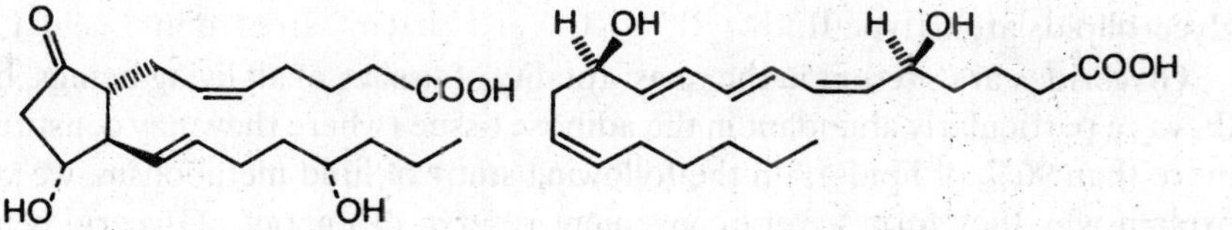

Prostagladine PGE_2 Leukotrien B_4

FIG. 5-1. — Structure of prostaglandines and leukotriens.

6. – Other close compounds

Besides the fatty acids, one finds aldehydes and fatty alcohols, such as for example, palmitaldehyde, stearaldehyde, olealdehyde and the corresponding primary alcohols. These compounds are rarely in free state, but are part of the structure of glycerophospholipids (see II-2) or cerides (see IV). Medium-chain linear aldehydes play a role of pheromone in insects.

II. – GLYCEROLIPIDS

1. – Glycerides

These compounds are obtained by esterification of the alcohol groups of glycerol by fatty acids; there are mono-, di- and triglycerides. Moreover, glycerides may differ by the nature and position of esterified fatty acids. To indicate the position, the carbon atoms of glycerol are denoted 1, 2 and 3.

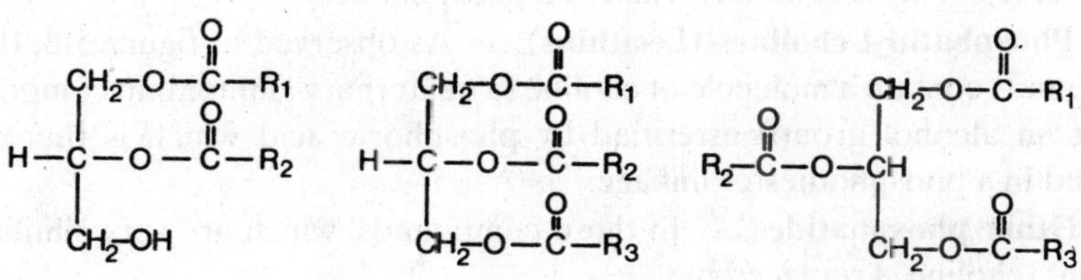

A B (I) B (II)

$R_1 = -(CH_2)_{14}-CH_3$ $R_2 = -(CH_2)_7-CH = CH-(CH_2)_7-CH_3$

$R_3 = -(CH_2)_{16}-CH_3$

FIG. 5-2. — Structures of di- and triglycerides.

Thus, the compound A of figure 5-2 is 1-palmitoyl 2-oleyl glycerol, compound B is 1-palmitoyl 2-oleyl-3-stearoyl glycerol.

When the fatty acids esterified in position 1 and 3 are different (as in the compound B), a centre of asymmetry appears in carbon 2 and one can therefore have the isomers I and II represented in figure 5-2. Most of the natural glycerolipids are of type II.

Glycerides are present in the quasi-totality of tissues of all living beings, but they are particularly abundant in the adipose tissue (where they may constitute more than 90% of lipids). In the following study of lipid metabolism, we will explain why they form a very convenient reserve of energy. Glycerides are generally present in cells in liquid state as cytoplasmic inclusions.

2. – Glycerophospholipids

Also called *phosphatides* they are the most numerous representatives of the large family of phospholipids.

They are found in high concentrations in the cellular and subcellular membranes of all living organisms. Only viruses called "non-enveloped" (viruses which do not incorporate in their structure, membrane elements of the host cell) are free from phosphatides. Some phosphatides are good emulsifying agents (lysophosphatides, lecithins).

A. – *Diacylphosphatides*

a) **Phosphatidic acids.** – They result from the esterification of glycerol by two fatty acids and phosphoric acid; the latter having only one of its acidic OH esterified, imparts an acid character to the molecule (see fig. 5-3). They exist in small quantities in free state and play an important role in the biosynthesis of glycerophospholipids, the structure of which derives from that of phosphatidic acids (see fig. 5-21). As in the case of glycerides, the molecule is asymmetric. The 2 fatty acids of glycerol have the same orientation as observed in glycerides (fig. 5-2, II). This is generally true of all phosphatides.

b) **Phosphatidyl cholines (Lecithins).** – As observed in figure 5-3, these compounds contain a molecule of choline (a quaternary ammonium compound having an alcohol group) esterified by phosphoric acid which is therefore involved in a phosphodiester linkage.

c) **Other phosphatides.** – In these compounds, which are very similar to lecithins, choline is replaced by:

– ethanolamine in *phosphatidyl ethanolamines*

– serine in *phosphatidyl serines*

These two types of lipids were earlier called "cephalins".

Choline can also be replaced by some alcohols, like glycerol in *phosphatidylglycerols* (abundant in some micro-organisms and plants) or inositol, a cyclic polyalcohol, in *phosphatidyl inositols.* Important derivatives of the latter phospholipids are phosphatidylinositol-4, 5-diphosphates. The structure of some of these lipids are represented in figure 5-3.

Mention may also be made of *diphosphatidyl-glycerol* or *cardiolipid*, a phosphatide which is specifically located in the mitochondria in mammals. It is formed by the union of 2 molecules of phosphatidic acid the phosphate atoms of which are linked through a molecule of glycerol.

B. – Alkenylphosphatides (Plasmalogens)

They differ from diacyphosphatides in that the fatty acid bound in position 1 of the glycerophosphate is replaced by a fatty aldehyde, bound by an ethylenic ether-oxide linkage (see fig. 5-3).

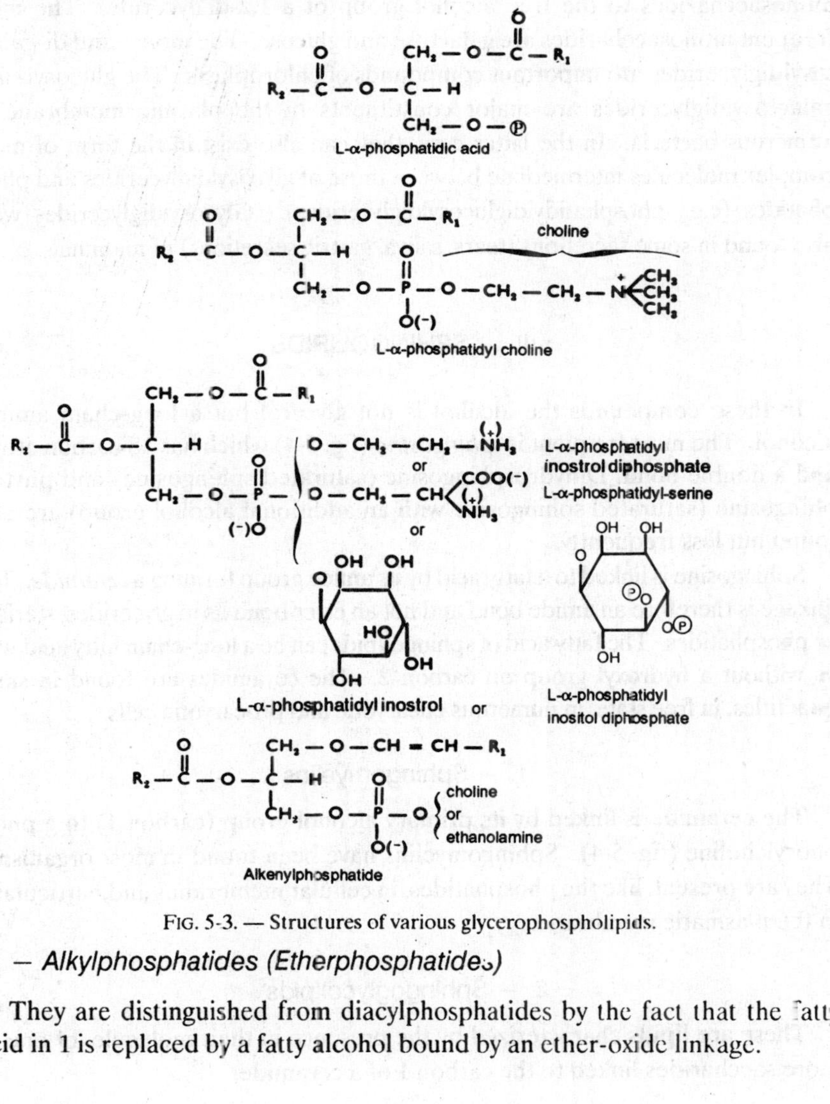

FIG. 5-3. — Structures of various glycerophospholipids.

C. – Alkylphosphatides (Etherphosphatides)

They are distinguished from diacylphosphatides by the fact that the fatty acid in 1 is replaced by a fatty alcohol bound by an ether-oxide linkage.

Like the diacylphosphatides, alkenyl- and alkyl phosphatides are found in a wide range of organisms from the unicellular to mammals. However, the percentage of these compounds varies considerably from one species to another and in the same species from one tissue to another. An important derivative of alkylphosphatides is PAF acether (Platelet Activating Factor). This compound, synthesized particularly by the platelets, has a very high aggregation activity and plays an important role in the formation of "platelet clots".

3. – Glycosyldiglycerides

They result from the binding of one or several (up to 10) molecules of monosaccharides to the free alcohol group of a 1,2-diglyceride. The most frequent monosaccharides are galactose and glucose. The mono- and di-galactosyldiglycerides are important compounds of chloroplasts. The glucosyl- and galactosyldiglycerides are major constituents of the plasmic membrane of numerous bacteria. In the latter case, they can also exist in the form of more complex molecules intermediate between those of glycosyldiglycerides and phosphatides (e.g., phosphatidyldiglucosyldiglycerides). Glycosyldiglycerides were also found in some secretions (tears, saliva, gastric secretions) of mammals.

III. – SPHINGOLIPIDS

In these compounds the alcohol is not glycerol but a long-chain aminoalcohol. The most frequent is *sphingosine* (fig. 5-4) which has 18 carbon atoms and a double bond. Dihydrosphingosine (saturated sphingosine) and phytosphingosine (saturated sphingosine with an additional alcohol group) are also found but less frequently.

Sphingosine is linked to a fatty acid by its amine group forming a *ceramide*. The linkage is therefore an amide bond and not an ester bond as in glycerides, sterides or phosphatides. The fatty acid of sphingolipids can be a long-chain fatty acid with or without a hydroxyl group on carbon 2. The ceramides are found in small quantities, in free state, in numerous eucaryotic and procaryotic cells.

1. – Sphingomyelins

The ceramide is linked by its primary alcohol group (carbon 1) to a phosphorylcholine (fig. 5-4). Sphingomyelins have been found in most organisms. They are present, like the phosphatides, in cellular membranes and particularly in the plasmatic membrane.

2. – Sphingoglycolipids

These are lipids characterized by the presenee in their molecule, of one or more saccharides linked to the carbon 1 of a ceramide.

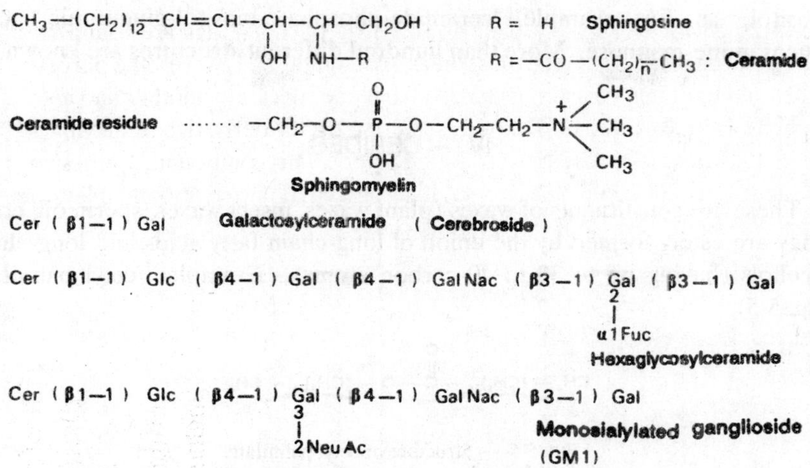

FIG. 5-4. — Structure of sphingolipids. The abbreviated nomenclature of monosaccharides is indicated in page 152.

A. – Galactolipids

In the case of galactocerebrosides, the monosaccharide fixed on the ceramide is galactose. Galactose can be esterified by a molecule of sulphuric acid. The compounds are then called *sulphatides*. In mammals, *galactocerebrosides* and sulphatides are mainly located in the renal tissue and nervous tissue (myelin sheath). They are rarely found in organisms other than the vertebrates.

B. – Neutral glycolipids

One or several (up to about ten) sugars are bound to the ceramide. In vertebrates, the first sugar is glucose. The compounds are then spoken of as *glucocerebrosides*. In addition to glucose, the most frequently found monosaccharides are galactose, mannose, fucose, glucosamine and galactosamine. Neutral glycolipids particularly the glucocerebrosides are present in a very large number of organisms ranging from the procaryotes to mammals where they can form up to 90% of lipids.

C. – Gangliosides

Their structure is that of a glucosylceramide to which are bound one or several molecules of galactose, N-acetyl galactosamine, N-acetylglucosamine or fucose (see fig. 5-4). The most characteristic sugar of gangliosides is however neuraminic acid (or *sialic acid*) (see fig. 4-11) in the N-acetylated or N-glycosylated form. They are found only in vertebrates. The brain is particularly rich in gangliosides.

D. – Phytoglycolipids

Besides the conventional neutral glycolipids, most plants, yeasts and fungi include more complex glycolipids which generally contain phosphorus and inositol, as for example, ceramide-phosphate-inositol-glucuronic acid-glucosamine-mannose. More than hundred different structures are known.

IV. – CERIDES

These are constituents of waxes (plant waxes, insect waxes, sperm oil, etc.). They are esters formed by the union of long-chain fatty acids and long-chain alcohols (having up to 30 to 40 carbon atoms). Example: cetyl palmitate (fig. 5-5).

$$CH_3-(CH_2)_{14}-\overset{O}{\overset{\|}{C}}-O-(CH_2)_{15}-CH_3$$

FIG. 5-5. — Structure of cetyl palmitate.

V. – HYDROCARBONS

The general structure of these compounds is $CH_3-(CH_2)_n-CH_3$. They are sometimes branched or unsaturated. They are found in small concentrations in most living organisms.

VI. – POLYISOPRENIC LIPIDS

1. – Polyisoprenic hydrocarbons

A very large number of compounds present especially in plants, are formed by the polymerisation of *isoprene* units; this is the case (as may be seen in figure 5-6) with *squalene* (an intermediate of the biosynthesis of sterols), *carotenes* and other terpenes like *limonene*.

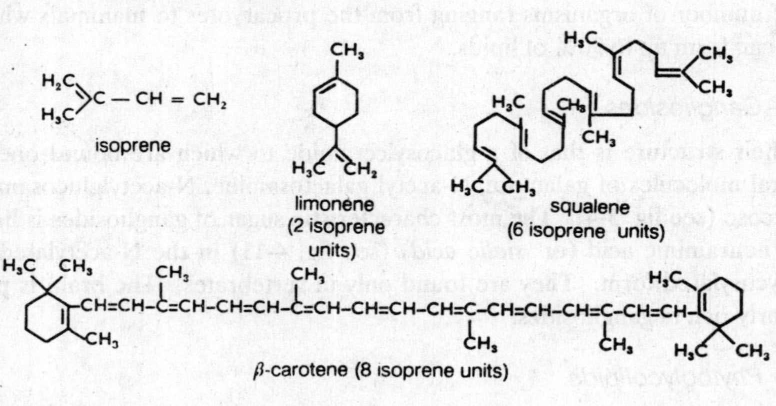

FIG. 5-6. — Structures of isoprene and various hydrocarbons derived from isoprene.

In addition to these terpenes, there are other polyisoprenes in plants; the best known is rubber, formed by the condensation of thousands of isoprene units.

One may also cite a group of linear polyisoprenes, having in average 20 isoprenic units, like the *dolichols* whose general structure diagram is as follows:

$$H-\left[CH_2-\underset{\displaystyle CH_3}{\underset{|}{C}}=CH-CH_2\right]_{15\ \text{to}\ 21}-CH_2-\underset{\displaystyle CH_3}{\underset{|}{CH}}-CH_2-CH_2\,OH$$

These compounds exist either free, or in the form of phosphoric esters (dolichols monophosphates) or pyrophosphoric esters (dolichols diphosphates) combined with a mono- or polysaccharide. The latter are intermediates acting in the synthesis of N-glycosylated proteins (see figs. 4-47 and 4-48).

A compound similar to dolichols is *phytol*:

$$H-\left[CH_2-\underset{\displaystyle CH_3}{\underset{|}{CH}}-CH_2-CH_2\right]_3-CH_2-\underset{\displaystyle CH_3}{\underset{|}{C}}=CH-CH_2\,OH$$

This compound is part of the structure of chlorophyll (see fig. 3-13). Phytanic acid, obtained by oxidation of the alcohol group of phytol, is one of the main fatty acids found in lipids of halophile and thermophile bacteria.

Some lipids can be considered as derivatives of isoprene (steroids, carotenoids, quinones with isoprenic side chain). These compounds are studied in the following.

2. – Sterols and steroids

These compounds derive from a polycyclic ring, called ring of cyclopentanophenanthrene. The stereochemistry and nomenclature problems of steroids are rather complex; we will not examine them in detail here. As far as stereochemistry is concerned we will envisage the main possibilities of isomerism with the maximum possible simplification; and to denote the steroids we will use the terms adopted by usage instead of the nomenclature based on structure.

The study of the polycyclic ring shows that there are 6 asymmetric carbons (indicated by asterisks in figure 5-7). Additional asymmetric carbons are formed by the introduction of substituents in the cycle. The quasi-totality of sterols and their derivatives have a substituent in 10 or 13 (generally a methyl group). Methyl groups 18 and/or 19 have the same orientation in space with respect to the plane in which the polycyclic ring is located. These 2 methyl groups serve as reference base. Any substituent which is situated on the same side as the methyl groups in 18 or 19 with respect to the plane of the molecule is said to be in position *cis*, called "β" and represented by a solid valence line; substituents situated on the other side of the plane of the molecule with respect to the methyl groups in 18 or 19 are said to be in position *trans*, called "α" and represented by a dotted valence line. α and β isomers of sterols and their derivatives are known only on carbons 3, 5, 7 and 17. Owing to the arrangement of cycles B, C, D in space and the existence of methyl groups 18 and 19, there is only one possibility of isomerism (β for 11 and α for 12) in carbons 11 and 12.

Steroids represent a very large and varied family of compounds. Their biological activities are very diverse and it is found that often, small variations of the structure or nature of the substituents result in major modifications of biological activity. There are:

A. – Sterols

This is a very important group of lipids found in practically all eucaryotes. More than hundred sterols are known of which we may cite *cholesterol* (see fig. 5-7) which is the principal sterol of vertebrates, *ergosterol* which is a natural precursor of vitamin D, (see fig. 5-7), stigmasterol, sitosterol, important sterols in plants.

By esterification of the alcoholic group by fatty acids, sterols give *sterides*. In general, in normal physiological conditions, the quantity of sterides in a given tissue is very small compared to free sterols (blood is an exception to this rule). Sterides exist only as traces in biological membranes. Their accumulation in the latter is pathological (atheroma). Plants contain an appreciable part of their sterols conjugated with a saccharide like glucose (sterylglucosides). The saccharide is linked by its reducing group to the alcohol group in position 3 of the sterol.

B. – Derivatives of sterols

***a*) Bile acids.** – The two main bile acids are *cholic acid* and *deoxycholic acid.* Their solubility in aqueous medium is extremely low. They are found in the bile, conjugated with *glycine* or *taurine* (the latter derives from cysteine by oxidation of the SH group and decarboxylation, see fig. 7-19), thus forming glycocholic, glycodeoxycholic, taurocholic and taurodeoxycholic acids (see fig. 5-7). Bile acids can be salified by monovalent ions (Na, K), thus forming *bile salts.*

***b*) Steroid hormones.**

TESTICULAR HORMONES, CALLED MALE SEX HORMONES.

Examples: *testosterone* (see fig. 5-7), androsterone.

OVARIAN AND PLACENTARY HORMONES, CALLED FEMALE SEX HORMONES.

– *Estrogen hormones*.

They are phenolsteroids (cycle A is aromatic).

Examples: estradiol (see fig. 5-7), estrone, estriol.

– *Luteal hormone.*
Example: *progesterone* (see fig. 5-7).

HORMONES OF THE ADRENAL CORTEX.
Examples: *Corticosterone* (see fig. 5-7), cortisol, aldosterone.

c) **Hormones of insects.** – A large group of sterol derivatives is represented by the insects "pupation" hormones. These compounds are analogous to the steroid hormones of mammals. The most important is *ecdysone* (fig. 5-7). The other compounds differ by the number and position of hydroxyl groups.

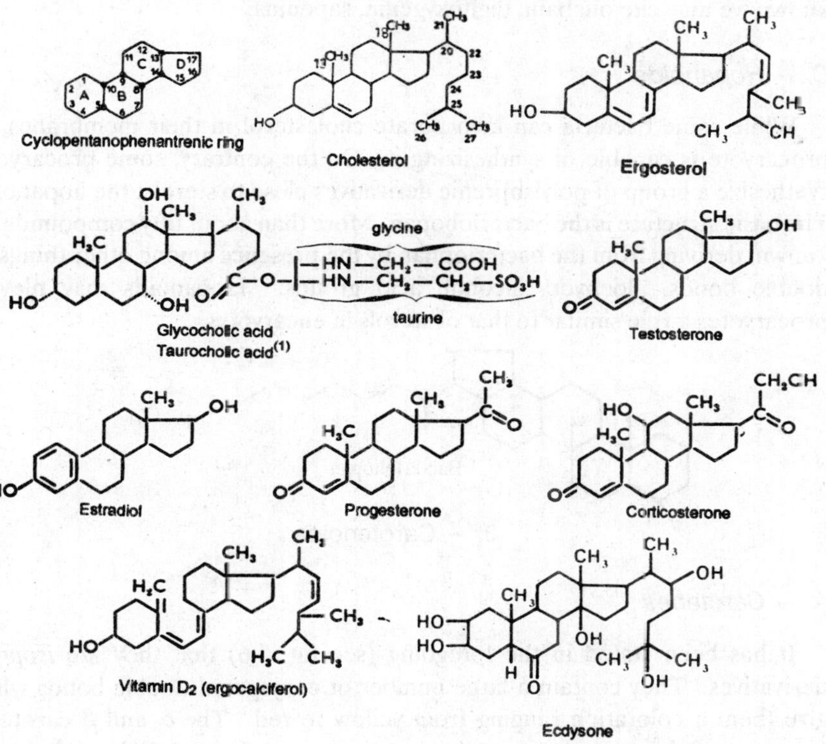

(1) The structures of glycodeoxycholic and taurodeoxycholic acids are identical except for the fact that in position 7 there is a H atom instead of a OH group.

FIG. 5-7. — Structures of various steroids.

d) **Vitamin D.** – There are several, very similar compounds having the same vitamin action; one of them is *vitamin* D_2 or *ergocalciferol* (fig. 5-7). It must be noted that cycle B is open and therefore the cyclopentanophenantrene ring characteristic of steroids is no longer present; however, the study of vitamin D in the chapter on sterols is justified by the fact that they derive from some sterols (especially ergosterol) by simple ultra-violet irradiation. Vitamin D is necessary for the proper formation of bones and teeth because it controls the phospho-calcium metabolism. These vitamins D are actually the precursors of biologically active compounds which are derivatives hydroxylated in position 24 or 25.

e) **Steroid alkaloids and heterosides.** – They are represented by a large number of compounds (more than hundred) synthesized by plants. They derive from the molecule of steroids, generally by the introduction of new groups (acid, alcohol, amine...). Some are combined with a sugar (glucose, galactose, arabinose, rhamnose...) linked to an alcohol or acid group of the cycle. The majority of these are pharmacologically active. Among the best known we may cite ouabain, digitoxygenin, saponins.

C. – Hopanoids

While some bacteria can incorporate cholesterol in their membranes, no procaryote is capable of synthesizing it. On the contrary, some procaryotes synthesize a group of polyisoprenic derivatives close to sterols: the hopanoids. The basic structure is the bacteriohopan. More than about fifty compounds are known, deriving from the bacteriophan by the presence among other things, of double bonds, aldehyde, alcohol, acid groups. Hopanoids may play in procaryotes a role similar to that of sterols in eucaryotes.

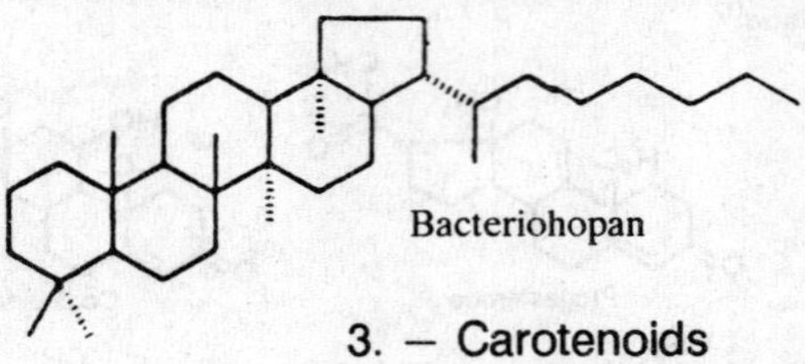

Bacteriohopan

3. – Carotenoids

A. – Carotenes

It has been stated in the foregoing (see fig. 5-6) that they are *isoprene* derivatives. They contain a large number of conjugated double bonds which give them a coloration ranging from yellow to red. The α and β carotenes (pigment of the carrot) are cyclized at the two ends (and differ only by the position of one double bond of the ring), whereas γ-carotene has only one ring (see fig. 5-8) and *lycopene* (pigment of the tomato) is not cyclized at all.

B. – Xanthophylls

These pigments derive from carotenes by oxidation and have hydroxyl groups on the rings.

C. – Vitamin A

Compounds with vitamin A activity can be placed next to the carotenes because they derive directly from them, as shown in fig. 5-8. Actually, the animal organism splits carotene giving rise to *retinal*, aldehyde of vitamin A which can then be reduced to alcohol. It must be noted that vitamin A can result only from one of the halves of α-carotene or γ-carotene molecule (the left hand half in figure 5-8), because the other half, in the former case (α-carotene), has the double bond of the ring in a different position, and in the latter case, has no ring at all (γ-carotene); on the contrary, β-carotene could theoretically yield 2 molecules of vitamin A, but in reality the initial scission does not occur at the centre of the molecule, so that one molecule of β-carotene also gives only one molecule of vitamin A.

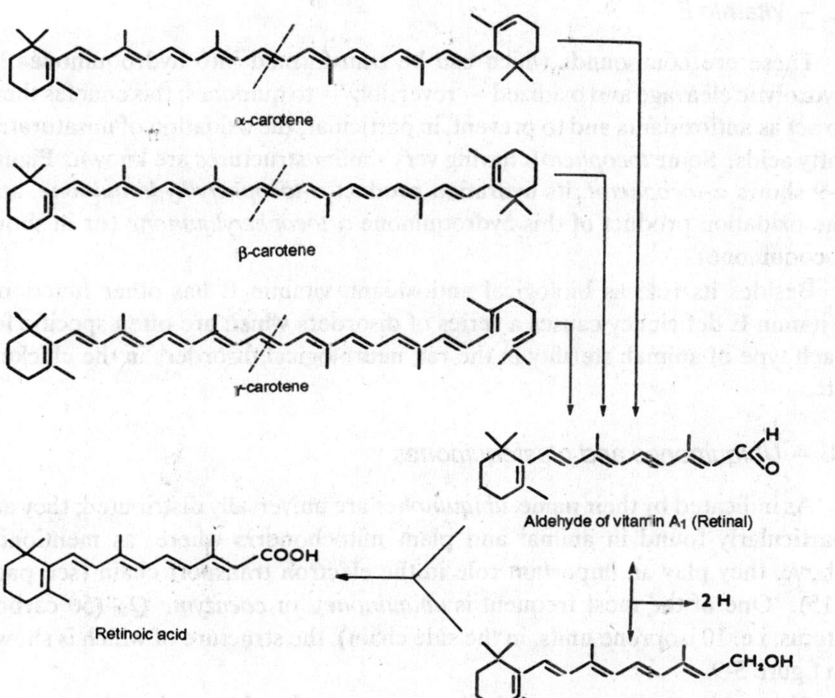

FIG. 5-8. — Structures of carotenes and vitamin A_1.

Vitamin A has several roles. It influences the growth of the animal and protects epithelial tissues. A derivative of vitamin A, retinoic acid (see fig. 5-8) is a modulator of cellular growth. Its role in the protection of epitheliums seems to be correlated with the fact that retinol derivatives (retinol pyrophosphoryl monosaccharide) are, like dolichols, important intermediates in the synthesis of some glycoproteins. Its best known role is that of a co-factor in the process of vision: vitamin A binds opsin, a protein of the retina, to give the visual pigment called rhodopsin.

4. – Isoprenic chain quinones

We are also listing under this heading, substances which can readily give quinones (tocopherols or vitamins E). All the substances examined in this paragraph can undergo the reversible transformation *quinone ⇔ hydroquinone.* In two of them (ubiquinone and plastoquinone) this transformation appears to be the basis of their biological activity, but in others (vitamins E and K) it is not yet known whether this transformation has any relation with their physiological action.

A. – Vitamin E

These are compounds which can be transformed into hydroquinones by hydrolytic cleavage and oxidized – reversibly – to quinones; this enables them to act as antioxidants and to prevent, in particular, the oxidation of unsaturated fatty acids. Some *tocopherols* having very similar structures are known. Figure 5-9 shows *α-tocopherol*, its hydration product *α-tocopherylhydroquinone*, and the oxidation product of this hydroquinone *α-tocopherylquinone* (or in short, tocoquinone).

Besides its role as biological antioxidant, vitamin E has other functions. Vitamin E deficiency causes a series of disorders which are often specific for each type of animal: sterility in the rat, neurological disorders in the chicken, etc.

B. – Ubiquinones and plastoquinones

As indicated by their name, *ubiquinones* are universally distributed; they are particularly found in animal and plant mitochondria where, as mentioned above, they play an important role in the electron transport chain (see page 115). One of the most frequent is *ubiquinone*$_{50}$ or *coenzyme* Q_{10} (50 carbon atoms, i.e. 10 isoprene units, in the side chain), the structure of which is shown in figure 5-9.

Plastoquinone has a very similar structure (fig. 5-9) and participates in electron transport in chloroplasts (see fig. 3-13).

C. – Vitamin K

As may be observed in figure 5-9, the structure of vitamin K_1 or *phylloquinone* is also very similar; it is a naphtoquinone with a chain of 4 isoprene units called phytyl residue, grafted on it. *Phylloquinone* is present in plants where it plays a role of electron acceptor in processes related to photosynthesis. *Menaquinones*, or vitamins K_2, differ from vitamin K_1 by the number of isoprene units of the side chain (there may be up to 10 units), and the number of double bonds present in this chain. They are found in bacteria. Vitamin K_2 is the active form of vitamin K in mammals (phylloquinone is converted into

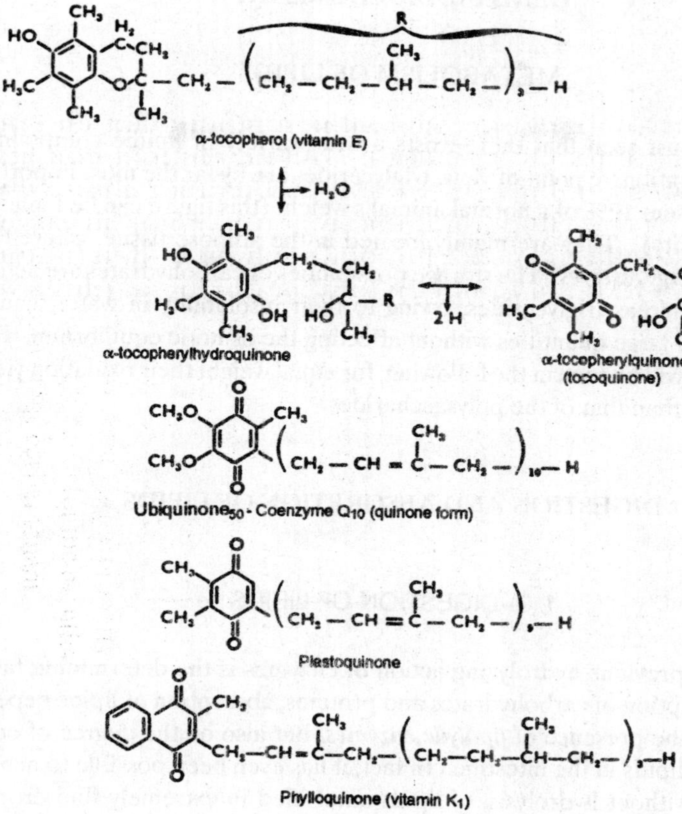

FIG. 5-9. — **Structures of some biologically active natural quinones having an isoprene chain.**

menaquinone in the liver). Vitamin K is also the coenzyme of an enzyme catalyzing the carboxylation of glutamic residues of proteins. The carboxylation is for example, necessary for the activation of a serum factor permitting the synthesis of *prothrombin*, a substance indispensable for blood-clotting. This explains the anti-hemorragic action of vitamin K.

This last paragraph devoted to isoprenic chain quinones (and substances readily leading to such quinones), included vitamins E and K. On the contrary, ubiquinone is not a vitamin for mammals : the organism can synthesize the ring from tyrosine; as for the side chain, it results from the condensation of C_5 units which are intermediate in the synthesis of cholesterol (see fig. 5-22).

In this study of isoprenic lipids we followed a classification based on the structural characteristics of various compounds and we thus found vitamins among the derivatives of steroids (D), carotenoids (A) and among isoprenic chain quinones (E and K), but we must bear in mind that these vitamins are often grouped under the name *liposoluble vitamins*.

METABOLISM OF LIPIDS

We have just seen that there exists a wide variety of lipidic compounds. From the quantitative point of view, triglycerides are by far the most important as they represent 10% of a normal animal's weight (this figure can be higher in cases of obesity). They are mainly located in the adipose tissue, where they constitute *energy reserves*. The storage possibilities of carbohydrates are actually limited, whereas triglycerides, owing to their insolubility in water, can be stored in very large quantities without affecting the osmotic equilibrium. Furthermore, as will be seen in the following, for equal weight their oxidation yields more energy than that of the polysaccharides.

DIGESTION AND ABSORPTION OF LIPIDS

I. – DIGESTION OF LIPIDS

Whereas previous hydrolyzing action of enzymes is the determining factor for the absorption of carbohydrates and proteins, absorption of lipids depends not only on the presence of *lipolytic enzymes*, but also on the degree of emulsification of lipids in the intestine. In fact, it has even been possible to achieve absorption without hydrolysis, of lipids emulsified in extremely fine droplets thanks to a synthetic detergent, much more efficient than the natural emulsifiers. These natural agents are the *bile salts* (especially sodium glycocholate) which, with the help of other emulsifiers like the phospholipids and the mono- and diglycerides (resulting from the partial hydrolysis of triglycerides) will permit the *absorption* of a mixture of partially hydrolyzed lipids. But in the first instance, the action of bile salts already manifests itself during *digestion*; they stabilize the pancreatic lipase and prevent its inactivation in the intestinal lumen.

a) **The pancreatic lipases.** – They are esterases which specifically hydrolyze the glycerol esters. The bonds in position 1 and 3 are the first to be broken, thus transforming a triglyceride into a 2-monoglyceride. The 2-monoglycerides are isomerized spontaneously to 1-monoglycerides which are then hydrolyzed. The pancreatic lipase does not act on lipids in solution, but only on micelles of triglycerides.

b) **The cholesterol-esterase.** – This enzyme hydrolyzes the ester linkage of sterides, form in which dietary cholesterol is generally found, thus liberating sterol and the fatty acid.

c) **The phospholipases.** – There are 4 types.

Phospholipase A_1 detaches the fatty acid in the position 1 from glycerophospholipids.

Phospholipase A_2 detaches the fatty acid in the position 2 from glycerophospholipids.

In both cases one obtains a phospholipid having only one fatty acid, called *lysophosphatide* (in the case of a lecithin, the degradation by a phospholipase A_1 or A_2 gives lysolecithin).

Lysophosphatides are very powerful surface active agents capable of causing serious hemolysis. This may explain, at least partially, the toxic action of certain snake venoms which contain high concentrations of phospholipase A.

Phospholipase A_2 plays a very important role in the biosynthesis of eicosanoids. This enzyme liberates from phosphatides, arachidonic acid, precursor of eicosanoids. The activators of the synthesis of these hormones are all activators of phospholipase A_2. The same enzyme can hydrolyze alkyphosphatidylcholines to alkyl-2-lyso derivatives. The alcohol group liberated is acetylated yielding PAF acether (see page 218).

The successive action of phospholipases A_1 and A_2 (or in the reverse order) detaches the 2 fatty acids from the molecule of the phospholipid to give a glyceryl-phosphoryl-derivative (glyceryl-phosphoryl-choline in the case of a lecithin).

Phospholipase C hydrolyzes the ester linkage between phosphoric acid and glycerol, thus liberating the prosphorylated nitrogenous base (e.g., phosphorylcholine) and a diglyceride.

Phospholipase D splits the ester linkage between the nitrogenous base (choline, ethanolamine, etc.) and phosphoric acid, and thus liberates the base and phosphátidic acid.

d) **The phosphodiesterases.** – As indicated by their name, these enzymes are specific of the phosphodiester linkages and by acting on glyceryl-phosphoryl-choline for example (obtained, as mentioned above, by action of phospholipases A on a lecithin) they will:

— either break the ester linkage between phosphoric acid and choline, thus liberating choline and glycerophosphate,

— or break the ester linkage between phosphoric acid and glycerol, thus liberating glycerol and phosphorylcholine (see fig. 5-3).

e) **The phosphomonoesterases (or phosphatases).** – They complete the hydrolysis of the products liberated by the phosphodiesterases; for example, from phosphorylcholine they liberate choline and phosphoric acid.

II. – ABSORPTION OF LIPIDS

The product passing through the intestinal mucosa is therefore a mixture of lipids more or less hydrolyzed (in fact, an emulsion containing fatty acids and also mono-, di-, triglycerides, phospholipids in more or less hydrolyzed state, sterols, etc.).

Glycerol and the short-chain fatty acids (up to 10 carbon atoms) will proceed to the liver through the blood vessels (particularly the portal vein). All other long-chain fatty acids as well as the mono- and diglycerides contribute to the reconstitution of triglycerides in the intestinal mucosa; these *neosynthesized triglycerides* will be brought to the liver through the lymphatic vessels and then the blood stream, in the form of lipoprotein particles, called *chylomicrons*; the latter are of about 1 μ diameter and consist of 98% lipids (mainly triglycerides) and 2% proteins.

The lipids which are not absorbed are found in the faeces, together with lipids (especially steroids) discharged into the intestine by the bile, and lipids excreted through the intestinal wall.

The transport by the blood stream, of the lipids synthesized in the liver to various tissues, particularly the adipose tissue (storage site of triglycerides) or the muscle (where fatty acids are utilized for energy production) takes place in the form of lipoproteins. The latter are complexes whose protein and lipid contents vary according to the particles, which enables a separation of the various lipoproteins according to their density. After reaction with specific cellular receptors, the lipoproteins are re-arranged partly thanks to the action of the lipoprotein-lipase which catalyzes the hydrolysis of the triglycerides present in the lipoprotein, with liberation of fatty acids; these can then be metabolized in the cells as explained below.

METABOLISM OF FATTY ACIDS

I. – OXIDATION OF FATTY ACIDS

1. – Oxidation of even-carbon, saturated straight chained fatty acids

A. – Activation of fatty acids

Before they are oxidized, fatty acids must be *activated*. This activation requires energy (supplied by ATP) and takes place in 2 steps, as may be seen in figure 5-10.

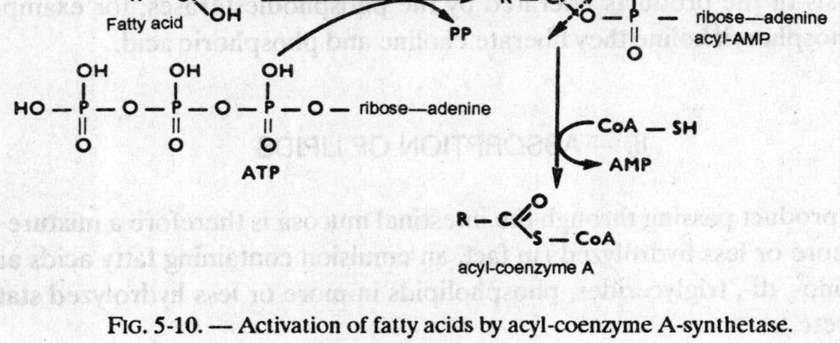

FIG. 5-10. — Activation of fatty acids by acyl-coenzyme A-synthetase.

In a first step, ATP reacts with the fatty acid to form a mixed anhydride (with liberation of pyrophosphate). This step is very similar to the reaction of activation of amino acids which will be studied in the chapter on the biosynthesis of proteins (see fig. 6-39). Then, in a second step, the acyl-AMP formed reacts with coenzyme A-SH to give a thio-ester, the *acyl-coenzyme A* (with liberation of AMP).

B. – Reactions of β-oxidation

The acyl-coenzyme A thus formed can then follow the pathway of *β-oxidation* which comprises the following reactions (see fig. 5-11).

– dehydrogenation under the action of an acyl-coA-dehydrogenase, a FAD enzyme, with formation of a α-β double bond in *trans* configuration;

– hydration of this double bond, catalyzed by *enoyl-coA hydratase* (*crotonase*), with formation of a β-hydroxylated derivative of L-configuration;

– dehydrogenation by *L-β-hydroxyacyl-coA-dehydrogenase*, a NAD^+ enzyme, with formation of a β-keto derivative;

– by intervention of a molecule of coenzyme A, detachment of a two carbon fragment in the form of acetyl-coA, catalyzed by *β-ketothiolase*. This is thiolysis or "thioclastic scission." There remains an acyl-coA having 2 carbon atoms less than the starting acyl-coA (or fatty acid). This set of 4 reactions can be summarized as follows:

$$CH_3(CH_2)_n—CO—S—coA + FAD + NAD^+ + coA—SH \longrightarrow$$
$$CH_3—(CH_2)_{n-2}—CO—S—coA + FADH_2 + NADH + H^+ + acetyl—coA$$

Then the acyl-coA formed (having 2 carbon atoms less than the starting fatty acid) will, in its turn, undergo the same series of 4 reactions to give an acetyl-coA and a new acyl-coA (having 4 carbon atoms less than the starting fatty acid) and so on. As proposed by Lynen, β-oxidation may be represented by a helix, each turn of which corresponds to a shortening of the fatty acid by 2 carbon atoms (liberated in the form of acetyl-coA) and a departure of 4H (see fig. 5-13).

The β-oxidation of fatty acids in an *intra-mitochondrial* process. The molecules of acetyl-coA formed will be used either in anabolic processes (biosynthesis of fatty acids and isoprenic lipids), or in ketogenesis, or they will be completely oxidized by the Krebs cycle. The first two possibilities will be studied in the discussion which follows in this chapter; the reactions of the Krebs cycle were examined in the previous chapter (see fig. 4-38).

Besides this β-oxidation, there exists a second pathway: the peroxysomal β-oxidation which takes place in the peroxysomes. The reactions are identical to those of the mitochondrial β-oxidation. Only the first enzyme is different. It transfers directly the 2 H atoms removed from the acyl-coenzyme A during the formation of the double bond, to cytochrome a_3 with formation of H_2O_2. In

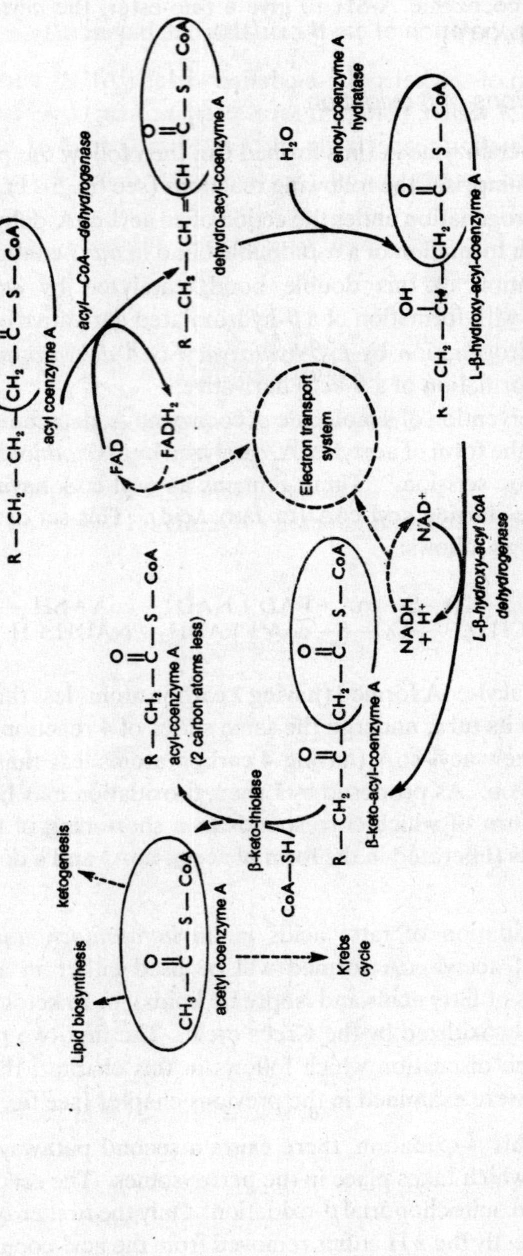

FIG. 5-11. — The 4 reactions constituting one turn of the β-oxidation of fatty acids.

plants, peroxysomal β-oxidation is the most important catabolism process of fatty acids. This system also exists in some tissues of mammals like the liver or kidney, and in yeasts, moulds and ciliates.

C. – *Energy balance of the β-oxidation of fatty acids*

Each turn of the helix of β-oxidation yields $FADH_2$ and NADH + H^+ the reoxidation of which, thanks to the electron transport chain, allows (see page 122) the formation respectively, of 2 and 3 molecules of ATP, i.e. a total of 5 ATP per turn (or per unit of 2 carbon atoms detached from the molecule of fatty acid). The acetyl-coA is then oxidized by the Krebs cycle with formation (as mentioned on page 195) of 12 ATP. *The complete oxidation of one unit of 2 carbon atoms therefore yields 17 ATP.* Referring to the example of stearic acid (see fig. 5-12) which has 18 carbon atoms, one can see that its complete oxidation will allow the formation of (8 × 5) + (9 × 12) i.e. 148 ATP, and after deducting the ATP required for the preliminary activation of the fatty acid, there is a gain of 147 ATP.

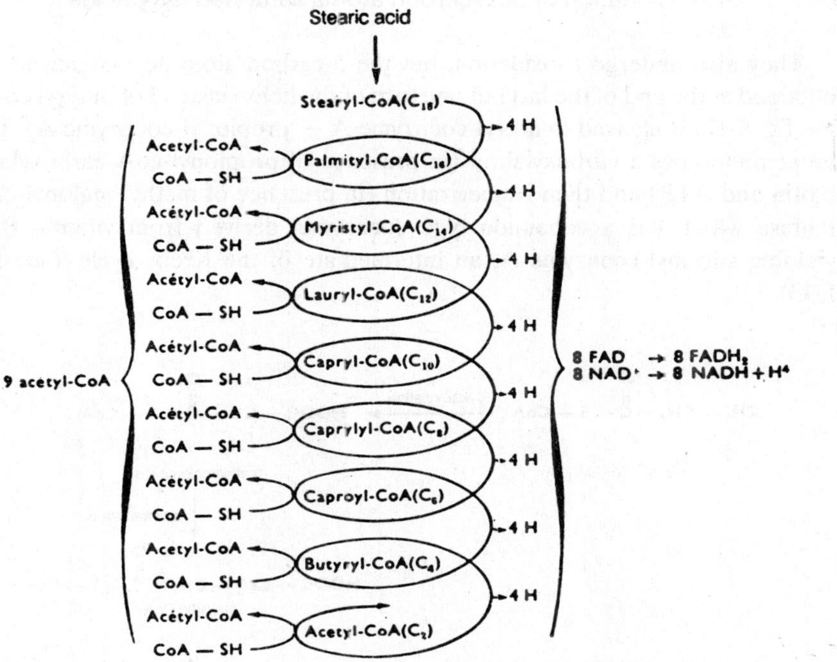

FIG. 5-12. — β-oxidation of stearic acid (Lynen's helix).

The energy yield per carbon atom is therefore of the order of 8 ATP (147/18) in the oxidation of fatty acids and 6 (38/6) in the oxidation of glucose (see page 195). Moreover, triglycerides contain many more carbon atoms per unit weight than the polysaccharides and can therefore constitute larger energy reserves for a much smaller weight.

At the end, let us point out that the enzyme implied in the β-oxidation of fatty acids are mainly located in the mitochondria, just like the enzymes responsible for electron transport and oxidative phosphorylation. One can therefore make the same observation as in the case of Krebs cycle (which is also an intra-mitochondrial process) : it appears that in the mitochondria there is a concentration of enzymatic systems which permits on the one hand, the oxidation of the energy reserve substances, and on the other hand, the formation of ATP, which must enhance the efficiency of these energy transfers.

However, due to intra-mitochondrial location of β-oxidation, the acetyl-coenzyme A which is required for the synthesis of fatty acids or sterols, has to leave the mitochondrion. It comes out in the form of *citrate* formed by reaction of acetyl-coenzyme A with the oxalacetate (see fig. 4-38). In the cytoplasm, citric acid is cleaved by ATP-citrate lyase which, in presence of a molecule of coenzyme A and ATP, yields one molecule of acetyl-coenzyme A and one molecule of oxalacetate (ATP being hydrolyzed to ADP + P_i).

2. – Oxidation of odd-carbon atoms saturated fatty acids

They also undergo β-oxidation, but the 5-carbon atom acyl-coenzyme A, obtained at the end of the last but one turn of the helix (instead of butyryl co-A, see fig. 5-12) is cleaved to acetyl-coenzyme A + propionyl-coenzyme A. The latter undergoes a carboxylation (in presence of propionyl-coA-carboxylase, biotin and ATP) and then isomerization (in presence of methyl malonyl-coA mutase which has a cobamide type coenzyme, derived from vitamin B_{12}), yielding succinyl-coenzyme A, an intermediate of the Krebs cycle. (see fig. 5-13).

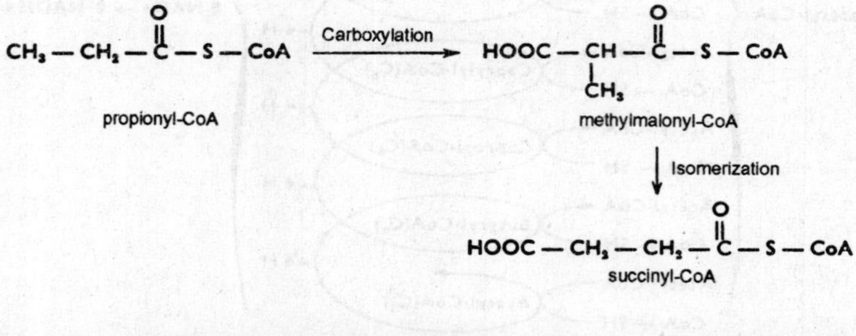

FIG. 5-13. — Transformation of propionyl-coenzyme A into succinyl-coenzyme A.

3. – Oxidation of unsaturated fatty acids

The catabolism of unsaturated fatty acids takes place by a modification of the β-oxidation consecutive to the position of double bonds. We shall take the

example of linoleic acid which is the most complex case (see fig. 5-15). The numbering of carbon atoms is indicated below each formula:

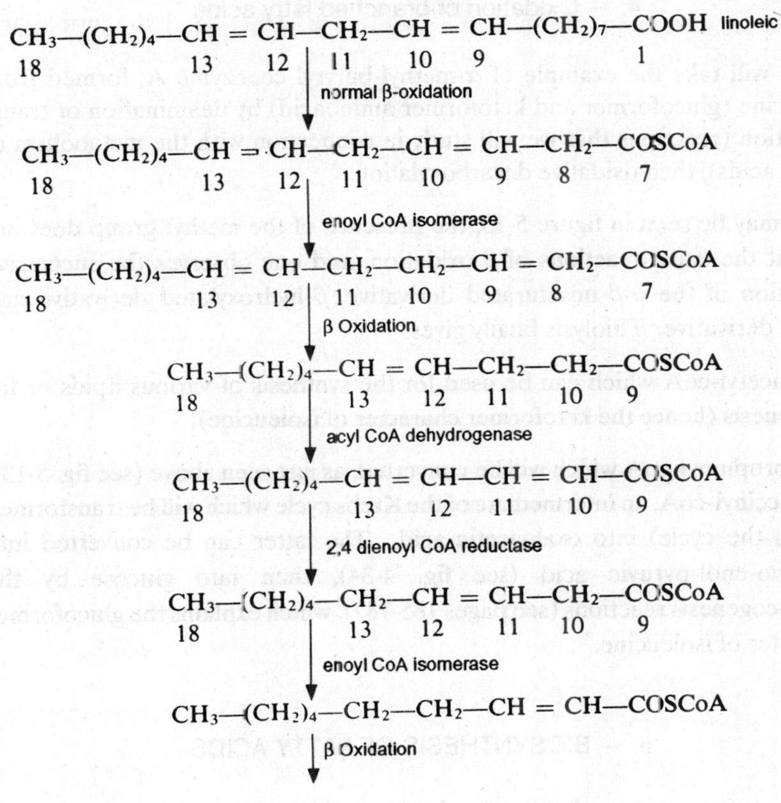

FIG. 5-14.—β-oxidation of unsaturated fatty acids.

The reaction begins with a conventional β-oxidation till formation of an acyl coA with 12 carbon atoms having a double bond *cis* in β-γ. Thanks to an isomerase, the latter is converted into a double bond *trans* placed in α-β. The acyl coA obtained will therefore continue to be catabolized by the conventional reactions of β-oxidation yielding a fatty acid with 10 carbon atoms. Acyl coA dehydrogenase introduces a double bond. After the action of 2,4 dienoyl coA reductase which has NADPH as coenzyme, the 2 double bonds are replaced by a single one localized in β-γ. The intervention of enoyl coA isomerase permits

the introduction of this latter compound into the normal cycle of β-oxidation. In the case of a fatty acid like oleic acid (18:1(9)), only the first 3 reactions of figure 5-14 come into play. This catabolism pathway is present in mitochondria and peroxysomes of the liver. It was found in mammals. An analogous system exists in fungi and bacteria.

4. – Oxidation of branched fatty acids

We will take the example of α-methyl-butyryl-coenzyme A, formed from isoleucine (glucoformer and ketoformer aminoacid) by deamination or transamination (reactions that we will study in connection with the metabolism of amino acids), then oxidative decarboxylation.

As may be seen in figure 5-15, the presence of the methyl group does not prevent the usual reactions of β-oxidation, and one observes the successive formation of the α-β-unsaturated derivative, β-hydroxylated derivative and β-keto derivative. Thiolysis finally gives:

– acetyl-coA which can be used for the synthesis of various lipids or for ketogenesis (hence the ketoformer character of isoleucine);

– propionyl-coA which will be converted, as just seen above (see fig. 5-13), into succinyl-coA, an intermediate of the Krebs cycle which will be transformed (within the cycle) into oxaloacetic acid. The latter can be converted into phospho-enol-pyruvic acid (see fig. 4-34), then into glucose by the neoglucogenesis reactions (see pages 185-187), which explains the glucoformer character of isoleucine.

II. – BIOSYNTHESIS OF FATTY ACIDS

1. – Synthesis of saturated fatty acids

It must be pointed out at the very outset that the biosynthesis of fatty acids does not generally take place by the reactions – in the reverse direction – of β-oxidation; the latter are indeed reversible in mammals, except the one catalyzed by acyl-coA dehydrogenase, but there is a dehydroacyl-coenzyme A-reductase, a NADPH enzyme, which can permit reduction on the double bond; it however appears that this mode of formation of fatty acids is of relatively limited importance. But the reverse pathway of β-oxidation is of great physiological interest, as it permits the elongation of pre-existing, medium-chain fatty acids, leading to stearic acid (C18), one of the principal saturated fatty acids of tissues, and to long-chain (C20 to C26) fatty acids. This system is intra-mitochondrial. However, in some organisms the short-chain fatty acids can be synthesized by the reactions – in the reverse direction – of β-oxidation.

In mammals, the major pathway of the biosynthesis of fatty acids is an extramitochondrial process (cytosolic and/or microsomal). To produce fatty acids from the precursor which is acetyl-coA, the cells must be able to *reduce the ketone groups*: this will be achieved thanks to NADPH; they must also be able *to form C–C bonds* in order to condense acetyl radicals : although the methyl group of acetyl-coA is capable of binding to a carbonyl (see for example the binding of acetyl-coA to acetoacetyl-coA in figure 5-24, or to oxaloacetate for entry into the Krebs cycle, in figure 4-38), this is not the reaction used for obtaining chains of fatty acids; the cells use a more reactive intermediate, malonyl-coA.

The synthesis of *malonyl-coenzyme A* consists in the binding of a molecule of CO_2 to a molecule of acetyl-coA, catalyzed by *acetyl-coA-carboxylase*, a biotine enzyme, in presence of ATP, according to the mechanism described in figure 5-16. This is an example of CO_2 binding which can be carried out by living beings (we have seen, in figure 4-34, another example in connection with the transformation of pyruvic acid into oxaloacetic acid by pyruvate-carboxylase).

In procaryotes, the acyl groups of acetyl-coA and malonyl-coA are transferred, respectively by an acetyl-transferase and a malonyl-transferase, to a small protein (M.W. # 9 000) called *Acyl Carrier Protein* or *ACP*. The prosthetic group of this protein is *phosphopantetheine* bound by an ester linkage between its phosphate group and the hydroxyl of a serine of the protein. Phosphopantetheine bears a close resemblance to coenzyme A; the sulphydryl group is again the active part in the binding and transfer of acyl groups (see fig. 5-17).

Acetyl-ACP and malonyl-ACP then react with *acyl-synthetase* (see below), responsible for the synthesis of fatty acids. Transfer of acyl groups to the polyenzymatic complex takes place, without any free intermediate at any time.

In the case of mammals and numerous other eucaryotes studied, the process is simplified. There is no intervention of any acyl group carrier protein (ACP). The acetyl-coA and malonyl coA (the latter being synthesized by acetyl-coA carboxylase) react directly with *acyl-synthetase*. This enzyme is a multienzymatic complex of M.W. = 2.3×10^6, which possesses "binding" sites ending by a –SH group: a cysteine radical or a phosphopantetheine radical. A transfer of acetyl and malonyl radicals therefore takes place from a –SH group (that of the coenzyme A) directly to another (that of the synthetase). The first condensation can take place as indicated in figure 5-18. It must be noted that this condensation is accompanied by a decarboxylation affecting the CO_2 previously bound by the action of acetyl-coA-carboxylase, which is therefore not incorporated in the fatty acids.

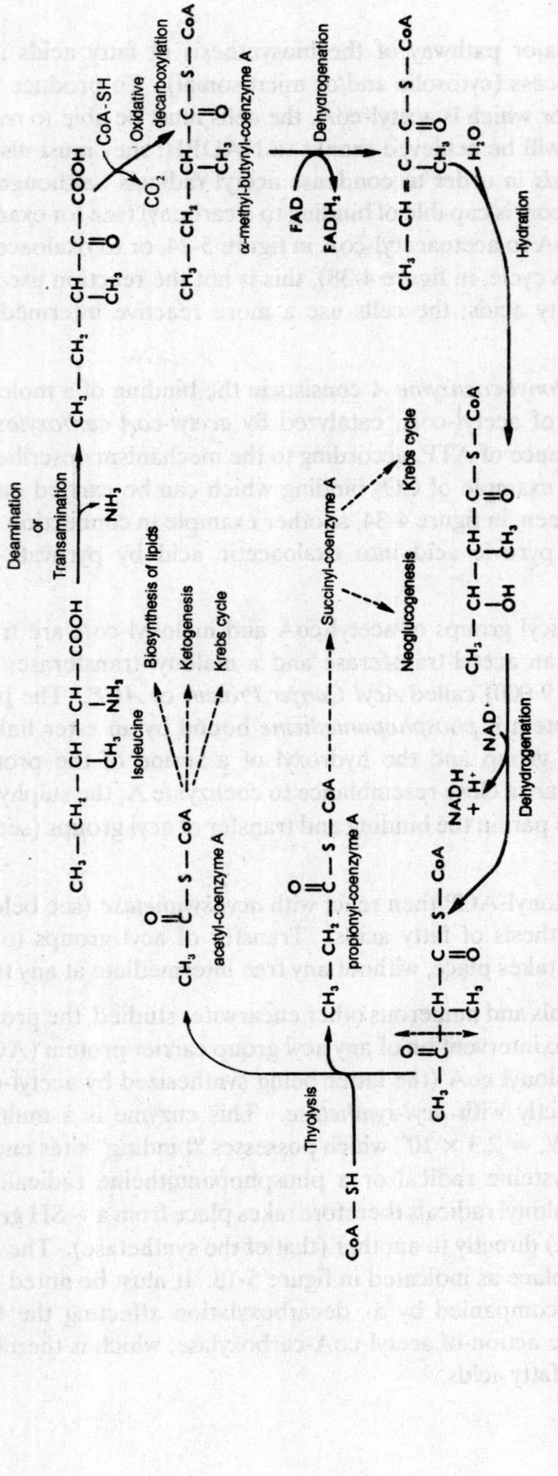

FIG. 5-15. — Oxidation of a branched fatty acid. Example of α-methyl-butyryl-coenzyme A.

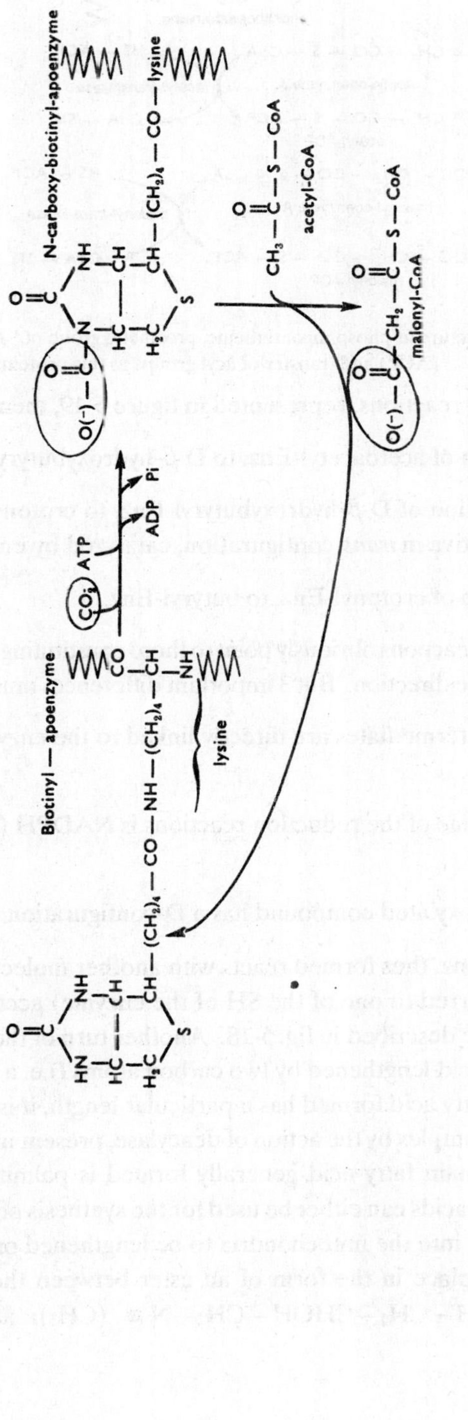

FIG. 5-16. — Carboxylation of acetyl-coenzyme A to malonyl-coenzyme A.

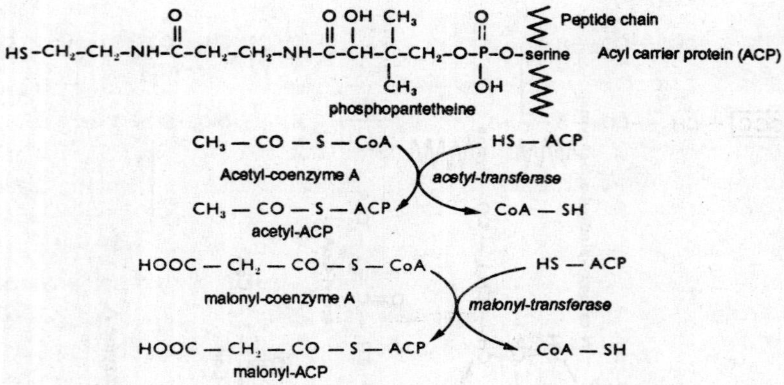

FIG. 5-17. — Structure of phosphopantetheine, prosthetic group of "Acyl Carrier Protein" (ACP) and transfer of acyl groups to this protein.

The following reactions, represented in figure 5-19, then take place:

– a reduction of acetoacetyl-Enz. to D-β-hydroxybutyryl-Enz.,

– a dehydration of D-β-hydroxybutyryl-Enz. to crotonyl-Enz., an α-β unsaturated derivative in *trans* configuration, catalyzed by enoyl-hydratase

– a reduction of crotonyl-Enz. to butyryl-Enz.,

This series of reactions obviously point to those constituting one turn of β-oxidation, in the reverse direction. But 3 important differences must be noted:

– here the intermediates are directly linked to the enzyme (and not to the coenzyme A),

– the coenzyme of the reduction reactions is NADPH (and not $FADH_2$ or NADH),

– the β-hydroxylated compound has a D-configuration (not L).

The butyryl-Enz. thus formed reacts with another molecule of malonyl-coA (which is transferred to one of the SH of the enzyme) according to a process similar to the one described in fig. 5-18. Another turn of the helix will produce a chain of fatty acid lengthened by two carbon atoms (i.e. a chain in C_6) and so on. When the fatty acid formed has a particular length, it is liberated from the polyenzymatic complex by the action of deacylase, present in the acylsynthetase complex. The main fatty acid generally formed is palmitic acid (C_{16}). The synthesized fatty acids can either be used for the synthesis of glycerides or other lipids, or carried into the mitochondria to be lengthened or catabolized. This transport takes place in the form of an ester between the alcohol group of *carnitine*: $COOH-CH_2-CHOH-CH_2-N \equiv (CH_3)_3$ and the fatty acid,

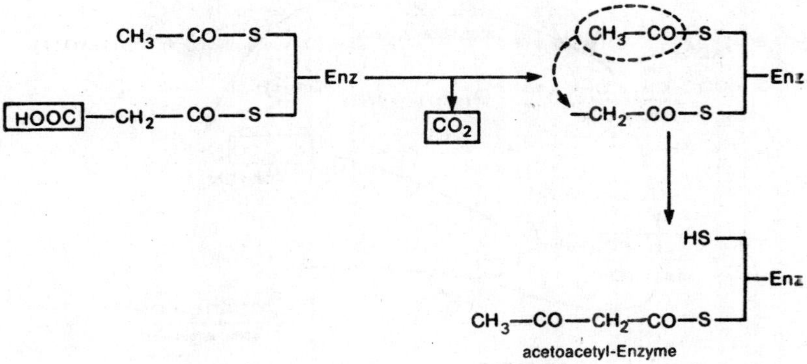

FIG. 5-18. — First condensation of the biosynthesis of a fatty acid.

called acylcarnitine. The esterification reaction is catalyzed by carnitine palmityl transferase.

Plants possess a double system. The cytosolic system uses acetyl-coenzyme A. The system located in the chloroplast uses acetyl-ACP. The biosynthesis reactions are similar to those described above. Palmitic acid *and* stearic acid are formed. The latter is lengthened (probably in the endoplasmic reticulum) by a system requiring malonyl coA.

2. – Synthesis of unsaturated fatty acids

A. – *Monounsaturated fatty acids*

There are 2 systems, one anaerobic present in some bacteria (*E.coli*), the other aerobic present in all other cells.

The anaerobic synthesis is carried out by the enzymatic complex synthesizing the saturated fatty acids with the following variant: the 10 carbon atom β-hydroxyacyl-ACP will be dehydrated (see fig. 5-19) to give simultaneously a α, β-dehydroacyl-ACP (C_{10}, Δ^2) and a β, γ-dehydroacyl-ACP (C_{10}, Δ^3). Only the former will be reduced by NADPH + H^+, the latter maintains its double bond and will be again lengthened in the conventional manner. One will therefore obtain successively:

$$C_8 \text{ (saturated)} + \text{malonyl-ACP} \begin{cases} \nearrow C_{10}, \Delta^2 \longrightarrow C_{10} \longrightarrow C_{12} \longrightarrow C_{14} \text{ etc.} \\ \longrightarrow C_{10}, \Delta^3 \longrightarrow C_{12}, \Delta^5 \longrightarrow C_{14}, \Delta^7 \\ \qquad\qquad \longrightarrow C_{16}, \Delta^9 \longrightarrow C_{18}, \Delta^{11} \end{cases}$$

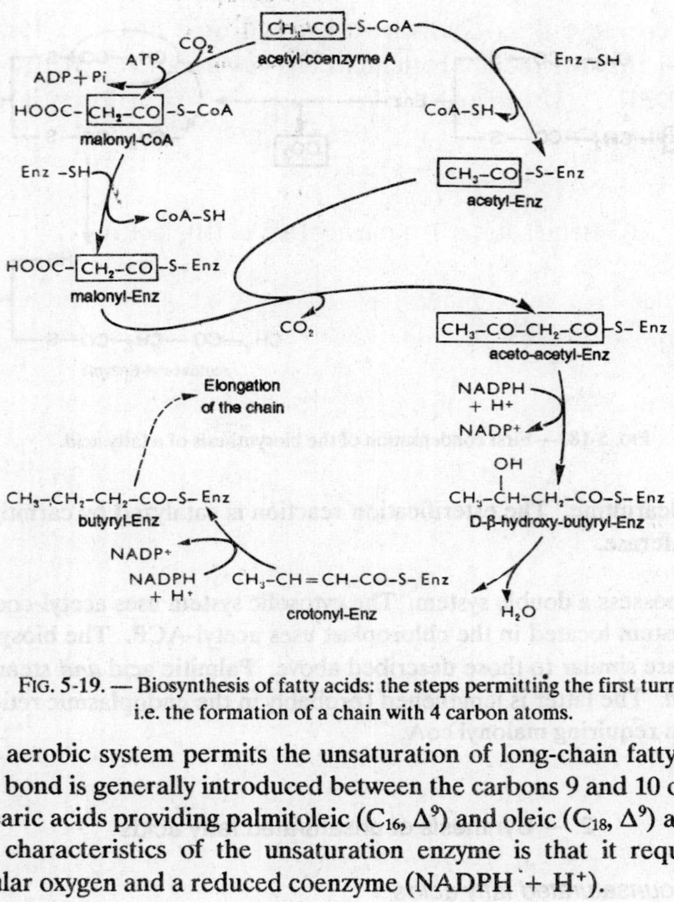

FIG. 5-19. — Biosynthesis of fatty acids: the steps permitting the first turn, i.e. the formation of a chain with 4 carbon atoms.

The aerobic system permits the unsaturation of long-chain fatty acids. A double bond is generally introduced between the carbons 9 and 10 of palmitic and stearic acids providing palmitoleic (C_{16}, Δ^9) and oleic (C_{18}, Δ^9) acids. One of the characteristics of the unsaturation enzyme is that it requires both molecular oxygen and a reduced coenzyme (NADPH + H^+).

B. — Polyunsaturated fatty acids

As far as polyunsaturated fatty acids are concerned, only non-bacterial microorganisms and plants are capable of synthesizing linoleic acid (C_{18}, $\Delta^{9,12}$) and α-linolenic acid (C_{18}, $\Delta^{9,12,15}$) by unsaturation of oleic acid. Some insects synthesize linoleic acid. The synthesis of oleic and linoleic acids takes place in the endoplasmic reticulum, whereas that of linoleic acid takes place in the chloroplasts and seems to be linked with chlorophyll synthesis. Linoleic and linolenic acids which are not synthesized by several groups of animals (numerous insects, mammals...) are called *essential fatty acids*. Contrary to plants, animals can introduce new double bonds in these two essential fatty acids to give polyunsaturated fatty acids like arachidonic acid (C_{20} $\Delta^{5,8,11,14}$) or docosahexaenoic acid (C_{22} $\Delta^{4,7,10,13,16,19}$). This biosynthesis is microsomal. It takes place by a series of reactions wherein unsaturations and elongations alternate [ex: 18:2 (9,12) $\longrightarrow$ 18:3 (6,9,12) $\longrightarrow$ 20:3 (8,11,14) $\longrightarrow$ 20:4

(5,8,11,14) ⟶ 22.5 (7,10,13,16) ⟶ 22.6 (4,7,10,13,16)]. An identical diagram is operative if one starts from linolenic acid 18:3 (9,12,15). The unsaturation enzymes require both molecular oxygen and a reduced coenzyme (NADPH). Lengthening takes place by the pathway involving malonyl-coenzyme A.

3. Regulation of the metabolism of fatty acids

The fact that the pathway of biosynthesis of fatty acids is different from the pathway of oxidation allows – as in the case of the biosynthesis and degradation of glycogen, or in glycolysis and neoglucogenesis – an independent regulation of the 2 processes. These regulation mechanisms will not be studied here, but some important factors may be mentioned. The biosynthesis of fatty acids requires *NADPH* which is mainly provided by the oxidation of glucose in the pentose-phosphates cycle (see fig. 4-43). It also requires energy and therefore the presence of ATP, supplied by the oxidation of carbohydrates (or, in plants, by photosynthesis); if ATP concentration decreases (and therefore ADP concentration increases), biosynthesis slows down, but on the contrary β-oxidation is stimulated, which will lead to a rise in ATP concentration.

The reaction catalyzed by *acetyl-coA-carboxylase* is the limiting step of the biosynthesis of fatty acids. This enzyme is activated by citric acid or insulin, but inhibited by glucagon or fatty acids, whether they be the terminal products of acylsynthetase (feedback inhibition mechanism) or of exogenous origin, for example nutritional. An accumulation of fatty acids may also result from a deficiency of *L-α-glycerophosphoric acid* which – as will be seen in the following paragraph (see fig. 5-21) – is the compound to which the acyl-coA bind in the biosynthesis of glycerides and glycerophospholides; but this compound is formed from triosesphosphates (see fig. 4-32). It is important to note that the 3 factors we have just mentioned (NADPH, ATP, L-α-glycero-phosphoric acid) are provided for their major part, by the metabolism of carbohydrates, which underlines the close relations existing between carbohydrates and lipids in respect of metabolism and regulation (not forgetting the important link represented by acetyl-coA).

Synthesis and catabolism of fatty acids are 2 competitive mechanisms which are regulated, at least in mammals. In order to penetrate into the mitochondrion, the fatty acids must be in the form of acyl carnitine. Malonyl coA, an intermediate in the biosynthesis, is a powerful inhibitor of carnitine palmityl transferase (see p. 241), thus blocking the β oxidation. When synthesis stops, the fatty acids can be esterified by carnitine and penetrate into the mitochondrion where they will be catabolized.

METABOLISM OF OTHER LIPIDIC COMPOUNDS

I. – METABOLISM OF GLYCEROLIPIDS

1. – Synthesis and degradation of glycerophospholipids

The common precursor of all phosphatides is *phosphatidic acid* (see fig. 5-3). There are two pathways for carrying out the biosynthesis of this lipid. In the first, glyceraldehyde-3-phosphate, which is one of the intermediates of glycolysis (see fig. 4-26) is first reduced in presence of NADH, to L-α-glycerophosphoric acid. (A specific kinase also permits, in presence of ATP, the phosphorylation of free glycerol). Glycerophosphoric acid will react successively with two molecules of acyl-coenzyme A, in presence of an acyl transferase, to give *phosphatidic acid* (see fig. 5-21). The second pathway starts from dihydroxy-acetone-phosphate which reacts with an acyl-coenzyme A and gives *acyl-dihydroxy-acetone-phosphate*. The ketone group is then reduced in presence of NADPH and the compound obtained will lead, after binding of a second molecule of acyl-coenzyme, to phosphatidic acid (see fig. 5-21). This second pathway permits the specific binding to glycerophosphate of saturated fatty acids in the position 1 and unsaturated fatty acids in the position 2.

The biosynthesis of non-nitrogenous phosphatides (phosphatidyl-glycerol, phosphatidyl-inositol, cardiolipid) takes place from phosphatidic acid. The latter first yields CDP-diglyceride (which may also be written CMP-P-diglyceride) after reaction with cytidine-triphosphate or CTP[1]. *Phosphatidyl-inositol* (see fig. 5-21) is obtained after the transfer of the "P-diglyceride" fraction to inositol. A similar reaction gives *phosphatidyl-glycerol*, precursor of the cardiolipid.

The biosynthesis of nitrogenous phosphatides (*phosphatidyl-choline, phosphatidyl-ethanolamine, phosphatidyl-serine*) begins with a hydrolysis of phosphatidic acid under the influence of a specific phosphatase (see fig. 5-21). The diglyceride thus formed will then react with a derivative of choline (or ethanolamine) formed as follows: choline (or ethanolamine) is first phosphorylated by ATP under the influence of a specific kinase. Then, the phosphoryl-choline (or phosphoryl-ethanolamine) thus formed reacts with CTP (see fig. 5-20) to give *cytidine-diphospho-choline* (or cytidine-diphospho-ethanolamine) called *CDP-choline* (or CDP-ethanolamine). Writing compounds in the form CMP-P-choline (or CMP-ethanolamine), affords a better understanding of what takes place during the reaction with the diglyceride in the formation of phosphatidyl-cholines and phosphatidyl-ethanolamines.

---

[1] Structures of cytidylic nucleotides and particularly that of CTP are given in the chapter on nucleic acids (see fig. 6-7).

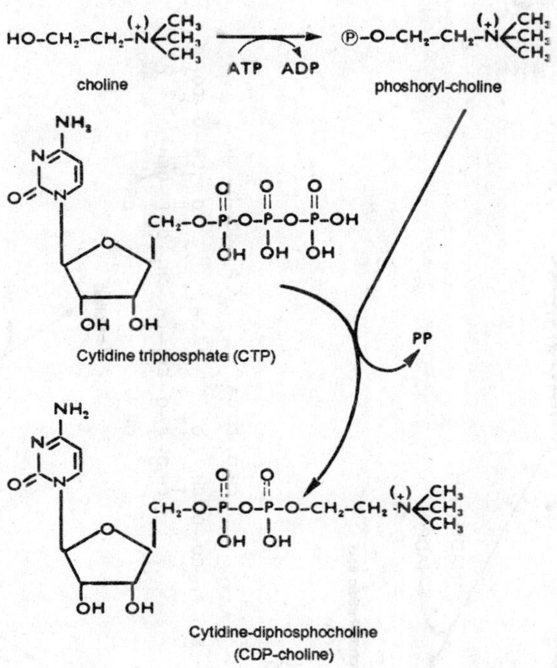

FIG. 5-20. — Formation of CDP-choline.

Indeed, the last step (see fig. 5-21) consists of a transfer, from CMP-P-choline (or CMP-P-ethanolamine), of phosphoryl-choline (or phosphoryl-ethanolamine) to the diglyceride, thus forming phosphatidyl-choline (or phosphatidyl-ethanolamine). Phosphatidyl-serine is on the contrary synthesized from phosphatidyl-ethanolamine by an exchange of the two bases. Lastly, there is a pathway which permits the passage from phosphatidyl-ethanolamine to phosphatidyl-choline by the binding to the amino group of phosphatidyl-ethanolamine of 3 methyl groups originating from S-adenosyl-methionine (see fig. 5-21).

Phosphatidylserine is synthesized by a base exchange reaction. Phosphatidylserine can be retransformed into phosphatidyl ethanolamine by decarboxylation.

A deficiency in choline or methionine (which are not synthesized by mammalian cells) will lead to a marked decrease in the synthesis of phosphatidylcholine in the liver. As a result, the diglycerides used for the synthesis of this phosphatide are then transformed into triglycerides which accumulate (hepatic steatosis).

Phosphatidylinositol and cardiolipid (phosphatide very abundant in procaryotes) are formed from CDP diglycerides (see fig. 5-21). In bacteria, phosphatidylserine is exclusively synthesized by the reaction CDP-diglyceride + serine ⟶ CMP + phosphatidylserine. The latter is decarboxylated to

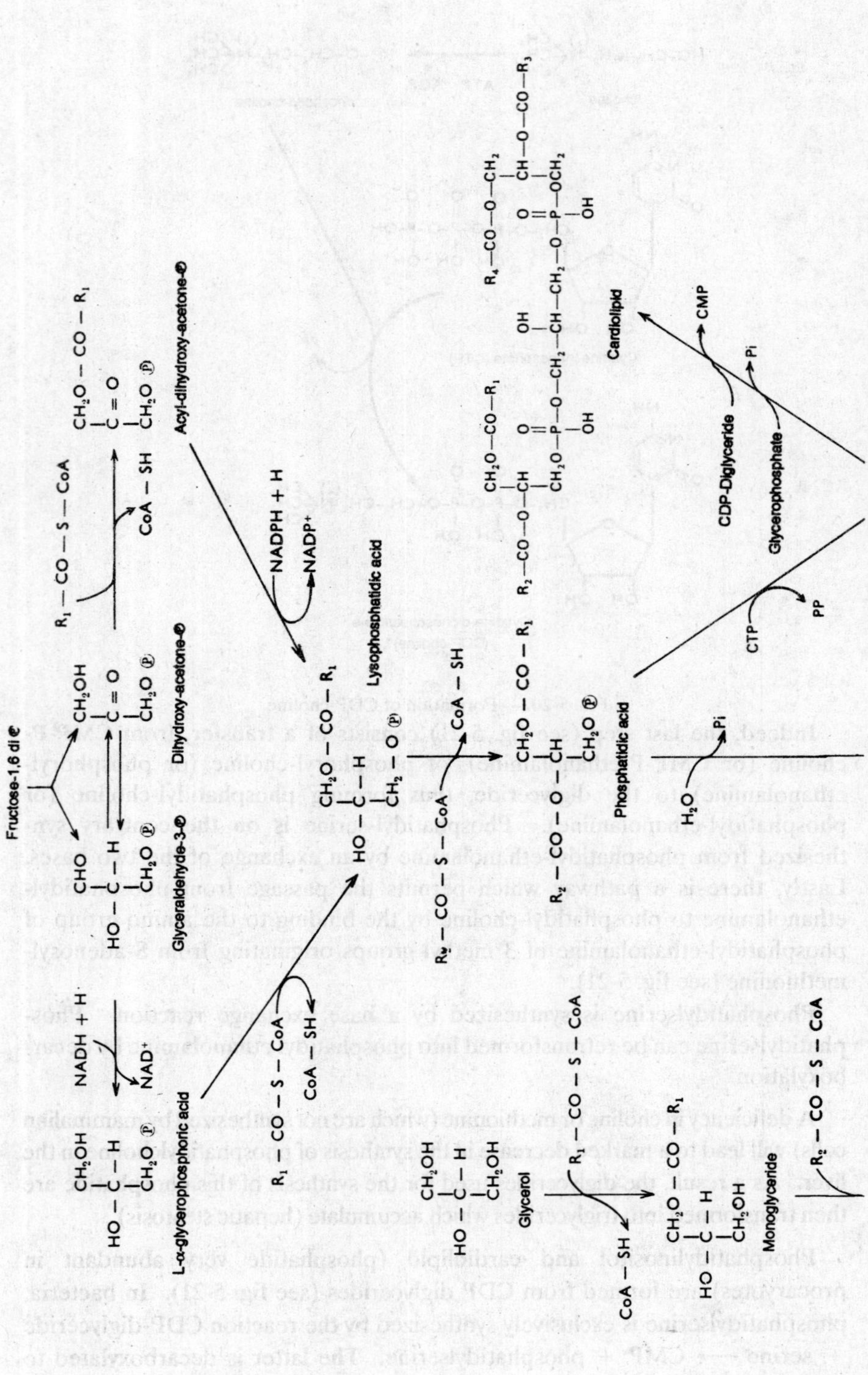
Fructose-1,6 di ℗
Glyceraldehyde-3-℗
Dihydroxy-acetone-℗
Acyl-dihydroxy-acetone-℗
L-α-glycerophosphoric acid
Lysophosphatidic acid
Phosphatidic acid
Glycerol
Monoglyceride
CDP-Diglyceride
Glycerophosphate
Cardiolipid

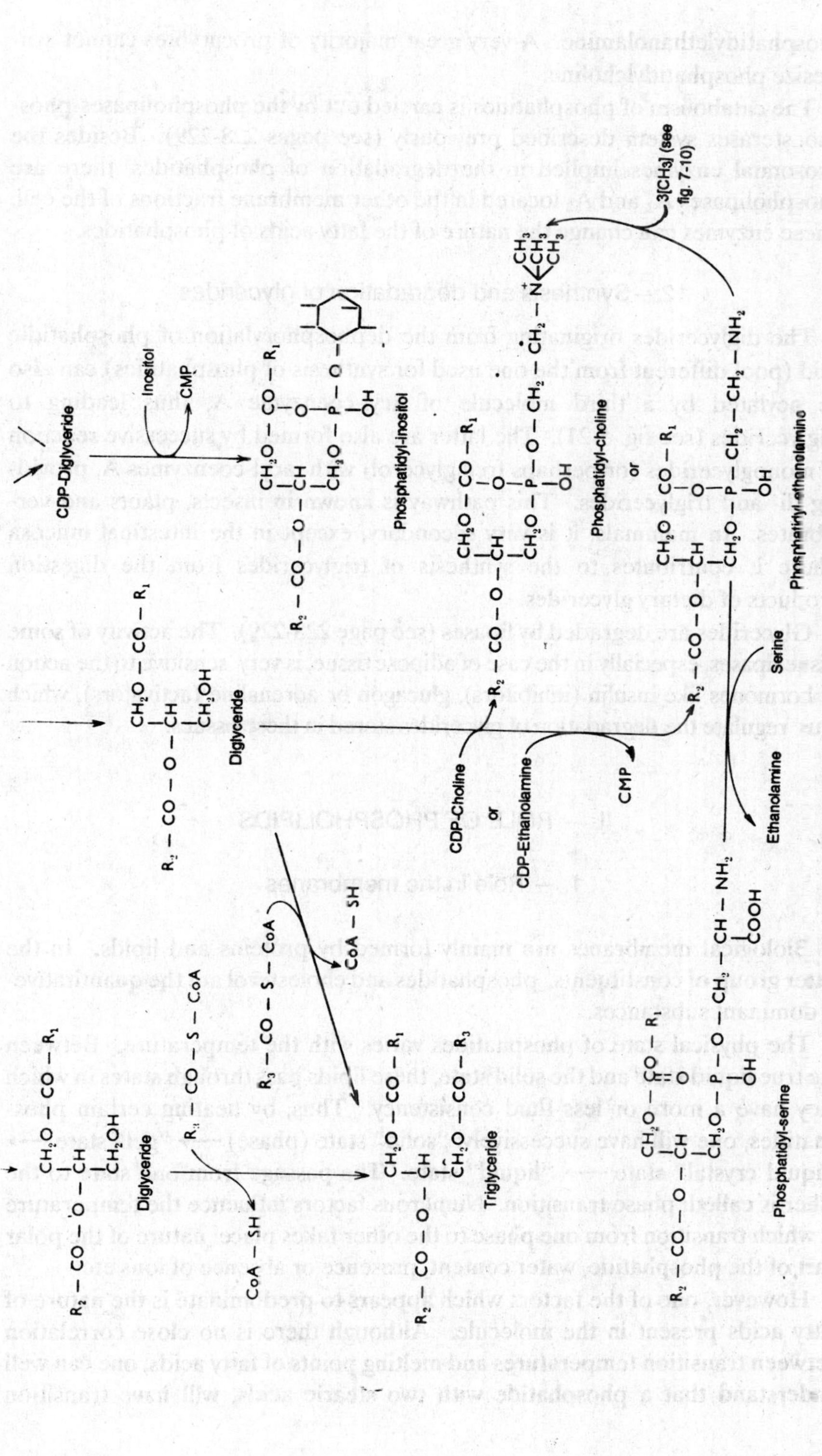

FIG. 5-21. — Biosynthesis of glycerides and glycerophospholipids.

phosphatidylethanolamine. A very great majority of procaryotes cannot synthesize phosphatidylcholine.

The catabolism of phosphatides is carried out by the phospholipases-phosphoesterases system described previously (see pages 228-229). Besides the lysosomial enzymes implied in the degradation of phosphatides, there are phospholipases A_1 and A_2 located in the other membrane fractions of the cell. These enzymes can change the nature of the fatty acids of phosphatides.

2. – Synthesis and degradation of glycerides

The diglycerides originating from the dephosphorylation of phosphatidic acid (pool different from the one used for synthesis of phosphatides) can also be acylated by a third molecule of acyl-coenzyme A, thus leading to triglycerides (see fig. 5-21). The latter are also formed by successive reaction of monoglycerides (or perhaps free glycerol) with acyl-coenzymes A, providing di- and triglycerides. This pathway is known in insects, plants and vertebrates. In mammals, it is very secondary, except in the intestinal mucosa where it contributes to the synthesis of triglycerides from the digestion products of dietary glycerides.

Glycerides are degraded by lipases (see page 228-229). The activity of some tissue lipases, especially in the case of adipose tissue, is very sensitive to the action of hormones like insulin (inhibitors), glucagon or adrenaline (activators), which thus regulate the degradation of glycerides stored in these tissues.

II. – ROLE OF PHOSPHOLIPIDS

1. – Role in the membranes

Biological membranes are mainly formed by proteins and lipids. In the latter group of constituents, phosphatides and cholesterol are the quantitatively dominant substances.

The physical state of phosphatides varies with the temperature. Between the true liquid state and the solid state, these lipids pass through states in which they have a more or less fluid consistency. Thus, by heating certain phosphatides, one will have successively: "solid" state (phase) ⟶ "gel" state ⟶ "liquid crystal" state ⟶ "liquid" state. The passage from one state to the other is called: phase transition. Numerous factors influence the temperature at which transition from one phase to the other takes place: nature of the polar part of the phosphatide, water content, presence or absence of ions etc.

However, one of the factors which appears to predominate is the nature of fatty acids present in the molecule. Although there is no close correlation between transition temperatures and melting points of fatty acids, one can well understand that a phosphatide with two stearic acids, will have transition temperatures much higher than the same phosphatide having 2 arachidonic acids (see page 214). In almost all the biological membranes studied, the arrangement of fatty acids in the phosphatides (presence of a saturated fatty acid and an unsaturated fatty acid) is such that the latter are, at the "biological" temperatures (between 20 and 40°), in a gel or "liquid crystal" state which represents a more or less fluid state. It would appear, at least in mammals that the phosphatides of a particular composition in fatty acids could play a special role (for example, like some molecules of phosphatidyl-choline of the lung which have both their fatty acids saturated and which play a role of surface active lipids). Cholesterol serves as a "buffer". It "stabilizes" the biological membranes in a fluid state. It acts by fluidizing the structure which are too rigid (for example, myelin which would "crystallize" in absence of cholesterol) or by rigidifying those excessively fluid.

Studies, using numerous physical techniques, have shown that biological membranes have the same phase transitions as the lipids they contain. On measuring the gel – "liquid crystal" transition temperature, it is observed that it is similar for the membrane and the lipids isolated from this membrane. It may therefore be concluded that the membrane lipids are the main agents responsible for the physical state of the membrane.

This physical state is of great importance in certain phenomena. Membranes are permeability barriers. One may well imagine that the movement of numerous molecules which must cross the membrane (whatever the mechanism permitting this passage) will be more or less easy depending on the degree of fluidity of the membrane. Similarly, certain enzymes change their conformation during their action on a substrate. An excessively rigid environment can inhibit this change in conformation and therefore inhibit enzymatic activity.

Numerous organisms, especially the unicellular ones, are capable of growing at very different temperatures (e.g., 10 to 40°). These organisms possess systems which can adjust the fluidity of their lipids, i.e. of their membrane, according to the temperature. In particular, the activity of enzymes which unsaturate the fatty acids is very sensitive to temperature. The same is true of the enzymes responsible for the incorporation of fatty acids in phosphatidic acids (see fig. 5-21) and therefore in phosphatides. Variations can thus exist in the structure of phosphatides incorporated in the membranes; this maintains the fluidity compatible with proper cellular functioning.

2. – Anchoring of (glyco) proteins

One of the modes of anchoring of (glyco)proteins in cellular plasmatic membranes takes place thanks to a phosphatide: phosphatidyl inositol. The structure of the anchoring complex would be: phosphatidyl inositol – (ethanolamine-monosaccharides) – protein, the diacyl-glycerol of the phosphatide being inserted in the membrane.

Several proteins (acetylcholinesterase, 5′ nucleotidase for example in mammals and also a surface glycoprotein in *Trypanosoma brucei*, a proteinase of *Leishmania major*,...) are attached to the membrane in this manner.

Simpler anchoring systems are also known. A fatty acid (generally, myristic acid) is linked by an amide group to an α-amino group of the N-terminal residue. This is the case for example, of NADH: cytochrome b_5 reductase or several viral proteins. If the N-terminal residue of protein is cysteine, it is linked by amide linkage with palmitic acid as previously, but in addition by thioether linkage with a diacylglycerol. This is the case of the penicillinase of the *Bacillus* or the receptor of transferrin for example.

3. — Cycle of inositol triphosphate

The binding to the membrane receptor of some compounds including neurotransmitter (acetylcholine, adrenaline, dopamine), hormones (vasopressin...), mitogens (growth factors), as well as various other compounds like thrombin (protease) or the platelet activation factor (PAF), leads to the activation of a phospholipase C which specifically hydrolyzes phosphatidyl-inositol diphosphate (Phyl IP2) to diglycerides and inositol triphosphate (IP3). The diglycerides formed can activate a protein kinase (kinase C). Inositol triphosphate brings about the mobilization of intracellular calcium and induces the activation of Ca^{++} – dependent enzymes, particularly those with their activity modulated by calmoduline, an activator protein sequestering calcium. These phenomena are very rapid. IP3 is inactivated by phosphatases which hydrolyse it to inositol. Phosphatidyl-inositol diphosphate is resynthesized from inositol and glycerides or phosphatidic acid (see diagram). In view of its role, inositol triphosphate is considered to be a secondary mediator.

In some cases, the diglyceride formed, instead of activating a kinase C, will be hydrolyzed by a phospholipase A with liberation of arachidonic acid which can either be transformed into prostaglandines, or directly activate adenylate cyclase with formation of cyclic AMP. This cycle of inositol is also present in

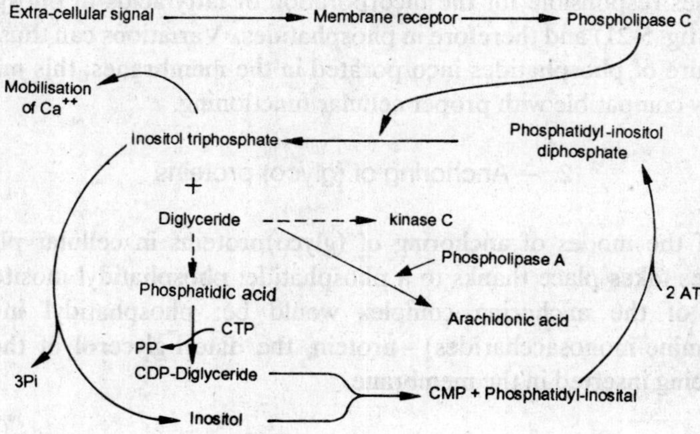

plants, but its exact role is not yet known. Light would be one of the activators. As in mammals, the liberated IP_3 is a mobilizing agent of the intracellular Ca^{++}.

A simpler cycle exists in cells of mammals. Stimulation by compounds like the *Epidermal Growth Factor* (EGF), a growth factor, activates a phospholipase C which hydrolyzes phosphatidylcholine with liberation of diglyceride which will activate a kinase C.

III. – BIOSYNTHESIS OF STEROIDS

1. – Biosynthesis of cholesterol

The main steps of the biosynthesis of cholesterol are diagrammatically represented in figure 5-22. The first reaction consists of the condensation of 2 molecules of acetyl-coA. It is the reverse of the reaction which takes place during the last turn of the helix in β-oxidation (see fig. 5-11). Then a third molecule of acetyl-coA binds to the acetoacetyl-coenzyme A thus formed which gives β-hydroxy-β-methyl-glutaryl coenzyme A (HMG coA). This binding of an acetyl-coenzyme A to a carbonyl group is similar to the reaction permitting the entry of acetyl-coenzyme A in the Krebs cycle by condensation on oxaloacetic acid. The reduction of the acid group (engaged in a thioester linkage) to alcohol, catalyzed by HMG coA reductase, gives *mevalonic acid*. It must be noted that all the carbon atoms of cholesterol originate from acetyl-coenzyme A.

A pyrophosphate group will then bind to the primary alcohol group of mevalonic acid. Mevalonyl-pyrophosphate will react with a third molecule of ATP; this reaction gives an unstable compound which decomposes spontaneously, losing the tertiary alcohol group and the free carbonyl group. An isoprene derivative with 5 carbon atoms is formed, *isopentenyl-pyrophosphate* which can be isomerized to *dimethyl-allyl-pyrophosphate*. The condensation of 2 fragments in C_5 gives *geranyl-pyrophosphate* (C_{10}), and after the binding of a third fragment in C_5, *farnesyl-pyrophosphate* (C_{15}) is obtained. The dimerization of the latter leads to *squalene* (C_{30}).

In vertebrates, the cyclization of squalene by squalene-oxidocyclase takes place by a series of reactions requiring molecular oxygen and a reducing coenzyme like NADPH, and leads to *lanosterol*. The passage from lanosterol to cholesterol takes place through several parallel pathways. The most important intermediates are desmosterol and 7-dehydrocholesterol, immediate precursors of cholesterol.

The isoprene derivatives with 5 carbon atoms are the precursors of dolichols, of the side chain of ubiquinone and vitamin K, of the isopentenyl group of some tRNAs (see page 263).

In vertebrates, the biosynthesis of cholesterol is microsomal. It is regulated by a feedback inhibition mechanism by a metabolite of cholesterol (most

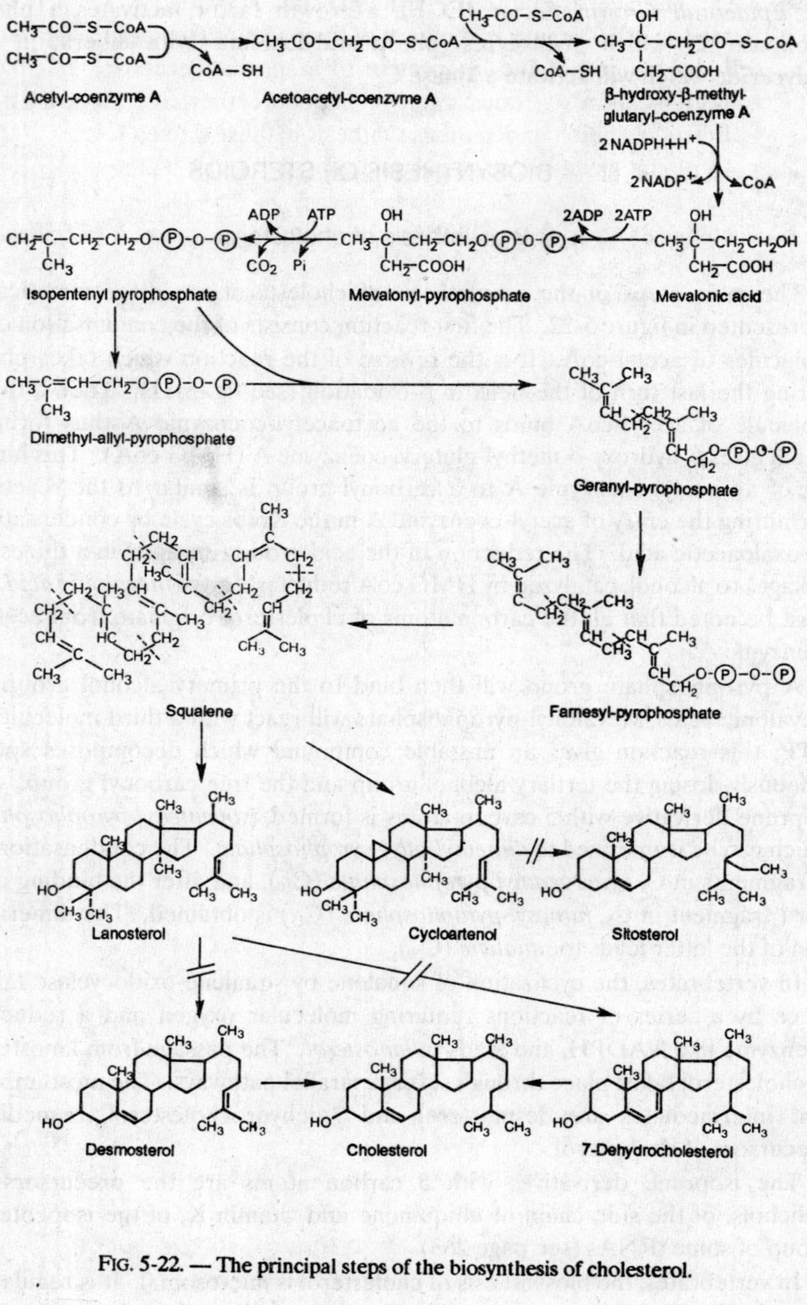

FIG. 5-22. — **The principal steps of the biosynthesis of cholesterol.**

probably the 25 hydroxycholesterol) acting on the HMG coA reductase. This feedback inhibition is never total to always permit the synthesis of polyisoprenoids important for other metabolisms, like the dolichols (see page 221). The liver is one of the principal sites of synthesis. Cholesterol is then carried to other organs in the form of lipoproteins (see page 230). It enters the cells by binding of the lipoprotein to a specific receptor. In physiological conditions the exogenous input of hepatic cholesterol to various tissues is sufficient to inhibit endogenous synthesis in these tissues.

In plants, the cyclization of squalene takes place similarly as in vertebrates except for the last step. One obtains cycloartenol, precursor of other sterols. Cycloartenol-oxidocyclase also requires O_2 and a reducing coenzyme. A part of sterols are glycosylated in presence of UDP-glucose, before being incorporated into the membranes. In plants, there are two synthesis sites. Chloroplastic synthesis leads to squalene and permits the formation of terpenes, carotenoids, xantophylls, polyisoprene chains of plastoquinones, vitamins K and E. In cytosol, on the contrary, the synthesized squalene is cyclized, yielding the various plant sterols. There is apparently no feedback inhibition mechanism in these organisms.

Insects are capable of synthesizing squalene, but they cannot cyclize it. They use the sterols (animal or plant) present in their food and are capable of metabolizing them to cholesterol, precursor of hormonal derivatives like ecdysone.

Procaryotes also synthesize squalene. Squalene-hopene cyclase which carries out the cyclization (see page 224) is an enzyme which does not require oxygen.

2. – Biosynthesis of bile acids

The transformation of cholesterol into bile acids takes place by two distinct processes. The shortening of the side chain first requires the oxidation of one of the terminal methyl groups with formation of a carboxyl. A process of β-oxidation of branched fatty acids (see fig. 5-15) gives a derivative with 24 carbon atoms, possessing a carboxylic group ending the side chain. At the same time hydroxyls (in α conformation) are introduced on the carbons 7 or 12. The transformation of the alcohol group 3 β of cholesterol into 3 α takes place by oxidation of alcohol to ketone and then reduction to a new alcohol group (see fig. 5-23).

3. – Formation of other steroids

As shown by the diagram of figure 5-23, cholesterol is the starting point of the synthesis of various steroids:

– *progesterone* is secreted by the corpus luteum, the placenta and the cortex of the adrenal gland, and acts mainly in the uterus to permit implantation and gestation;

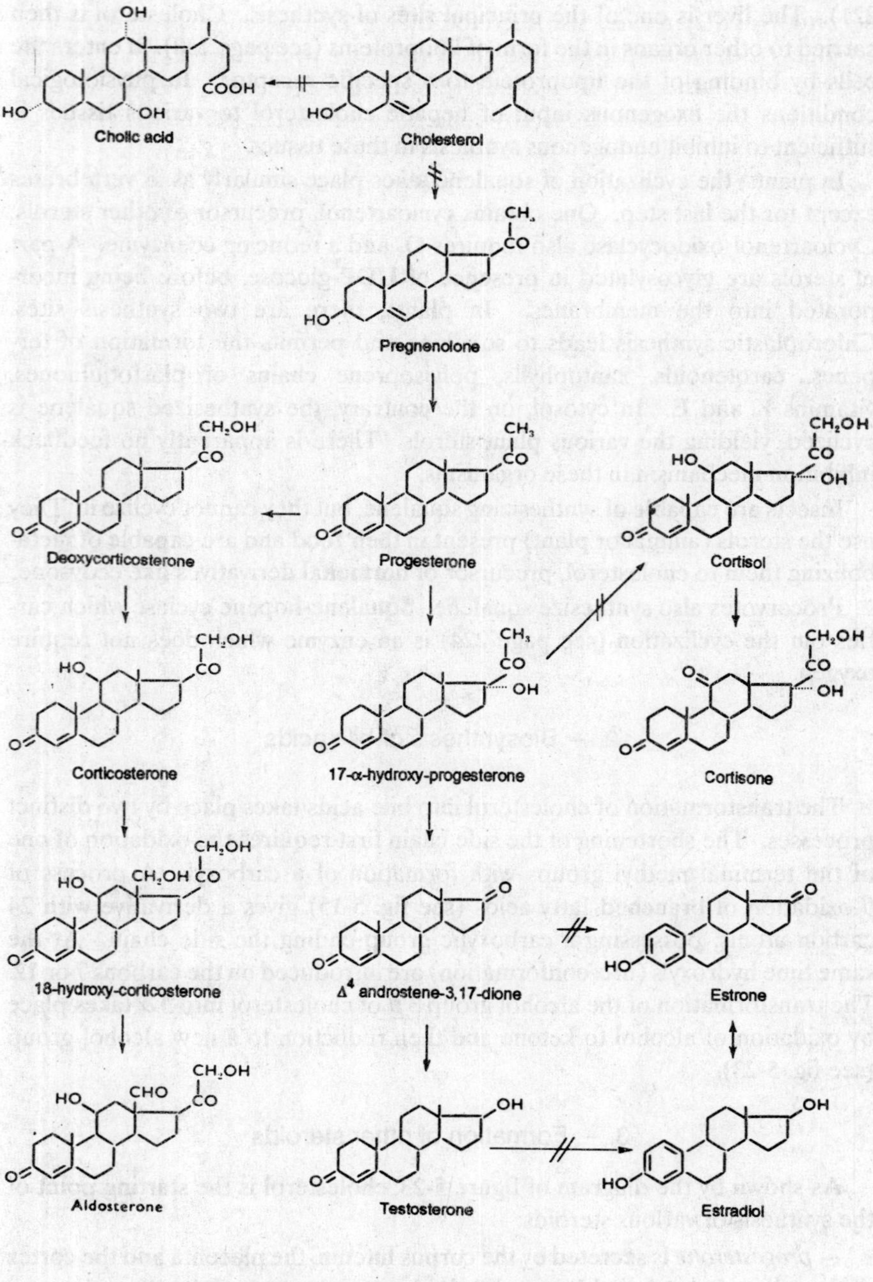

FIG. 5-23. — Formation of various steroids from cholesterol. (simplified diagram; all the intermediates are not shown).

— *aldosterone*, a hormone secreted by the adrenal cortex, which permits the reabsorption of sodium (and secondarily of chlorine and water) in the kidney, hence its name, *mineralocorticosteroid*;

— *cortisol* and *cortisone*, also secreted by the adrenal cortex, sometimes called glucocorticosteroids, because they stimulate protein catabolism and neoglucogenesis in the liver (they are therefore hyperglycemic). They also act on conjunctive and limphoid tissues by depressing membrane permeability and opposing the inflammatory processes (which explains their use in therapeutics). The synthesis of cortisol (and therefore of cortisone) is stimulated by the corticotropic hormone of the anterior lobe of the pituitary gland or ACTH (*A*dreno *C*ortico *T*ropic *H*ormone);

— *testosterone*, secreted mainly by the testicles, is responsible for the various male sexual characters;

— *estrogenic hormones* (estradiol and estrone), responsible for the various female sexual characters, synthesized mainly in the ovary and placenta, and characterized — from the structural point of view — by a phenolic ring.

It may be observed that it is relatively easy to pass (in few steps) from progesterone to other hormones having very different physiological properties; in other words, in this family of steroid hormones, small structures modifications correspond to large differences in biological activity.

We mentioned that the synthesis of cortisol and cortisone by the adrenal glands is influenced by ACTH, a hormone of anterior lobe of the pituitary gland; we must indicate that the secretory activities of ovaries and testicles are also controlled by the *hormones* of the anterior lobe of the pituitary gland called *gonadotropins*, like FSH (*F*ollicule *S*timulating *H*ormone) and LH (*L*uteinizing *H*ormone), for example. Moreover, this anterior lobe also secretes other stimulines, like the growth hormone or *somatotropic hormone*, and *thyreotropic hormone* (TSH), which stimulates the synthesis of thyroid hormones by the thyroid gland.

Chemists sometimes classify the steriod hormones according to the number of carbon atoms contained in their molecules and thus distinguish:

— C_{18} hormones (estradiol, estrone);

— C_{19} hormones (testosterone);

— C_{21} hormones (progesterone and most of the hormones secreted by the cortical part of adrenal glands, the mineralocorticoids like aldosterone as well as the glucocorticoids like cortisol and cortisone).

Vitamins D are formed by the opening of the B cycle due to ultraviolet light, either from ergosterol (vitamin D_2), or from 7-dehydro-cholesterol (which gives vitamin D_3).

IV. — KETOGENESIS

In vertebrates, the acetyl-coenzyme A originating from the β-oxidation can leave the mitochondrion in the form of citrate (see fig. 4-38). However a

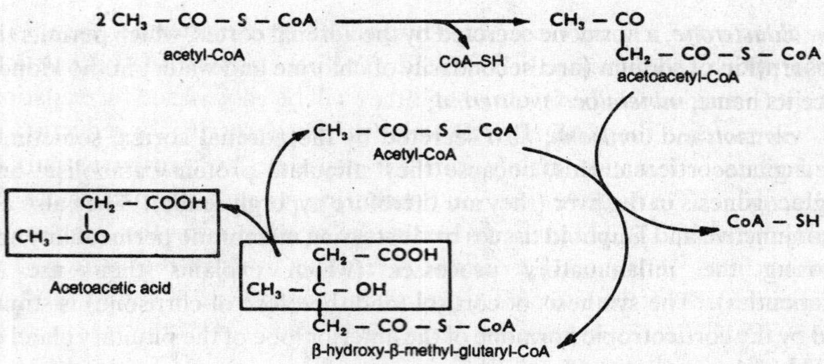

FIG. 5-24. — Formation of acetoacetic acid by the hydroxymethyl-glutaryl-coenzyme A cycle.

notable part of this acetyl-coA is used for the synthesis of *ketone substances*: acetoacetic acid, β-hydroxy-butyric acid and acetone. This process, called *ketogenesis*, takes place in the mitochondrion. The first steps are similar to those of cholesterol synthesis (see fig. 5-24). The β-hydroxy-β-methyl-glutaryl-coenzyme A formed is split into acetyl-coenzyme A and free acetoacetic acid.

The acetoacetic acid can be

— either spontaneously decarboxylated to *acetone:*

$$CH_3-CO-CH_2-COOH \longrightarrow CO_2 + CH_3-CO-CH_3$$

— or reduced to β-hydroxybutyric acid:

$$CH_3—CO—CH_2—COOH + NADH + H^+ \longrightarrow CH_3—CHOH—CH_2—COOH + NAD^+$$

Acetoacetic acid and β-hydroxybutyric acid, which are produced in large quantities, are utilized as energy source by various tissues like muscles, kidney and the brain. Ketone substances do indeed diffuse freely across the cell membranes. Furthermore, these tissues, contrary to the liver, are capable of reactivating acetoacetic acid by the following reaction:

$$\text{Acetoacetic acid} + \text{Succinyl} - \text{coA} \longrightarrow \text{Succinic acid} + \text{Acetoacetyl} - \text{coA}$$

The latter can be split by thiolysis into two molecules of acetyl-coA which will enter the Krebs cycle (see fig. 5-11). Ketogenesis is a very important physiological process. It is exclusively hepatic. The regulation of ketogenesis depends mainly on the concentration of 2 hormones: insulin and glucagon. In mammals a decrease in blood glucose concentration depresses insulin and increases glucagon. A change in the concentration of these hormones has the following main consequences in the liver: arrest of glycolysis (a very minor process in the liver), increase of neoglucogenesis, inhibition of the synthesis of fatty acids at the level of acetyl-coA carboxylase (see pages 237 and 239). The same hormonal change brings about in the adipose tissue, a hydrolysis of triglycerides by action on the hormone-dependent lipase (see page 245). The fatty acids thus liberated will be transported to the liver. The inhibition of acetyl-coA carboxylase depresses malonyl-coenzyme A concentration. The latter compound is a repressor of acylcarnitine synthetase (see page 240), enzyme which will therefore be activated. The penetration of fatty acids into the mitochondrion will be favoured and they will be catabolized by oxidation. This will increase the NADH/NAD ratio which will result in a reduction of oxaloacetate (originating mainly from the aspartate by transamination) to malate, which can leave the mitochondrion (and will participate in neoglucogenesis). The reduction of oxaloacetate will prevent the transformation into citrate of the acetyl-coenzyme A resulting from the β-oxidation. The only possible fate of the latter will therefore be the synthesis of ketone substances which will freely leave the hepatocyte to be captured by the other tissues capable of utilizing them.

V. — METABOLISM OF SPHINGOLIPIDS

Sphingosine results from the condensation of palmityl-coenzyme A with serine. The next step consists in the binding of a fatty acid which gives a ceramide. This ceramide can react either with CDP-choline to form a sphingomyelin, or with UDP-galactose to give a cerebroside, or with other UDP-monosaccharides to yield the other sphingoglycolipids (the monosaccharides are transferred one after the other by a specific enzyme). The sialic acid of gangliosides is transferred by reaction of a glycolipid with CMP-sialic acid (CMP-NeuAc). These reactions are diagrammatically represented in figure 5-25. The catabolism of these compounds takes place by reverse reactions catalyzed by specific osidases at each step.

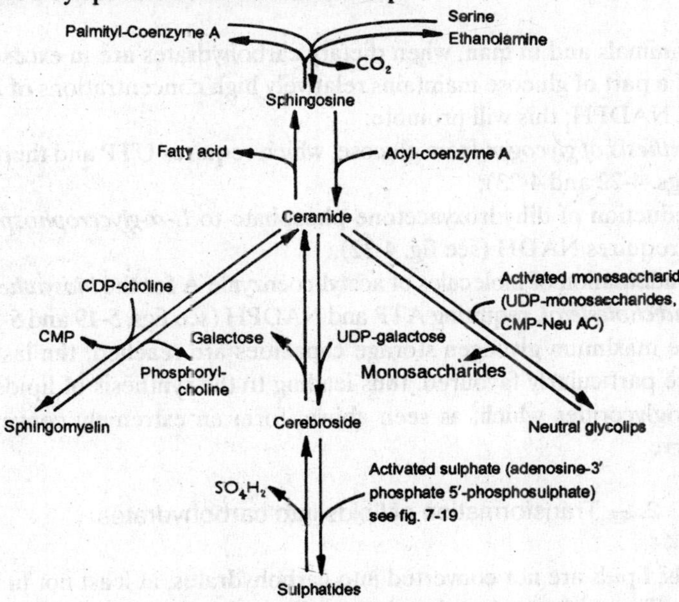

FIG. 5-25 — Biosynthesis and catabolism of sphingolipids.

The glycosyltransferases and glycosidases involved in this metabolism are different from those involved in the metabolism of glycoproteins. In the case of glycolipids, the substrates being insoluble in aqueous medium, the enzymes require protein activators in order to act.

Let us note that the synthesis and catabolism of glycosyldiglycerides take place according to a similar diagram, the primary acceptor being a diglyceride and not a ceramide.

VI. – ROLE OF GLYCOLIPIDS

Sphingolipids are mainly localized in the plasmatic cellular membrane. Numerous glycolipids and gangliosides have antigenic properties, at least in mammals. Hexaglycosylceramide represented in figure 5-4 is for example, a determinant of B group blood. Some gangliosides are cellular receptors for bacterial toxins (GM1, represented in figure 5-4, is the receptor of cholera toxin). Other gangliosides modulate the accessibility of receptors, like that of the thyreotropic hormone (TSH), or the secretion of growth factors (lymphokins) by the lymphocytes.

VII. – INTERRELATIONS BETWEEN CARBOHYDRATES AND LIPID METABOLISMS

1. – Transformation of carbohydrates into lipids

In most animals and in man, when dietary carbohydrates are in excess, the oxidation of a part of glucose maintains relatively high concentrations of ATP, NADH and NADPH; this will promote:

– the *synthesis of glycogen* from glucose, which requires UTP and therefore ATP (see figs. 4-22 and 4-23);

– the reduction of dihydroxyacetone-phosphate to *L-α-glycerophosphoric acid*, which requires NADH (see fig. 4-32).

– the condensation of molecules of acetyl-coenzyme A for the *biosynthesis of fatty acids and cholesterol*, requiring ATP and NADPH (see figs. 5-19 and 5-22).

When the maximum glycogen storage capacities are reached, the last two pathways are particularly favoured, thus leading to the synthesis of lipids and especially triglycerides which, as seen above, form an extremely convenient energy reserve.

2. – Transformation of lipids into carbohydrates

In animals, lipids are not converted into carbohydrates, at least not in large proportions. The oxidative decarboxylation of pyruvic acid to acetyl-coenzyme

A is indeed not reversible. It is true that the carbon atoms of acetyl-coenzyme A (originating for example from the oxidation of fatty acids) can, after entering the Krebs cycle, be found in oxaloacetic acid, then in phosphoenolpyruvic acid and be incorporated in glucose thanks to neoglucogenesis (see figs. 4-34 and 4-35), but a mere glance at the diagram of figure 4-38 will show that this incorporation takes place only after one complete turn of the cycle, (comprising 2 decarboxylations) and one cannot therefore speak of actual synthesis of glucose.

On the contrary, in plants this conversion takes place, especially in grains rich in lipids, which can be rapidly transformed into carbohydrates during germination; it also takes place in some microorganisms. The key-reaction of this transformation is the condensation of acetyl-coenzyme A on glyoxylic acid, catalyzed by *malic acid synthetase*; here also, acetyl-coenzyme A binds by its methyl group to a carbonyl (as it happens during its condensation on oxaloacetic acid for its entry into the Krebs cycle and as in the synthesis of β-hydroxy-β-methyl-glutaryl-coenzyme A). As for *glyoxylic acid*, it results from the scission of isocitric acid by *isocitratase*. These two reactions are represented in figure 5-26.

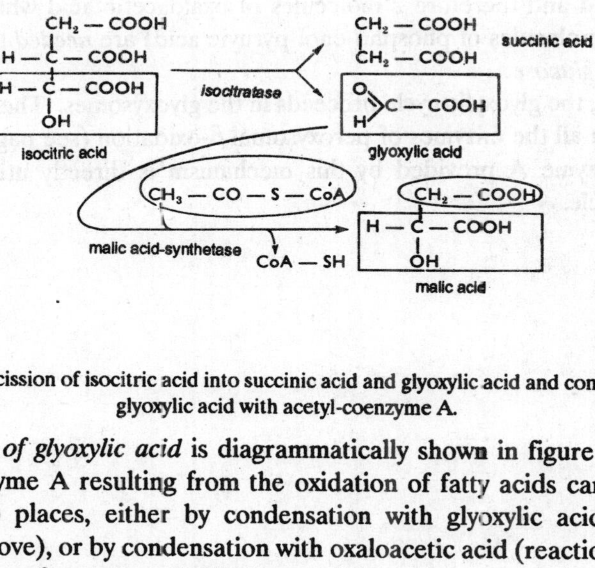

FIG. 5-26. — Scission of isocitric acid into succinic acid and glyoxylic acid and condensation of glyoxylic acid with acetyl-coenzyme A.

The *cycle of glyoxylic acid* is diagrammatically shown in figure 5-27. The acetyl-coenzyme A resulting from the oxidation of fatty acids can enter the cycle at two places, either by condensation with glyoxylic acid (reaction described above), or by condensation with oxaloacetic acid (reaction identical to the reaction of entry into the Krebs cycle).

Each turn of the cycle sees the formation of one molecule of *succinic acid* at the cost of 2 molecules of acetyl-coenzyme A. Succinic acid is oxidized to oxaloacetic acid by a series of reactions which we examined while studying the Krebs cycle (see fig. 4-38), and oxaloacetic acid can give phospho-enol-pyruvic acid by decarboxylation and phosphorylation (see fig. 4-34); the latter acid can be converted into glucose-6-phosphate by the pathway of neoglucogenesis (see

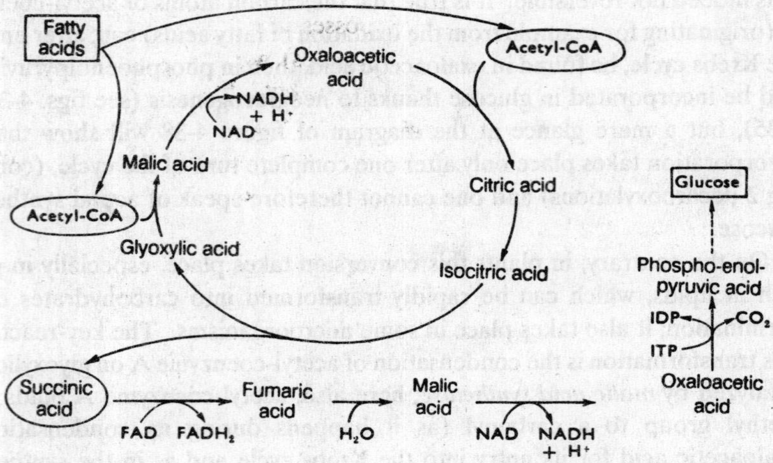

FIG. 5-27. — Transformation of acetyl-coenzyme A into glucose by the glyoxylic acid cycle followed by other reactions already studied.

fig. 4-35). *Four molecules of acetyl-coenzyme A* (these will give 2 molecules of succinic acid and therefore 2 molecules of oxaloacetic acid which will yield $2CO_2$ + 2 molecules of phospho-enol-pyruvic acid) are *needed to obtain one molecule of glucose*.

In plants, the glyoxylic cycle proceeds in the glyoxysomes. These structures also contain all the enzymes of peroxysomal β-oxidation (see page 231). The acetyl-coenzyme A provided by this mechanism is directly utilized by the glyoxylic cycle.

CHAPTER 6

STRUCTURE AND METABOLISM OF NUCLEIC ACIDS GENETIC CODE AND BIOSYNTHESIS OF PROTEINS

STRUCTURE OF NUCLEIC ACIDS

Nucleic acids are macromolecules present in all living cells, either in free state, or in combination with proteins to form nucleoproteins. Besides, the simplest viruses consist exclusively of a molecule of nucleic acid enclosed in a protein shell. We will see in the following that nucleic acids are of fundamental importance because *they are either carriers of genetic information, or agents enabling the expression of this information.*

Nucleic acids are polymers consisting of units called *nucleotides*; they are hence called *polynucleotides*. It may be recalled here that nucleotides are not only compounds leading to nucleic acids on polymerization, but also substances which can play an important role in intermediate metabolism, say, for energy storage (ATP), or as coenzymes (NAD^+, $NADP^+$, FAD, coenzyme A; see figs. 2-16, 2-17, 2-18), or for the synthesis of polysaccharides or of lipids (see for example, the role of UDP glucose in the synthesis of glycogen in figure 4-22, or the role of CDP-choline in the synthesis of phosphatidyl-choline in figure 5-21, or in the control of various processes (see for example, the role of cyclic AMP as relay of hormonal action pages 481-486).

Nucleotides differ from monomers which constitute the other macromolecules already studied (glucose of starch or glycogen, amino acids of proteins) in that they can themselves be further hydrolyzed into their 3 constituents : *nitrogenous heterocyclic base, pentose and phosphoric acid.* We will study these constituents, at least the first two, in some more details, but we may state at the outset that two types of nucleic acids are distinguished depending on whether the pentose is deoxyribose or ribose.

1) *The deoxyribonucleic acids* (DNAs) which are mainly found in the chromosomes in the nucleus of plant and animal cells; in procaryotes (bacteria, blue-green algae) it is also DNA which forms the chromosomes. Some viruses,

especially bacterial viruses (or bacteriophages) and animal viruses have a genome consisting of DNA. Furthermore, DNA is also found in mitochondria of plant and animal cells and in chloroplasts of photosynthetic organisms.

2) *The ribonucleic acids* (RNAs), mainly found in the cytoplasm of cells. There are various types of ribonucleic acids (ribosomal ribonucleic acids, transfer ribonucleic acids, messenger ribonucleic acids) involved in the expression of genetic information; we will study them in detail later in this chapter. Besides, some viruses, especially plant viruses and also some animal viruses and bacteriophages, have a genome consisting of RNA.

I. – PENTOSES

In ribonucleic acids, the sugar is *β-D-ribose*, in deoxyribonucleic acids it is *β*-2-deoxy-D-ribose (see fig. 6-1). These 2 sugars are of the D-series due to their configuration on carbon 4 which is the carbon next to the one carrying the primary alcohol group (see pages 141-142). They are represented in the *β*-form because it is in this form that they are found in nucleotides (and therefore in nucleic acids). Let us point out that 2-O-methyl-pentoses are sometimes found in some ribonucleic acids.

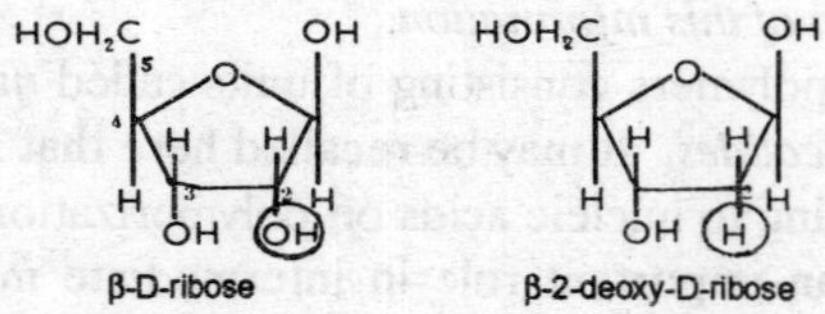

FIG. 6-1. — Structures of ribose and deoxyribose.

II. – NITROGENOUS BASES

All nitrogenous bases derive from 2 heterocyclic bases, purine and pyrimidine; we therefore have :

1. – Purine bases or substituted purines

The two principal purine bases found in deoxyribonucleic acids as well as ribonucleic acids are *adenine* and *guanine* (see fig. 6-2). There are also other purine bases, but they are found only in some nucleic acids (especially the transfer ribonucleic acids or *t*RNAs) and in small quantities, hence their name, *rare bases* or *abnormal bases* or *minor bases* (see fig. 6-2); they are often methylated derivatives of the two major bases, but there are some having more complex structures (for example, isopentenyl-adenine, which is a cytokinine with hormonal activity in plants); about fifty minor bases (purine and pyrimidine bases) are known at present.

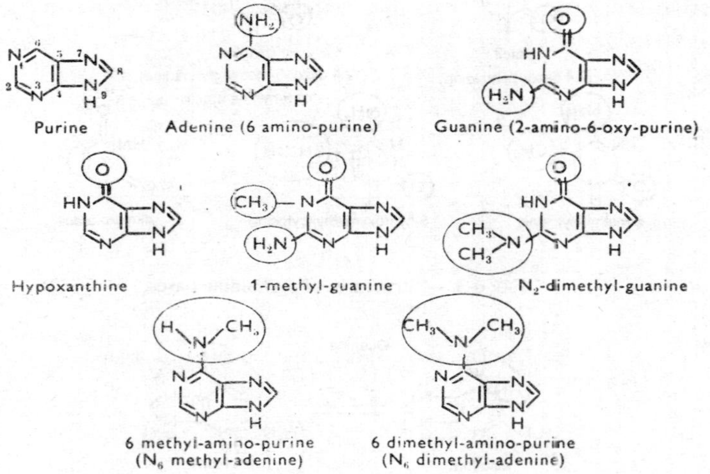

FIG. 6-2. — Structures of purine bases.

2. – Pyrimidine bases or substituted pyrimidines

Cytosine and *uracil* are found in ribonucleic acids; *cytosine* and *thymine*, in deoxyribonucleic acids. One also finds, particularly in the deoxyribonucleic acids of some bacteriophages, abnormal bases like 5-hydroxymethylcytosine (see fig. 6-3) instead of cytosine (in even T phages: T_2, T_4, T_6), or hydroxymethyluracil instead of thymine (in some phages infecting *B. subtilis*). Among the rare bases found in the transfer ribonucleic acids, one must cite dihydrouracil and thymine (the latter is a normal base in deoxyribonucleic acids and a rare base in ribonucleic acids).

3. – Tautomerism of bases

The structure of oxygen-containing bases is given in figures 6-2 and 6-3 in the *keto* (or lactam) form. This is the common form at the physiological pH, but as shown by figure 6-4, there is equilibrium between this form and the *enol* (or lactim) form.

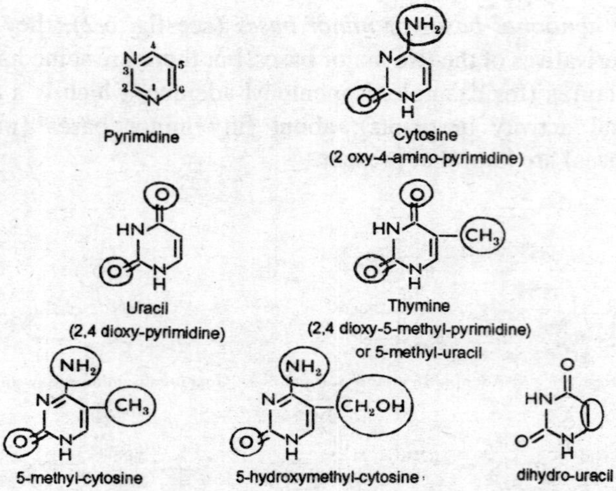

FIG. 6-3. — Structures of pyrimidine bases.

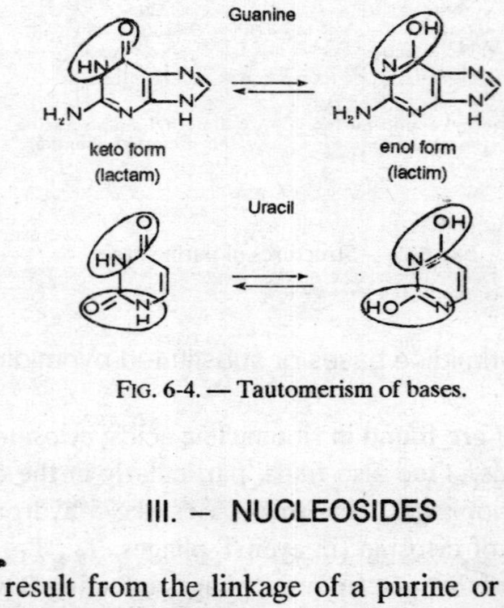

FIG. 6-4. — Tautomerism of bases.

III. – NUCLEOSIDES

Nucleosides result from the linkage of a purine or pyrimidine base with ribose or deoxyribose. This linkage joins nitrogen 9 of the purine base, or nitrogen 1 of the pyrimidine base with carbon 1′ of pentose[(1)]. Pentose is in the form of furanose and the glycosidic linkage is in the β-form. With ribose one has *ribonucleosides* (or ribosides); with dexoyribose, *deoxyribonucleosides* (or deoxyribosides). The structures of these two types of nucleosides are illustrated by the examples of figure 6-5.

(1) The carbon atoms of pentose are denoted 1′, 2′, 3′ etc. to avoid confusion with the numbering of the atoms of the base.

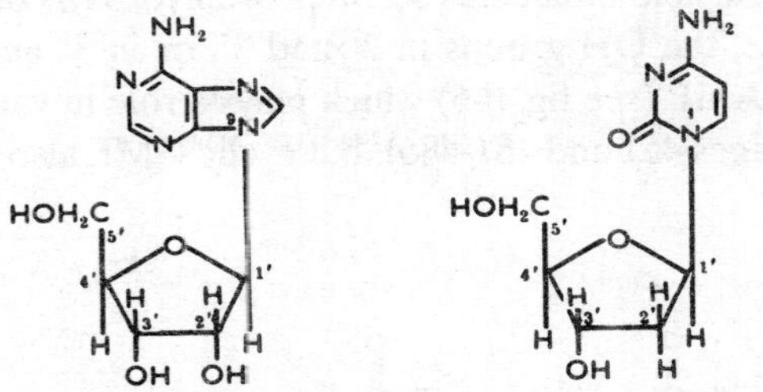

Adenosine
(Adenine ribonucleoside)

Deoxycytidine
(cytosine deoxyribonucleoside)

FIG. 6-5. — Structures of nucleosides.

The table below indicates the nomenclature of the main nucleosides. It will be noted that purine nucleosides have the suffix "osine" and the pyrimidine nucleosides, the suffix "idine".

Base	*Ribonucleoside*	*Deoxyribonucleoside*
Adenine	Adenosine	Deoxyadenosine
Guanine	Guanosine	Deoxyguanosine
Uracil	Uridine	Deoxyuridine
Cytosine	Cytidine	Deoxycytidine
Thymine	Thymine ribonucleoside or ribothymidine (rare)	Deoxythymidine(1)

(1) "Deoxythymidine" is the correct appellation but one still finds frequently the term "thymidine" which dates back to the time when it was thought that thymine existed only in DNA (and therefore necessarily linked to deoxyribose), so that the prefix "deoxy" was understood.

IV. – NUCLEOSIDES-MONOPHOSPHATES

Nucleotides are the phosphoric esters of nucleosides. Depending on the nature of the pentose one will have *ribonucleotides* (or ribotides) and *deoxyribonucleotides* (or deoxyribotides). Mild hydrolysis of a nucleotide can be carried out in two ways:

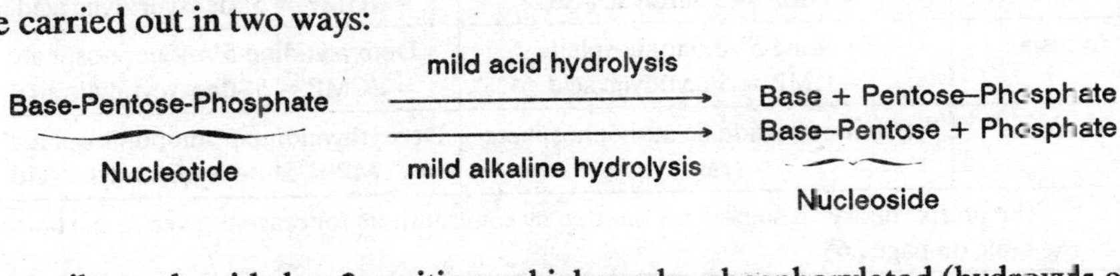

A ribonucleoside has 3 positions which can be phosphorylated (hydroxyls of carbons 2′, 3′ and 5′) while a deoxyribonucleoside can be phosphorylated only in two places (3′ and 5′). These various isomers of nucleosides-monophosphates exist (see examples of figure 6-6). Besides, there are cyclic nucleotides where the phosphoric acid molecule esterifies two hydroxyls of the pentose at a time (for example, the OH groups in 2′ and 3′, or in 3′ and 5′); the most important is cyclic AMP (see fig. 6-6) which plays a role in various regulation mechanisms (see pages 451 and 481-486), but cyclic GMP also exists.

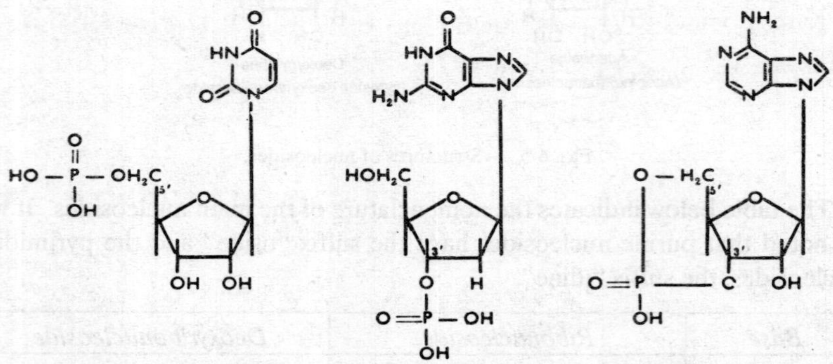

FIG. 6-6. — Structures of nucleosides-monophosphates.

The nomenclature of the principal nucleosides-monophosphates is indicated in the table below; these are 5′ isomers, but the other isomers are denoted in identical manner by simply replacing 5′ by 2′ or 3′ respectively.

Base	*Ribonucleoside-5′-monophosphate*	*Deoxyribonucleoside-5′-monophosphate*
Adenine	Adenosine-5′-monophosphate = AMP = 5′-adenylic acid	Deoxyadenosine-5′-monophosphate = *d*AMP = 5′-deoxyadenylic acid
Guanine	Guanosine-5′-monophosphate = GMP = 5′-guanylic acid	Deoxyguanosine-5′- monophosphate = *d*GMP = 5′-deoxyguanylic acid
Uracil	Uridine-5′-monophosphate = UMP = 5′-uridylic acid	Deoxyuridine-5′-monophosphate = *d*UMP = 5′ deoxyuridylic acid.
Cytosine	Cytidine-5′-monophosphate = CMP = 5′ cytidylic acid	Deoxycytidine-5′-monophosphate = *d*CMP = 5′-deoxycytidylic acid
Thymine	Thymine riboside-5′-monophosphate (rare)	Deoxythymidine-5′-monophosphate(1) = *d*TMP = 5′-deoxythymidylic acid(1)

(1) The prefix "deoxy" is sometimes omitted by some authors for reasons given at the bottom of the table on page 265.

V. NUCLEOSIDES-DI- AND TRIPHOSPHATES

A second phosphate group can be bound to the phosphate of a nucleoside-5′-monophosphate, thus forming a *ribo-* or a *deoxyribonucleoside-5′-diphosphate* (ZDP or dZDP); a third phosphate can also be added to the second, forming a *ribo-* or a *deoxyribonucleoside-5′-triphosphate* (ZTP or *d*ZTP). Examples of this type of nucleotides are represented in figure 6-7 (they are sometimes called nucleosides-polyphosphates, although the number of phosphate groups bound to the nucleoside does not exceed 3, barring very rare exceptions).

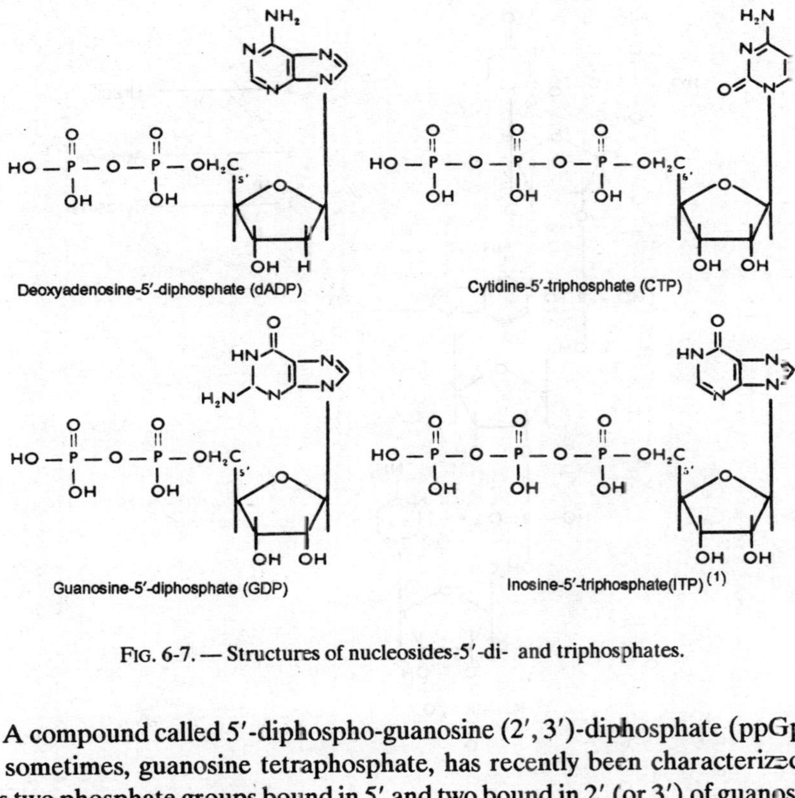

FIG. 6-7. — Structures of nucleosides-5′-di- and triphosphates.

A compound called 5′-diphospho-guanosine (2′, 3′)-diphosphate (ppGpp), or sometimes, guanosine tetraphosphate, has recently been characterized; it has two phosphate groups bound in 5′ and two bound in 2′ (or 3′) of guanosine. This compound accumulates in some auxotrophic mutants of *E.coli* when the latter are deprived of the essential aminoacid, and protein synthesis and formation of the two ribosomal RNAs of high molecular weight are inhibited.

(1) While names of nucleosides and nucleotides are generally derived from the corresponding bases (see tables on pages 265 and 266), we have here an exception to this rule: the base corresponding to the nucleoside called inosine (and the derived nucleotides) is called hypoxanthine (see fig. 6-2).

VI. PRIMARY STRUCTURE OF NUCLEIC ACIDS

In deoxyribonucleic acids as in ribonucleic acids, the nucleotides are joined by *3′-5′ phosphodiester* bonds; in other words each phosphate group (except those situated at the end of chains) is esterified to the 3′ hydroxyl group of a pentose and to the 5′ hydroxyl group of the next pentose. This mode of binding between nucleotides is demonstrated by titration experiments which reveal that in the polynucleotide each phosphate group has only one free acid OH, and by the fact that enzymatic hydrolysis of nucleic acids yields either nucleosides-5′-monophosphates, or nucleosides-3′-monophosphates, depending on the enzyme used.

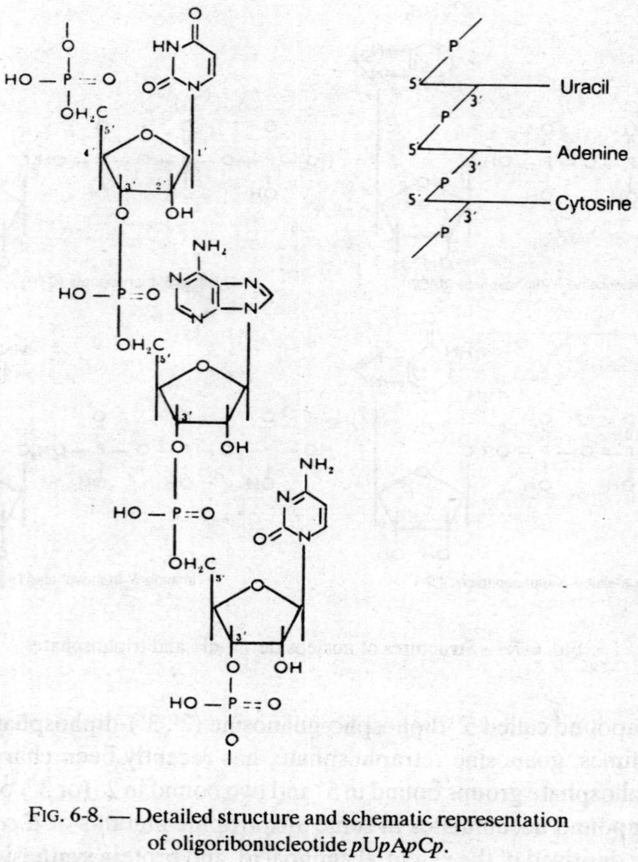

FIG. 6-8. — Detailed structure and schematic representation of oligoribonucleotide *pUpApCp*.

As shown by figure 6-8, the polyribonucleotide chain therefore consists of alternating ribose and phosphate residues. This chain is absolutely identical for all ribonucleic acids, only its length can vary. The DNA chains have the same structure, with deoxyribose in place of ribose. It is observed that the bases do not participate in the chain linkages but it should not be concluded that they are unimportant. On the contrary, the order in which the bases follow one another along the chain, called *base sequence* (or nucleotide sequence) is characteristic of a nucleic acid (just as the sequence of amino acids is characteristic of a protein), and we will see in this chapter that genetic information in fact resides in the base sequence.

Figure 6-8 shows the detailed structure of a fragment of RNA, and also its schematic representation (where a single line denotes pentose) which reflects the main features and is much more convenient to write. This oligoribonucleotide fragment can also be denoted – very explicitly – as follows: *pUpApCp*. It will be noted that in this mode of writing, a "*p*" to the left of the capital letter designating the nucleoside signifies that the phosphate group is bound to the 5′ carbon of the pentose, and a "*p*" to the right of the capital letter represents a phosphate bound to the 3′ carbon of the pentose of the nucleoside considered.

VII. DETERMINATION OF NUCLEOTIDE SEQUENCES

The determination of the base sequence of a nucleic acid is much more difficult than the determination of the sequence of amino acids of a protein, because there are only 4 types of monomers instead of 20. This explains why one began with the determination of the sequences of transfer ribonucleic acids (more than 200 sequences are known); these do offer two great advantages: they are relatively short and have – in addition to the 4 normal bases – a few rare bases (acting as reference points), which considerably facilitates the determination of their primary structure. In the meantime, the structure of other types of ribonucleic acids (ribosomal ribonucleic acids, viral ribonucleic acids), has been determined or is under study, and presently, the nucleotide sequences of deoxyribonucleic acids are also being studied. In practice, to determine the base sequence of an RNA or DNA, several selective hydrolyses are carried out and the fragments obtained are analyzed. We will now see briefly the principal modes of hydrolysis of ribonucleic acids and deoxyribonucleic acids.

1. – Hydrolysis of ribonucleic acids and deoxyribonucleic acids

A. – Alkaline hydrolysis

Figure 6-9 illustrates the mechanism of this hydrolysis; it is seen that it implies the intermediate formation of a 2′ 3′ cyclic phosphate; this is the reason why the rupture of all phosphodiester bonds of the RNA molecule – by scission between the phosphate group and carbon 5′ of the adjacent ribose – leads to the formation of not only nucleosides-3′-phosphates, but also nucleosides-2′-phosphates, although the hydroxyl of carbon 2′ is not at all involved in the internucleotide linkages of the intact RNA chain.

Deoxyribonucleic acids are resistant to alkaline hydrolysis, because there is no hydroxyl group at the position 2′ to permit the reactive mechanism illustrated by figure 6-9.

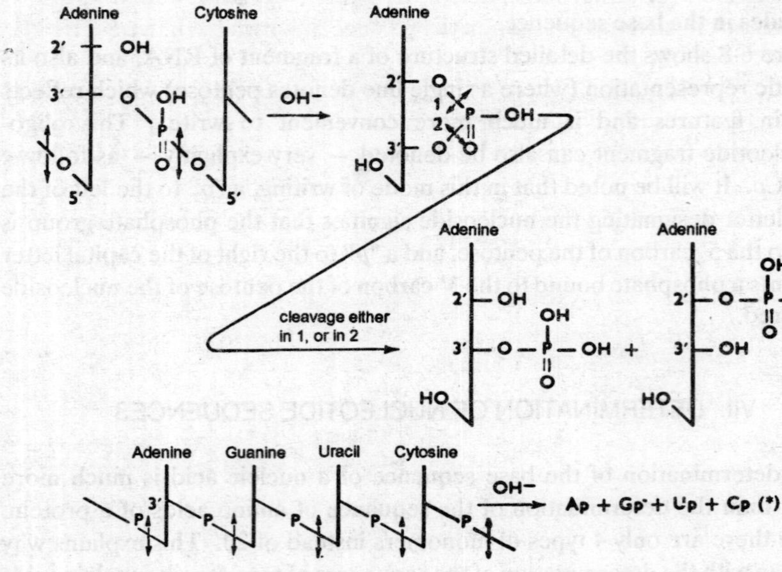

(*) This representation denotes the nucleosides-3′-phosphates, but these are accompanied by nucleosides-2′-phosphates.

FIG. 6-9. — Alkaline hydrolysis of ribonucleic acids.

B. — Hydrolysis by the pancreatic ribonuclease (RNase)

This enzyme is an *endonuclease,* which splits the phosphodiester bonds within the RNA chains; it acts in the same manner as the alkalis, by the same mechanism (which explains that deoxyribonucleic acids are resistant), but only when the nucleotide situated on the 3′ side is a *pyrimidine nucleotide,* as shown in figure 6-10, and it liberates nucleosides-3′-phosphates and oligonucleotides-3′-phosphates.

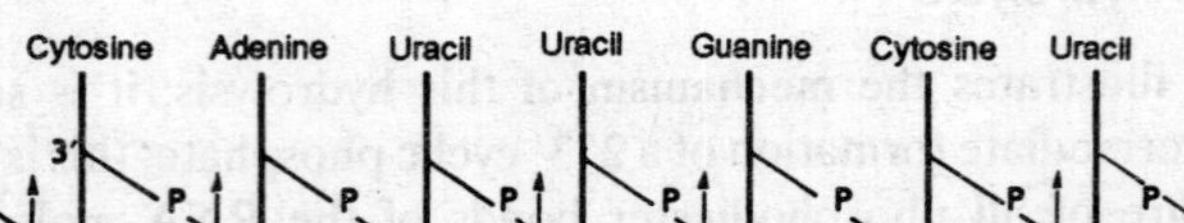

FIG. 6-10. — Hydrolysis of ribonucleic acids by pancreatic RNase.

C. – Hydrolysis by the ribonuclease T_1

This RNase has a specificity different from the one above; it is an endonuclease which splits the phosphodiester bonds of ribonucleic acids when a *guanylic* nucleotide (or an inosinic nucleotide, a rare nucleotide found in the transfer ribonucleic acids) is present on the 3′ side, as shown by figure 6-11.

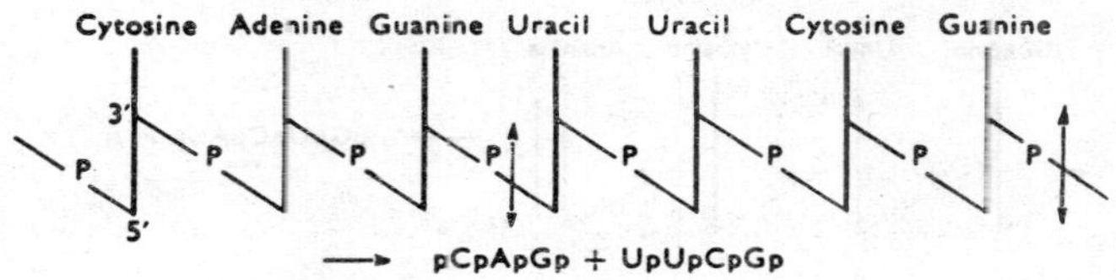

FIG. 6-11. — Hydrolysis of ribonucleic acids by RNase T_1.

D. – Hydrolysis by the ribonuclease U_2

It is an endonuclease which splits the phosphodiester bonds of ribonucleic acids when an adenylic or guanylic nucleotide is present on the 3′ side. Used on fragments resulting from the hydrolysis by RNase, T_1, it will specifically split at the adenylic nucleotides. Besides, conditions have been determined, permitting the enzyme to split the ribonucleic acids preferentially at the adenylic nucleotides.

E. – Hydrolysis by the snake venom phosphodiesterase

Contrary to what we have seen so far, this enzyme splits the phosphodiester bonds between the phosphate group and the carbon 3′ of the adjacent ribose, thus liberating nucleosides-5′-monophosphates (see fig. 6-12). Another difference: this is an exonuclease which attacks the oligo- and poly-nucleotides progressively from their 3′ end by detaching one nucleoside-5′-monophosphate at a time. It attacks the deoxyribonucleic acids as well as the ribonucleic acids.

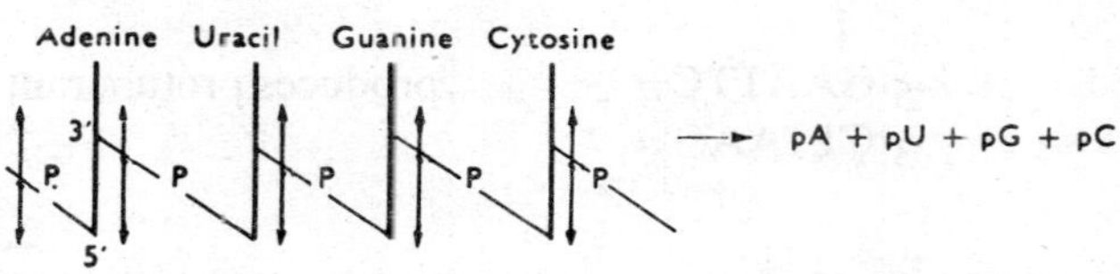

FIG. 6-12. — Hydrolysis of ribonucleic acids by snake venom phosphodiesterase.

F. – Hydrolysis by the spleen phosphodiesterase

This is also an exonuclease but it acts in a direction opposite to the previous one. Attacking the ribo- and deoxyribo-polynucleotides at their 5′ end, it liberates sequentially nucleosides-3′-monophosphates.

G. – Action of alkaline phosphatase of E.coli

The alkaline phosphatase of *E.coli* and the other phosphatases are – contrary to the enzymes studied till now, which were all phosphodiesterases – *phosphomonoesterases* which can detach the phosphate group only at the end of *a chain* as shown by figure 6-13.

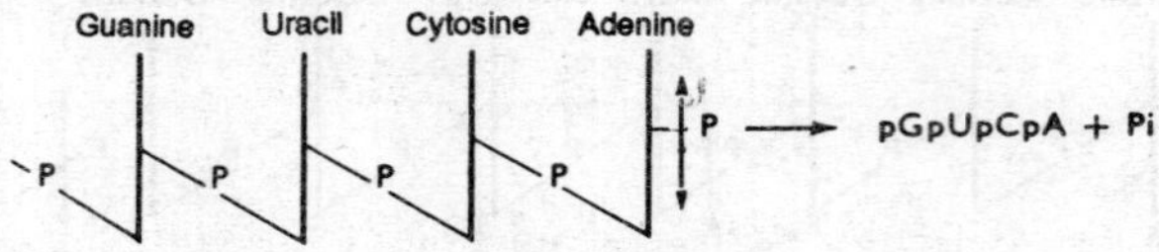

FIG. 6-13. — Action of phosphatase.

H. – Action of deoxyribonucleases ("DNases")

The first DNases known were less specific than the RNases. Two endonucleases were particularly studied:

– the pancreatic DNase (I) which generates oligodeoxyribonucleotides-5′-monophosphates;

– the acid DNase (II), which yields oligodeoxyribonucleotides-3′-monophosphates.

But, more recently *restriction endonucleases* were discovered, they are present in diverse bacterial species where they act in degrading any foreign DNA which would have penetrated in the cell (while the homologous DNA is protected by methylation of a base on each strand at the site of the cleavage). These restriction enzymes cleave the molecules of double-stranded DNA at very special nucleotide sequences, characterized by a symmetry and present in a very limited number in the DNA molecules. As examples, are given below the sequences recognized by three restriction endonucleases commonly used and extracted respectively from *E.coli* (Eco R I), *Haemophilus aegyptus* (Hae III) and *Providencia stuarti* (Pst I), as well as the places where the cleavages take place (indicated by arrows):

Eco R I	5′G↓AATTC3′ 3′CTTAA↑G5′	produces protuberant 5′ends
Hae III	5′GG↓CC3′ 3′CC↑GG5′	produces blunt ends
Pst I	5′CTGCA↓G3′ 3′G↑ACGTC5′	produces protuberant 3′ends

It is interesting to note that the cleavage by a restriction endonuclease like Eco R I or Pst I produces fragments possessing *cohesive ends* (i.e. single-stranded ends having complementary sequences), which then allows the joining of two fragments of DNA of different origins but which were cleaved by the same restriction enzyme and therefore possess the same cohesive ends. This property is used in experiments of genetic recombination *in vitro*, particularly to introduce deoxyribonucleic acids of diverse origins in phages or plasmids (see pages 380 to 388).

2. – Determination of the sequence of an oligoribonucleotide

To illustrate how one can determine a nucleotide sequence with the help of carefully chosen enzymes, we will take the example of the trinucleotide G*p*A*p*-U*p*. As shown by figure 6-14, *phosphatase* can be used first; it detaches the terminal phosphate at the 3′ end of the fragment. Then follows an *alkaline hydrolysis* which will yield a nucleoside (uridine) easily separable from the 2 nucleotides also liberated; we can now say that there was a *uridylic nucleotide* at the 3′ end of the original trinucleotide. Besides, a *quantitative analysis of nucleotides* (and nucleoside) obtained by alkaline hydrolysis reveals the base composition of the oligonucleotide. In this case, the trinucleotide G*p*A*p*U yields G*p*, A*p* and U in equal quantities, whereas for instance the tetra-nucleotide G*p*A*p*A*p*U would have given 1G*p* and 1U but 2 A*p*. The quantitative

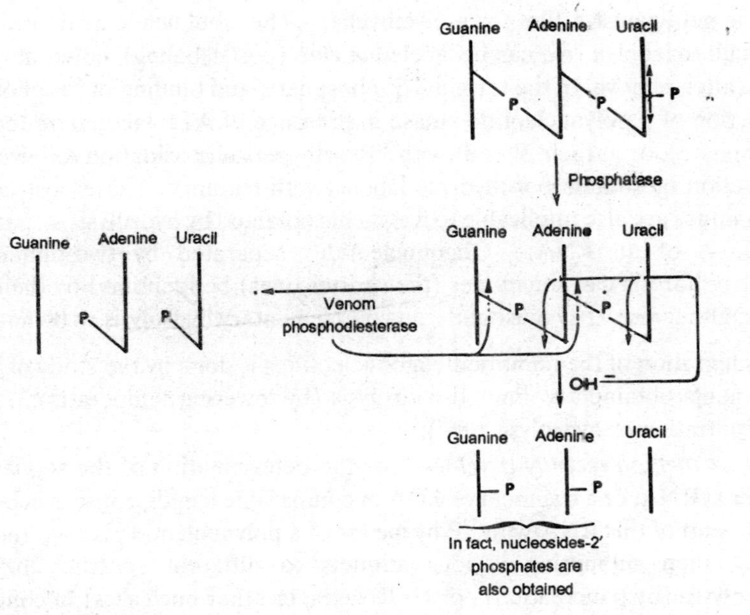

FIG. 6-14. — Determination of the sequence of a trinucleotide (this is a fragment which can be obtained by action of pancreatic RNase on a RNA).

analysis of bases is carried out in some cases (after perchloric hydrolysis). The action of *snake venom phosphodiesterase* on another aliquot of the trinucleotide previously treated by the phosphatase, will yield a nucleoside (guanosine) easily separable from the 2 nucleotides liberated; this will indicate that there is a guanylic nucleotide at the 5′ end of the trinucleotide. Knowing the nucleotide of each end, it is deduced that the adenylic nucleotide is in the middle and that the trinucleotide has the sequence *GpApUp*.

3. – Determination of the sequence of ribonucleic acids

Three types of methods are presently used for determining the primary structures of ribonucleic acids.

– *Enzymatic hydrolysis* with the help of pancreatic, T_1 or U_2 ribonucleases, or another nucleases, followed by fractionation of the oligonucleotides obtained, either by *column chromatography* (which requires rather large quantities of material, but permits subsequent characterization of nucleotides – especially the rare nucleotides – by their absorption spectrum in ultra violet light), or by *two-dimensional electrophoresis* on cellulose acetate, ion exchange paper, or polyacrylamide gel (which necessitates lesser quantities, but requires the use of ribonucleic acids labeled with ^{32}P – which is relatively easy in microorganisms, but poses problems in higher organisms – because the identification of oligonucleotides is done by autoradiography by applying the paper or the gel against a film for a given time). The ribonucleic acids which are difficult to label *in vivo* can be labeled *in vitro* (post-labeling), either at their 5′ end (after removal of the terminal 5′ phosphate, and binding of ^{32}P phosphate by action of a polynucleotide kinase in presence of ATP labeled on the third phosphate), or at their 3′ end (with ^{32}P or by periodic oxidation followed by a reduction by sodium borohydride labeled with tritium). These post-labeling techniques are also applicable to fragments obtained by hydrolysis – partial or total – of an RNA. Oligonucleotides separated by two-dimensional electrophoresis can sometimes (the shorter ones) be identified by their position. Otherwise, they must undergo complementary hydrolysis to be analyzed.

Integration of the identified oligonucleotides is done by the study of longer fragments obtained by limited hydrolysis (by lowering temperature, enzyme concentration or hydrolysis time).

– *A method recently developed* for the determination of the sequence of either a *t*RNA, or a fragment of RNA of comparable length, consists in labeling the 5′ end of this RNA with ^{32}P by means of a polynucleotide kinase and ATP ($^{32}P_\gamma$), then submitting various aliquots to different specific enzymatic hydrolyses (by pancreatic, T_1 or U_2 RNases, or other nucleases) in conditions under which there will be statistically only one cleavage per chain. The fragments thus obtained will be fractionated according to their size by electrophoresis on polyacrylamide gel, and only the fragments carrying the 5′

end labeled with ^{32}P (*p) will be revealed by autoradiography. At the same time, a partial alkaline hydrolysis which can cleave all phosphodiester bonds (but causes statistically only one cleavage per chain) serves as reference. The sequence of a fragment (which can reach 70 nucleotides or more) can thus be read off directly on the gel. The principle of this reading is illustrated in the following diagram, in the case of a pentanucleotide.

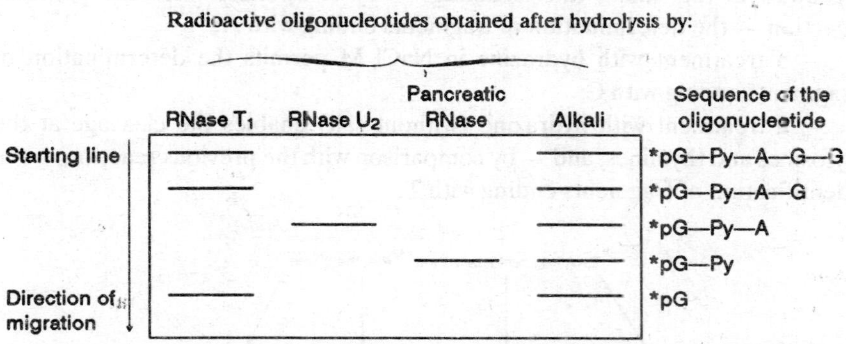

Diagram showing the radioactive oligonucleotides (labeled with ^{32}P in 5′) that would be obtained by subjecting the pentanucleotide *pG-Py-A-G-G to various partial hydrolyses and fractionating the resulting oligonucleotides by gel electrophoresis followed by an autoradiography. Other oligonucleotides are produced by diverse hydrolyses (this is evident especially in the case of alkaline hydrolysis) but only those having ^{32}P in 5′ (indicated by *p) are revealed by autoradiography.

– *Study of a strand of viral RNA synthesized in vitro* synchronously, in present of the RNA replicase, the complementary strand serving as template and the 4 radioactive ribonucleoside-triphosphates (see page 344). Interruption of the synthesis after very short times (a few seconds) gives radioactive newly synthesized fragments (whose lengths increase as reaction time is extended) the sequence of which can be determined. This method was used for studying the sequence of the RNA the Qβ phage.

4. – Determination of the sequence of deoxyribonucleic acids

In recent years, methods were developed for determining DNA sequences.

These methods are based on the preparation of specific fragments with the help of restriction endonucleases (see page 272) and the cloning of these fragments (see pages 380-388) which provides large quantities of the DNA to be sequenced. The analysis of these fragments can be made by the method described by Maxam and Gilbert, a method very similar to the one described above for the determination of the sequence of ribonucleic acids. The principle is actually identical, except that various chemical hydrolyses are used (instead of enzymatic hydrolyses). Four aliquots of a DNA fragment (labeled with ^{32}P at one end) are subjected to four different reactions, in conditions under which there will be statistically, only one cleavage (or a small number of cleavages) per chain:

— a treatment with dimethylsulphate which methylates particularly guanine in N7 (and to a lesser degree, adenine in N3); under the effect of heat, guanine is detached and the polynucleotide chain is broken there. One therefore obtains fragments ending at the point where there was a G in the chain;

— a treatment with dimethylsulphate in acid medium detaches the 2 purines, and therefore yields fragments ending where there were adenines and guanines in the chain, thus enabling — by comparison with the previous reaction — the determination of fragments ending with A;

— a treatment with hydrazine in NaCl M permits the determination of fragments ending with C;

— a treatment with hydrazine (without salt) enables the cleavage at the cytosines and thymines, and — by comparison with the previous reaction — the identification of fragments ending with T.

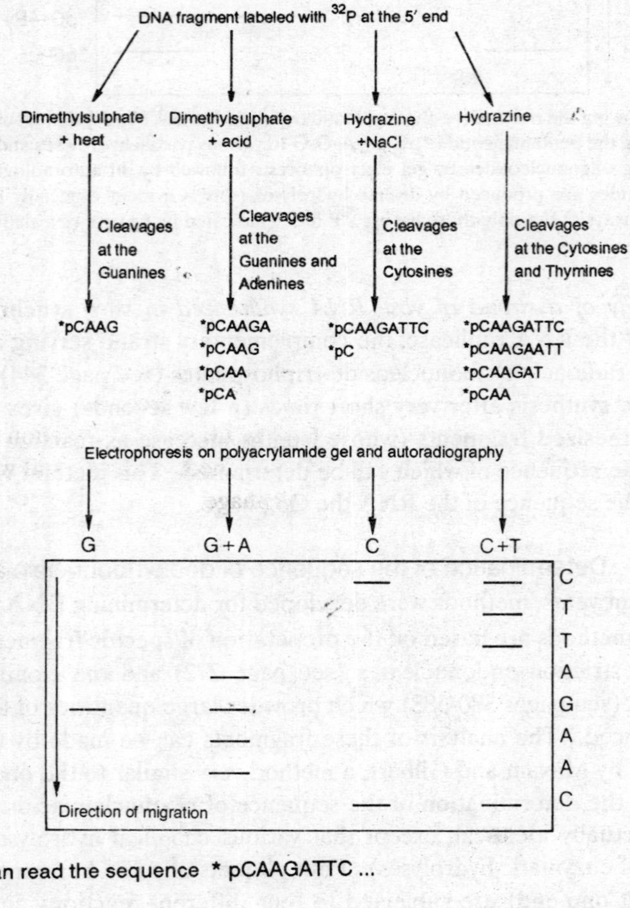

DETERMINATION OF THE SEQUENCE OF DNAS BY THE MAXAM AND GILBERT METHOD

The fragments produced by these four reactions are subjected concurrently to an electrophoresis which separates them according to their size. The shortest fragment is found at the bottom of the gel and, knowing that the size of two successive fragments differs only by one nucleotide, the sequence of the DNA can be read off after autoradiography, as shown by the example of the above diagram.

Presently, one can separate, by electrophoresis on a polyacrylamide gel of 40 cm length, fragments ranging from 1 to about 300 nucleotides (or more, by using longer gels). One may verify that no error has been committed by sequencing the two strands and checking that the complementary is perfect. To establish the sequence of a longer DNA region (one gene can contain several thousands of nucleotides, especially if it has introns), the sequence of adjacent fragments is determined. To make sure that small fragments were not lost (two restriction sites recognized by a given enzyme can be relatively near one another), one must sequence overlapping fragments, obtained by means of various restriction endonucleases, and the order of which has been established.

One can also determine the sequence of deoxyribonucleic acids by the dideoxynucleotides method or Sanger's method: to a single-stranded DNA (which will serve as template) is hybridized a short complementary primer (having a 3′OH), from which DNA polymerase I will polymerize the four deoxyribonucleoside-triphosphates (dNTP) and thus synthesize a DNA strand complementary to the template. In fact, four independent polymerization reactions are performed by introducing in each, one of the four dideoxyribonucleoside-triphosphates (ddNTP), having neither an OH in 3′, nor in 2′, and whose incorporation in the DNA chain in place of the corresponding dNTP will therefore bring about the (prematured) termination of the chain (because there is no OH in 3′ to enable the formation of the next phosphodiester bond).

Each reaction mixture therefore contains the 4 dNTP (of which one is labeled on the Pα) and only one ddNTP. When the analogue is ddATP for example, a random incorporation of ddATP (and therefore, a random termination of the chain) can be obtained by adjusting the ddATP/dATP ratio; there will be finally, a series of DNA fragments of different sizes, having all the primer at their 5′ end and a ddA in 3′. After the 4 polymerization reactions (each with a different ddNTP) 4 series of fragments are obtained, each series having at the 3′ end, respectively ddA, ddG, ddC, ddT. These fragments are then separated, by electrophoresis on polyacrylamide gel, according to their size (resolution is such that one can distinguish between two fragments of sizes differing only by one nucleotide), and after autoradiography one can read the sequence on the gel, proceeding from the bottom of the gel (where is found the shortest fragment) to the top, as shown by the diagram. This gives the sequence of the newly synthesized DNA (which is complementary to the strand used as template) from the primer, in the direction 5′ $\longrightarrow$ 3′.

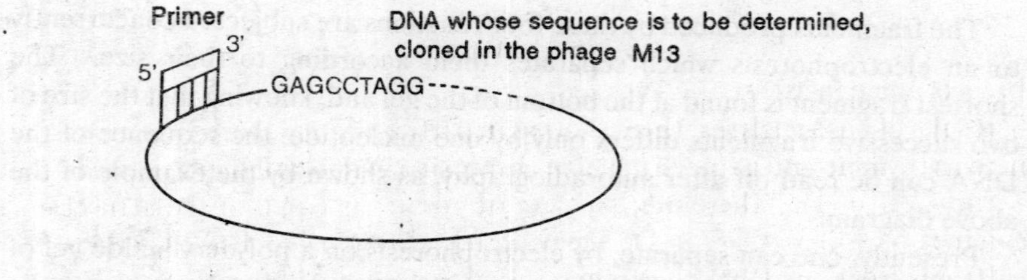

Polymerization in presence of the above template, of DNA polymerase I (in fact, one uses a part of this enzyme, called Klenow fragment, having no 5′→3′ exonuclease activity which would hydrolyze the primer; this would perturb the analysis of the fragments) and substrates: dATP ($^{32}P\alpha$), dGTP, dCTP, dTTP and one of the four dideoxyribonucleoside-triphosphates:

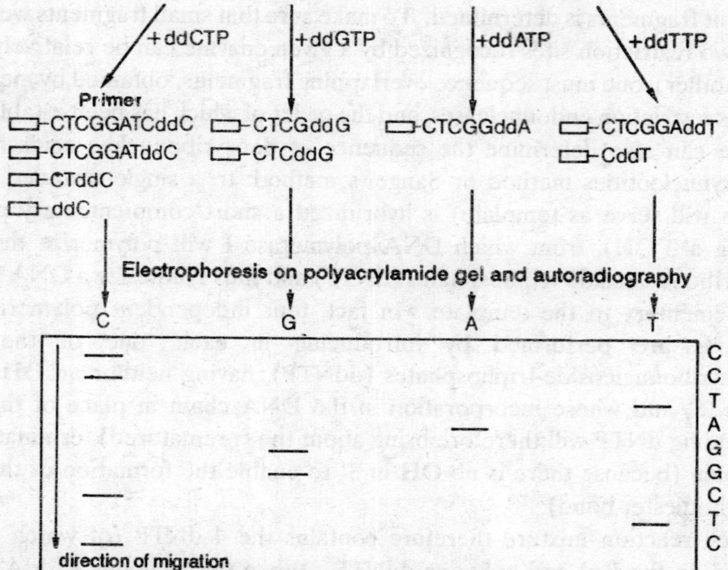

The sequence of the newly formed strand can be read off:

5′ CTCGGATCC3′, and one can deduce from it the sequence of the strand used as template

DETERMINATION OF THE SEQUENCE OF DNAS BY SANGER'S METHOD

VIII. – SECONDARY STRUCTURE OF DEOXYRIBONUCLEIC ACIDS

Some observations made on deoxyribonucleic acids in solution suggested that DNA molecules – as they could be extracted from cells – had structural characteristics different from those expected in molecules having only the primary structure we just described (see fig. 6-8); on the contrary, after heating, the DNA molecules behaved like simple flexible chains. It therefore appeared that native DNA has a secondary structure and it must be said to the credit of Crick and Watson that they proposed in 1953, a model which explained the various properties of the DNA. Crick and Watson established their model on the basis of some facts:

– results of base analyses which showed in particular, that there is as much adenine as thymine and as much guanine as cytosine in deoxyribonucleic acids (A/T and G/C = 1). Therefore, the sum of purines is equal to the sum of pyrimidines (A+G = C+T). On the contrary, the ratio A+T/G+C varies considerably from one DNA to another; it is characteristic of the species, and is called the dissymmetry ratio (if it is greater than 1, the DNA is rich in A–T, if it is less than 1 the DNA is rich in G–C);

– results of X-ray diffraction studies (carried out by Wilkins) which suggested a helicoidal configuration;

– experimental results which suggested that the polydeoxyribonucleotide chains were held together by hydrogen bonds.

Let us describe the important characteristics of this model illustrated in figure 6-15.

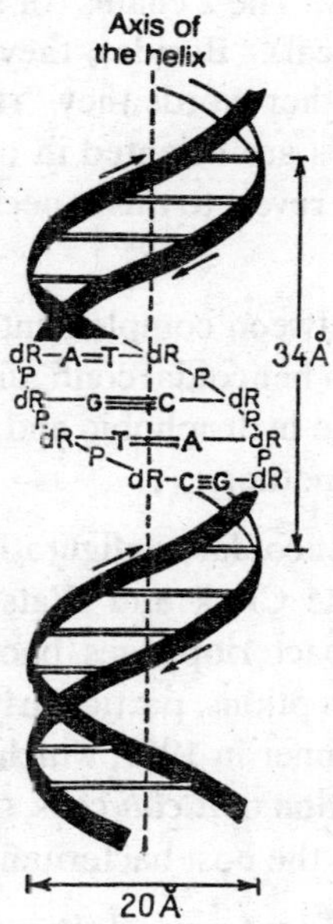

FIG. 6-15. — The double helix of the DNA. Each strand is constituted by the alternate linking deoxyribose (dR) — phosphate (P). The pairs of complementary bases A—T and G—C form a stacking of planes perpendicular to the axis of the helix.

The DNA molecule consists of two polydeoxyribonucleotide chains twisted around one another in a double helix.

The bases of each of these chains are projected towards the centre (i.e. towards the axis of the helix); the planes formed by the atoms of these bases are parallel to one another and perpendicular to the axis of the double helix. These planes are 3.4 Å apart and since the pitch of the helix is 34 Å, there will be 10 planes per turn.

In fact, these planes consist of, not one base but one base of each chain (i.e. one pair of bases) held together by hydrogen bonds. While, in principle any base when free, can exchange one or several hydrogen bonds with any other base, the configuration of the helix is such that a stable pairing – obtained by the exchange of more than one hydrogen bond – can take place only between adenine and thymine on the one hand (2 hydrogen bonds) and between guanine and cytosine on the other (3 hydrogen bonds). Adenine and thymine are said to be *complementary bases*, as also guanine and cytosine (see fig. 6-16).

One now understands why the A/T and G/C ratios are equal to 1; as a matter of fact, to one adenine in a chain, must necessarily correspond one thymine in the other chain (and conversely), and to one guanine in one chain must necessarily correspond one cytosine in the other chain (and conversely). Hence, the base sequence in one of the chains automatically determines the sequence in the other chain and we will see that this is of great importance in the study of the replication of the DNA. The 2 chains (or strands) of a DNA molecule are *complementary* (not identical). Besides, they are *antiparallel*, as may be observed in figure 6-16; in other words they "run" in opposite direction (their 3′-5′ phosphodiester bonds are oriented in opposite directions); this can be demonstrated and we shall revert to this aspect a little later in this chapter (see fig. 6-30).

The hydrogen bonds between complementary bases are partly responsible for holding the 2 chains in a helicoidal configuration, but some other forces are also involved, especially the hydrophobic and electronic interactions between the bases stacked on one another.

This *double-stranded* helicoidal configuration is a general characteristic of deoxyribonucleic acids; the Crick and Watson model is thus valid for the deoxyribonucleic acids of bacteriophages, bacteria, plants and animals. There are however some rare exceptions, particularly the DNA of the bacteriophage $\varphi\times174$, isolated by Sinsheimer in 1959, which has a single-stranded structure (however, during the infection of *E.coli* cells, the double-stranded form, called *replicative form*, appears in the host bacterium).

The DNAs can in fact, adopt several three-dimensional conformations and change reversibly from one form to the other.

The B form of the helix is characterized by the presence of a large groove (width 12 Å, depth 8.5 Å) and a small groove (width 6 Å, depth 7.5 Å), because the glycosidic bonds of a base pair are not diametrically opposed. The capacities of these 2 grooves in respect of the donor and acceptor atoms of hydrogen bonds are not the same. The large groove is more accessible to proteins interacting with specific sequences of the DNA. We have seen in the foregoing that, according to the Crick and Watson model, the double helix comprises 10 base pairs per turn, such that from one residue to the following on the chain, there is a rotation of 36°, but in reality this rotation angle can vary

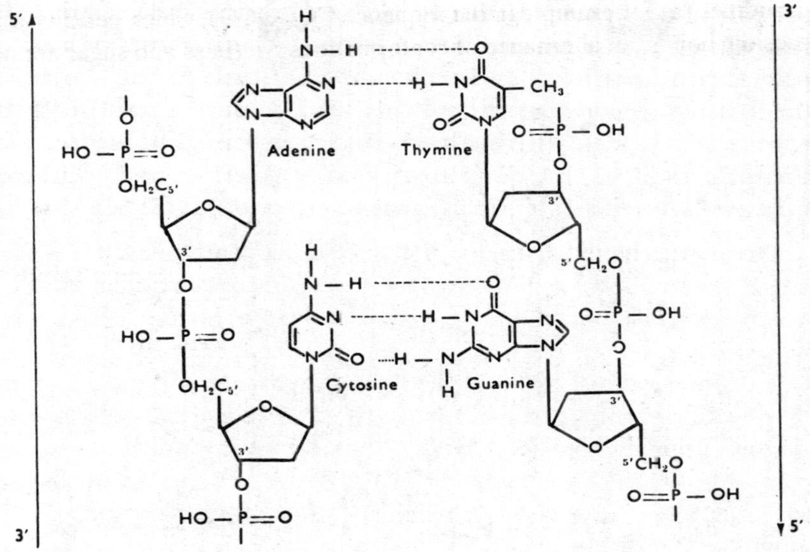

FIG. 6-16. — Hydrogen bonds between the complementary bases of a fragment of double-stranded DNA molecule.

from 28° to 42°. Furthermore, the 2 bases of a pair are not absolutely coplanar, they are turned with respect to one another like the blades of an aircraft propeller which increases the stacking of bases in each strand. Also, the base pairs are more or less inclined with respect to the neighbouring pairs. All these localized variations of the structure of the double helix depend on the nucleotide sequence, and can be recognized by a protein in search of its target region on the DNA. Lastly, the B helix can be deformed (curved, folded back, or twisted to form a super-helix), which permits for example the formation of covalent circles or the twisting of the DNA around proteins (for example, to form the nucleosomes, (see page 283).

Form A is, like form B, a double helix twisted on its right, but it is wider and the base pairs are inclined at 19° instead of being perpendicular to the axis of the helix. These differences are due to the fact that the carbon atom in 3′ of the furanose ring is at 0.5Å from the plane formed by the 4 other atoms of pentose (C−3′ endo). The phosphate groups can bind less water molecules than in the form B, and the small groove is practically non-existent.

Form Z is on the contrary, a double helix turning to its left. It owes its name to the fact that the phosphates of the chain form a zigzag. Form Z has only one deep groove.

The existence of form Z is facilitated by an alternance of purines and pyrimidines (as for example in the sequence CGCGCG), which permits at the glycosidic bonds, an alternance of conformations *syn* (base and sugar are near

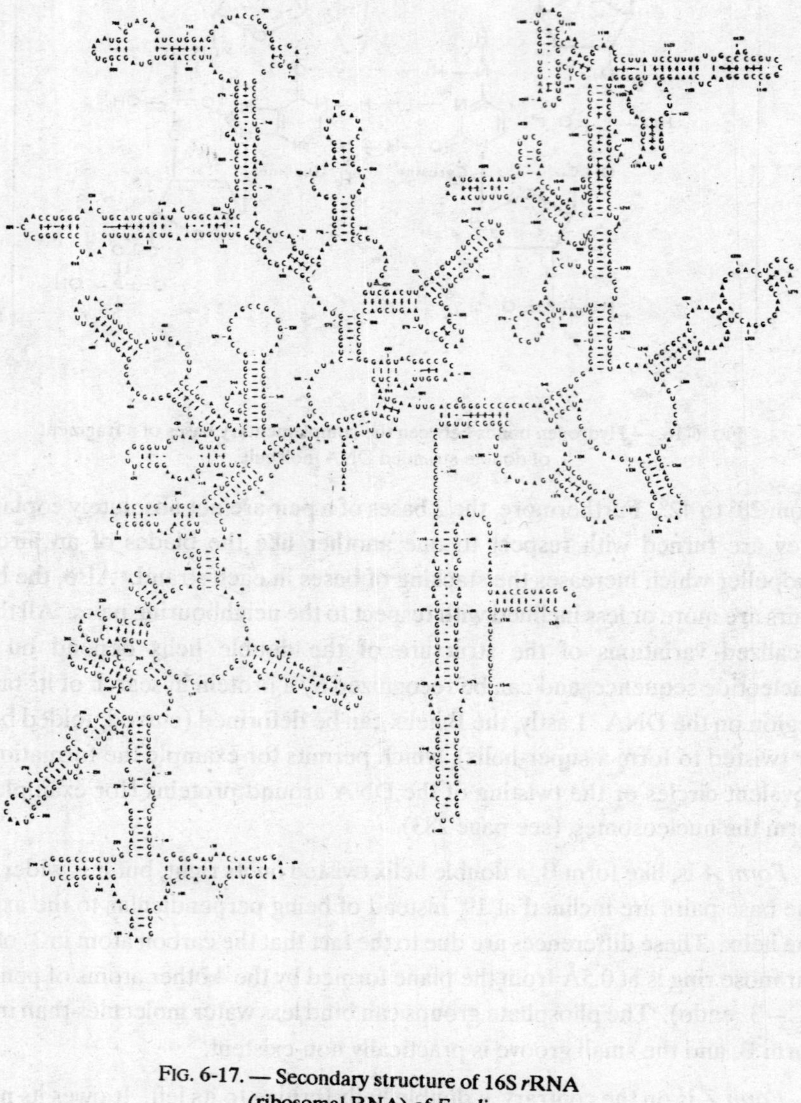

FIG. 6-17. — Secondary structure of 16S *r*RNA (ribosomal RNA) of *E.coli*.

one another, on the same side) and *anti* (base and sugar are distant, because they are on opposite sides). While in forms A and B of the DNA all glycosidic bonds are *anti*, in form Z the purines are *syn* and the pyrimidines are *anti*. Although the major part of the DNA is in form B, in procaryotes as well as in eucaryotes, the specific antibodies of form Z enable one to detect regions of DNA-Z in the chromosomes of eucaryotes. The formation of DNA-Z is facilitated by the methylation of C-5 of cytosine (a modification observed in eucaryotes and which favours hydrophobic interactions) and by the negative supercoiling of native molecules of DNA (which promotes the separation of the 2 strands of the helix in view of replication and transcription).

IX. – SECONDARY STRUCTURE OF RIBONUCLEIC ACIDS

Hydrogen bonds can also be formed between adenine and uracil on the one hand and guanine and cytosine on the other. But first, *these bonds are formed between 2 bases of the same chain* (whereas in the DNA the hydrogen bonds are interchain), and moreover *these bonds do not present the regularity characteristic of deoxyribonucleic acids* (where the complementarily of bases is total, i.e. the bases are paired all along the molecule). However, after such pairing, the RNA chains can take in some zones, a helicoidal configuration, sometimes called pseudo-helicoidal (see figs. 6-17 and 6-44).

But it must be mentioned that there are *double-stranded replicative forms of some viral ribonucleic acids which have a regular double helix structure* similar to that of deoxyribonucleic acids; actually, the single-stranded RNA which constitutes the genome of the virus serves, during the infection, as template for the synthesis of a complementary strand (see fig. 6-38) and double-stranded structures are then observed transitorily.

X. – PHYSICO-CHEMICAL PROPERTIES OF NUCLEIC ACIDS

The size of nucleic acids varies immensely. The smallest ribonucleic acids are the *t*RNAs which comprise about 80 nucleotides; their molecular weight is about 30 000. Ribosomal ribonucleic acids are larger; they contain several thousands of nucleotides and their molecular weight can exceed 1 million (see page 353). The RNA of the tobacco mosaic virus has about 6 500 nucleotides, which corresponds to a molecular weight greater than 2 million; the polycistronic messenger ribonucleic acids have comparable dimensions.

Deoxyribonucleic acids are much larger. A small DNA like the *circular* single-stranded DNA of phage $\varphi \times 174$ already has 5.5×10^3 nucleotides and a molecular weight of 1.7 million, but that of the phage T_2 comprises 2×10^5 pairs of nucleotides which corresponds to a molecular weight of 130 million.

The chromosome of *E.coli* is a *circular* double-stranded molecule[1] comprising about 4×10^6 pairs of nucleotides; its molecular weight is therefore of the order of 2 to 3×10^9. In fact, the extraction of intact DNA molecules (from bacterial, plant or animal cells) is a rather delicate operation; cleavages occur easily and in the past, one often attributed molecular weights much lower than the real ones. Cells of eucaryotes have a variable number of chromosomes according to the species. Each chromosome seems to consist of only one molecule of DNA (containing about 10^8 pairs of nucleotides), associated with basic proteins, the histones, to form chromatin, consisting of a flexible chain of repetitive units, the *nucleosomes*. The nucleosome, the fundamental packing unit of the DNA, contains about 200 pairs of nucleotides coiled around a nucleus consisting of 8 molecules of histones. The nucleosomal chain is itself folded up in such a manner that finally, the length of a metaphasic chromosome is about 8000 times shorter than that of the DNA it contains. Non-histone proteins (called "acid") are also attached to the chromatin. Some of them most probably play a role in the regulation of the genetic expression at the transcriptional level.

On the other hand, it must be noted that the deoxyribonucleic acids found in cellular organelles (mitochondria, chloroplasts) differ by their size and base sequence from the nuclear DNA present in the same cell.

The molecular weights of deoxyribonucleic acids are determined either upon the observation of molecules by electron microscopy (it is estimated that a length of 1 μ corresponds to a molecular weight of 10^6 for a single-stranded structure, or 2×10^6 for a double-stranded structure), or from the sedimentation constant of the DNA, obtained by comparing its rate of sedimentation in sucrose gradient (from 5 to 20%) to that of a DNA of known sedimentation constant. (Then, with the help of an empirical equation, the molecular weight is calculated from the value of S).

Very often, for preparative of analytical purposes, one uses the sedimentation equilibrium (see page 38) in a CsCl gradient established during the centrifugation; this gives a density gradient ranging generally from 1.65 to about 1.75 g/ml. The DNA is concentrated in a band at the place where the density of CsCl solution is equal to its own and this density is generally determined by comparison with a DNA of known density, centrifuged in the same gradient. The density of a DNA at the sedimentation equilibrium increases with greater percentages of G-C pairs of the DNA (because G-C pairs are more dense then A-T pairs).

The DNA solutions have a very high *viscosity* due to the considerable length and relative rigidity of the double helix. The *denaturation* of DNA molecules can be followed up with the help of viscosity measurements (see below). On the other hand, due to the presence of purine and pyrimidine bases, they *absorb*

---

(1) In other words, without ends. In the bacterium, this circular molecule does not have the form of a geometric circle, but a coiled structure.

in ultra-violet light (just like the RNA solutions) with a maximum around 260 nm[1]. When a solution of native DNA is heated one observes a decrease of viscosity, and an increase of it *o*ptical *d*ensity (O.D.) at 260 nm (between 30 and 40%). This phenomenon is called *hyperchromic effect* or *hyperchromicity*. It is due to the separation of the 2 strands of the double helix (spoken of as the *melting* of DNA), by rupture of the interchain hydrogen bonds, and since the G≡C bonds are stronger then the A = T bonds, the higher the GC content of a DNA, the higher will be its melting point. The curve of O.D. as a function of temperature has a sigmoidal form. It permits the determination of a transition point, T_m, which corresponds to the temperature at which the DNA molecules are half-denatured. A DNA whose strands are thus separated, either by heat, or under the action of other physical or chemical agents (e.g., urea) is called *denatured*. The DNA passes from an ordered structure (double helix) to a disordered structure (random coil). The DNA solution can be cooled again in 2 ways:

– rapidly: the 2 strands remain separated (the DNA remains denatured);

– slowly: this permits a re-association of the 2 strands (the AT and GC bonds are formed again) and the initial helix is reconstituted. The DNA is thus *renatured* and recovers its biological properties (for example the transforming activity in the case of some bacterial deoxyribonucleic acids).

Heating followed by slow cooling forms the basis of an important technique of study of deoxyribonucleic acids and ribonucleic acids: *hybridization*.

XI. – HYBRIDIZATION

Hybridization is a method universally used in all studies relating to nucleic acids. It is used, either to detect, identify or quantify a given molecule of RNA or DNA, or to compare the sequences of two molecules or two populations of DNA or RNA molecules.

Practically, the molecule or set of DNA or RNA molecules to be used for the detection or measurement (called "probe") is labeled with a radioactive isotope (frequently ^{32}P), then the probe and the molecules to be studied are mixed, heated beyond T_m and incubated at a temperature slightly lower than T_m. In these conditions, the complementary molecules can associate again. The non-hybridized molecules are then eliminated, either by chromatography on hydroxyapatite, or by hydrolysis using a nuclease specific of single-stranded nucleic acids. Very often the DNA or RNA to be studied is attached on a nitrocellulose membrane. This membrane is incubated with the probe, the excess of which is then simply eliminated by washing the membrane.

---

(1) In practice, DNA and RNA solutions are titrated (when they have been isolated and purified) by spectrophotometry, by measuring absorption in U.V. at 260 nm.

The radioactivity selected by chromatography on hydroxyapatite, or resistant to nucleases, or remaining on the nitrocellulose membrane, permits the detection of the presence, or the measurement of the quantity of RNA or DNA sequences complementary to the probe.

BIOSYNTHESIS OF NUCLEIC ACIDS

There are two steps in the biosynthesis of nucleic acids:

– the biosynthesis of purine and pyrimidine ribo- and deoxyribo-nucleosides-5′-triphosphates;

– the polymerization of these precursors into RNA or DNA.

BIOSYNTHESIS OF NUCLEOSIDES-5′-TRIPHOSPHATES

Among animals, some amino acids cannot be synthesized and must therefore be supplied through diet (essential amino acids). On the contrary, purine and pyrimidine nucleotides can be synthesized "*de novo*", from precursors like CO_2, NH_3, formiate, glycine etc., and this is true of animals and plants as well as bacteria (except some mutants lacking an enzyme catalyzing one of the reactions leading to nucleotides). This does not prevent the cells from using, for the biosynthesis of their nucleic acids, the already formed purines and pyrimidines which may be provided to them (e.g., through food).

Purine and pyrimidine nucleotides have a common precursor, *ribose-5-phosphate*, the formation of which will not be discussed here again since we already mentioned while studying the pentose-phosphates cycle that the passage from glucose-6-phosphate to ribose-5-phosphate is possible, either by the *oxidative pathway* (necessitating $NADP^+$), or by the *pathway of interconversions*; these reactions are schematically represented in figure 4-43).

Ribose-5-phosphate binds on its carbon 1 a pyrophosphate group from ATP and thus transforms (see fig. 6-18) into *5-phosphoribosyl-1-pyrophosphate* or

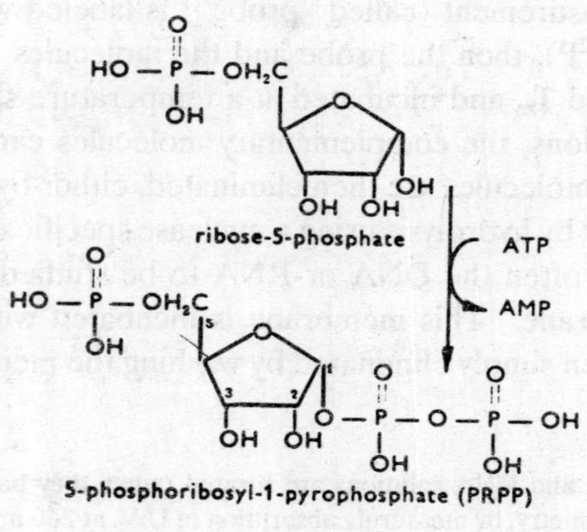

FIG. 6-18. — Formation of PRPP, precursor of purine and pyrimidine nucleotides.

PRPP, a precursor which will provide the "ribose-5-phosphate" part in the *de novo* biosynthesis of purine and pyrimidine nucleotides, as well as in the synthesis of nucleotides from preformed purines or pyrimidines.

I. – BIOSYNTHESIS OF PURINE RIBONUCLEOSIDES-5'-TRIPHOSPHATES

1. – *De novo* biosynthesis of IMP

Figure 6-19 shows the series of 11 reactions leading from ribose-5-phosphate to IMP (or inosine-5'-monophosphate, or inosinic acid), the base of which, we may repeat, is called hypoxanthine (see figs. 6-2 and 6-7). Names of intermediate compounds and enzymes catalyzing the various reactions are not important here; we will confine ourselves to some remarks on some of the steps.

Reaction 2 is subjected to *feedback inhibition*; in other words, the enzyme is inhibited by purine nucleotides which are the final products of this metabolic series; such a mechanism, inhibiting the first reaction of the metabolic pathway when the terminal products are in excess, is obviously very economical for the cell because it prevents a useless energy expenditure. It is indeed the first reaction of the metabolic pathway leading specifically to purine nucleotides which is subjected to feedback inhibition, because the previous reaction, leading from ribose-5-phosphate to PRPP, permits the formation of a compound (PRPP) which is a precursor of the biosynthesis of purine and pyrimidine nucleotides and it would be undesirable that it should be inhibited by an excess of purine nucleotides only. Besides, it must be noted that at this stage there is a change of configuration at carbon 1 of ribose, since there is passage from configuration α (PRPP) to configuration β (phospho-ribosylamine) which is that of nucleotides.

Reaction 4 and reaction 10 consist of a transfer of formyl group, from N_5-N_{10} methenyl-tetrahydrofolic acid (reaction 4) and N_{10}-formyl-tetrahydrofolic acid (reaction 10). But we have seen (see pages 73-74) that sulphamides are structural analogues of para-aminobenzoic acid, the compound used by bacteria to synthesize folic acid. Sulphamides therefore prevent the formation of the tetrahydrofolic acid required for the biosynthesis of purine nucleotides which is blocked and this explains the bacteriostatic effect of sulphamides.

Reactions 8 and 9 consist of the input of a nitrogen atom provided by aspartic acid which is thus transformed into fumaric acid. A similar mechanism is involved in the transformation of IMP into AMP which we will study in the next paragraph (see fig. 6-21) and in the transformation of citrulline into arginine in the ureo-genesis cycle (see fig. 7-30).

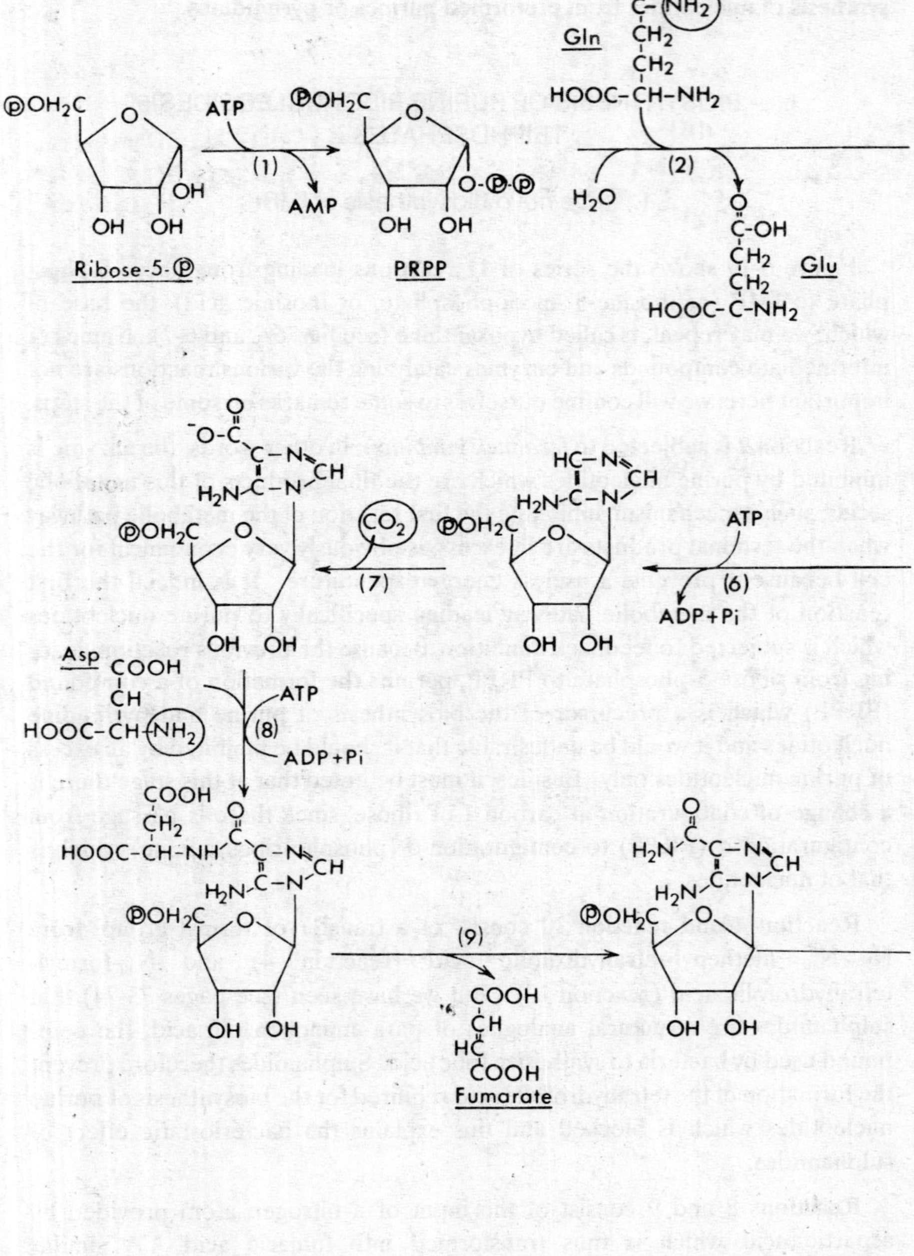
ATP
(1)
AMP
Ribose-5-Ⓟ
PRPP
Gln
(2)
H2O
Glu
CO2
(7)
ATP
(6)
ADP+Pi
Asp
ATP
(8)
ADP+Pi
(9)
Fumarate

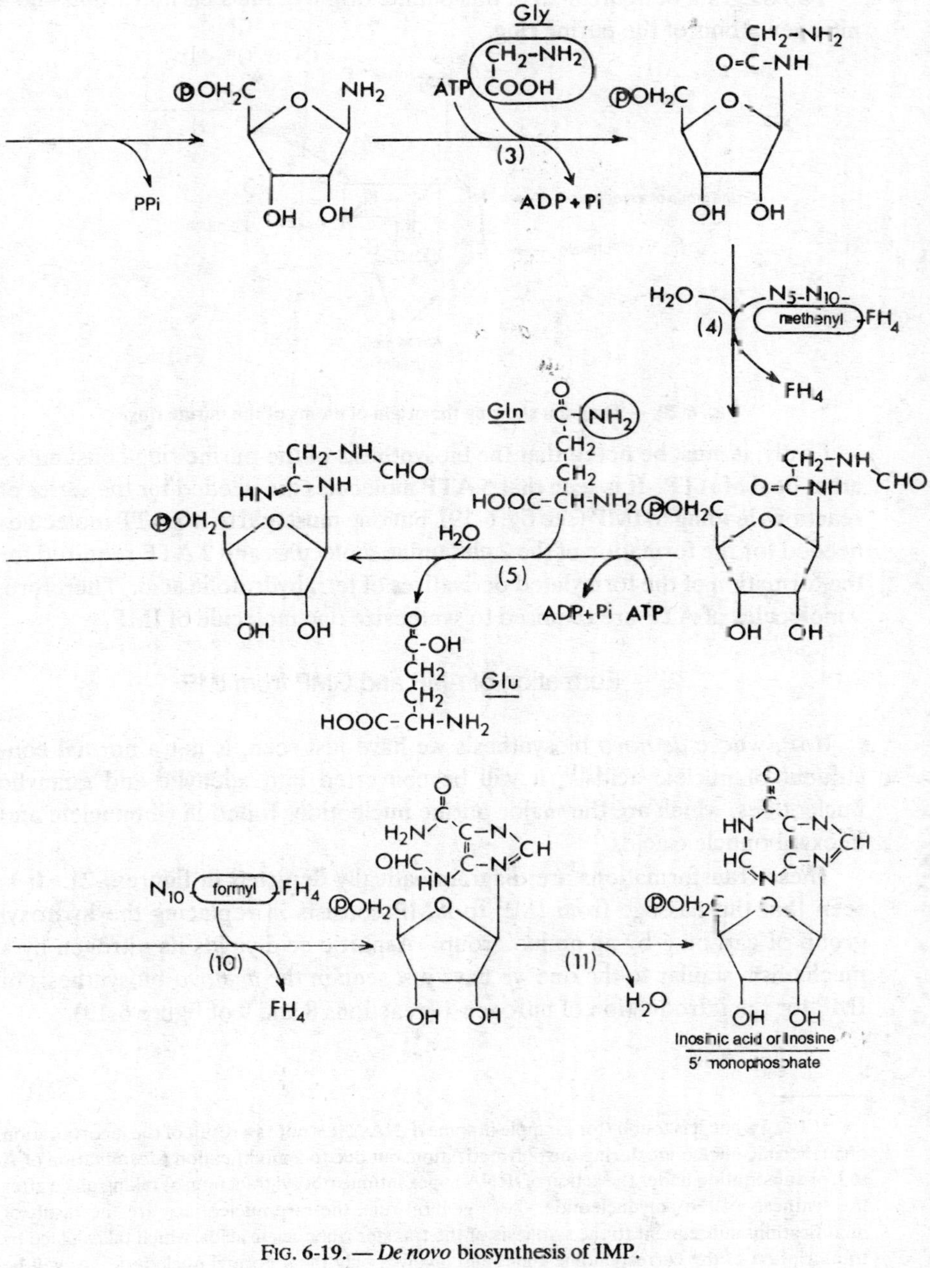

FIG. 6-19. — *De novo* biosynthesis of IMP.

This series of reactions can be summarized as follows:

ribose-5-phosphate + glycine + aspartate + 2 glutamine
+ 2 formiate + $CO_2 \longrightarrow$ IMP + 2 glutamate + fumarate.

The diagram of figure 6-20 points out the origin of the 5 carbon atoms and 4 nitrogen atoms of the purine ring.

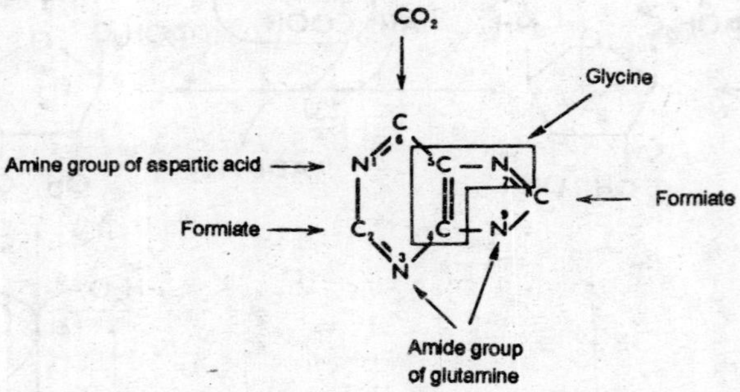

FIG. 6-20. — Diagram showing the origin of atoms of the purine ring.

Lastly, it must be noted that the biosynthesis of the purine ring consumes a great deal of ATP. It is seen that 5 ATP molecules are needed for the series of reactions leading to IMP (see fig. 6-19), but one must add to it 2 ATP molecules needed for the formation of the 2 glutamine molecules and 2 ATP required for the formation of the formylated derivatives of tetrahydrofolic acid. Therefore, 9 molecules of ATP are required to synthesize one molecule of IMP.

2. – Formation of AMP and GMP from IMP

IMP, whose *de novo* biosynthesis we have just seen, is not a normal constituent of nucleic acids(1); it will be converted into adenylic and guanylic nucleotides, which are the major purine nucleotides found in ribonucleic and deoxyribonucleic acids.

These transformations are diagrammatically depicted in figure 6-21. It is seen that the passage from IMP to AMP consists in replacing the hydroxyl group of carbon 6 by an amino group. Aspartic acid yields its nitrogen by a mechanism similar to the one we have just seen in the *de novo* biosynthesis of IMP for the introduction of nitrogen 1 (reactions 8 and 9 of figure 6-19).

---

(1) Even when it is found (for example in some *t*RNAs), it is not as a result of the incorporation of an inosinic nucleotide during the polymerization, but due to a modification (deamination of A to I, or substitution under the action of *t*RNA hypoxanthine ribosyl transferase) taking place after the synthesis of the polynucleotide. As a general rule, the rare nucleotides are the result of modifications subsequent to the synthesis of the transfer ribonucleic acids, which takes place by transcription of the corresponding genes and involves only the 4 normal nucleotides, as will be seen in the following part of this chapter.

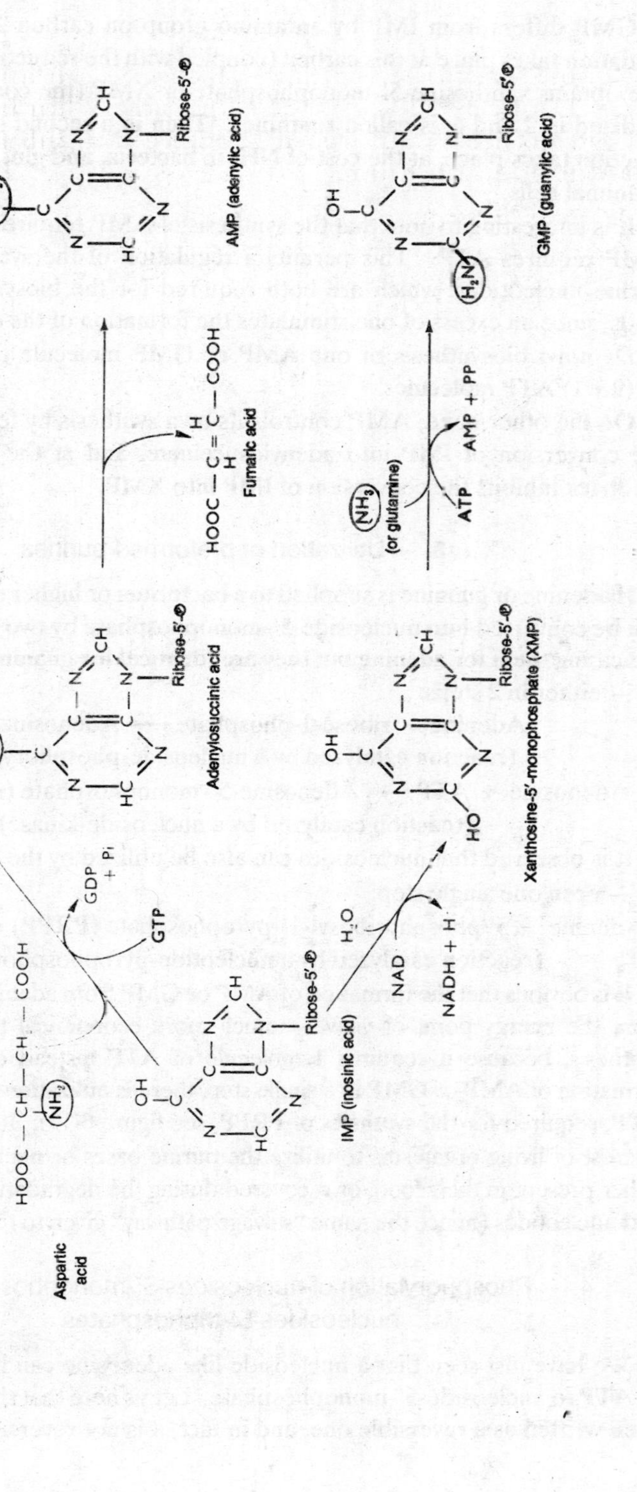

FIG. 6.21. — Formation of AMP and GMP from IMP.

GMP differs from IMP by an amino group on carbon 2. In a first step, oxidation takes place at this carbon (coupled with the reduction of NAD^+) and one obtains xanthosine-5′-monophosphate or XMP (the corresponding base, oxidized in 2 and 6, is called xanthine). Then in a second step an amination reaction takes place, at the cost of NH_3 in bacteria, and glutamine (amidic N) in animal cells.

It is interesting to note that the synthesis of AMP requires GTP and that of GMP requires ATP. This permits a regulation of the synthesis of the two purine nucleotides, which are both required for the biosynthesis of nucleic acids, since an excess of one stimulates the formation of the other.

De novo biosynthesis of one AMP or GMP molecule therefore requires 10 (9 + 1) ATP molecules.

On the other hand, AMP controls its own synthesis by feedback inhibiting the conversion of IMP into adenylosuccinate, and at the same time GMP feedback inhibits the conversion of IMP into XMP.

3. – Utilization of preformed purines

If adenine or guanine is supplied to a bacterium or higher organism, the base can be converted into nucleoside-5′-monophosphate by two processes (we are indicating them for adenine but they are identical for guanine):

– either in 2 steps:

$$\text{Adenine} + \text{ribose-1-phosphate} \longrightarrow \text{Adenosine} + P_i$$

(reaction catalyzed by a nucleoside-phosphorylase)

$$\text{Adenosine} + \text{ATP} \longrightarrow \text{Adenosine-5'-monophosphate (AMP)} + \text{ADP}$$

(reaction catalyzed by a nucleoside-kinase).

It is observed that nucleosides can also be utilized by the cells;

– or in one single step:

$$\text{Adenine} + 5' \text{ phosphoribosyl-1'-pyrophosphate (PRPP)} \longleftrightarrow \text{AMP} + \text{PP}$$

(reaction catalyzed by a nucleotide-pyrophosphorylase).

It is obvious that the formation of AMP or GMP from adenine or guanine is – from the energy point of view – much more economical than their *de novo* synthesis, because it requires 1 molecule of ATP instead of 10 (even in the formation of AMP or GMP in a single step, there is utilization of one molecule of ATP, required for the synthesis of PRPP, see figure 6-19). It is therefore in the interest of living organisms to utilize the purine bases or nucleosides, which are either present in their food, or recovered during the degradation of nucleic acids and nucleotides (hence the name "salvage pathway" given to these reactions).

4. – Phosphorylation of nucleosides-5′-monophosphates into nucleosides-5′-triphosphates

We have just seen that a nucleoside like adenosine can be phosphorylated by ATP to nucleoside-5′-monophosphate. Let us note that the reaction has not been written as a reversible one, and in fact, it is not reversible, because there is rupture of a phosphoanhydride bond and formation of an ester linkage, so that the equilibrium is very strongly in favour of the ester formation (see pages 104 and 107).

On the contrary, the transphosphorylation reactions that we will see now are reversible because there is rupture and formation of the same type of bond (phosphoanhydride bond).

A. – Reactions catalyzed by nucleosides-5′-monophosphates-kinases

The best known enzyme of this group is AMP kinase (or adenylate kinase or myokinase) which catalyzes the reaction:

$$\text{AMP} + \text{ATP} \rightleftarrows \text{ADP} + \text{ADP (2 ADP)}$$

i.e. phosphorylation of a nucleoside-5′-monophosphate into diphosphate.

B. – Reactions catalyzed by nucleosides-5′-diphosphates-kinases

As indicated by their name, these enzymes catalyze the phosphorylation of a nucleoside-5′-diphosphate into triphosphate. For example, the following reaction may take place:

$$\text{GDP} + \text{ATP} \leftrightarrows \text{GTP} + \text{ADP}$$

This shows how, from AMP and GMP, formed either by *de novo* biosynthesis, or from preformed purines, can be synthesized the corresponding ribonucleosides-5′-triphosphates which, as we will see later in this chapter, are the precursors required for the synthesis of ribonucleic acids.

It should be noted that the 3 classes of transphophorylations we have just considered exist not only in the "ribo" series but also in the "deoxyribo" series and that they also exist for the pyrimidine compounds. In reality that following reactions can take place[1]:

Purine or pyrimidine ribonucleoside

$$\text{(Z)} \rightarrow \text{ZMP} \leftrightarrow \text{ZDP} \leftrightarrow \text{ZTP}$$

Purine or pyrimidine deoxyribonucleoside

$$(d\text{Z}) \rightarrow d\text{ZMP} \leftrightarrow d\text{ZDP} \leftrightarrow d\text{ZTP}$$

These reactions are therefore also involved in the formation of pyrimidine ribonucleotides and purine and pyrimidine deoxyribonucleotides, but we will not revert to this question, and while studying the synthesis of these nucleotides we will assume the reactions known.

---

(1) Having seen above (page 292 and fig. 6-21) that the letter X represents xanthosine, it can be understood why one rather uses the letters N, Y or Z to signify "any nucleoside".

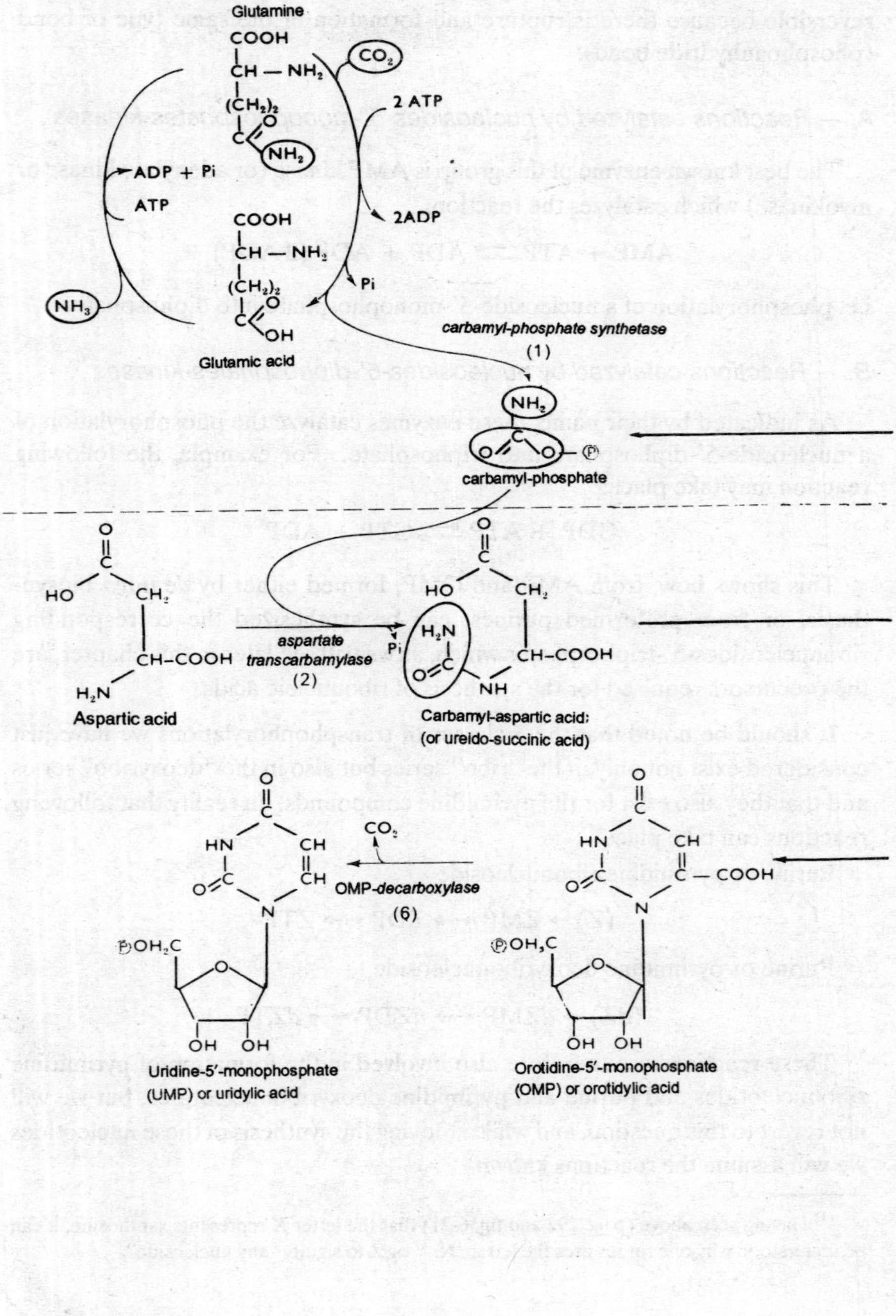
Glutamine
CO2
2 ATP
ADP + Pi
ATP
2ADP
Pi
NH3
Glutamic acid
carbamyl-phosphate synthetase
(1)
carbamyl-phosphate
aspartate
transcarbamylase
(2)
Aspartic acid
Carbamyl-aspartic acid:
(or ureido-succinic acid)
CO2
OMP-decarboxylase
(6)
Uridine-5′-monophosphate
(UMP) or uridylic acid
Orotidine-5′-monophosphate
(OMP) or orotidylic acid

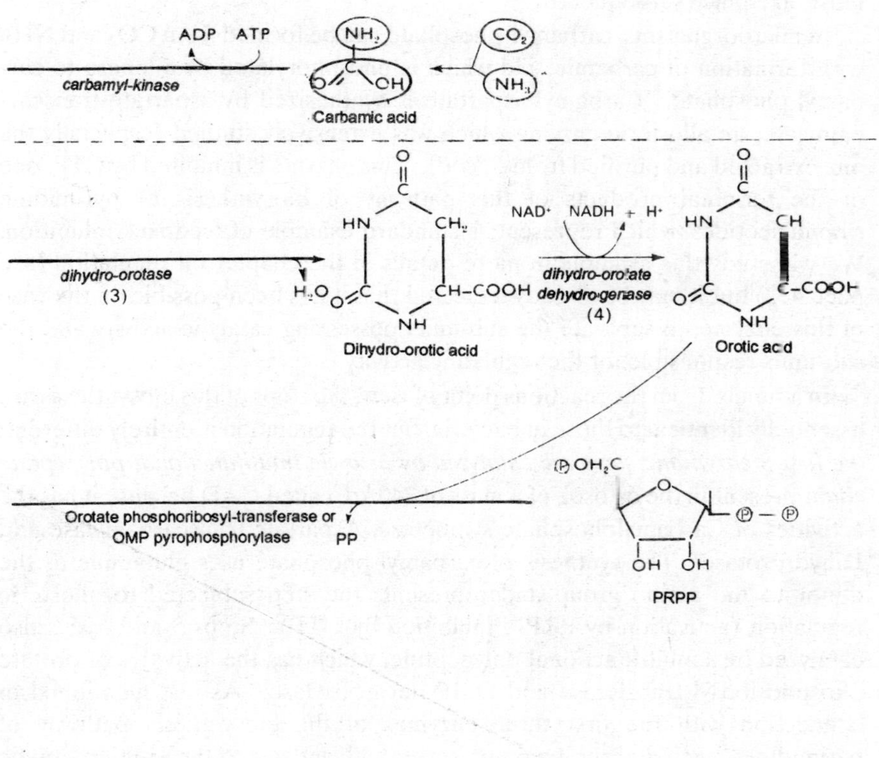

FIG. 6-22. — *De novo* biosynthesis of UMP. The reactions of this biosynthesis pathway are represented below the dotted line (above this line can be seen the 2 modes of formation of carbamyl-phosphate).

II. – BIOSYNTHESIS OF PYRIMIDINE RIBONUCLEOSIDES-5 - TRIPHOSPHATES

1. – *De novo* biosynthesis of UMP

The series of reactions leading to UMP is shown in figure 6-22. It is observed that it is *much shorter* than the one leading to the biosynthesis of purine nucleotides. It also differs in that the *"ribose-5'-phosphate" part is introduced in the molecule only at the last but one step*, when the pyrimidine ring is already formed (whereas in the biosynthesis of purine nucleotides, ribose-5'-phosphate is the starting point on which are grafted the constitutive atoms of the purine ring and therefore, an aliphatic ribonucleotide is obtained right at the start). Another difference lies in the fact that the *de novo* biosynthesis of pyrimidine nucleotides leads to UMP, *one of the four major nucleotides of ribonucleic acids*, whereas IMP is not a nucleotide frequently found in nucleic acids. However, it may be noted that in both cases, *de novo biosynthesis leads to a nucleotide deprived of any amino group*, and that the amination reactions must take place subsequently.

In microorganisms, carbamyl-phosphate can be formed from CO_2 and NH_3, with formation of carbamic acid which is phosphorylated by a kinase to carbamyl-phosphate. Carbamyl-aspartate is synthesized by aspartate-transcarbamylase, an allosteric enzyme which was extensively studied (especially the one extracted and purified from *E. coli*). This enzyme is inhibited by CTP, one of the terminal products of this pathway of biosynthesis of pyrimidine ribonucleotides, which represents a standard example of feedback inhibition. We will study this example in more details in the chapter on regulation (see page 472) but it may be already indicated that it has been possible, in the case of this enzyme, to separate the sub-units possessing catalytic activity and the sub-units responsible for the regulating activity.

In animals, from the reactions point of view, the steps of this biosynthesis are essentially identical to those in bacteria, but the regulation is entirely different; *the first 3 enzymatic steps are catalyzed by a single multifunctional polypeptide chain* present in the cytosol, of a mass of 240 kd, called CAD because it has the activities of Carbamylphosphate synthetase, Aspartate transcarbamylase and Dihydrorotase. The synthesis of carbamyl phosphate uses glutamine as the donor of the amino group and represents the step subjected to allosteric regulation (activation by PRPP, inhibition by UTP). Steps 5 and 6 are also catalyzed by a multifunctional polypeptide, which has the activities of orotate phosphoribosyl transferase and OMP decarboxylase. As just mentioned in connection with the first three enzymes of the biosynthesis pathway of pyrimidines, such a grouping offers several advantages to the multienzymatic complex itself (whose formation is simpler and more effective than if it would require joining by non-covalent bonds, different enzymes which were synthesized separately), as well as to the metabolic pathway because the substrates can thus be effectively transferred from one catalytic site to the other. In the chapter on lipids we have seen another multienzymatic complex, acyl-synthetase (see page 237). Some believe that the genes corresponding to these plurifunctional polypeptides result from rearrangements of various exons. This is, in animals, a striking example of *compartmentation* or *metabolic channeling* for two pathways having a common precursor. The carbamyl-phosphate required for the synthesis of urea (see fig. 7-30) is indeed produced by a mitochondrial enzyme characterized by different substrate (NH_3) and regulation.

2. – Utilization of preformed pyrimidines

Uracil can be converted into UMP by reactions identical to those we have seen in connection with the utilization of preformed purines

– either in 2 steps:

$$\text{Uracil} + \text{ribose-1-P} \longleftrightarrow \text{Uridine} + P_i$$

(reaction catalyzed by a nucleoside-phosphorylase)

$$\text{Uridine} + \text{ATP} \longrightarrow \text{UMP} + \text{ADP}$$

(reaction catalyzed by a nucleoside-kinase)

– or in 1 step:

$$\text{Uracil} + \text{PRPP} \longleftrightarrow \text{UMP} + \text{PP}$$

(reaction catalyzed by a nucleotide pyrophosphorylase)

Cytosine on the contrary, is not incorporated as such; it is generally deaminated into uracil, which is transformed into UMP as we have just seen. However cytidine can be incorporated after phosphorylation into CMP.

3. – Formation of uridylic and cytidylic ribonucleotides

With the help of the kinases studied in connection with the formation of purine nucleotides, UMP (formed *de novo*, or from uracil) can be phosphorylated into UDP and then UTP. The latter will be aminated into CTP, as shown by figure 6-23, thanks to NH_3 (in microorganisms) or amidic nitrogen of glutamine (in higher organisms).

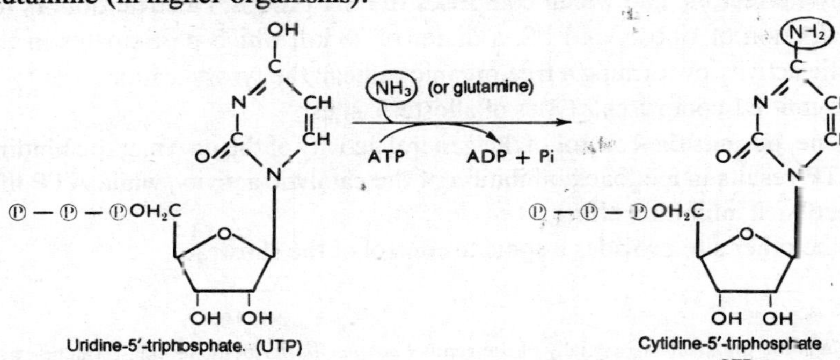

FIG. 6-23. — Amination of UTP into CTP.

III. – FORMATION OF PURINE DEOXYRIBONUCLEOSIDES-5′-TRIPHOSPHATES

The deoxyribonucleotides are simply formed by reduction of the corresponding ribonucleotides. Thus reduction of carbon 2′ of ribose, takes place in *E.coli* and animal tissues at the ribonucleosides-5′-diphosphates stage[(1)], at the cost of a protein of low molecular weight, *thioredoxin*, comprising 2 thiol groups; thioredoxin will thus be oxidized, but it can be again reduced at the cost of NADPH. A mutant of *E.coli*, without thioredoxin, but yet capable of forming deoxyribonucleotides, enabled the characterization of a protein similar to thioredoxin, *glutaredoxin*, which, after oxidation during the reduction of ribonucleotides, is reduced by glutaredoxin reductase, an enzyme whose prosthetic group is glutathione. Oxidized glutathione (disulphide form) will be finally reduced by NADPH. All these reactions, schematically represented in figure 6-24, concern the *reduction of purine as well as pyrimidine ribonucleosides-diphosphates*. In the case of purine derivatives, one thus obtains *d*ADP and *d*GDP which will be phosphorylated, by a nucleoside-diphosphate kinase, to *d*ATP and *d*GTP, the purine precursors required for the synthesis of deoxyribonucleic acids.

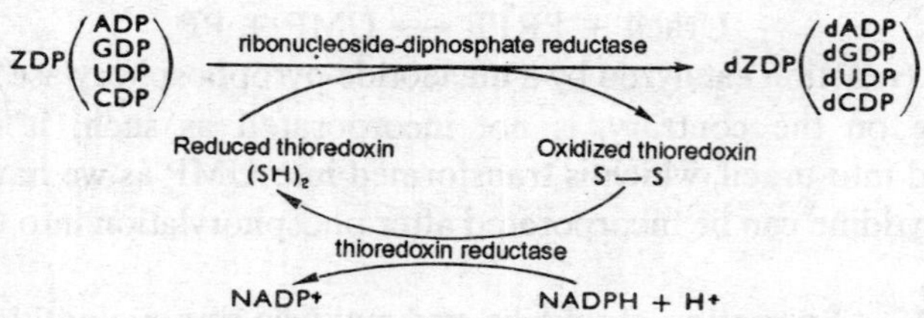

FIG. 6-24. — System enabling the reduction of ribonucleosides-5′-diphosphates to deoxyribonucleosides-5′-diphosphates.

Ribonucleoside diphosphate reductase comprises two sub-units: B1, a dimer of 160 kd which carries the binding sites of substrates (ZDP) and allosteric effectors, and which comprises the SH groups, electron donors for the reduction of ribose, and B2, a dimer of 78 kd, which participates in the catalytic activity by forming a free organic radical: the tyrosyl cation.

Sub-unit B1 comprises 2 types of allosteric sites:

– one, permits the control of the general activity of the enzyme; the binding of *d*ATP results in feedback inhibition of the catalytic activity, while ATP lifts this feedback inhibition effect;

– the other site provides a specific control of the substrate:

(1) In some organisms (especially of the genus *Lactobacillus*) ribonucleotide-reductase acts on ribonucleosides-5′-triphosphates.

The reduction of pyrimidine nucleotides, UDP and CDP, is activated by the binding of *d*ATP on this site.

Reduction of GDP is favoured by the binding of *d*TTP, which also inhibits the reduction of pyrimidine nucleotides.

The resulting formation of *d*GTP stimulates the reduction of ADP;

It appears that ribonucleotide reductase can exist in different conformations, each having specific reducing properties, and that this system of allosteric regulation provides a balanced production of the 4 deoxyribonucleotides required for the synthesis of DNA.

IV. – FORMATION OF PYRIMIDINE DEOXYRIBONUCLEOSIDES-5'-TRIPHOSPHATES

UDP and CDP can be reduced ((just like ADP and GDP) to *d*UDP and *d*CDP. The latter will be phosphorylated into *d*CTP, one of the pyrimidine precursors required for the synthesis of deoxyribonucleic acids. The other precursor is not *d*UTP but *d*TTP, and we will now see the reactions involved in the *biosynthesis of thymidylic deoxyribonucleotides*.

1. – Formation of *d*UMP

A. From uridylic deoxyribonucleotides

*d*UDP can

– either lose a phosphate group and thus give *d*UMP

– or be phosphorylated into *d*UTP, but the latter is not a precursor of deoxyribonucleic acids, and it is degraded into *d*UMP + PP.

B. – From cytidylic deoxyribonucleotides

*d*CMP can be deaminated into *d*UMP by deoxycytidylate aminohydrolase (or *d*CMP deaminase) which catalyzes the following reaction:

$$d\text{CMP} + H_2O \longrightarrow d\text{UMP} + NH_3$$

This is the reverse reaction of the amination reaction of UTP into CTP which we mentioned in the foregoing, with the difference that here the base is linked to a deoxyribose-5'-monophosphate and not to a ribose-5'-triphosphate. The enzyme has a very narrow specificity and does not deaminate deoxycytidine, *d*CDP, *d*CTP, cytosine, cytidine, CMP, although the structures of these compounds are closely related to that of *d*CMP.

2. – Methylation of *d*UMP into *d*TMP

This reaction catalyzed by *(deoxy) thymidylate* synthase takes place by the action of $N_5 - N_{10}$ methylene tetrahydrofolic acid which on the one hand, yields its one carbon atom group and on the other hand acts as a reductant and is thus finally found in the form of *dihydrofolic* acid. Tetrahydrofolic acid will be regenerated by *dihydrofolate reductase* which catalyzes the following reaction:

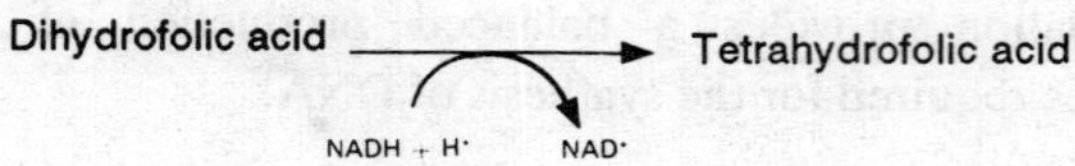

Generally, cancerous cells multiply rapidly; they are therefore subjected to a significant synthesis of DNA which requires especially *d*TMP in large quantities. This is why thymidylate synthase and dihydrofolate reductase are choice targets in the treatment of cancer by chemotherapy.

Methotrexate and *aminopterin*, which are analogues of dihydrofolate, are powerful competitive inhibitors of dihydrofolate reductase, and methotrexate is used in the treatment of various rapid growth tumours; but this compound kills the cells in rapid division, malignant or otherwise; it therefore has secondary toxic effects (in bone marrow, capillary follicles, epithelial cells of the intestine, etc.). On the other hand, it was observed that mammalian cells cultured in presence of methotrexate become resistant to this substance; mutated cells appear which are affected either in the uptake of methotrexate or in the affinity of dihydrofolate reductase for methotrexate (which is decreased), or because of an excess production of dihydrofolate reductase (1 000 times more enzymes) due to an amplification of the corresponding gene (some cells can have up to 100 copies of the gene coding for this enzyme). The treatment must therefore use rather high doses of methotrexate to prevent the appearance of mutant cells increasingly resistant, due to genic amplification.

Fluorouracil is used as anti-cancerous agent because it is converted *in vivo* into fluorodeoxyuridylate (F-*d*UMP) which irreversibly inhibits thymidylate synthase. At the beginning of the reaction F-*d*UMP behaves as the normal substrate (*d*UMP) and methylene tetrahydrofolate can fix itself on the C-5 of the analogue. But the enzyme cannot detach F^+ from the C-5 of F-*d*UMP (as it detaches H^+ from the C-5 of *d*UMP), with the result that the reaction cannot proceed. It is a *suicide-inhibition* because the enzyme transforms a substance (which first behaves as a substrate) into an inhibitor which blocks the reaction.

3. – Utilization of thymine and deoxythymidine

Lastly, it should be mentioned that preformed purines and pyrimidines can be incorporated in the corresponding deoxyribonucleotides, but only through deoxyribonucleosides (there is no equivalent of PRPP for deoxyribose). One can have for example:

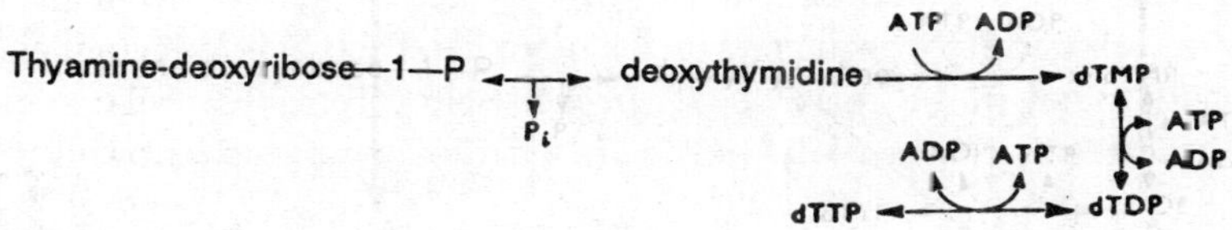

Figure 6-25 gives a general diagram recapitulating the main steps of the biosynthesis of precursors of ribonucleic acids (purine and pyrimidine ribonucleosides-5′-triphosphates) and deoxyribonucleic acids (purine and pyrimidine deoxyribonucleosides-5′-triphosphates).

NUCLEIC ACIDS AND GENETIC INFORMATION

The synthesis of nucleic acids (DNAs and RNAs) from nucleosides-5′-triphosphates is characterized by the necessity of conserving the genetic information intact. The DNA is indeed the carrier of this information which is transmitted from generation to generation; this notion is now well established, but it was long believed that this information was carried by proteins. It would therefore be appropriate to recall the important observations and experiments which led to the correct conclusion that DNA is actually responsible for the hereditary characters.

I. – PROOFS OF THE ROLE OF DNA AS CARRIER OF GENETIC INFORMATION

We will only mention a few arguments. Some of them are relative to notions which cannot be developed here and form the subject of more specific studies in Microbiology and Genetics.

1. *In higher organisms, there is a correlation between the quantity of DNA contained in a cell and the quantity of genetic information*. Thus, the quantity of DNA contained in the nuclei of diploid, somatic cells (having $2n$ chromosomes), is twice greater than that contained in the nuclei of haploid, germinal cells called the gametes (having n chromosomes).

2. *A correlation can be observed between the genetic relationship of certain species and the analogy of base composition of their deoxyribonucleic acids or even, the degree of homology of the nucleotide sequences of their deoxyribonucleic acids* (measured by hybridization, a technique, the principle of which was explained on page 285).

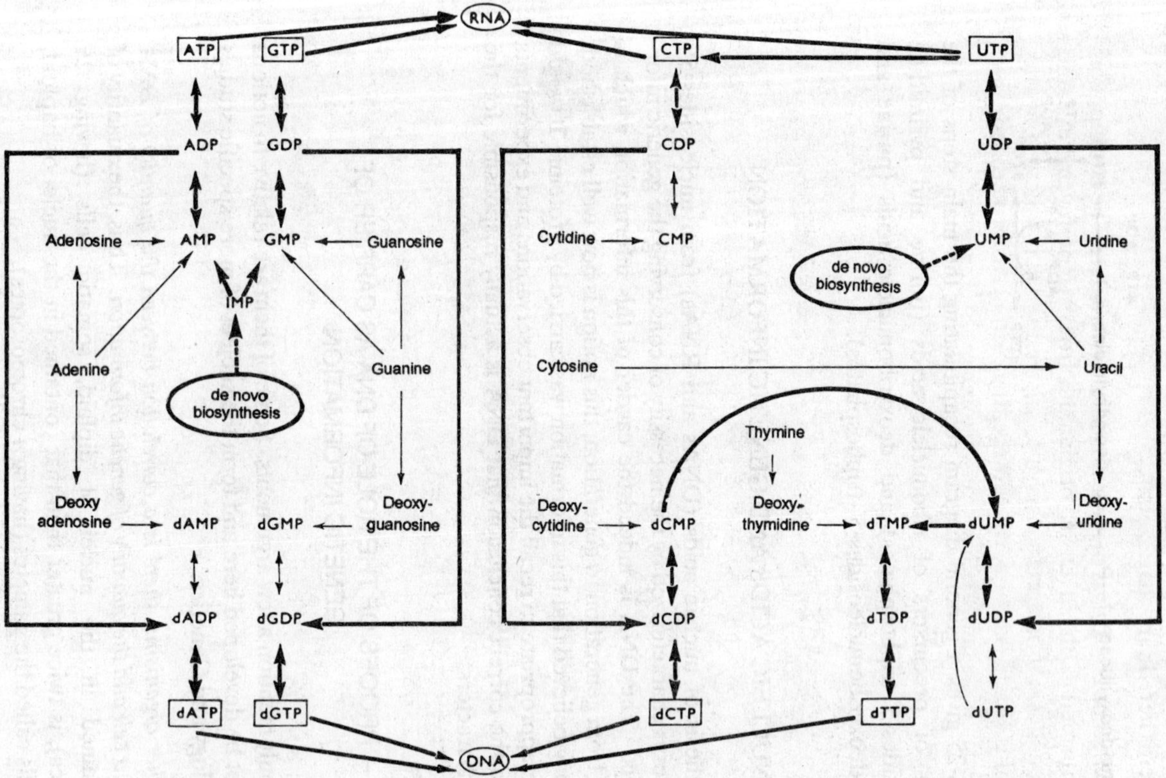

FIG. 6-25. — Recapitulative diagram of the biosynthesis of purine and pyrimidine ribonucleosides-5′-triphosphates and purine and pyrimidine deoxyribonucleosides-5′-triphosphates. (Red arrows indicate the reactions following *de novo* biosynthesis; black arrows show the utilization of preformed purines and pyrimidines).

3. *Agents* (physical or chemical) *which can cause mutations* (changes of hereditary characters) *are those capable of altering the deoxyribonucleic acids.*

4. In microorganisms, it was observed that the DNA can permit the transfer of one or several hereditary characters:

– *Transformation.* – This is the transfer of one character. The addition for example of purified, isolated DNA, originating from streptomycin-resistant bacteria, to a culture of bacteria sensitive to this antibiotic, is followed by the appearance of resistant bacteria, which were therefore transformed and have acquired a new genetic information, transmissible to their progeny.

– *Transduction.* – The transfer of several characters, thanks to a transducing phage DNA carrying several bacterial genes, when it penetrates into in a new bacterial cell.

– *F-duction.* – It is the transfer of a small number of genes, present together with the sex factor F in a fragment of DNA called *episome*, which passes from a donor cell to a receptor cell.

– *Conjugation.* – Entering in direct contact, a male bacterial cell can transfer a part or the whole of its chromosome to a female cell. This conjugation can be interrupted and a *correlation is observed between the quantity of DNA having penetrated the receptor cell and the number of characters transferred.*

5. – **Phage and viral infection.** – *The infection of a bacterial cell by a phage results from the penetration of the phage DNA only* and it can be verified that the proteins of the phage (specifically labeled with ^{35}S for example) do not penetrate. The infection can also be brought about with the help of purified DNA, in cases of phages as well as some viruses. As complete phage or viral particles will appear in the cell, it must be concluded that the information required for their formation is carried by the DNA. This conclusion is further supported by the study of the *lysogenic* pathway where the DNA of the phage is integrated into the bacterial DNA, transmitted from generation to generation with the bacterial genome, and can suddenly start the infectious process (lytic pathway) after remaining for several generations without manifesting itself.

One thus arrived at the conclusion that the DNA is the carrier of genetic information; in other words the genes, responsible for the hereditary characters are DNA fragments and the information is contained (coded) in the nucleotide sequence of these fragments. We must now examine how these genes are expressed, and determine the nature of compounds which are characteristic of a given organism.

II. – PROTEIN NATURE OF THE PRODUCTS OF GENE EXPRESSION. EFFECTS OF MUTATIONS

These specific compounds, whose synthesis depends on genetic information, are mostly *proteins*. For example, in microorganisms there are numerous mutants which differ from the *wild* type strain because an enzyme is absent (due to a mutation having affected the corresponding gene), with the result that they can no longer perform the reaction in question and the metabolic pathway (synthesis of an amino acid or synthesis of purine nucleotides, for example) is blocked.

In higher organisms particularly in man, numerous examples are presently known where the protein modified after a mutation could be identified (which in fact, enables the management of hereditary metabolic diseases caused by these mutations). In the chapter on carbohydrates, we have already mentioned *galactosemia* (see page 184) caused by the absence of phosphogalactose-uridyl-transferase. In phenylpyruvic oligophrenia or *phenylketonuria*, it is phenylalanine-hydroxylase which is absent, blocking the normal pathway of catabolism of this amino acid and thus causing the accumulation of the toxic transformation products (see fig. 7-25). The absence of another enzyme prevents the formation of melanic pigments from dihydroxy-phenylalanine (see fig. 7-26) and causes *albinism*, a disease characterized by the non-pigmentation of the skin and hair. Similarly, *hemophilia* is due to the absence of a protein factor required for the coagulation of blood.

An enzyme is said to be absent when the functional enzyme is lacking, in other words, the corresponding catalytic activity is not detectable in the cell. Often, the polypeptide chain is present, but it is modified after the mutation, with the result that it is biologically inactive; but very often it conserves – at least partly – the antigenic properties of the normal protein, which enables the detection of its presence and its titration by immunological methods (with the help of an immune serum containing the corresponding specific antibody obtained by injecting the animal with the purified normal protein). The modification can be very slight as shown by the example of *sickle-cell anemia*.

The hemoglobin of individuals affected by this disease is very slightly different from normal hemoglobin. Adult human hemoglobin or Hb-A consists of 2α-globin chains having 141 amino acids, and 2β-chains having 146 amino acids (each chain is linked to a prosthetic group called *heme*). The analysis of amino acid sequences of polypeptide chains of the Hb-S hemoglobin contained in the red blood cells of patients suffering from sickle-cell anemia (which are sickle-shaped and easily distinguishable from normal red blood cells which are spherical) revealed that the only difference lies at the sixth amino acid (from the N-terminal end) of the β-chain which is a valine, whereas it is a glutamic acid in the case of the normal hemoglobin Hb-A. Since there are 2β-chains per molecule of hemoglobin, 2 amino acids are therefore substituted out of a total of 574; this may seem to be a very slight difference, but the consequences can be tragic, because the capacity of this modified hemoglobin to bind oxygen is reduced and individuals affected by this disease risk to die very young especially homozygous individuals who can synthesize only abnormal hemoglobin (Hb-S/Hb-S). The prognosis is less serious for heterozygous individuals

(Hb-A/Hb-S) who are capable of synthesizing both normal and abnormal hemoglobins.

It is clear that *the genes – DNA fragments, carrying the genetic information coded in the form of the nucleotide sequence – contain instructions enabling the cells to polymerize the amino acids in a very precise order and thus synthesize specific proteins*. A mutation in a given gene generally results in the synthesis of a modified protein. This modification may have no influence on the biological properties of the protein (the replacement of a neutral amino acid by another neutral amino acid at a position in the chain which participates neither in the constitution of the catalytic site of the enzyme, nor in the establishment of the three-dimensional conformation of the protein, may not produce any visible effect).

We have seen above, a few cases where a mutation was followed by the disappearance – or at least alteration – of a biological activity essential for the metabolism of the organism. But a mutation can also have a beneficial effect and confer an advantageous character on the mutated organism; around the year 1850, Darwin made a very significant contribution in advancing a theory of the evolution of species based on *natural selection of favourable mutations*. Evolution may therefore be regarded as a series of mutations resulting in modifications of proteins, and in fact one can watch the course of evolution of species by following up the evolution of the amino acid sequence of a given protein (e.g., cytochrome *c*, or hemoglobin).

III. – TRANSFERS OF INFORMATION

The DNA is the carrier of genetic information for all organisms, except RNA-containing phages and viruses (they are mostly plant viruses, but also animal viruses like that of poliomyelitis) which have no DNA and where the RNA is the carrier of genetic information. It is clear that this information must be conserved during cell division and that is why this division is preceded by a synthesis of DNA which was named DNA *duplication* or *replication*, because an exact replica of the genetic information contained in the mother cell must be produced, so that the two daughter cells can inherit this information (i.e. a qualitatively and quantitatively identical information).

This information permits the synthesis of specific proteins in conformity with the rule "one gene → one enzyme" or rather "one gene → one polypeptide chain" since one enzyme can be formed by different polypeptide chains[1]. But the information carried by the DNA is not used as such for protein synthesis; it must be first transcribed into messenger RNA, thus called because it is in its

[1] There are exceptions to this rule and it must be said that some genes specify ribonucleic acids (ribosomal RNAs, transfer RNAs) which are DNA transcription products and will not — contrary to messenger ribonucleic acids — be subsequently translated into proteins (but participate in this process, as will be seen in the following part of this chapter).

turn, carrier of the genetic message; it is the messenger RNA which controls the ordered incorporation of amino acids in proteins.

There are therefore different *transfers of informations*:

1. DNA ⟶ DNA transfer. This is the *replication* or *duplication* required for the conservation of genetic information;

2. DNA ⟶ RNA transfer. This is the *transcription*, permitting the passage of information from the DNA to various ribonucleic acids.

3. RNA ⟶ Protein transfer. This is the *translation* allowing the synthesis of specific proteins, thanks to the information contained in the messenger ribonucleic acids.

Besides, in the case of RNA-containing viruses, there are two additional transfer possibilities:

4. RNA ⟶ RNA transfer. The synthesis of RNA using the viral RNA as template has a double objective: it supplies the genomic RNA thus allowing the course of the infectious cycle (replication of the ribonucleic genome) and it provides the *messenger ribonucleic acids* coding for the viral proteins required for the infectious process (it is the equivalent of the transcription process).

5. RNA ⟶ DNA transfer. It is the prerequisite for a ribonucleic genome to be integrated in the deoxyribonucleic genome of the host cell. Furthermore, it appears that for some viruses causing tumors (oncogenic viruses), the ribonucleic genome must be first put in the form of DNA so as to be then transcribed like the genome of the host cell.

We are now ready to take up the study of the biosynthesis of nucleic acids. We have seen how precursors (ribo- and deoxyribonucleosides-5′-triphosphates) are formed and we know that the synthesis of deoxyribonucleic acids and ribonucleic acids must meet the requirement of a faithful transfer of genetic information.

BIOSYNTHESIS OF DEOXYRIBONUCLEIC ACIDS[1]

I. – FUNDAMENTAL MECHANISMS OF THE SYNTHESIS OF DEOXYRIBONUCLEIC ACIDS

We have just seen that the biosynthesis of the DNA must consist in a replication or duplication. The study of the DNA structure as proposed by Crick and Watson (see fig. 6-15) gives an idea of the mechanism of this replication. Each nucleotide of the template DNA chain need only be recognized by a complementary nucleotide before the latter is incorporated at the end of the chain in formation. To this end, the double helix must therefore be

(1) Revised by M. Wright and J.-M. Rossignol

open, at least partly and transitorily, to permit the pairing of the 2 complementary nucleotides by hydrogen bonding. The nucleotide thus selected is then incorporated in the chain being synthesized and 2 molecules identical to the starting molecule will be finally obtained (see fig. 6-26).

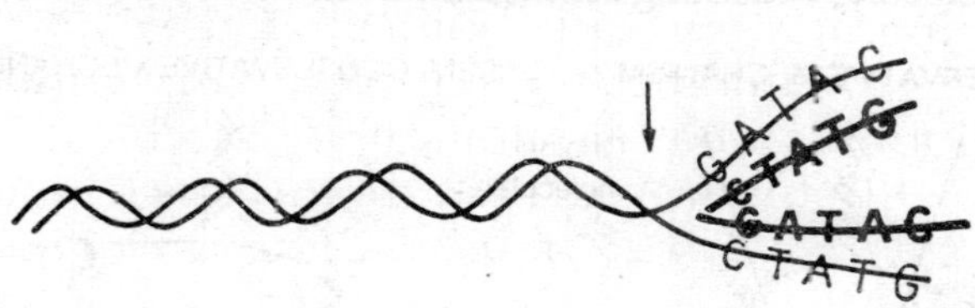

FIG. 6-26. — Model of DNA duplication: each strand of the original molecule (black lines) serves as template for the synthesis of a complementary strand (red lines). The arrow indicates the replication point (or fork).

1. – The replication is semi-conservative

In fact, two mechanisms of replication can be imagined (fig. 6-27):

– the conservative mechanism, where one of the strands of the original molecule serves as template for the synthesis of a complementary strand; this newly formed strand serves in its turn as template for the synthesis of a strand which is complementary to it. Therefore, after a replication cycle, the starting molecule remains intact (hence the name of this mechanism) and a new molecule consisting of 2 newly synthesized strands is formed;

– the semi-conservative mechanism, thus named because each strand of the original molecule serves as template for the synthesis of a complementary strand, with the result that after a duplication cycle, there are 2 molecules of hybrid DNA, consisting of one strand of the original molecule paired with a newly formed strand.

The experiments of Meselson and Stahl could show that DNA replication takes place according to the semi-conservative mechanism. They cultivated *E.coli* for several generations in a mixture containing $^{15}NH_4Cl$ as the only source of nitrogen, with the result that the DNA synthesized was heavy[(1)]. At a given time (time 0), they transferred the culture to a mixture containing $^{14}NH_4Cl$. Then, at regular intervals, they analyzed by centrifugation in cesium chloride gradient, the DNA extracted from the bacteria. This technique permits the separation of molecules as a function of their density; therefore, in the experiment described here, the denser the DNA, the more it will migrate to the bottom of the tube.

Meselson and Stahl obtained the following results:

– at time 0 : a single band, corresponding to the heavy DNA;

– after one generation in the mixture containing ^{14}N: a single band having a density intermediate between the density of the heavy DNA and that of the light

---

(1) ^{15}N is a nitrogen isotope heavier than the usual isotope, ^{14}N, because its ring contains an additional neutron.

DNA, which means that the DNA molecules consist of an original heavy strand (^{15}N) and a newly synthesized light strand (^{14}N);

– after 2 generations in the mixture containing ^{14}N: two bands of equal intensity, one corresponding to the hybrids (heavy-light), the other corresponding to the molecules consisting of two light strands.

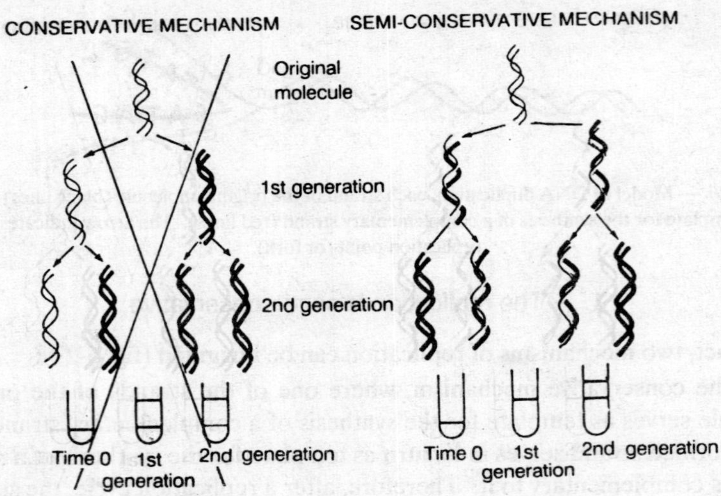

FIG. 6-27. — Conservative and semi-conservative mechanisms of replication of DNA. (black lines represent the heavy strands of the original molecule; red lines correspond to the newly synthesized light strands). The results obtained by Meselson and Stahl are in conformity with the predictions in the case of a semi-conservative replication (distances between bands have been exaggerated so that the differences may appear more distinctly).

These results are in conformity with those expected in the case of a semi-conservative replication, as may be observed in figure 6-27, which also indicates the results that would have been observed if the replication had taken place in the conservative mechanism (it is noted that in this case one would never observe the formation of hybrids of intermediate density).

Experimental results therefore exclude the conservative mechanism (see fig. 6-27).

2. – The polymerization of nucleotides is carried out by an enzyme

Although Watson and Crick had suggested in 1953, that the precursors of deoxyribonucleic acids can pair and bind one another without any enzymatic action, in 1958 an enzyme capable of polymerizing the 2′-deoxyribonucleosides-5′-triphosphates was isolated, by Kornberg from *E.coli* (DNA polymerase I). Subsequently, DNA polymerases were found in all bacteria and extracts of animal or plant cells in which a DNA synthesis could be observed.

3. – The DNA polymerase needs a DNA template and a primer

The substrates of polymerization are compulsorily 2′-deoxyribonucleosides-5′triphosphates (*d*NTP). As shown by figure 6-28, the DNA polymerase

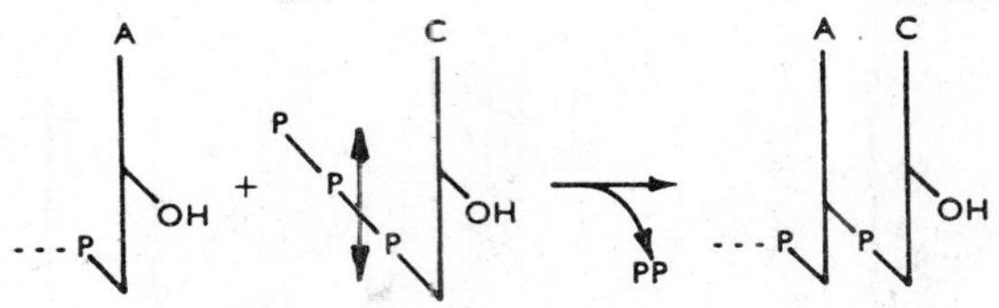

FIG. 6-28. — Elongation of a DNA chain. (The red arrow indicates the position where cleavage will take place during polymerization).

catalyzes the lengthening of polydeoxyribonucleotide chains (the term elongation is generally used) in the direction 5′→3′.

It must be noted that:

(1) the phosphodiester bond forms between the 3′ OH group of the growing chain and the 5′-phosphate group of the nucleotide to be incorporated. The growing chain is called primer;

(2) as already mentioned above, each new deoxyribonucleotide to be incorporated is selected as complementary to the nucleotide situated just opposite, on the chain to be copied. This chain is therefore called template.

Therefore, no known DNA polymerase can synthesize DNA without a primer and a template(1), hence the exact name of this enzyme: DNA-dependent-DNA polymerase. Moreover, the DNA polymerases cannot synthesize the primer; the latter must be provided to them, hence the importance of priming which we will discuss again.

4. – Polymerization takes place only in one direction: 5′→3′

A DNA polymerase catalyzes the following reaction (see fig. 6-28):

$$(d\text{NMP})_n + d\text{NTP} \xrightarrow[\text{Enzyme}]{\text{DNA, Mg}^{++}} (dNMP)_{n+1} + \text{PPi}^{(2)}$$

The enzyme has two substrates. One is the couple template-primer 3′OH. The other substrate is compulsorily a deoxyribonucleosides-5′-triphosphate

Some points could be confirmed by studying the nearest neighbour's frequency. This is done by introducing in the polymerization reaction, the 4 *d*NTP, one of them labeled with ^{32}P on the α phosphate group (the one bound to carbon 5′ of pentose). At the end of the reaction, the DNA formed is subjected to an enzymatic hydrolysis by micrococcal DNase and phosphodiesterase. In this way, the phosphodiester bonds are specifically split between carbon 5′ of deoxyribose and phosphate, and the deoxyribonucleosides-3′-monophosphates are liberated quantitatively. Hence, as may be seen in figure 6-29, the ^{32}P which was incorporated in the DNA when it

---

(1) Except the DNA polymerase I in certain special conditions called "*de novo* synthesis".

(2) *d*NMP = deoxyribonucleoside-5′-monophosphate
PPi = inorganic pyrophosphate.

Deoxyadenosine-5′-triphosphate (dATP) labeled with ^{32}P on Pα. (The arrow shows the place where cleavage will occur during the polymerization).

Hydrolysis of the DNA cleaving the phosphodiester bonds between P and $C_{5'}$ (arrows). It is seen that ^{32}P initially carried by the dATP will be found in the form of dYMP (i.e. in the nucleosides 3′-monophosphates which were adjacent to the adenylic nucleotides in the polymer).

FIG. 6-29. — Principle of the analysis of nearest neighbour's frequency.

was bound in position 5′ of a nucleoside-triphosphate, is found, after hydrolysis, attached in position 3′ of the adjacent nucleotide (i.e., the ^{32}P has shifted from the nucleotide which carried it, to its nearest neighbour in the chain).

Separating the 4 deoxyribonucleosides-3′-monophosphates and determining their radioactivity, one can calculate the frequency at which each base was adjacent to a given base.

This type of study has led to three important conclusions:

– the frequencies of the nearest neighbours of the template DNA and newly synthesized DNA are identical, which confirms that the DNA added to the reaction does serve as a template;

– any sequence of the DNA template is copied, without preference nor specificity;

– the polarity of the newly synthesized chain is the opposite of that of the template chain. The growing chain and the template chain are also said to be "anti-parallel". This conclusion is based on the fact that the equal frequencies of dinucleotide sequences which may be expected in the case of opposite polarities are verified by experiments, while those expected in the case of identical polarity of the 2 strands are contradicted by experimental results (see fig. 6-30).

II. – STEPS OF REPLICATION

The replication of the DNA can be schematically divided in three steps: initiation, elongation and termination.

1. – The initiation step

As seen above, no DNA polymerase can perform any *de novo* synthesis (contrary to RNA polymerases which can initiate by placing a complementary

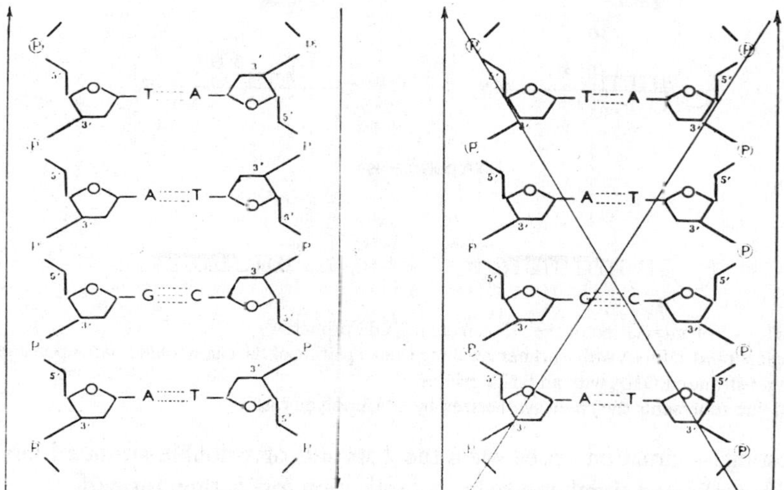

TpA (0.012) = TpA (0.012)
ApG (0.045) = CpT (0.045)
GpA (0.065) = TpC (0.061)
Opposite polarities

TpA (0.012) ought to be equal to ApT (0.031)
ApG (0.045) ought to be equal to TpC (0.061)
GpA (0.065) ought to be equal to CpT (0.045)
Identical polarity

FIG. 6-30. — Nearest neighbour's frequency. Experimental results (values between brackets) confirm the predictions made in the case of opposite polarities of the 2 strands, and exclude the identical polarity.

ribonucleoside-5′-triphosphate opposite to the DNA to be transcribed). The primer 3′OH which will serve as start point for the synthesis of DNA must therefore be supplied to the DNA polymerase during the initiation step.

On the other hand, replication starts at a precise point of the DNA called origin of replication. When the DNA is double-stranded, it is first necessary to open out the 2 strands in order to carry out the actual initiation step. We will again examine this mechanism.

The 3′OH can be supplied in various ways. In a very large number of cases it is a short RNA fragment either synthesized by a RNA polymerase or a DNA primase (*E.coli, B.subtilis*, phages M13, φX 174, T4, T7, eucaryotic cells, polyoma virus or Simian virus SV40) or vestige of a larger RNA (virus of hepatitis B). In the case of single-stranded DNA, the primer can also be formed by one end doubling back upon itself (hair-pin structure); this is postulated for the parvoviruses (see fig. 6-31). Lastly, it may be a deoxyribonucleoside-5′-monophosphate bound covalently to a protein; the 2 best known cases are, on the one hand, the adenoviruses, and on the other, a phage of *B.subtilis* called φ29.

2. — The elongation step

The elongation steps poses more or less problems depending on whether the DNA is single-stranded or double-stranded. The most complex — and most

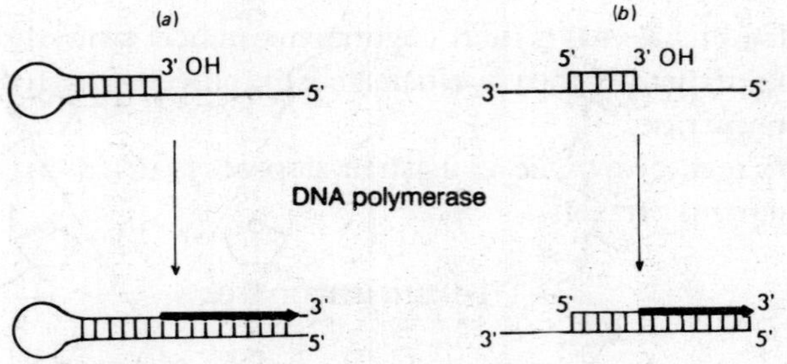

FIG. 6-31. — Polymerization of the DNA from a 3'OH primer
(*a*) Single-stranded DNA with a primer resulting from a pairing of the chain folded back upon itself.
(*b*) Single-stranded DNA with an RNA primer.
The red line represents the chain synthesized by DNA polymerase.

interesting – situation arises when the 2 strands of a double-stranded DNA must be replicated simultaneously. A replication fork is then formed.

We have seen in the foregoing (fig. 6-30), that the 2 chains of the double helix are of opposite polarities. But we know that polymerization takes place strictly in the direction 5′→3′. This implies that the DNA synthesis can take place on one strand continuously in the direction 5′→3′, but, in order to copy the other strand, it can take place only in the direction opposite of that of the progression of the fork.

To solve this dilemma it was tried to find a mechanism which would allow progression in the direction 3′→5′, for example the polymerization of nucleosides 3′ triphosphates. One could never find such precursors, nor a DNA polymerase capable of polymerizing nucleosides 3′ triphosphates synthesized chemically.

The question therefore remained unanswered: how does synthesis take place in the direction 3′→5′. The answer was given by experiments carried out on the replication of phage T4, by the Okazaki team; by adding a precursor of DNA synthesis, like thymidine (deoxythymidine) in a highly radioactive form and for a very short time, it is possible to label only the DNA which has just been synthesized. This method shows that the newly synthesized DNA is in the form of nucleotide chains of 1 000 to 2 000 residues (Okazaki fragments). These fragments are then bound to one another to form a continuous chain in the direction 5′→3′. The same type of experiment could be reproduced with various phages, bacteria and eucaryotic cells. In eucaryotes however, these fragments are definitely shorter in size (100 to 200 nucleotides).

Therefore, the replication fork has, in fact, an asymmetric structure (see fig. 6-32). One of the chains is synthesized in the direction 5′→3′ in a continuous manner (leading strand). The other, although advancing as a whole in the direction 3′→5′, consists of fragments synthesized individually in the direction 5′→3′; this is a discontinuous synthesis (lagging strand).

Although these results eliminate one difficulty they require a solution to the problem of the primer (initiation of polymerization). Here again, it could be shown that the primer was a short oligoribonucleotide, synthesized by DNA primase. This primer will then be eliminated, the gap thus created being filled by a DNA polymerase.

As will be seen below, the elongation step necessitates the coordinated action of numerous proteins.

3. – The termination step

Studies on the synthesis of DNA were first focussed on the elongation step, then, more recently, on the initiation step which in fact, permits the control of the process. Little is therefore known presently on the termination of replication. There are two important problems:

– how does the replication cycle terminate? Why replication does not continue? Why newly synthesized molecules do not serve immediately as templates for a new replication cycle?

– how newly synthesized DNA chains separate? This point seems easy to solve for a linear genome but it poses major topological problems for circular genomes.

III. – REPLICATION IN *E.COLI*

The replication of numerous procaryotes was studied in recent years. The mechanisms of the replication of phages T4, T7 and λ are now well known, while those of *B. subtilis* are beginning to be understood. However, the most extensively studied system remains *Escherichia coli*, thanks to the combination of various types of approaches.

1. – Experimental approaches to the study of replication

In a first step, the isolation and then enzymological study of DNA polymerase I gave a better understanding of the mechanism of polymerization of nucleotides. In a second step, the isolation of thermosensitive conditional mutants, i.e. incapable of replicating their DNA in certain temperature conditions, enabled the identification of the proteins which play a role of primary importance in replication. It could thus be shown that the "replicative" enzyme is not DNA polymerase I, but DNA polymerase III(1).

A third type of approach consists in studying the replication *in vitro* of DNA phages whose genome is circular, single-stranded and relatively short (5 to 6 000 nucleotides). For their replication, these phages (M13, φX174, G4) use

(1) There is still another polymerase, DNA polymerase II which seems to be the enzyme involved in the repair processes, i.e., when a fragment of altered DNA must be replaced by a fragment of normal DNA.

mainly the proteins of the *E.coli* replication machinery. It was thus possible to show, for example, the existence of the primase, responsible for the synthesis of not only the RNA primer allowing the replication of phage G4, but also the RNA primers of Okazaki fragments in the bacterium.

Lastly, more recently, using genetic engineering techniques it was possible to obtain large quantities of enzymes relatively scarce in the cell, or to create template deoxyribonucleic acids having the desired sizes or sequences. Major advances could be made thanks to the construction of a plasmid carrying the origin of the replication of the chromosome of *E.coli* (oriC), and then the isolation of a semi-purified cell-free fraction capable of replicating this DNA *in vitro*. Presently, there are at least 25 known proteins which participate in the replication of the chromosome in *E.coli*.

2. – Initiation at the origin of replication

There is an important point to note: the fact that *E.coli* has only one origin of replication provides the bacterium an easier control of the replication of the chromosome. However, practically nothing is known at present, of the processes responsible for the origin of replication being active or at rest.

On the contrary, knowledge is being acquired on the proteins involved in the initiation at the origin of replication. In fact, these can be divided in two classes:

(a) proteins involved in the initiation;

(b) "specificity proteins" which prevent the initiation process at sites other than oriC.

The specific binding of the product of gene *dna*A to oriC represents the first step of initiation. Complexed with ATP, the protein *dna*A will open the double helix, permitting the formation of the initiation complex. The separation of the two strands of the DNA molecule by the product of *dna*B will introduce new constraints which will cause twists of the whole DNA molecule upon itself (supercoils). These supercoils are eliminated by DNA gyrase. This enzyme belongs to the category of topo-isomerases which are enzymes capable of causing a transitory and reversible rupture of the phosphodiester bond. One may imagine a situation where the two DNA fragments rotate round one another. The DNA is thus "ready", the ribonucleotide primer can then be synthesized, probably by the primase. All the proteins playing a role in this step are sometimes defined as a complex called *orisome*. Some of them specifically recognize oriC (gyrase, *dna*A) or interact with the other proteins of the orisome (*dna*B, gyrase).

Lastly, other proteins called "specificity proteins" act in preventing initiation at sites other than oriC (protein HU) or eliminating RNA primers made at illegitimate sites (RNase H, an enzyme which specifically degrades the RNA strand of a RNA/DNA hybrid).

3. – Progression of the replication fork

Before considering the events which permit the progression of the replication fork, it would be useful to briefly describe here, the principal agent of replication in *E.coli*: the DNA polymerase III holoenzyme.

As mentioned above[1], there are three DNA polymerases in this bacterium. These three enzymes differ by a few biochemical criteria which permit a distinction between them in semi-purified extracts or during the purification of the enzymes. The DNA polymerase III holoenzyme is responsible for the synthesis of the two DNA strands at the replication fork. This enzyme is composed of at least 7 sub-units: three of them form the core enzyme; the 4 others are auxiliary sub-units. The core enzyme comprises sub-unit α responsible for the polymerization, sub-unit ε responsible for the $3' \longrightarrow 5'$ exonuclease activity (see in the following, the paragraph: "Fidelity of replication") and a third, θ, whose role is still unknown. The four other sub-units (τ, γ, β and δ) confer on the core enzyme a greater efficiency during the polymerization and control of fidelity.

The mechanisms of the elongation step are schematically represented in fig. 6.32. We mentioned above that replication is continuous on one strand (see fig. 6.32 part *e*) and discontinuous on the other. One must continue to unwind the two strands of the DNA from one another and hence the presence of a DNA helicase is necessary (see fig. 6.32, part *a*). It is not yet known which helicase is really implied in this process (product of the gene dnaB or gene rep?). As described in the initiation step, one must eliminate the supercoils thus formed, with the help of *a* topoisomerase, probably *DNA gyrase* (see fig. 6.32, part *b*). The unwinding of the two strands of the DNA molecule brings about the formation of single-stranded zones. These single-stranded zones must be stabilized (prevention of the re-association of the two strands of the DNA molecule) and protected from the action of enzymes capable of degrading them (DNase). These functions are carried out by a protein called single-stranded DNA binding protein, generally known by its abbreviation SSB (see fig. 6.32, part *c*).

The ribonucleotidic primers of Okazaki fragments are synthesized by the primase (see fig. 6.32, part *d*). This enzyme may be distinguished from RNA polymerase by its resistance to rifampicin. It synthesizes short RNA fragments (about 10 nucleotides) without any particular sequence specificity. These primers are then extended by the DNA polymerase III holoenzyme. This therefore leads to the formation of a chain which contains at its $5'$ end, an RNA fragment covalently bonded to the DNA fragment (see fig. 6.32, part *f*). But we know that the chromosome contains no ribonucleotide when the replication is completed. It is therefore necessary to eliminate the ribonucleotide primers and then fill the gaps thus formed. These two functions are carried out by DNA

---

(1) It appears that all living organisms have several DNA polymerases, with the exception of viruses and phages.

polymerase I, which eliminates the ribonucleotide primers thanks to its 5′ → 3′ exonuclease activity (which is not specific for DNA) and simultaneously performs the polymerization of the missing DNA fragment using as a primer the 3′ end of the next Okazaki fragment, i.e. the nearest on the side of the replication fork (see fig. 6.32, part *g*).

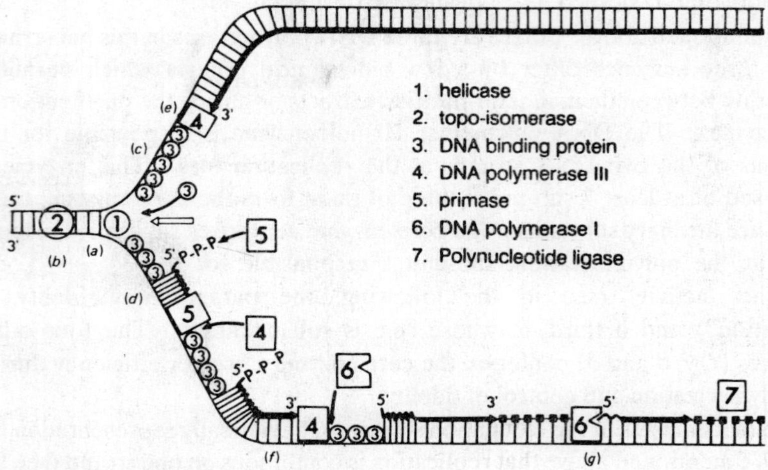

FIG. 6-32. — Diagram of the replication fork in *E.coli.*
The white arrow indicates the direction of displacement of the fork. This is a simplified diagram: the number of proteins was deliberately reduced and the respective sizes of various parts of the replication fork have not been taken into consideration.
——— DNA matrix
〰 RNA synthesized by the primase
——→ DNA synthesized by DNA polymerase III
■ ■ ■ DNA synthesized by DNA polymerase I

The synthesis of this strand is therefore discontinuous, and it is then in the form of juxtaposed DNA fragments. It is therefore necessary to bind these fragments together covalently. This is accomplished by DNA ligase, which is capable of forming a phosphodiester bond between a 3′-OH and a 5′-monophosphate, provided that these groups are maintained near one another thanks to hydrogen bonds exchanged between the newly synthesized DNA strand and the complementary parental strand. The DNA ligase cannot achieve the closure if the 5′-monophosphate belongs to a ribonucleotide which compulsorily implies the removal of RNA primers from the Okazaki fragments.

It must be emphasized that the situation described here is still a model, very probable, but not proved. More recently, other hypotheses – complementary – were advanced, in order to take into consideration, particularly the fact that, *in vivo*, the replication fork progresses relatively fast (500 to 1 000 nucleotides per second). For a better coordination of the action of the various enzymes, the latter could be physically associated in a multi-enzymatic complex: the replisome (see fig. 6.33). This hypothetical structure could contain 2 molecules of DNA polymerase III, one or several DNA helicases and the primosome (the set of proteins permitting the initiation of chains). To enable this structure to replicate the 2 strands concomitantly, one must imagine that the strand which will be copied discontinuously winds on itself at 180°; this allows the two

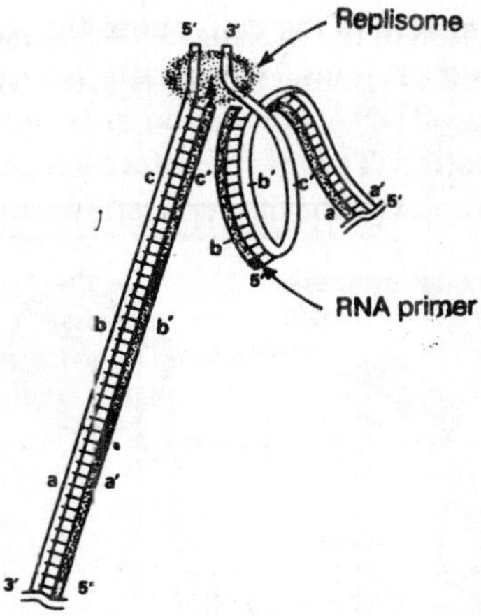

FIG. 6-33. — Diagram showing a concomitant replication of two DNA strands by the replisome.

parental strands to have the same polarity locally, at the fork. By such a mechanism, not only a molecule of polymerase (or a dimer of polymerase) would copy the two chains simultaneously, but it would also remain attached to the primosome which progresses in a direction opposite to that of the elongation on the discontinuous synthesis strand.

As far as termination is concerned, it is known that in *E.coli*, there is a genetically silent termination region where the two replication forks slow down. Going over this region could be the signal of cell division.

4. – New hypotheses on the replication in *E.coli*

As regards initiation of replication, it appears that the bacterial membrane plays an important role, on one hand because the protein *dna*A is associated with this membrane, and on the other hand, because the activity of the complex *dna*A-ATP is modified by a membrane phospholipid: cardiolipin.

At the level of elongation, according to a recent hypothesis, the DNA polymerase III holoenzyme is in the form of an asymmetric dimer. This asymmetry would be due to the mutually exclusive presence of two partly identical sub-units. The half of the dimer comprising the γ sub-unit would perform continuous synthesis and the other half comprising the τ sub-unit would carry out discontinuous synthesis.

IV. – REPLICATION IN EUCARYOTES

Although substantial progress was made during the last decade, all replication mechanisms in eucaryotes are not yet understood because:

(a) replication is an event which occurs much more rarely than in bacteria. Initiation of replication is therefore highly controlled; this complicates the understanding of this process;

(b) the DNA is not naked, in the cell, but in the form of chromatin, i.e. a combination of DNA and proteins. Chromatin is organized in nucleosome: about 200 pairs of bases of DNA are wound around an octamer of histones (basic proteins of chromatin). There is therefore a higher level of organization resulting in a greater complexity for the replication (see fig. 6-34).

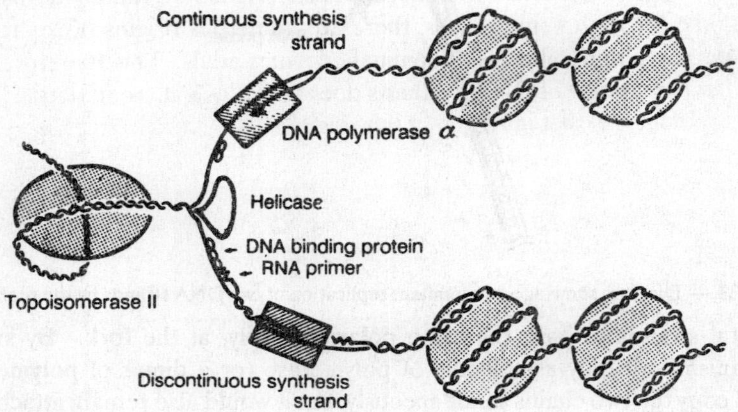

FIG. 6-34. — Replication fork in eucaryotes. The figure shows the winding of the DNA in the nucleosomes, and some of the proteins which act at the replication fork.

Mainly 3 experimental approaches were followed.

First of all, enzymes identical to those implied in bacterial replication were sought out, identified and very often purified. It is however difficult to know whether the protein has not been degraded during the purification as long as the gene coding for this polypeptide has not been isolated. On the other hand, it is only in some cases, by using appropriate biological systems, that it has been possible to correlate an activity showed *in vitro* to a precise role *in vivo*.

Recently, the existence of conditional mutants in some yeasts (*Saccharomyces cerevisiae* and *Schizosaccharomyces pombe*) has given much information on the role of DNA polymerases as well as on the role of topoisomerases (see below).

The third approach consists in studying the replication of animal DNA viruses, like the monkey virus SV 40 (Simian Virus 40). As a matter of fact, except for one protein coded by their genome, antigen T, these viruses utilize the enzymatic machinery of the cell to replicate their DNA.

Based on the results obtained by these three approaches we will now summarize the present knowledge on the subject.

1. — Steps of replication

The long chromosomal DNA molecules (10^8 base pairs) probably replicate from multiple origins of replication. By means of electron microscopy one can indeed observe "replication eyes", i.e. open regions of the DNA molecule, which may correspond to two replication forks progressing in opposite directions. If radioactive precursors are incorporated in the DNA being synthesized, one can watch the course of the phenomenon by autoradiography. These experiments confirm that there are numerous origins of replication, and that DNA synthesis is bidirectional from these origins. In rapidly developing tissues, like the embryonic tissues, there are numerous origins of replication which decrease in number as the animal becomes adult. This therefore confirms that the control of DNA synthesis does take place at the initiation stage. The present view is that initiation at the origin of replication takes place using an RNA.

In the case of SV 40, initiation of replication implies the recognition of a specific sequence by a viral protein, antigen T, followed by a local unwinding of the double helix thanks to the DNA helicase activity of this protein. This situation is similar to the one described previously for *E.coli*.

On the strand of discontinuous synthesis, the Okazaki fragments are much smaller than in procaryotes (about 200 nucleotides). This difference in size could be due to the nucleosomal structure of chromatin. These Okazaki fragments contain at their 5′ end, a short RNA fragment (about 10 nucleotides) which is synthesized by the eucaryotic primase. This primer is then elongated by DNA polymerase α. Unlike in *E.coli*, DNA polymerase and primase are combined into a complex which can be isolated *in vitro* (see below). The elimination of RNA primers could be made by RNase H. The filling of gaps thus created would be accomplished either by DNA polymerase α, or DNA polymerase β, the fragments being then bound to one another by a polynucleotide ligase. According to results obtained recently with SV40, the replication of the strand of continuous synthesis would be carried out by the DNA polymerase δ (see below).

Very little information is available on termination; some experiments were carried out with SV40 whose replication is bidirectional. Examining molecules replicated to 90%, one observes that there exists a termination zone, in which the two replication forks stop at about 500 nucleotides from one another, probably due to the effect of topological constraints. It then appears that an unwinding step (achieved by a topo-isomerase II) is required for the separation of the two daughter-strands. This has been confirmed in *S.pombe* thanks to a heat sensitive mutant of DNA topoisomerase II.

2. – The DNA polymerases

There are probably four DNA polymerase in eucaryotes: DNA polymerase α, DNA polymerase β, DNA polymerase γ and DNA polymerase δ. They can be distinguished by their physiochemical properties and sensitivity to various drugs.

DNA polymerase α or DNA polymerase I in *S.cerevisiae*, was purified in a native form. It comprises at least 4 sub-units. One of them is responsible for the polymerase activity. The primase activity is carried by one (or two) different sub-unit(s). One can therefore speak of a DNA polymerase α-primase complex, active *in vitro*.

The existence of DNA polymerase δ (DNA polymerase III in yeast) was confirmed only recently. It is a multimeric enzyme which has an exonuclease activity $3' \longrightarrow 5'$. It is stimulated by cycline, a protein whose quantity is maximal during the DNA synthesis phase.

The role of DNA polymerase β has not yet been established accurately. This enzyme is found in all cells, including those at a terminal differentiation stage (neurone, spermatozoid). Moreover, it presents a constant level of activity during the phases of the cell cycle. It is generally admitted that it has a role in the phenomena of DNA repair, especially in those taking place outside a phase of DNA synthesis.

DNA polymerase γ is localized in the mitochondria. An exonuclease activity $3' \longrightarrow 5'$ is probably associated to it. It probably replicates the mitochondrial DNA, whose replication is chronologically and enzymatically autonomous from the replication of the nuclear DNA.

3. – Other proteins involved in replication

As mentioned above, there are other proteins playing a role at the replication fork. They can be classified into proteins with enzymatic activity, proteins having a structural role and lastly modulator proteins, i.e. proteins which modify some enzymatic activities or make them specific.

It is rather easy to identify proteins having an enzymatic activity capable of playing a role in replication. It was thus possible to isolate a topo-isomerase of type I, one of type II, one RNase H, two DNA ligases and recently a DNA helicase. However, although it could be shown that the activities of topo-isomerase II or ligase I increase in parallel with the DNA synthesis, the role of these diverse enzymes in the replication process remains to be determined (see below).

Numerous proteins having affinity for the single-stranded DNA were partially purified from various tissues, but it could not be proved that any of them played a role in replication.

The situation is even more complex for modulator proteins, because, they are *a priori*, present in very small quantities in the cell. However, it has been possible to isolate factors capable of increasing *in vitro*, the affinity between RNA primers and the DNA polymerase isolated from the same tissue.

4. Developments and perspectives

As in the case of *E.coli* it is envisaged that the replication of the two chains of DNA in eucaryotes takes place in a coordinated manner thanks to a complex formed by the DNA polymerase α and DNA polymerase δ. This attractive hypothesis is however yet to be proved.

Cloning of genes coding for the catalytic sub-units of DNA polymerases has already been carried out for several enzymes obtained from different cells. It was thus possible to show some common sequences at the protein level, sequences which would probably correspond to the binding sites of the template, *d*NTPs etc. It was also possible to classify the various DNA polymerases in two families. At present, the larger family comprises only the DNA polymerases which perform the replication of the genomic DNA with the exception of DNA polymerase III of *E.coli* which derives from a different ancestor.

V. – FIDELITY OF THE REPLICATION

The DNA being the carrier of genetic information it is essential that this information remains correct. But the DNA can be damaged by the action of chemical or physical agents. The cell therefore possesses repair systems which can maintain the integrity of genetic information.

It is also of utmost importance that during the replication, the newly synthesized DNA be an exact copy, correct to one nucleotide, of the template DNA. Replication must therefore be a faithful process i.e. it should make no errors. If such errors did exist, they could be detected by the repair systems but the latter would not be able to distinguish between the strand carrying the correct information and the one carrying the erroneous information. It is therefore necessary to have a corrector activity during the synthesis.

1. – Corrector activity of DNA polymerases

Given that the genome of *E.coli* comprises about 4×10^6 base pairs, the admissible rate of error for a totally correct replication is one erroneous base for every 10^6, or even 10^7, base pairs. The mere complementarity of bases will not ensure such a low rate of error because the bases can tautomerize (see pages 263-264). Their enol form (unusual in physiological conditions) can pair with a normally non complementary base: e.g., a cytosine with adenine. It is believed that the frequency of these enolic forms is 10^{-4} to 10^{-5} base pairs. It may be noted therefore that without a correction mechanism, the fidelity of replication would be 10^{-4} or 10^{-5} and not 10^{-7}.

The correction mechanism was extensively studied in *E.coli*. It is based on the fact that the DNA polymerases need both a primer and a template. In other words, the polymerase will not be able to bind a new nucleotide to the 3′OH of a nucleotide not paired to the template DNA. If a base in its enol form has been incorporated, its return to the keto form results in a non-pairing.

Therefore, the DNA polymerase can no longer catalyze the formation of a phosphodiester bond with the following nucleotide. Its 3′ ⟶ 5′ exonuclease activity eliminates the ill-paired nucleotide, thus liberating a 3′OH utilizable by the DNA polymerase. A mutation of sub-unit ε, carrying the 3′ ⟶ 5′ exonuclease activity, generates a DNA polymerase III which makes many more errors.

For eucaryotes, the problem of fidelity of replication becomes even more crucial, because the genome of a mammal for example, contains about 3×10^9 base pairs. The fidelity of replication must therefore be equal to, or greater than 10^{-9}. The existence of an exonuclease activity 3′ ⟶ 5′ associated with DNA polymerase δ and perhaps with DNA polymerase α explains this fidelity to a large extent. It is generally admitted that the presence of auxiliary proteins and the very nature of the eucaryote chromosome permit this very low rate of error. In this sense, it has been observed that the error rate of DNA polymerases increases when the enzymes are in experimental conditions in which they synthesize the DNA in a relatively continuous manner (they are then said to be processive). This suggests that in their active conformation these enzymes are very faithful.

2. – Fidelity and mechanism of replication

As just seen, the fact that a DNA polymerase has an absolute requirement for a primer 3′OH paired with the strand of the template DNA to synthesize a new DNA chain, allows the corrector activity to fully play its role.

It may also be observed that if one of the two strands of the DNA were copied by a DNA polymerase acting in the direction 3′ ⟶ 5′ that copy would be less faithful. As a matter of fact, in this case, the growing chain would carry the triphosphate and the nucleotide to be added would carry the 3′OH. After the hydrolysis of an ill-paired nucleotide, there would be no more triphosphate available on the growing chain to permit the addition of a new nucleotide.

If the first nucleotide incorporated were a deoxyribonucleoside triphosphate, it could never be corrected by the exonuclease activity. At the end of the synthesis, the ribonucleotide primer will be systematically recognized and eliminated; as a result, the fact that the RNA polymerases are not very faithful (10^{-4} to 10^{-5}) loses its importance.

One can realize that the very complexity of the replication mechanism (necessity of a primer, synthesis only in the 5′ ⟶ 3′ direction, ribonucleic nature of the primer) provides a corrector activity at the precise moment of replication, and sufficient fidelity for maintaining the exact genetic information.

VI. – RNA-DEPENDENT DNA POLYMERASE (REVERSE TRANSCRIPTASE)[1]

The existence of an RNA-dependent DNA polymerase was established in the case of various RNA-containing oncogenic viruses. The reaction catalyzed by this enzymes necessitates Mg^{2+} or Mn^{2+} and the four usual deoxyribonucleosides triphosphates. *In vitro*, the viral ribonucleic acids or other natural ribonucleic acids (such as some *m*RNAs) or some single-stranded synthetic polyribonucleotides can be used as templates[2]. The presence of a primer is necessary, and the DNA synthesized during brief reaction periods is complementary to the RNA which serves as template. For long reaction periods, the proportion of DNA complementary to the RNA decreases. The appearance of double-stranded DNA indicates that the DNA transcribed from the RNA serves as template for the synthesis of a complementary DNA strand. The process can be summarized as follows:

$$\text{viral RNA} \longrightarrow \text{RNA}\cdot\text{DNA} \longrightarrow \text{DNA}\cdot\text{DNA}$$

The double-stranded DNA thus synthesized can be integrated into the genome of the host cell. This integration can take place during productive infections resulting in the multiplication of the virus (the integrated viral DNA being used as template for the synthesis of viral RNA molecules by a DNA-dependent RNA polymerase), as well as during the transformation phenomenon (when a cell acquires properties of a cancerous cell). The presence of the viral genome integrated in the form of DNA was indeed detected in infected cells as well as in transformed cells.

Azidothymidine (AZT) is an inhibitor of reverse transcriptase, used in the treatment of AIDS at present.

BIOSYNTHESIS OF RNAs

The polymerization of ribonucleosides-5′-triphosphates to RNA is a process called *transcription*, because the genetic information carried by the DNA is transcribed into RNA. This transfer of information is based on the same principle as the duplication of DNA, i.e. the pairing of complementary bases (we will see that this pairing also ensures the fidelity of translation). We

---

(1) The term *reverse transcriptase* underlines the fact that this enzyme catalyzes the reverse reaction of the one catalyzed by the DNA-dependent RNA polymerase (or transcriptase) discovered earlier and which we will examine in the following (see page 324).

(2) Reverse transcriptase is used in numerous laboratories for preparing from various ribonucleic acids (especially messenger ribonucleic acids), the complementary deoxyribonucleic acids (*c*DNAs) which can be either used as probes (radioactive) for detecting and titrating, by hybridization, the corresponding messenger ribonucleic acids synthesized *in vitro* or *in vivo* (e.g, at different stages of development) or introduced into a vector (e.g., plasmid) for cloning. (see page 383).

will first study the synthesis of various types of cellular RNAs (ribosomal RNAs, transfer RNAs, messenger RNAs) which are transcription products of the corresponding genes; then we will consider the particular case of the synthesis of ribonucleic acids of RNA-containing viruses and phages.

I. – POLYNUCLEOTIDE-PHOSPHORYLASE

Isolated by Grunberg-Manago and Ochoa in 1955, polynucleotide phosphorylase is the first known enzyme capable of catalyzing the formation of polyribonucleotides. It catalyzes the following reaction:

$$n_1\text{ADP} + n_2\text{GDP} + n_3\text{CDP} + n_4\text{UDP} \longrightarrow [(\text{AMP})n_1, (\text{GMP})n_2, (\text{CMP})n_3, (\text{UMP})n_4] + (n_1+n_2+n_3+n_4)\text{P}_i$$

This reaction does not require the presence of DNA as template, and the base composition of the product formed in the reaction generally reflects the relative proportions of ribonucleosides-5′-diphosphates added to the reaction mixture. This is not compatible with the requirements of a faithful transcription of the genetic information from DNA to RNA, and the idea that this enzyme could be responsible for the biosynthesis of ribonucleic acids had to be abandoned. Although it is widely distributed among bacteria, its physiological role is still unknown. It perhaps enables the degradation of cellular ribonucleic acids since it also catalyzes the phosphorolysis reaction (whence its name) at the intracellular phosphate concentrations. In any case, this enzyme is extremely useful because it permits *in vitro* synthesis of numerous homo- and copolymers depending on whether one or several ribonucleosides-diphosphates are introduced in the reaction mixture: with only UDP, polyuridylic acid (poly U) is obtained; with only ADP, one obtains polyadenylic acid (poly A); ADP + UDP will give poly AU; with the 4-nucleosides-diphosphates one will have poly AGUC, etc. The polynucleotides obtained in this manner were used in numerous physico-chemical works and in the first studies on the deciphering of the genetic code (see page 362).

II. – DNA-DEPENDENT RNA-POLYMERASE (TRANSCRIPTASE)(1)

As indicated by its name, this is the enzyme permitting transcription. It was discovered in 1960 by Hurwitz in *E.coli* and by Weiss in the rat liver; it catalyzes the following reaction:

$$n_1\text{ATP} + n_2\text{GTP} + n_3\text{CTP} + n_4\text{UTP} \xrightarrow[\text{Enzyme}]{\text{DNA, Mg}^{++}} [(\text{AMP})_{n1}, (\text{GMP})_{n2}, (\text{CMP})_{n3}, (\text{UMP})_{n4}] + (n_1+n_2+n_3+n_4)\text{PP}$$

(1) Revised by P. CHAMBON and C. KEDINGER.

The presence of the *4 ribonucleosides-5'-triphosphates* is necessary for the transcription of DNA to take place. The polymer formed is a polyribonucleotide identical to the natural ribonucleic acids. The DNA necessary for the reaction serves as the template; in other words, the base sequence of this DNA controls the base sequence of the RNA synthesized. This is shown by the analysis of the base composition and study of the nearest neighbours frequency of the DNA introduced in the reaction and the RNA synthesized. Besides, if instead of introducing a natural DNA, a synthetic copolymer, poly *d*AT, is added to the reaction mixture, the enzyme catalyzes the synthesis of the complementary copolymer, poly UA, even if the 4 ribonucleosides triphosphates are present in the mixture. Lastly, the DNA introduced in the reaction and the newly synthesized RNA can form, after heating and slow cooling, hybrids which are the sign of a complementarity. The polarity of the synthesized RNA is the opposite of that of the DNA strand which served as template (see fig. 6-35).

As shown by fig. 6-35, the ribonucleic acids are synthesized in the direction $5' \longrightarrow 3'$. It is observed that the first nucleotide incorporated maintains its triphosphate group, contrary to the nucleotides incorporated subsequently; therefore, by introducing a nucleoside triphosphate labeled on its γ phosphate ($^{32}P_{\gamma}$-ATP for example) in the reaction mixture, one can determine the number of polyribonucleotide chains synthesized in the reaction which begin with an adenylic nucleotide, by simply measuring the radioactivity incorporated in the polymer (as only the adenylic nucleotides incorporated at the beginning of the

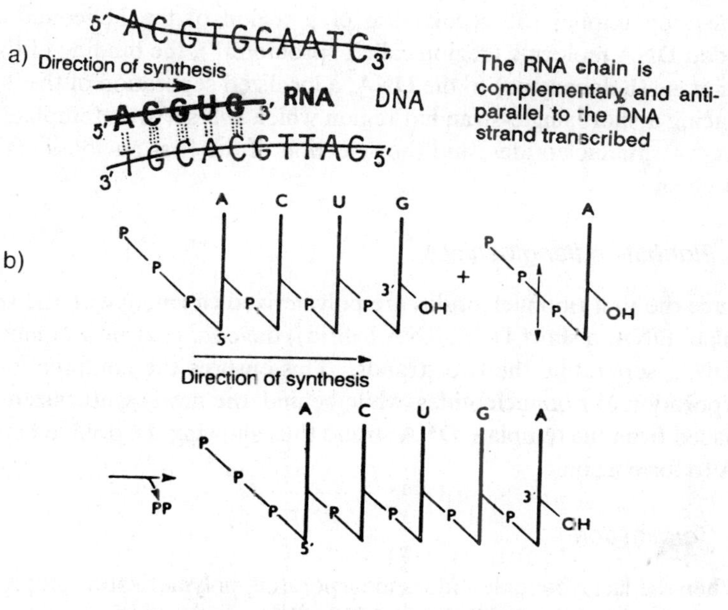

FIG. 6-35. — Mechanisms of the transcription.
(a) Diagrammatic representation of asymmetric transcription.
(b) Diagram showing how the polyribonucleotide chain elongates by one unit.

chain conserve their β and γ phosphate groups). The use of ATP, GTP, CTP and UTP labeled on γ-phosphate showed that RNA chains have a purine at their 5′ end, i.e., they begin with a pppA or pppG. This peculiarity also confirmed that the ribonucleic acids are actually synthesized in the direction 5′ ⟶ 3′; if some $^{32}P_{\gamma}$-ATP is introduced at the very first moments of incubation and then "chased" by adding an excess of non-radioactive ATP, radioactivity (due to the triphosphate group) is found at the 5′ end of chains formed in the reaction, which proves that the 5′ ends were constituted at the beginning of the reaction when the radioactive ATP was present (if the chains were formed in the direction 3′ ⟶ 5′, the 5′ ends would be constituted later, when non-radioactive ATP is present in the mixture, and they would not be radioactive). The use of 3′ deoxy ATP (labeled on adenine) confirmed that the RNAs are indeed synthesized in the direction 5′ ⟶ 3′. This analogue is found incorporated in the ribonucleic chains thanks to its 5′ triphosphate group (but any subsequent polymerization is then blocked), whereas if the elongation took place from 3′ to 5′, it could not be incorporated due to the absence of OH in 3′.

1. – Steps of the transcription

Several steps can be distinguished in the transcription.

A – *Initiation*

This step implies the recognition of a region of the helicoidal double-stranded DNA molecule (region called "promoter"), the binding of the RNA polymerase (holoenzyme) to the DNA, a localized separation of the 2 strands producing a short single-stranded region which will serve as template for the pairing of ribonucleotides, and the selection of the first nucleotide(s) of the RNA chain.

B. – *Elongation (lengthening)*

Once the first ribonucleotides are polymerized (forming with the strand of template DNA, a short DNA-RNA hybrid), the core enzyme will move along the DNA, separating the two strands. This ensures the continuation of the incorporation of ribonucleotides, while behind, the newly synthesized RNA is displaced from the template DNA strand thus allowing the double helix of the DNA to form again.

C. – *Termination*

When the last ribonucleotide is incorporated, polymerization stops, and the transcription complex is dissociated; this liberates the enzyme and the newly formed RNA (after dissociation of the DNA-RNA hybrid) so that the DNA can resume its double stranded structure. This step implies the recognition of a DNA sequence called "terminator".

2. – The bacterial RNA polymerase

A. – Structure

In the bacterial cell, a single RNA polymerase permits the synthesis of ribosomal RNAs, transfer RNAs and messenger RNAs. In *E.coli* the complete enzyme (or holoenzyme) has a molecular weight of about 450 000 and consists of four types of polypeptide chains, namely *two α chains* (mol. weight : 40 000), *one β chain* (mol. weight: 150 000), *one β' chain* (mol. weight: 155 000) and *one σ chain* (mol. weight: 70 000). The core enzyme of $\alpha_2\beta\beta'$ structure, can perform the elongation of RNA chains, but the factor (or sub-unit) σ is indispensable for the initiation of the transcription in a specific site on a double-stranded DNA (see below). Eucaryotic cells, on the contrary, possess several RNA polymerases which differ by their localization, structure, properties and the type of RNA whose synthesis they catalyze (see page 333).

B. – Mode of action of the enzyme

Whereas the core enzyme can bind to any DNA sequence (called weak binding site) forming a closed complex (enzyme-DNA) where the DNA remains double-stranded, the σ factor notably modifies the affinity of RNA polymerase for the DNA; consequently, the holoenzyme has a reduced capacity to recognize the weak binding sites (the complexes formed will be less stable), but it has, on the contrary, the property of recognizing specific binding sites, the promoters, and binds to them strongly (in fact, more or less strongly, which will determine the efficiency of different promoters in the initiation of transcription), forming an open complex, thus called because of the limited opening of the double helix. Out of 4×10^6 base pairs of the DNA of *E.coli* (see page 283), there are about 2 000 promoters.

This binary complex soon becomes a ternary complex (DNA-enzyme-nascent RNA chain) after formation of the first phosphodiester bond, then the σ factor detaches itself from the complex, and the core enzyme continues the polymerization (elongation). The transitory association of the σ factor with the enzyme decreases the stability of weak complexes (formed between the core enzyme and any region of the DNA) and, on the contrary, increases the stability of strong complexes formed by the complete enzyme at the promoters. But this high stability would block the reaction, if the dissociation of σ did not modify the enzyme which thus recovers (by again becoming core enzyme) its general affinity for the DNA (without specificity vis-a-vis particular sequences) enabling it to accomplish elongation.

The core enzyme moves along the template and the open region of the double helix accompanies this movement. It is considered that the enzyme covers about 60 DNA base pairs and that the region where the helix is open comprises 17 base pairs. It is believed that the hybrid formed (between the template DNA strand and the synthesized RNA strand) covers about 12 base pairs and has only a very transitory existence because following the passage of the enzyme, the RNA strand is displaced and the DNA double helix (duplex) is reconstituted. Transcription is a very rapid process: in *E.coli*, at 37°, the *m*RNAs are synthesized at the rate of 40 to 50 nucleotides per second.

It must be noted that RNA polymerase has no exonuclease activity and cannot therefore have a corrector activity, contrary to what takes place for the DNA polymerase (see page 321). Therefore, the fidelity of transcription is less than that of replication (it is estimated that one error occurs for 10^4 or 10^5 bases, which is 10^4 times more than in the case of replication). This relatively high rate of error is however acceptable, on one hand because the errors are not transmitted to the progeny, and on the other hand because each gene is transcribed into numerous RNA molecules and the presence of a few defective molecules is not significant.

C. – Role of various sub-units

Little information is presently available on the architecture of the RNA polymerase of *E.coli* and the roles of various sub-units. It is believed that the β chain is involved in the binding of nucleosides-triphosphates because it could be labeled by some analogues of these substrates and because the origin of the β-chain determines – in reconstitution experiments – the sensitivity to some antibiotics like rifamycins (especially rifampicin) which inhibit initiation by acting before the formation of the first phosphodiester bond, or streptolydigins which inhibit elongation. It is believed that the β' chain which is the most basic, acts by binding the template, because *in vitro* addition of heparin, a polyanion which binds to the β' sub-unit, blocks transcription, (heparin competes with the DNA for the binding of the enzyme). Addition of the σ factor changes the conformation of the core enzyme and its capacity to recognize the DNA (see above), but it is possible that sub-unit α also acts in the recognition of promoters because when *E.coli* is infected by the phage T_4, an arginine of the α chain undergoes an ADP-ribosylation and this modification is accompanied by a decrease of the affinity of the enzyme for the promoters previously recognized by the holoenzyme.

One might wonder why the bacterium needs a multimeric RNA polymerase of this size (mol. weight = 450 000), while some phages code for monomeric RNA polymerases of much smaller size (mol. weight = 110 000 in the case of RNA polymerases of phages T_3 and T_7), catalyzing the synthesis of RNA at a higher rate (200 nucleotides incorporated per second, at 37°) than the enzyme of *E.coli* (40 nucleotides per second). But whereas those phage polymerases recognize only some promoters on the DNA of the phage, the bacterial enzyme must catalyze the production of more than a thousand transcripts and its activity is modulated by numerous factors, of bacterial and of phage origin, specific of one or several operons, acting either positively or negatively; the complexity of the bacterial RNA polymerase may therefore be due to the necessary interactions with these diverse promoters and with different regulation factors rather than to the requirements of the catalytic activity proper.

3. – Asymmetry of the transcription

In vivo, transcription is an *asymmetric* process, i.e. for a given gene or group of genes, only one of the 2 strands of the DNA is transcribed as per the rule: 1 gene ⟶ 1 polypeptide chain. In presence of σ factor, it has also been possible to obtain an asymmetric transcription *in vitro*, which suggests that the signals permitting the RNA polymerase to recognize on the DNA the place where it must bind to begin the transcription have been preserved; this opinion is further supported by the fact that the RNAs formed *in vitro* can compete with those synthesized *in vivo*, during hybridization experiments, as would be expected if, in fact, they were actually the transcription products of the same genes. Thanks to advances made in the methods used for the determination of the nucleotide sequences of the DNA, it was possible to define, for a whole series of bacteriophage and bacterial genes, the structure of regions where the RNA polymerase binds specifically (promoter sites) before starting the RNA synthesis. The comparison of these sequences revealed structural patterns (motifs) common to the various promoter sites studied. In the same way, it was possible to find a typical nucleotide sequence playing the role of termination signal of the transcription (see fig. 8-4).

Let us add that in some cases, all the ribonucleic acids synthesized *in vivo* appear to result from the transcription of only one DNA strand (for example in the case of phage T_7), but in other cases they are the transcription products of one of the strands for some genes and of the other strand for other genes (for example in the case of phage λ).

4. – Sites involved in the initiation

We have seen that the RNA polymerase recognizes a region of DNA called "promoter" and binds to it, thus allowing transcription to be initiated and to proceed until the enzyme reaches the "terminator" sequence, thus producing a RNA molecule called "primary transcription product" or "primary transcript", which is mono- or polycistronic, and which is generally unstable, because it undergoes maturation processes.

A. – Starting point of the transcription

By convention, the starting point of the transcription is numbered +1, the bases situated upstream this point are numbered –1, –2, etc., the bases downstream are numbered +2, +3, etc. The starting point, i.e. the DNA base pair corresponding to the first nucleotide of the RNA, is identified as follows: the template DNA is denatured, it is hybridized with the RNA obtained by transcription *in vivo* or *in vitro*, the non-paired DNA (i.e. the non-transcribed strand as well as the regions of the template strand extending beyond – on both sides – the transcribed sequence) is degraded by nuclease S1, and the RNA involved in the hybrid is digested by alkaline hydrolysis: it is then possible to determine the sequence of the DNA which was involved in the hybrid and define the starting point of the transcription. This mapping technique with S1 nuclease was simplified by the use of single-stranded DNA fragments labeled at one end and corresponding to the presumptive transcription start region (see fig. 8-8). This starting point is part of the promoter, and is generally situated in the region of the DNA where the enzyme binds.

B. – Binding sites of E.coli RNA polymerase

The site where the RNA polymerase binds to form the initiation complex is part of the promoter; its sequence can be determined after binding the enzyme *in vitro* to a DNA fragment containing a promoter, digesting by a DNase the DNA not protected by the enzyme, and extracting and sequencing the protected DNA. This type of analysis provides fragments of 40 to 44 base pairs having at their centre the starting point of the transcription (which therefore extends from –20 to about +20), because the enzyme attached to the protected fragment can in fact initiate the transcription and synthesize a small RNA chain of 17 to 20 nucleotides, terminated when the RNA polymerase is released at the end of the DNA fragment.

The binding site of the RNA polymerase can also be studied by the "foot-printing" method (see diagram below) where the DNA fragment protected by the enzyme (and labeled on one strand at one of the ends) is subjected to a limited endonuclease digestion which will split all the accessible phosphodiester bonds (not those protected by the RNA polymerase), but causing statistically one cleavage per molecule. As in the method of the sequence gels (see pages 275-277), the fragments produced are separated according to their size by gel electrophoresis and for each ruptured bond, a band is obtained on the gel corresponding to the fragment extending from the labeled end to the cleavage site. The comparison of the bands obtained on the gel for the non-protected DNA fragment (where all the phosphodiester bonds can be ruptured) and the protected fragment, shows that some bands are missing on the gel corresponding to the protected fragment; this is the region where the polymerase bound to the DNA has prevented the endonuclease action (see diagram below). By combining foot-printing and sequencing it is possible to determine the position of the binding site and its sequence. Labeling the two DNA strands (in 2 separate experiments) one can even observe that the two strands are not protected with equal efficiency at the ends of the binding site, which suggests an asymmetric conformation of the polymerase bound to the DNA, in conformity with the asymmetry of the transcription itself. By this technique, the binding site appears slightly greater (upstream of the starting point) than by the method based on the recovery of the protected fragment (see above): it extends from about –50 to +20. Comparable results were obtained by studying the protection of the DNA (by the RNA polymerase) against an exonuclease attack, demarcating a promoter region extending from –44 to +20.

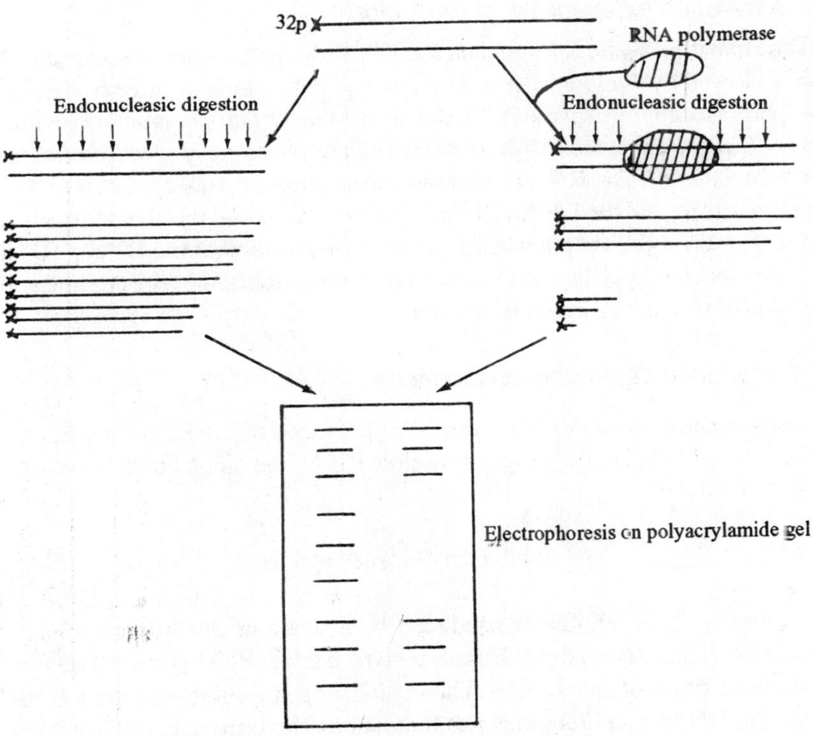

Determination of the position of the binding site of RNA polymerase on the DNA by the foot-printing method

The study of more than hundred procaryotic promoters revealed the existence, upstream the transcription initiation site, of common patterns (consensus) comprising 6 base pairs called respectively sequence –10 (or Pribnow box, or TATA box) and sequence –35.

	–35	17 base pairs	–10	+ 1 (initiation site)
5'———	TTGACA	————————	TATAAT	—————

Strong (effective) promoters, permitting frequent initiations of the transcription (every 2-5 seconds, for example in *E.coli*) have sequences very near the 2 consensus sequences, while weak promoters (permitting an initiation of the transcription of the gene only every 5 or 10 minutes in *E.coli*) present substitutions on their sequences −10 and −35. Site-directed mutagenesis experiments devised to modify *in vitro* one or several bases of these consensus sequences and introduce into a bacterial cell the gene flanked with a modified promoter, showed that the substitution of only one base, either in the sequence −10, or in the sequence −35, could bring about a considerable decrease or even loss of the promoter activity.

5. – Termination of the transcription

A. – Termination independent of rho factor

The termination signal is formed on the DNA by a palindromic sequence rich in G-C followed by a region rich in A-T (see fig. 8.7), which – after transcription – permits at the level of the RNA, the formation of *a hair-pin structure* (thanks to the G - C pairs which are relatively stable) followed by *a series of at least four or five uridylic residues*. The RNA polymerase slows down or pauses when it finds a hair-pin structure, and the DNA-DNA hybrid formed beyond this structure is less stable as it rests on dA-rU pairs which causes a dissociation of the DNA–RNA hybrid, a reassociation of the two DNA strands, a liberation of the core enzyme and a liberation of the newly formed RNA chain.

B. – Termination dependent on rho factor

In certain sites, termination requires the presence of ρ (rho), a hexameric protein which binds to the single-stranded RNA and detaches it from the DNA-RNA hybrid.

6. – Inhibition of transcription

Actinomycin D, an antibiotic produced by a strain of *Streptomyces*, has a phenoxazone ring carrying 2 cyclic peptides which is inserted between 2 successive GC base pairs of the DNA. This fixation on a double-stranded DNA prevents the DNA from functioning as template in the transcription, but a low concentration, it prevents neither the replication of the DNA nor the protein synthesis, which makes of actinomycin D *a specific inhibitor of the synthesis of new RNA molecules* in procaryotes and eucaryotes. This antibiotic has been used as an anti-cancerous agent, because it inhibits the growth of rapidly dividing cells.

Rifamycin, isolated from another strain of *Streptomyces*, and its derivative, *rifampicin*, inhibit the *initiation* of new RNA chains, but not the elongation of chains already being synthesized. It is the formation of the first phosphodiester bond which is blocked, and the sub-unit β of RNA polymerase is the target of the antibiotic.

7. – Transcription in eucaryotes

A. – Transcription and translation take place in different compartments

Whereas in procaryotes transcription and translation are coupled, i.e. translation of a *m*RNA can begin before its synthesis is completed, in eucaryotes transcription takes place in the nucleus while translation takes place in the cytoplasm, so that the RNAs (*r*RNAs, *t*RNAs, *m*RNAs) will have to migrate from the nucleus to the cytoplasm. The *heterogenous nuclear RNAs (hnRNAs)* are partly hydrolyzed by nucleases in the nucleus and partly converted into *m*RNAs. We shall see that this conversion (maturation) of the primary products of transcription (primary transcripts) into *m*RNAs implies some modifications at the 5′end (formation of a cap) and 3′end (addition of a tail of poly A) and the elimination of introns, with the result that sometimes the mature *m*RNA has only 10% of the size of the corresponding precursor. It must be noted that these processes of transport and maturation of *m*RNAs constitute steps during which may come into play mechanisms of regulation of the expression of genes, which obviously do not exist in procaryotes.

B. – Existence of 3 RNA polymerases

In eucaryotes, the mechanisms controlling transcription are varied (see pages 454-467) and the presence of several RNA polymerases (contrary to what takes place in procaryotes where the synthesis of all RNAs is carried out by a single polymerase) is an important aspect. Three DNA-dependent RNA polymerases are found in the nuclei of eucaryotic cells:

– RNA polymerase I (or A), responsible for the synthesis of about 60% of the cellular RNAs, localized in the nucleolus, catalyzes the synthesis of ribosomal RNAs (*r*RNAs) and is insensitive to α-amanitine (one of the toxins of *Amanita phalloides*);

– RNA polymerase II (or B), responsible for the synthesis of about 30% of the cellular RNAs, localized in the nucleoplasm, catalyzes the synthesis of *heterogenous nuclear* RNAs (*hn*RNAs) which are the precursors of messenger RNAs (*m*RNAs) and is inhibited by small concentrations of α-aminitine (of the order of 0.1 μg/ml);

– RNA polymerase III (or C), responsible for the synthesis of about 10% of cellular RNAs, also localized in the nucleoplasm, catalyzes the synthesis of transfer RNAs (*t*RNAs) and 5S RNA; the enzyme of animal cells is sensitive to α-amanitine at concentrations in the order of 500 μg/ml, while the enzyme of yeast or insect is insensitive.

The quaternary structure of these enzymes (molecular weight in the order of 600 000) is based on the same model as that of the bacterial polymerase (two large-sized sub-units of molecular weight 220 000 to 140 000 depending on the enzyme and several sub-units of smaller size). The total number of sub-units which constitute these molecules is however much greater in eucaryotes (about fifteen), and only the largest ones have preserved resemblance with those of archaebacteria as revealed by their immunological properties. Some of the small sub-units (two or three depending on the species), probably involved in the basic catalytic function (formation of phosphodiester bonds), are common to the three classes (A, B and C) of polymerases. The more intricate structure of RNA polymerases of eucaryotes most probably reflects the greater complexity of the tasks they perform. As will be seen in the following (chapter 8), the regulation of transcription also involves other protein factors which interact both with the DNA and the polymerases.

As in procaryotes synthesis of RNAs takes place in eucaryotes in the direction 5′⟶3′, by a nucleophile attack of the 3′ OH of the growing chain on the P_α of the following ribonucleoside 5′ triphosphate and no primer is necessary. The eucaryotic RNA polymerases also have no exonuclease activity, which excludes the possibility of correcting the errors occurring during the transcription.

C. – *Promoters of eucaryotes*

Upstream the eucaryotic genes coding for proteins was observed the presence of sequences (or elements in "*cis*") important for the transcription, especially a *TATA* box which is however situated farther away from the site of initiation of the synthesis of *m*RNAs than in procaryotes, generally at –25 (in yeast, the TATAAA sequence is situated further upstream). This TATA box is necessary for the promoter activity as shown by site-directed mutagenesis experiments, but in addition, also come into play other sequences situated even further upstream (between – 40 and 110) called *CAAT box and GC box*.

D. – *Protein factors stimulating transcription*

One already knows a number of proteins having the property of stimulating transcription by binding specifically to promoter sequences situated upstream the genes. These proteins are called transcription factors (or elements acting in "*trans*") and their binding sites are determined by "*foot-printing*" experiments (see pages 330-331). The characterization of these proteins often uses the gel retardation technique which enables one to watch the migration of a DNA fragment (comprising a promoter sequence) slowed down by the binding of a transcription factor (in comparison to the migration of the DNA fragment alone). For example, the following transcription factors were characterized:

– Sp 1, a protein of about 100 kdaltons; in mammals it is required for the transcription of genes whose promoters comprise a GC box (especially in the case of the DNA of SV40 virus);

– CTF or *CCAAT-binding Transcription Factor*, a protein of about 60 kdaltons which, in mammals binds to CAAT box;

– protein B which binds to TATA box in the drosophila;

– HSTF, or *Heat-Shock Transcription Factor* which, in the drosophila binds to the promoters of heat-shock genes upstream their TATA box. It is distinct from factor σ^{32} which combines with the RNA polymerase of *E.coli* to permit the recognition of bacterial heat shock genes (See page 453).

E. – Enhancers

These elements exert their stimulating activity even if they are situated several thousand base pairs away from the gene in question and they may be localized either upstream or downstream or even in the middle of the gene, on the coding DNA strand or on the non-coding strand. In general a given enhancer is active only in some cells, because it requires a protein which is expressed only in these cells. For example the enhancer permitting the hormonal action of glucocorticoids must bind the soluble hormone-receptor complex to be able to stimulate the transcription of a battery of genes. How this long distance effect of enhancers is exerted is not yet known exactly, but it is believed that the binding of regulator proteins could bring nearer regions of the DNA initially distant and permit the formation of an active complex of transcription initiation comprising RNA polymerase II.

8. Maturation of transcription products

A. – Maturation of rRNAs and tRNAs

In procaryotes as in eucaryotes, *r*RNAs and *t*RNAs are transcribed as *precursors* (longer than the functional RNAs) which will have to undergo a series of endo- and exo-nucleolytic cleavages to assume their final size.

For example in *E.coli* a primary transcript will supply the 3 *r*RNAs (16 S, 23 S and 5 S) and 1 or 2 *t*RNAs. Depending on the operon considered (there are 7*r*RNA operons in *E.coli*), there are 1 or 2 *t*RNA genes situated in the spacer between the genes of RNA 16 S and RNA 23 S. Other precursors contain several *t*RNA molecules.

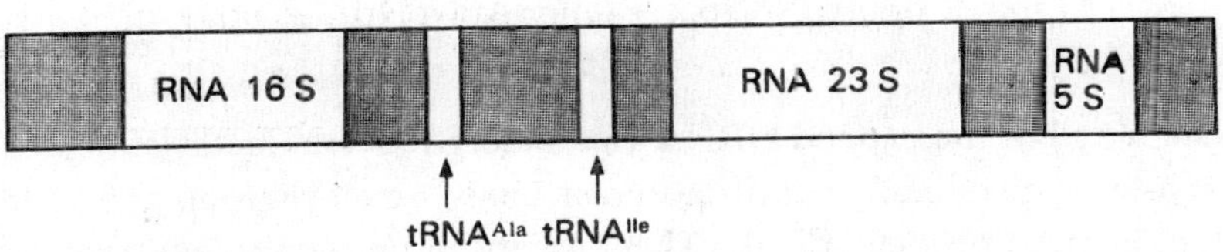

The maturation of *t*RNAs also implies the addition, under the effect of the *tRNA nucleotidyl transferase*, of the 3′ terminal CCA triplet (see pages 349-351), when the latter is not present (often it is not coded by the *t*RNA gene).

Furthermore, the precursors of *r*RNAs and especially of *t*RNAs undergo enzymatic modifications of which the simplest is methylation, either of the OH at position 2′ of ribose, or of the purine or pyrimidine base, thus producing the rare bases a few examples of which we mentioned in the foregoing (see pages 263-264). These methylations are catalyzed by *t*RNAs methyltransferases (using generally S-adenosyl-methionine as donor of the CH_3 group) which are extremely specific. This specificity concerns the nature of the base (an enzyme which methylates guanine does not methylate the 3 other bases), the position which is going to be methylated (different enzymes are responsible for the formation of 1 methylguanine, N_2 methylguanine, N_2 dimethylguanine and 7 methylguanine), and the localization of the base (of the 15 or 20 guanines of a *t*RNA molecule, only one will be methylated).

Lastly, some precursors of eucaryotic *r*RNAs and *t*RNAs (in the cytoplasm as well as in the organelles) contain introns which must be spliced out. It appears that this splicing is controlled more by the secondary structure of the mature RNA than by particular signals present at the ends of the intron (as in the case of splicing of introns in precursors of eucaryotic *m*RNAs, as will be seen below).

B. – *Maturation of eucaryotic mRNAs*

a) **Cap formation at the 5′end of *m*RNAs.** – In eucaryotes as in procaryotes, precursors of *m*RNAs begin with an adenylic or guanylic nucleoside triphosphate, but in eucaryotes this 5′end of the chain under synthesis is very rapidly blocked by the binding of a guanylic residue in reverse position. This nucleotide, called *cap*, is added by action of GTP on the purine nucleoside triphosphate, forming an unusual *5′–5′ triphosphate bond*. This cap stabilises the *m*RNAs by preventing their degradation at the 5′ end by phosphatases or exonucleases, and stimulates the translation of *m*RNAs by the systems of eucaryotic protein synthesis.

This reaction, catalyzed by a guanylyl transferase (a nuclear enzyme), creates a 5′–5′ triphosphate bond between the first nucleotide and the guanylic residue added:

$$\underset{5'}{\text{Gppp}} + \underset{5'}{\text{pppApNpNp...}} \rightarrow \underset{5'\;\;\;5'}{\text{GpppApNpNp...}} + \text{pp} + \text{p}$$

The 5′end of the capped *m*RNA can undergo several methylations: in all eucaryotes, a guanine-7-methyltransferase binds a methyl group at position 7 on the guanylic residue added. Then, except in unicellular eucaryotes, the ribose of the first nucleotide (the one which was first before the addition of the cap, i.e. the adenylic nucleotide in our diagram) is methylated at position 2′. Some *m*RNAs are also methylated at the position 6 of the adenine, or on the ribose of the next nucleotide (the first N). These possibilities of methylation are summarized in the diagram below:

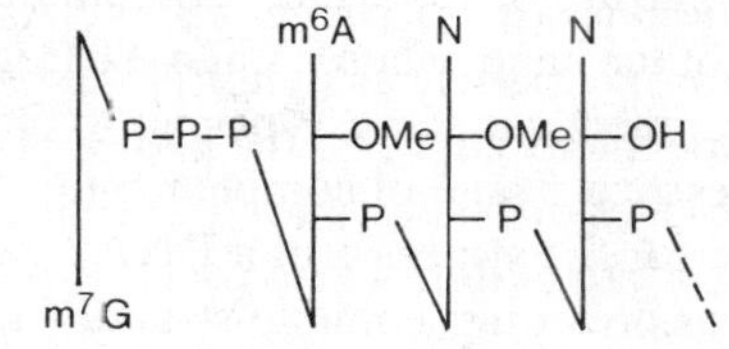

b) **Polyadenylation of the 3′end of *m*RNAs.** – The termination signals of transcription in eucaryotes are not fully understood, but some resemble those of procaryotes because they also have a hair-pin structure followed by a series of uridylic residues (see pages 332, 467 and fig. 8-7). The precursors of *m*RNAs are then cleaved at their 3′end by an endonuclease which recognizes a AAUAAA sequence situated at about twenty nucleotides upstream the cleavage point (and apparently a UGUGUU sequence situated at about twenty nucleotides downstream).

This cleavage is followed by the addition of several tens or even several hundreds (up to about 300) of adenylic residues at the 3′end of most eucaryotic *m*RNAs. Although cleavage and polyadenylation are correlated, the two reactions are catalyzed by different enzymes (endonuclease and poly A polymerase) within a complex which also involves a ribonucleoprotein particle containing a small nuclear RNA rich in uridine (RNA U1).

The *m*RNAs coding for histones are notable exception to this type of maturation. For these RNAs, the 3′ cleavage seems to imply the pairing of the 5′end of another small nuclear RNA (RNA U7) at a complementary sequence near the cleavage site. The cleavage is not followed by polyadenylation, but the existence of a palindromic sequence just upstream the 3′end of the RNA brings about the formation of a hair-pin type secondary structure at that place. It is probable that cap and polyadenylation or secondary structure provide a protection of 5′ and 3′ ends of *m*RNAs against exonucleases.

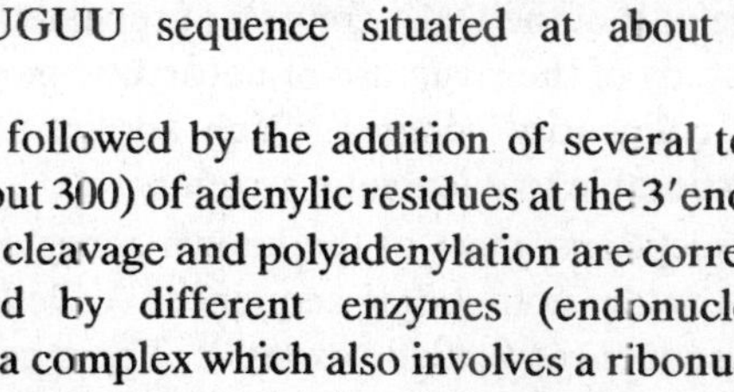

This poly A tail is of great practical interest. On the one hand, it enables the separation of *m*RNAs from other cellular RNAs (and especially from *r*RNAs) by using a column of oligo U-Sepharose (or oligo dT-Sepharose) to which the *m*RNAs are bound by pairing of their tail, then eluted by lowering the salt concentration. On the other hand, when a oligo dT fragment is hybridized to the tail of poly A, this fragment can serve as primer to the reverse transcriptase (see page 384) to synthesize a strand of complementary DNA (*c*DNA), which can be used as template for the synthesis of a DNA strand complementary to the first DNA strand, thus providing a double-stranded DNA corresponding to the *m*RNA and which can be cloned.

c) **Elimination of introns.** – Numerous works of fine genetic mapping and of nucleotide sequencing showed that in procaryotes the sequence of amino acids in the protein chain reflects exactly that of codons in the messenger RNA and that the latter is a copy strictly complementary to the DNA (gene-protein colinearity). While the quasi-totality of genome sequences are informational in procaryotes, less than 1/10 of the DNAs seem to be expressed in higher eucaryotes. A large part of the DNA consists of repeated sequences and is not transcribed. The study of the structure of eucaryotic genes also revealed that the "gene-protein colinearity" dogma which applies to procaryotes is not systematically transposable to eucaryotic systems.

But for a few exceptions, most of the known genes coding for proteins in higher eucaryotes contain non-coding sequences (called "introns") which interrupt the coding sequences (called "exons"). These genes are called "split" or "mosaic" genes. The length of introns present in precursors of eucaryotic *m*RNAs can vary from 50 to 10 000 nucleotides.

The precise demarcation of exons and introns could be carried out for numerous genes coding for proteins. The comparison of nucleotide sequences at the exon-intron junctions revealed consensus nucleotide sequences present there. Lastly a motif less strictly conserved but equally important is found inside the intron, 20-55 nucleotides away from its 3′ end; it is called "branch site" because it is involved in the formation of a loop (see below)

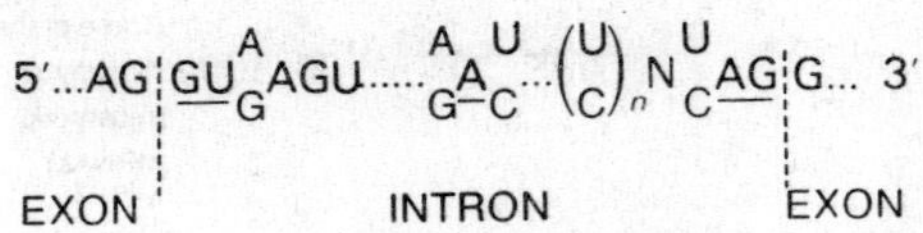

The most conserved nucleotides of these consensus sequences are underlined.

Site-directed mutagenesis experiments showed that these particular nucleotide sequences play the role of "signal" sequences indispensable for an adequate maturation process. On the contrary, the rest of the intron can be modified without affecting the effectiveness of the process of elimination of the intron; one can thus remove a part of the intron, or add sequences of varying lengths, provided one does not modify the "signal" sequences at the cleavage points at the 5′ and 3′ ends of the intron, nor the branch site and provided new "signal" sequences are not created inside the intron. The primary transcription products of split genes contain both the intron and exon sequences and constitute the pre-messenger-RNAs (pre-*m*RNAs) in the heterogeneous nuclear RNAs (h*n*RNAs). Before moving into the cytoplasm and after addition of the cap and the polyA tail these precursors undergo the *third phase* of maturation in which the intron sequences are eliminated. This operation called splicing takes place within ribonucleoprotein complexes (h*n*RNP) in which the h*n*RNAs are included right from their synthesis. The first step of this maturation (fig. 6-36A) is the cleavage of the RNA at the 5′ end of the intron and the appearance of a loop (or lariat) structure by formation of a 2′, 5′ phosphodiester bond between the 5′-terminal G of the intron and an internal adenylic residue of the intron. Then takes place a trans-esterification reaction where the 3′OH group of the free exon (donor site) attacks the phosphodiester bond of the second intron-exon junction (acceptor site), displaces the 3′ end of the intron and establishes the junction between the two exons, liberating the intron which carries the loop. The precision and efficiency of splicing are conditioned by the consensus sequences situated at the ends of the intron and by the loop connection sequence (or branch site) located inside the intron. The nucleus contains small RNAs (having less than 300 nucleotides) called *snRNAs*

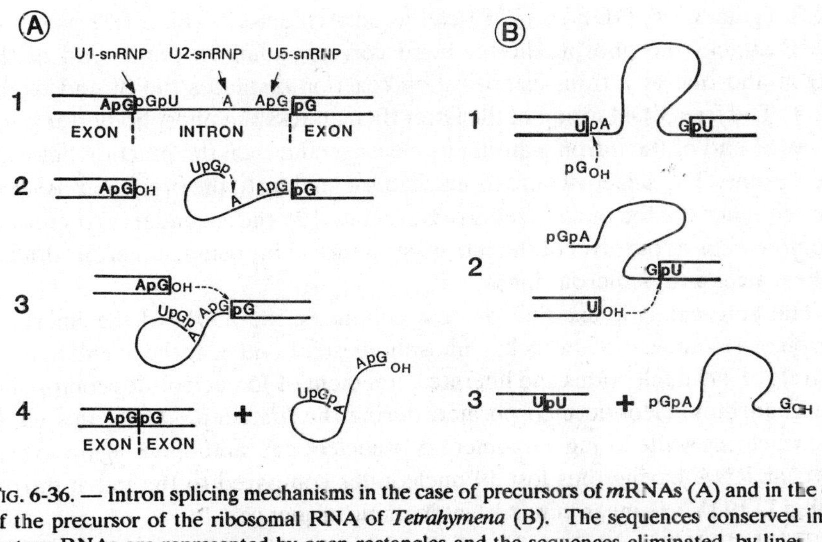

FIG. 6-36. — Intron splicing mechanisms in the case of precursors of *m*RNAs (A) and in the case of the precursor of the ribosomal RNA of *Tetrahymena* (B). The sequences conserved in the mature RNAs are represented by open rectangles and the sequences eliminated, by lines. The nucleotides participating in the reaction are indicated, as also the *sn*RNPs containing the small nuclear RNAs U1, U2 and U5. The *sn*RNP containing both RNAs U4 and U6 is not shown here because it does not interact directly with the RNA but with the other *sn*RNPs.

(small nuclear RNAs) associated with proteins to form complexes called *snRNPs* or *"snurps" (small nuclear ribonucleoprotein particles)* which play a role in the splicing of introns. In mammals, for example the U1-*sn*RNP binds at the cleavage site at the 5′ end of the intron (RNA U1 has a sequence complementary to that of this site), U2-*sn*RNP binds to the branch site and to the series of pyrimidines upstream the cleavage site at the 3′ end of the intron and U5-*sn*RNP recognizes the cleavage site at the 3′ end; a particle comprising the U4 and U6-*sn*RNPs is further added to form – with the precursor of *m*RNA – a large complex (of about 60S) called *spliceosome*. In patients suffering from lupus erythematosus, an auto-immune disease characterized by the formation of antibodies against various nuclear constituents, the excision of introns is blocked by antibodies specific of these ribonucleoproteins, thus confirming the important role played by *sn*RNPs in the process of elimination of introns.

It must be noted that anomalies in the excision of introns can cause hereditary diseases, like some *thalassemias* (characterized by the synthesis of abnormal hemoglobins); it was for example observed that a punctual mutation (G ⟶ A) about twenty nucleotides upstream the normal cleavage site at the 3′ end of the intron had created a new cleavage site (UUAG) on the pre-*m*RNA bringing about the synthesis of a defective hemoglobin.

In some cases, an auto-catalytic process has been described which needs no intervention of any protein. The best known example is that of the maturation of the precursor (6.4 kb) of the 26S *r*RNA of a protozoan, *Tetrahymena* (see fig. 6-36B), where the elimination of the intron needs, in addition to the RNA itself, GMP (guanosine, GDP or GTP) as the only cofactor. The 3′OH group of GMP attacks the phosphodiester bond corresponding to the 5′ end of the intron and due to a trans-esterification reaction displaces the 3′ end of the exon. The free 3′OH group of the latter then attacks the phosphodiester bond of the 3′ end of the intron, causing its cleavage and then the junction between the 2 exons. The selectivity of this mechanism leading to the precise excision of the sequences of the intron seems to be dictated by the secondary structure of the precursor at the level of the intron, structure comprising numerous double helical arms and numerous loops.

The auto-catalytic excision process continues: the 3′OH of the liberated intron of 414 nucleotides attacks a phosphodiester bond near the 5′ end to form a circle of 399 nucleotides and liberate a fragment of 15 nucleotides comprising the guanylic nucleotide incorporated during the first step. Then, this circle opens, closes while losing a fragment of 4 nucleotides, and opens again to give a linear RNA having thus lost 19 nucleotides compared to the initial intron, called L-19 IVS (Linear minus 19 intervening sequence).

This L-19 IVS is capable of catalyzing the conversion of pentacytidylate into shorter and longer oligonucleotides, and therefore has a nuclease and polymerase activity (the velocity of pentacytidylate hydrolysis by this RNA is

10^{10} greater than the hydrolysis velocity without catalyst). This RNA, as also the precursor of the *r*RNA, are therefore endowed with catalytic activity and were named *ribozymes*. Some believe that the existence of these riboenzymes points to the possibility of an auto-replication of RNAs without participation of proteins at the beginning of evolution.

It is interesting to note that the maturation of precursors of bacterial *t*RNAs calls upon a ribonucleoprotein, consisting of a protein of 20 000 daltons and a RNA (RNA M1) of 377 nucleotides, called *RNase* P and which causes an endonucleolytic cleavage generating the 5′end of mature *t*RNAs. The RNA M1 can by itself recognise the cleavage site and catalyze the specific cleavage at an appreciable velocity. This is therefore another example of RNA endowed with catalytic activity or ribozyme.

The auto-excision of the intron present in the precursor of *r*RNA of *Tetrahymena* implies the recognition and cleavage of the exon-intron junction (at the beginning, i.e. 5′ side of the intron) and the intron-exon junction (at the end of the intron), then the ligation of the two exons. Figure 6-37 shows the recognition of the exon-intron junction (at the 5′ end of the intron) and the position where cleavage takes place (indicated by an arrow). This recognition involves the nucleotides 22 to 27 of the intron which pair with the last 6 nucleotides of the exon.

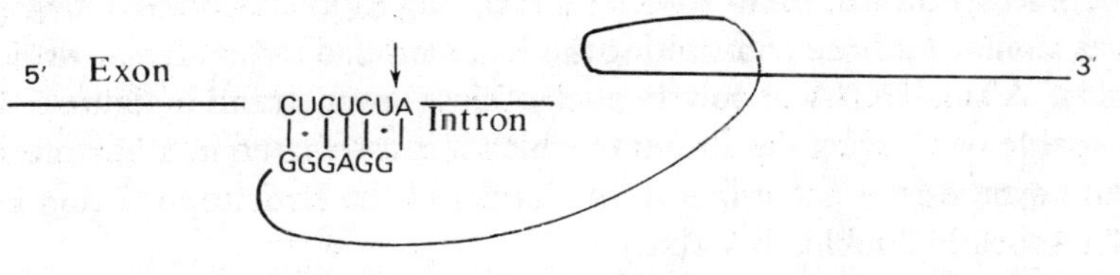

Fig. 6-37. — Recognition of the exon-intron junction at the 5′ end of the intron present in the precursor of *r*RNA of *Tetrahymena*.

The excised intron of the precursor of *r*RNA of *Tetrahymena* (this intron initially comprises 414 nucleotides) loses 19 nucleotides at its 5′ end to give L–19 IVS (Linear minus 19 intervening sequence). This fragment can (see fig. 6-38) cleave a RNA in a specific site, as a restriction endonuclease, thanks to the pairing between an internal sequence of this fragment and a complementary sequence on the RNA. This fragment of 395 nucleotides (414-19) is a ribozyme; it is found intact at the end of the reaction.

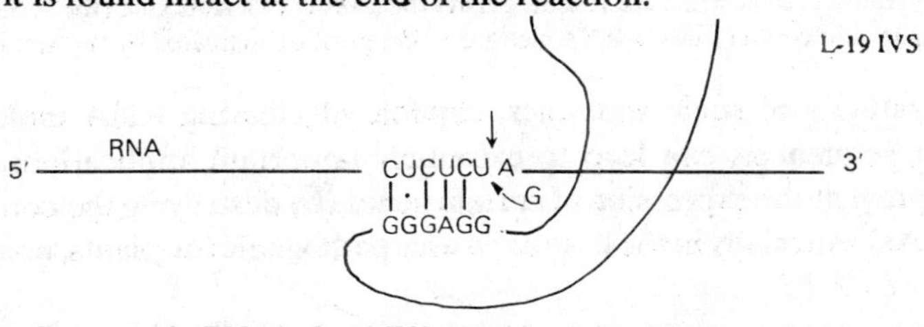

Fig. 6-38. — Cleavage of a RNA by L–19 IVS (resulting from the auto-excised intron from the precursor of *r*RNA of *Tetrahymena*. The RNA is cleaved after the CUCUCU sequence (arrow). A guanylic nucleotide is added at the 5′ end of A.

Figure 6-39 illustrates the auto-excision of the RNA (+ strand) of a virusoid of lucerne called LTSV (Lucerne Transient Streak Virus). This RNA of 324 nucleotides can take a hammerhead shape by pairing of complementary bases which permits the auto-excision between the nucleotides 167 and 168 (arrow).

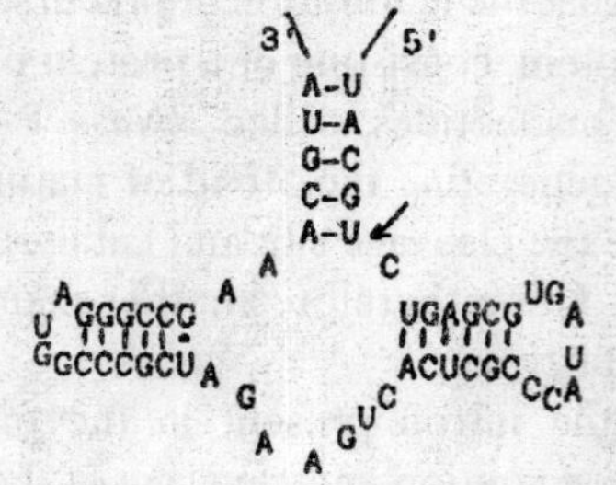

Fig. 6-39. — Auto-excision (at the place indicated by arrow) of the RNA of a virusoid (LTSV) having a hammerhead structure.

Such structures were observed in other plant virusoids, suggesting that the hammerhead shape is a sufficient condition to permit cleavage; this led researchers to synthesize small RNAs (in fact, oligoribonucleotides) having sequences similar to those constituting the hammerhead structures in viroids or virusoids. A small RNA of only 19 nucleotides (lower strand in figure 6-40) is thus capable of cleaving the strand to which it is associated in a hammerhead structure synthesized according to the model of the structure existing in the ASBV (Avocado Sunblotch Viroid).

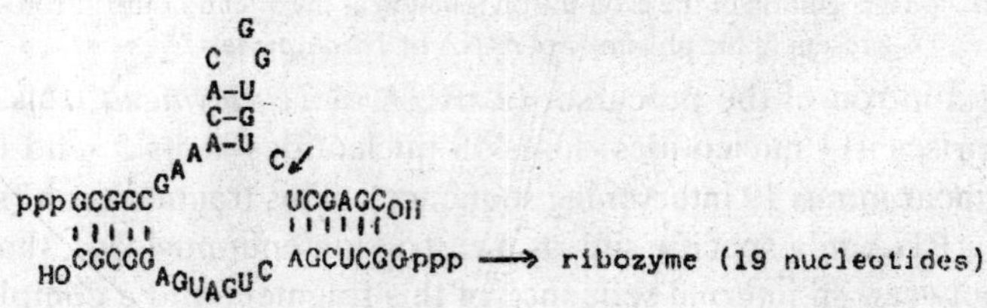

Fig. 6-40. — Hammerhead structure resulting from the pairing of a ribozyme (the RNA comprising 19 nucleotides) with a RNA cleaved at the position indicated by the arrow.

The synthesis of such ribozymes, capable of cleaving RNA molecules at particular sequences, can lead to extremely important applications because they can prevent the expression of certain genes (by destroying the corresponding *m*RNAs) especially in the case of viruses pathogenic for plants, animals and man.

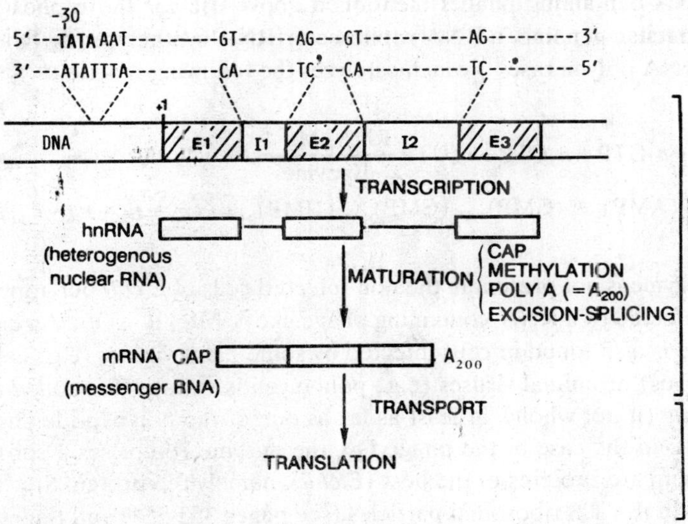

FIG. 6-41. — Important steps of the expression of genes coding for proteins in eucaryotes. Typical nucleotide sequences of eucaryotic genes and their position are represented schematically at the top (TATA sequence and limits of introns). E1, E2 and E3 represent 3 exons; I1 and I2 represent 2 introns of the same gene. The transcription product is represented by rectangles for the parts corresponding to exons (which after splicing will constitute the *m*RNA) and by lines for the parts corresponding to introns (which will be eliminated).

III. – RNA REPLICASE

Viruses (and RNA phages) use for the replication of their genome, one of the following 4 possibilities:

– viruses of the 1st group (like the virus of poliomyelitis, or phages of *E.coli*, such as f_2, MS_2, R_{17}, $Q\beta$) contain in their particles a strand of RNA (+) (by convention RNA + is the *m*RNA). During the infection, a (–) strand is synthesized, which serves a template for the synthesis of new (+) strands as we will see below.

– the reverse situation prevails in the 2nd group (e.g., comprising the virus of rabies) The virion contains a (–) strand of RNA which will serve as template for the synthesis of (+) RNA.

– the 3rd group of viruses (including the reoviruses) consists of double-stranded RNA viruses, in which the (±) duplex serves as template for the asymmetric synthesis of (+) strands.

– in the 4th group are found the retroviruses (like the Rous Sarcoma Virus) in which the (+) RNA of the virion permits the intermediate synthesis of a DNA which in its turn serves as template for the synthesis of new (+) RNAs.

The RNA-containing phages mentioned above use for the replication of their ribonucleic genome, a RNA replicase or RNA synthetase (or RNA-dependent RNA polymerase[1]) which catalyzes the following reaction:

$$n_1\text{ATP} + n_2\text{GTP} + n_3\text{CTP} + n_4\text{UTP} \xrightarrow[\text{Enzyme}]{\text{RNA, Mg}^{2+}}$$

$$[(\text{AMP})_{n1}, (\text{GMP})_{n2}, (\text{CMP})_{n3}, (\text{UMP})_{n4}] + (n_1 + n_2 + n_3 + n_4)\text{PP}$$

This enzyme is not present in the non-infected cells of *E.coli* but appears in the cells infected by a RNA-containing phage like f_2, MS_2, R_{17}, $Q\beta$. An enzyme of this type is also found in cells infected by some plant viruses (e.g, Tobacco Mosaic Virus) or animal viruses (e.g., poliomyelitis virus). It is *coded by the viral genome* (if not wholly, at least as far as one of the polypeptide chains is concerned): in the case of the phage $Q\beta$, the enzyme comprises 4 sub-units: three of them are proteins of the host (*E.coli*), namely the protein S_1 which is also found in the 30S ribosomal particles (see pages 353-354) and the elongation factors $EF-T_u$ and $EF-T_s$ (see page 359), while only the fourth sub-unit is coded by the RNA of the phage.

It may be noted that the reaction catalyzed is exactly the same as that catalyzed by the RNA polymerase of healthy cells (see page 324), only the template is different: it is not DNA (the action *in vitro* of DNase thus eliminates the synthesis of RNA by the DNA-dependent RNA polymerase, without affecting the activity of the viral RNA replicase), but *RNA*. Besides, the RNA replicase is extremely specific in the sense that it *requires the homologous RNA*; for example, the RNA replicase of phase $Q\beta$ does not accept as template, the RNA of phage MS_2 though the latter is rather close, and even less the various cellular RNAs. This specificity offers the virus the advantage of allowing the preferential replication of the viral RNA (and thus permits the infectious process), while this RNA is in an environment where the cellular RNAs very largely predominate.

The RNA synthesized is not complementary but identical to the viral RNA and one wondered whether some special mechanism is involved in this transfer of information. This is not so; the principle remains that of *pairing of complementary bases*, so that during the infection one has, in a first phase, synthesis of a (–) strand complementary to the (+) strand from viral particles, and formation of a duplex, or *replicative form*, consisting of a (+) strand and a (–)

---

[1] The term "polymerase" is not used in order to avoid confusion with the DNA-dependent RNA polymerase studied in the previous paragraph; one of the former two names is preferred.

strand in double helix (some believe that these duplexes constitute artefacts formed during the isolation of viral RNAs). The (–) strand then acts as template for the synthesis of numerous (+) strands (see fig. 6.42). The (+) strands function as messengers to direct the synthesis of viral proteins and after encapsidation they form the genome of new viral particles. They contain only 4 genes: the gene of the capsid protein, the genes of the maturation protein necessary to assemble RNA and protein to constitute the phage particles (the capsid comprises 180 molecules of the capsid protein which has a molecular weight of 14 kd, and 1 molecule of the maturation protein which has a molecular weight of 38 kd), the gene of one of the 4 sub-units of the RNA replicase (see above), and the gene of the protein responsible for the lysis of the bacterial cell (this gene overlaps on one hand the gene of the capsid protein and on the other hand the gene of the sub-unit of the replicase). With the help of the RNA replicase of Qβ, Spiegelman could obtain *in vitro* the *synthesis of RNA molecules identical to the RNA extracted from the phage; these RNA molecules synthesized in vitro were in their turn capable of acting as template in the reaction catalyzed by the enzyme* (we have seen that the latter is very demanding as far as its template is concerned) and furthermore, *they were capable of programming the formation of viral particles* in the protoplasts of *E.coli* (they therefore possessed infectious capacity). These were the first experiments which enabled the synthesis *in vitro* of a biologically active nucleic acid with the help of a purified enzyme; this synthesis *in vitro* of a ribonucleic genome was followed by that of a deoxyribonucleic genome (that of the phage φX174).

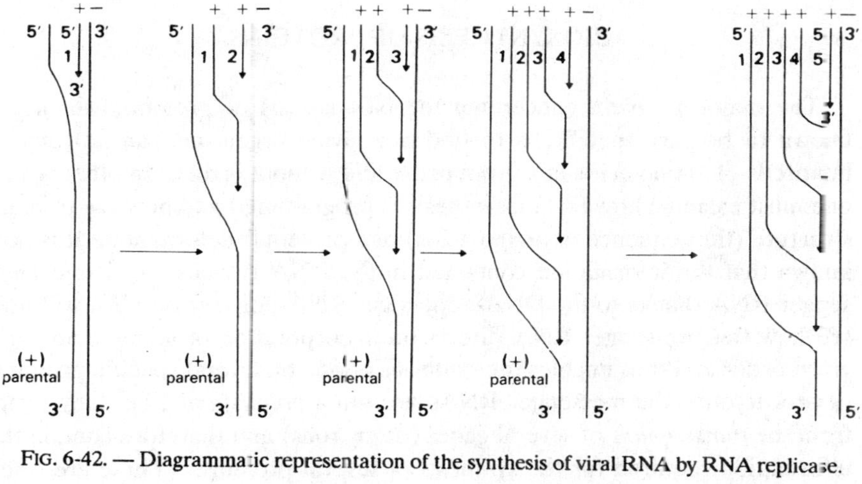

FIG. 6-42. — Diagrammatic representation of the synthesis of viral RNA by RNA replicase.

Even in the expression of such a small genome, there are *regulation* phenomena. On one hand, the replication of the (+) strand must not take place together with its translation (otherwise there would be collision between the RNA replicase moving along the (+) strand from 3′ towards 5′ and the ribosomes moving along the (+) strand from 5′ towards 3′). This is avoided by the inhibiting action of the RNA replicase of phage Qβ on the binding of ribosomes to the (+) strand as long as a sufficient quantity of (–) strands has not been synthesized. On the other hand, the 4 phage proteins are not synthesized in equal quantities, because much more capsid protein is required (180 molecules per phage particle); this is accomplished by the fact that the ribosomes bind much more strongly to the initiation site of the translation of the capsid protein than to the initiation sites of other proteins on RNA(+). Moreover, the capsid protein inhibits the synthesis of the sub-unit of RNA replicase by blocking the initiation site of the translation of this sub-unit.

As in the case of DNA, there are proofs that *the RNA is actually the genome of RNA-containing phages and viruses*; in some cases it was possible to cause the infection of the host cell by the RNA isolated and purified from viral particles. On the other hand, after separating RNA and protein from a virus, one can carry out reconstitution experiments in certain experimental conditions; in the case of the tobacco mosaic virus, for example, one can form a hybrid virus from the RNA of a given strain (x) and the protein of another strain (y); after infection of tobacco by this hybrid, one observes the appearance of viral particles the capsid of which is of the type x, i.e. of the type of the strain having provided the RNA, and not of the type of the protein present in the hybrid which started the infection; since it is the RNA which specified the characteristics of the viral progeny, it is therefore the actual carrier of the genetic information of the virus.

BIOSYNTHESIS OF PROTEINS

The major problem concerning the biosynthesis of proteins, which are known to be very specific, is to find how living organisms can polymerize hundreds of amino acids in a given order and without errors. In other words one must examine how this biosynthesis is programmed and how the primary structure (the sequence of amino acids) of a protein is determined. It is now known that the information contained in the DNA is transcribed into messenger RNA thanks to the DNA-dependent RNA-polymerase. We will now see how this messenger RNA directs the incorporation of amino acids in a given order and thus enables the synthesis of one or several specific proteins; in procaryotes, the messenger RNAs are often *polycistronic*, i.e. they result from the *transcription* of several genes (or cistrons) and therefore contain the information required for the synthesis of several proteins. (These are often enzymes which participate in the same metabolic pathway and whose corresponding genes are grouped in an *operon*). There are therefore two successive processes:

$$\text{DNA} \xrightarrow{\text{transcription}} m\text{RNA} \xrightarrow{\text{translation}} \text{Proteins}$$

Since the messenger RNAs are not only synthesized, but also translated in the direction 5′ ⟶ 3′ , there can be, in procaryotes, a coupling between transcription and translation, i.e. the translation of a *m*RNA can begin even before its synthesis is completed. In eucaryotes, such a coupling is not possible because the *m*RNAs are synthesized in the nucleus (where the DNA is localized) and must then migrate, in the form of ribonucleoprotein complexes called *informosomes* or *m*RNP (messenger ribonucleoprotein particles), to the cytoplasm where the major part of protein syntheses take place.

We will now study *translation*, i.e. the biosynthesis of proteins.

I. – ACTIVATION OF AMINO ACIDS AND FORMATION OF AMINOACYL-*t*RNAs

The first step of protein biosynthesis comprises two reactions (see fig. 6-43) catalyzed, for a given amino acid, by one and the same enzyme called *activation enzyme*, or aminoacyl-*t*RNA synthetase, or aminoacyl-*t*RNA ligase.

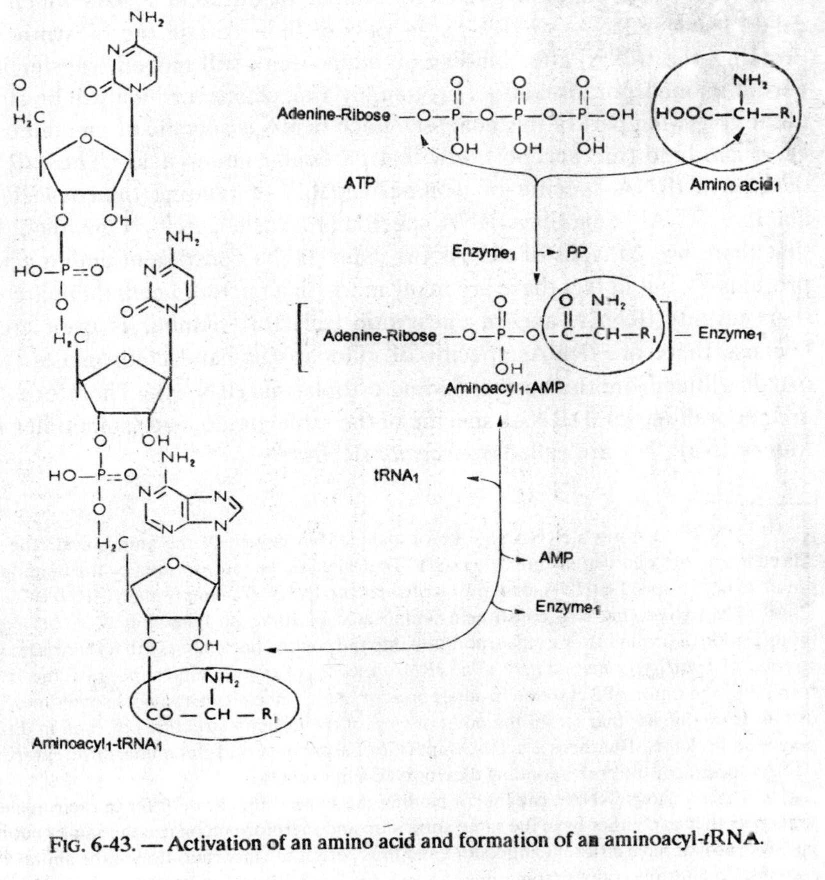

FIG. 6-43. — Activation of an amino acid and formation of an aminoacyl-*t*RNA.

In a first step, the L-amino acid is activated by ATP and forms an aminoacyl-AMP, by a reaction analogous to that of the activation of fatty acids where an acyl-AMP is formed (see fig. 5-10). A mixed anhydride is formed between the carboxyl and a phosphoric acidic OH. This compound remains bound to the enzyme; then, since it is very reactive, it will dissociate and thus allow the amino acid to bind (by its carboxyl, again) to a hydroxyl at the 3′ end of a particular RNA called transfer RNA or *t*RNA; an ester linkage is therefore formed and this compound is called *aminoacyl-tRNA* (see fig. 6-43). Depending on the *t*RNA, the amino acid is bound either to the OH at position 2′ or to the OH at position 3′ of the ribose of the 3′ terminal nucleotide.

1. – TRANSFER RNAs (*t*RNA)

The RNA present in the soluble part (non particulate) of the cytoplasm was originally called soluble RNA (*s*RNA), by opposition to the ribosomal RNAs present in the ribonucleoprotein particles, called ribosomes which will be referred to in the following. It was subsequently found that this *s*RNA is in reality a mixture containing a large number of different RNAs which were called transfer RNAs or *t*RNAs, in view of their role in the biosynthesis of proteins; the *t*RNA, after binding an amino acid, will indeed transfer it and permit its incorporation into a protein, by a mechanism which will be studied in the following part of this chapter. Each *t*RNA is specific of an amino acid, i.e. it can bind (or "accept") only that particular amino acid. Thus, *t*RNA^{ala} denotes a *t*RNA specific of alanine, capable of binding then transferring alanine, *t*RNA^{tyr} denotes a *t*RNA specific of tyrosine, etc[(1)]. One might think that there are 20 types of *t*RNAs (as many as the constituent amino acids of proteins[(2)]), but in fact there are many more: in a bacterial cell, there are more than seventy *t*RNAs, and in eucaryotic cells this number is even greater, because there are *t*RNAs specific of mitochondria and chloroplasts (which usually differ from the corresponding cytoplasmic *t*RNAs). Therefore, there are generally several tRNAs specific of the same amino acid (sometimes up to four or five): they are called *isoacceptor t*RNAs[(3)].

---

(1) *t*RNA^{ala} denotes a *t*RNA specific of alanine, but devoid of the amino acid (the terms "discharged" or "deacylated" are also used). To designate the ester formed by the binding of an amino acid to a specific *t*RNA, one must write alanyl-*t*RNA^{ala} (or simply alanyl-tRNA).

(2) Hydroxyproline is a constituent amino acid of some proteins, but it is formed by a modification of proline after its incorporation into the polypeptide chain, so that there is no *t*RNA specific of hydroxyproline. There is no *t*RNA specific of cystine either, because the latter is formed by the union of 2 cysteine residues present in the same chain at places sometimes rather distant from one another (from the point of view of the primary structure) or even in different polypeptide chains. But there are *t*RNAs specific of asparagine and glutamine, different from the tRNAs specific of the corresponding dicarboxylic amino acids.

(3) These various *t*RNAs, capable of binding the same amino acid, differ in their nucleotide sequence; they can either have the same anticodon and therefore recognize the same codon (see pp. 364-366), or have different anticodons and thus permit the incorporation of the amino acid in response to multiple codons (sometimes up to 6, see fig. 6-48) specifying the same amino acid.

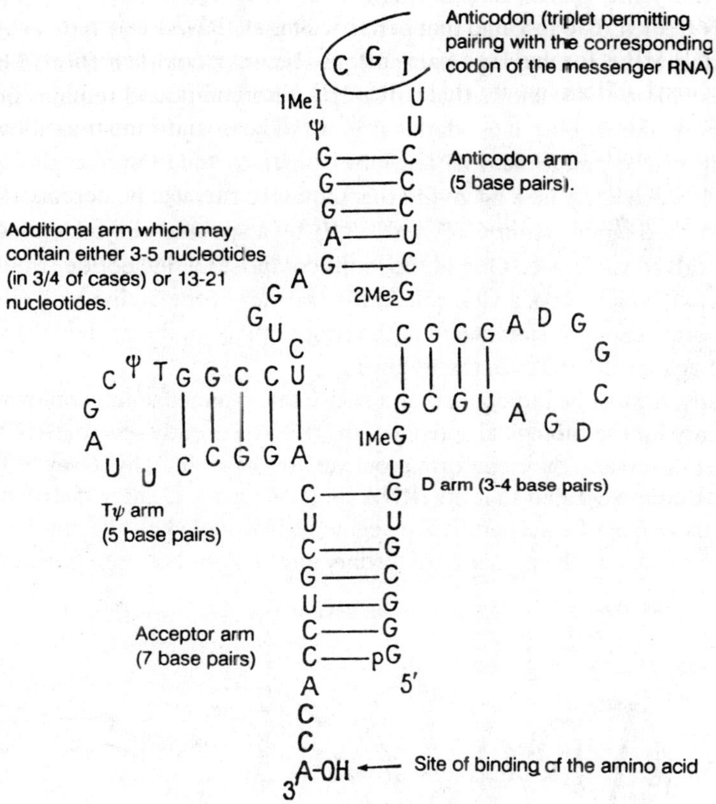

FIG. 6-44. — Primary and secondary structure (clover leaf-shaped) of yeast *t*RNA^{ala}, determined by Holley and co-workers (1965) and modified by Dirheimer and co-workers (1974). The phosphate groups (*p*) between the nucleosides are not shown in this diagram. Abbreviations for the rare nucleosides are as follows: D = dihydrouridine; 1 MeG = N_1 methylguanosine; 2 Me_2G = N_2 dimethylguanosine; 1MeI = N_1 methylinosine; I = inosine; T = thymine ribonucleoside; ψ = pseudouridine.

The *t*RNAs are small-sized ribonucleic acids, consisting of 70 to 90 nucleotides only, which corresponds to a molecular weight of about 30 000. They contain *rare nucleotides* (several tens of them are presently known), i.e. nucleotides which are unusual either by the nature of the base which is often a methylated base (see figs. 6-2 and 6-3), or even by the nature of the bond between the base and the sugar (in *t*RNAs one, for instance, finds pseudo-uridylic acid where C_1 of ribose is bound to C_5 of uracil and not to N_1). These modifications are catalyzed by various specific enzymes after the synthesis of the polyribonucleotide chain, as described in the study of processes of maturation of *t*RNAs (see pages 335-336).

The various RNAs differ by their nucleotide sequences. One presently knows several hundreds of complete sequence of *t*RNAs of very diverse origins (bacteria, yeast, plants, animals) and it is observed that there are common characteristics. It was found that in particular, all *t*RNAs can take a clover leaf secondary structure, having 4 pairing zones between complementary bases and 3 loops. Figure 6-44 shows that a *p*C*p*C*p*A terminal trinucleotide sequence is present at the 3′ end; it is identical in all *t*RNAs and indispensable for the binding of the amino acid as the latter binds to the ribose of the terminal adenylic nucleotide (see fig. 6-43); this sequence can also be degraded *in vitro*, then formed again (from CTP and ATP) by a specific enzyme called *t*RNA nucleotidyl transferase. One of the loops comprises a nucleotide triplet called *anticodon*, which plays a vital role in the transfer of the amino acid (i.e. in the incorporation of the amino acid at the proper place in the protein), and which will be again referred to in the following.

Lastly, it must be indicated that a native three-dimensional conformation is necessary for the biological activity of the *t*RNA; recently several *t*RNAs could be crystallized and their conformation can now be studied by X-ray diffraction. These studies revealed that the *t*RNA molecule has a L-shaped structure, due to the existence of 2 perpendicular segments in double helix having each about 10 base pairs, and the presence of tertiary interactions between the bases of the

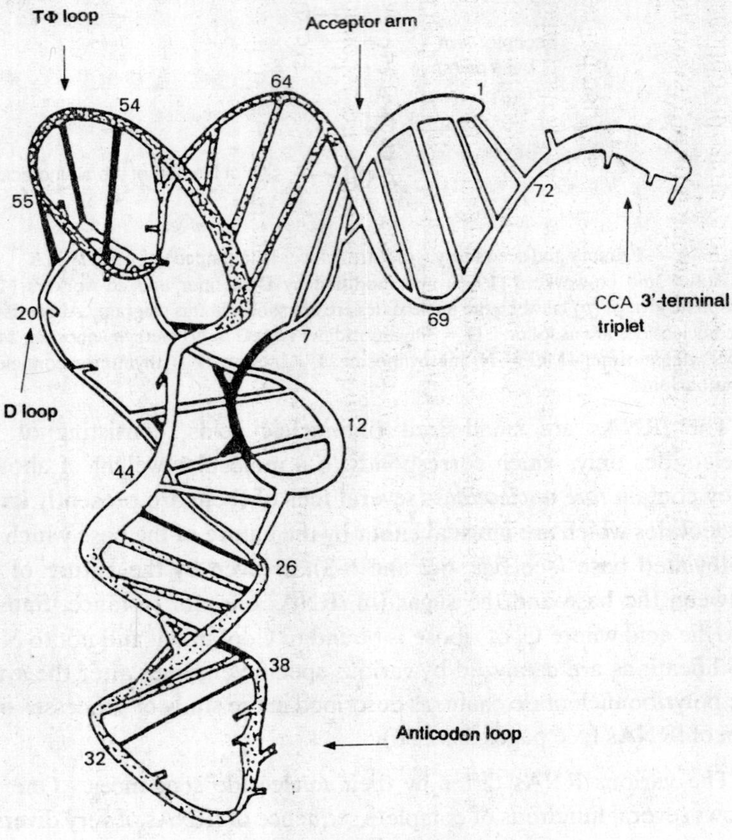

L-shaped structure of the *t*RNA molecule.

non-helicoidal regions. These interactions are often based on unusual hydrogen bonds between some bases (AC, AA, GG), between some bases and the ribose-phosphate chain, or between different regions of this chain (the 2′OH groups of riboses of the chain often participate in these hydrogen bonds). The anticodon and the binding site of the amino acid (i.e. the CCA 3′ terminal triplet) are at the ends of the L, at a distance of about 80 Å (see diagram on previous page).

2. – Activation enzyme or aminoacyl-tRNA synthetase

These two names recall the two reactions catalyzed by this enzyme (see fig. 6-43) which besides ATP must recognize, on one hand the corresponding amino acid (it can catalyze the activation of the latter only), and on the other hand, the corresponding *t*RNA (it can allow the binding of the amino acid only to a *t*RNA specific of this amino acid). A very high specificity is indeed necessary during this step in order to avoid errors in the biosynthesis of proteins, because once the aminoacyl-*t*RNA is formed, there is no longer any control mechanism in the cell to verify the nature of the amino acid and it is therefore no longer possible to reject an amino acid which would have been incorrectly bound; as will be seen in the following, the next steps of protein synthesis involve only the polynucleotide part of the aminoacyl-*t*RNA, with the result that an amino acid which would be bound to a *t*RNA other than the one corresponding to it, would be erroneously incorporated in the protein. This was shown by Chapeville and co-workers in the following manner: they bound cysteine to *t*RNA^{cys} thanks to the corresponding activation enzyme, then with the help of Nickel Raney they transformed cysteine into alanine without breaking the ester linkage and therefore obtained an alanyl-*t*RNA^{cys} (a compound which cannot normally form in the cell); they observed that from this compound alanine was incorporated in the polypeptides in place of cysteine (see fig. 6-45) in a cell-free system of protein synthesis programmed either by poly

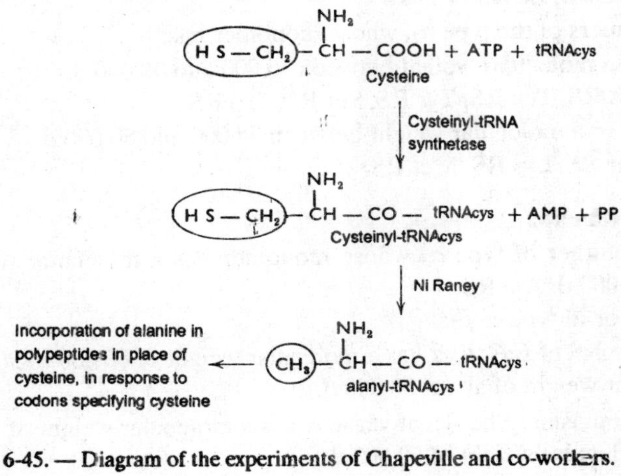

FIG. 6-45. — Diagram of the experiments of Chapeville and co-workers.

UG (where the ratio U/G was 5) permitting normally the incorporation of cysteine (codon UGU) but not that of alanine (GCN), or by globin *m*RNA (in this case, addition of ^{14}C-alanyl-*t*RNA^{cys} led to the formation of a radioactive trypsic peptide normally comprising cysteine).

In vivo, the aminoacyl-*t*RNA synthetases must therefore, either not make any errors or be able to correct them. They must, in particular be able to distinguish between two amino acids of very similar structures. Thus, tryosyl *t*RNA synthetase binds tyrosine 1 000 times more effectively than phenylalanine (which differs only by one hydroxyl less). But the discrimination by isoleucyl-*t*RNA synthetase between valine and isoleucine (which has only one $-CH_2-$ more implies a correction mechanism or "*proof-reading*". This additional $-CH_2-$ favours the activation of isoleucine compared to valine by a factor of about 200, but since in general, there is 5 times more valine than isoleucine *in vivo*, one would expect that valine is incorporated by error once out of 40 times in place of isoleucine, In fact, the frequency of such an error is only 1/3 000 because isoleucyl-*t*RNA synthetase is capable of correcting its own errors, i.e. in presence of *t*RNA^{ile}, the valyl-AMP formed is hydrolyzed (but not isoleucyl-AMP), thus preventing an erroneous aminoacylation (i.e. a mis-acylation) of *t*RNA^{ile}. Most aminoacyl-*t*RNA synthetases probably have a synthetic site which dces not accept amino acids larger than the corresponding amino acid, and a hydrolytic site which degrades the aminoacyl-AMPs smaller than the correct aminoacyl-AMP.

The number of sub-units and molecular weight vary widely from one aminoacyl-*t*RNA synthetase to another even within the same organism. One thus finds:

– monomers (α) of molecular weight between 50 000 and 60 000 d (daltons):

Cys RS, Glu Rs, Gln Rs, Arg Rs

or a molecular weight between 110 000 and 130 000 d:

Leu RS, Ile RS, Val RS

– dimers of the type α_2, whose monomer has:

– a molecular weight between 35 000 and 40 000 d:

His RS, Trp RS, Tyr RS, Ser RS, Pro RS

– or a molecular weight between 75 000 and 80 000 d:

Thr RS, Lys RS, Met RS

– tetramers:

– either of type α_4 whose monomer has a molecular weight of about 95 000 d: Ala RS

– or of type $\alpha_2\beta_2$:

In Phe RS of *E.Coli*, α has a molecular weight of 80 000 to 90 000 d; β has a molecular weight of about 30 000 d.

In cytoplasmic Phe RS of yeast, α has a molecular weight of 70 000 d and β has a molecular weight of 60 000 d.

II. – TRANSFER OF AMINO ACIDS FROM AMINOACYL-*t*RNAs TO PROTEINS

Amino acids will be incorporated into proteins from the aminoacyl-*t*RNAs. To study this process we must acquire a wider knowledge of ribosomes and the messenger RNA, which will form a complex with the aminoacyl-*t*RNAs.

1. – Ribosomes

These are ribonucleoprotein particles consisting of about 65% RNAs and 35% proteins in the case of *E.coli* and which may be obtained from cellular homogenates by simple centrifugation at high speed (105 000 × g). These particles are usually designated according to their sedimentation constant: the ribosomes present in bacteria (as also in chloroplasts) are called 70 S because such is their sedimentation constant when the Mg^{2+} ion concentration in the solution is 0.01 M. If this concentration is lowered to 0.001 M, the 70 S ribosomes dissociate into 50 S and 30 S particles, and this phenomenon is reversible if the Mg^{2+} ion concentration is again raised. The ribosomes of the cytoplasm of plant and animal cells are slightly heavier (80 S) and also dissociate into 60 S and 40 S particles when Mg^{2+} ion concentration is lowered.

The size of mitochondrial ribosomes varies according to their origin between 60 S in animals and about 75 S in yeast and plants.

The 50 S and 30 S, or 60 S and 40 S particles can themselves be dissociated, as shown by the diagram below, into ribosomal RNAs (*r*RNAs), and proteins (and also reconstituted from these components under certain conditions):

70 S Ribosome (bacterial)	30 S Particle 16 S RNA contains 1 541 nucleotides (molecular weight = 5.6×10^5) 21 proteins called S_1, S_2, S_3, etc. (S stands for "Small" and indicates that they are proteins of the small particle) 50 S Particle 23 S RNA contains 2 904 nucleotides (molecular weight = 1.1×10^6) 5 S RNA contains 120 nucleotides (molecular weight 3.5×10^4) 34 proteins called L_1, L_2, L_3 etc. (L stands for "Large" and indicates that they are proteins of the large particle)

There is one molecule of each protein per ribosome except for L_7 and L_{12} which are present with 2 molecules per ribosome.

*r*RNAs have a secondary structure comprising many short helicoidal regions; this was shown by studies of accessibility to chemical modifications and enzymatic digestions, as well as comparisons of the structures of *r*RNAs of various organisms enabling the detection of conserved characteristics. The secondary structure of 16 S *r*RNA is shown in figure 6-17.

80 S Ribosome (mammalian cytoplasmic)	40 S Particle 18 S RNA contains 1 900 nucleotides (molecular weight = 7×10^5) 33 proteins 60 S Particle 28 S RNA contains 4700 nucleotides (molecular weight = 1.8×10^6) 5.8 S RNA contains 160 nucleotides 5 S RNA contains 120 nucleotides 49 proteins

Intensive studies on these particles and their constituents are now in progress to determine the structure of these complexes (especially the nature of interactions between *r*RNAs and ribosomal proteins) and the respective roles of various constituents.

As regards studies on the architecture, they are based on the one hand, on experiments of assembly or reconstitution of ribosomal particles from the *r*RNAs and proteins (one can for example study the effects of the omission of a protein), on the other hand, on structural analyses using diverse methods, like the localization of proteins in the particle by observation of antibodies in electron microscopy, study of protein-protein or protein-RNA proximity with the help of cross-linking reagents or neutron diffusion by particles containing deuteriated proteins.

As for studies on the functions of the constituents of ribosomes they are also based on varied approaches like affinity labeling with analogues – particularly active – of compounds binding to ribosomes, chemical modification of proteins or RNAs in an attempt to establish correlations between structural modification and alteration of a function, analysis of mutations modifying a protein or a *r*RNA (for example, those responsible for resistance to an antibiotic, etc.). These studies could establish, on the ribosomes, a number of active centres implying some proteins or RNA regions and playing a role in the different steps of protein synthesis. For example, the P site (site of peptidyl-*t*RNA binding) seems to involve proteins L_2 and L_{27}, the binding of the 30 S particle to the *m*RNA seems to imply proteins S_1, S_{18} and S_{21} as well as the 3′ part of the 16 S *r*RNA (which pairs with a complementary sequence situated upstream the initiator codon AUG of the bacterial *m*RNAs), proteins L_2 and L_{16} seem to be involved in the peptidyl-transferase activity, proteins L_7/L_{12} (there are 2 copies of this dimer per ribosome, while all other ribosomal proteins are represented by only one copy) appear to be involved in GTP-ase activity, etc.

In systems where protein biosynthesis takes place, *in vitro* as well as *in vivo*, it is observed that the 70 S (or 80 S) ribosomes are attached to a chain of single-stranded RNA, thus forming aggregates called *polyribosomes* or *polysomes* which could be observed using the electron microscope. The number of ribosomes which can be present on the same RNA chain depends on several factors, especially the size of the *m*RNA and the efficiency with which the ribosomes bind to the *m*RNA to translate it. In optimal conditions for protein synthesis, there is in average one ribosome for every 80 nucleotides. In general, there are less than ten ribosomes on the eucaryotic *m*RNAs (which are usually monocistronic), while there can be several tens on procaryotic *m*RNAs (which are often polycistronic). These polysomes are the seat of proteins synthesis as we will see in the following, but we must first explain the nature and important characteristics of the RNA to which the ribosomes are bound and which is called messenger RNA.

2. – Messenger RNA (*m*RNA)[1]

The existence of *m*RNAs was not detected for a long time because they represent only a very small percentage (generally 2 to 4%) of the total cellular RNAs (*r*RNAs represent about 80% and *t*RNAs about 15% of the total RNA). *m*RNA molecules can be revealed by labeling techniques using a radioactive precursor (^{14}C–uracil, for example) which is allowed to act only for a very short time (this is called a pulse) which is about one minute for bacteria: at the end of this time, the *m*RNAs will be preferentially radioactive because their *turnover is extremely rapid.* In bacteria infected by a phage, one can observe, very soon after the infection, the appearance of RNA capable of hybridizing specifically to the DNA of this phage (but not to the DNA of the host-cell or to the DNA of other phages); this is therefore RNA transcribed from the phage DNA, carrying genetic information necessary for the expression of the genes of this phage, i.e. necessary for the infectious cycle, and which must direct the synthesis of proteins specific of the phage. By blocking transcription in cells (with the help of an appropriate antibiotic for example), one can study the life expectation of existing *m*RNA molecules either by measuring the capacity of the *m*RNA to program the synthesis of the corresponding proteins, or by measuring the capacity of the *m*RNA to hybridize to the DNA from which it is transcribed.

In bacteria, *the half-life of mRNA molecules* (period of time after which half is degraded) is about 1 to 2 minutes; this may appear short but considering that the time required for the synthesis of a polypeptide chain of average size is less than 10 seconds (the rate of protein synthesis is about 15 amino acids per second), one realizes that this life time is sufficient to enable a *m*RNA molecule to direct the synthesis of numerous protein molecules, especially as a series of ribosomes run along the *m*RNA, following one another at regular intervals and, as will be seen in the following (see page 356), each of them permitting the synthesis of a polypeptide chain. The degradation of *m*RNAs seems to proceed from 5′ to 3′ i.e. in the same direction as the synthesis of *m*RNAs and their translation. Such rapid synthesis and short life of bacterial *m*RNAs are necessary for the capability of adaptation that bacteria must possess to face

---

(1) Revised by C. KEDINGER.

changes of medium; they must be able to rapidly synthesize an inducible[(1)] enzyme when they need it and cease to synthesize it when it is no longer required. But the half-life of all *m*RNAs is not as short; in higher organisms in particular, there is no necessity of such rapid adaptation (the environment is more stable) and the *m*RNAs can exist for a longer time. The half-life of *m*RNAs of mammalian cells in culture is in the order of a few hours but can extend to 24 hours (i.e. the duration of the cell cycle). Furthermore, in the particular case of cells deprived of their nucleus either after a natural process (mammalian red blood cells), or after an experimental intervention (*Acetabularia*), and DNA is no longer present to serve as template for the synthesis of new *m*RNA molecules, the *m*RNAs can function (direct protein synthesis) for hours or even days.

The size of *m*RNAs is very variable because the size of the proteins whose synthesis they direct is itself very variable. A polycistronic *m*RNA can result from the uninterrupted transcription of about ten adjacent genes in a given operon, its molecular weight can be in the order of 3.5×10^6 (which corresponds to about 10^4 nucleotides) and it can thus be 3 times longer than the largest ribosomal RNAs (23 S RNA) and about 120 times longer than a *t*RNA.

Even in the case of monocistronic *m*RNA, the length is not limited to the coding region between the initiation codon (generally AUG) and the termination codon (UAG, UAA or UGA): upstream the initiation codon (i.e. on the 5′ side) there is a leader region, and downstream the termination codon (i.e. on the 3′ side) there is a tail region. Besides, in polycistronic *m*RNAs (i.e. in most bacterial *m*RNAs), there are intercistronic (or intergenic) regions which can be long (a few tens of nucleotides) or short (1 or 2 nucleotides only). In some cases there is on the contrary, an overlapping of two genes: the last nucleotide of the termination codon (UGA) of one of the genes is also the first nucleotide of the initiation codon (AUG) of the following gene.

3. – Mechanism of translation

As observed in bacteria, the polysomes consisting of 70 S ribosomes bound to a *m*RNA chain, are not static; on the contrary each ribosome moves along this chain in the direction $5' \longrightarrow 3'$. When a ribosome has moved a particular distance, a second ribosome binds and begins, in its turn, to move towards the 3′ end of the *m*RNA, and so on. This movement corresponds to the translation, the *deciphering of the message carried by the mRNA*, thus permitting protein synthesis.

---

(1) There are *constitutive enzymes*, always present in the cell because they are constantly synthesized, and *inducible (or adaptable) enzymes* which are synthesized only as and when needed; as will be seen in the chapter on regulations, the bacterium synthesizes the enzymes of the lactose operon (required for metabolizing lactose) only when it is made to grow in a medium containing lactose as the only source of carbon; and it synthesizes the enzymes of the tryptophane operon (permitting the formation of this amino acid) only when tryptophane is missing in the cell.

How is this message, coded in nucleic language, translated into protein language? How are amino acids specified (coded) to permit their incorporation in a correct order? This is the problem of the genetic code. If one groups in twos the 4 types of nucleotides found in the *m*RNAs it is clear that there are only 16 combinations (or doublets) possible, which is insufficient to code 20 amino acids. If on the contrary, *one groups the nucleotides in threes*, then *64 triplets are possible*, which is amply sufficient; there is even an excess of triplets and we will see that this enables most amino acids to be coded by several triplets. It could be shown that these triplets do not overlap (i.e. a nucleotide is never part of 2 triplets, it cannot, for example, be the last nucleotide of a triplet and the first of the following triplet), so that a polypeptide comprising *n* amino acids will be coded by a polyribonucleotide chain having 3*n* nucleotides. The *m*RNA must therefore be considered as a *succession of nucleotide triplets, or codons*. These codons are, so to say, the words of polynucleotide sentences; they will be translated into amino acids which are the words of protein sentences.

Considering a protein of average size, having a molecular weight of 10^5 and consisting of about 800 amino acids, one may calculate that the corresponding *m*RNA will have $800 \times 3 = 2\,400$ nucleotides, and the corresponding gene $2\,400 \times 2 = 4\,800$ nucleotides, which gives a molecular weight of $4\,800 \times 300$ (average molecular weight of a nucleotide) $= 1.5 \times 10^6$. Knowing that there are about 2 000 different proteins in a bacterial cell, one can calculate that $1.5 \times 10^6 \times 2\,000 = 3 \times 10^9$ daltons are required to code all these proteins, and one thus obtains the molecular weight of 3×10^9 indicated (page 284) for the DNA of *E.coli*.

The existence of a triplet code was confirmed by the study of mutations bringing about a *change in phase* of the translation *(frame-shift mutations)* and consisting of an addition or on contrary, a deletion (loss) of one or several nucleotides. It was indeed observed that if two mutations of opposite signs (e.g., an addition of one nucleotide and then a deletion of one nucleotide) take place sufficiently near one another (in the same gene), the correct translation is re-established after the second mutation, and if three mutations of the same sign (e.g., three additions of one nucleotide) take place in the same gene, the correct translation is re-established after the third mutation.

Translation is based on the same fundamental principle as the replication of DNA (or viral RNA) and the transcription of DNA into RNA: the *pairing of complementary bases*. It involves the *t*RNAs, or more exactly the aminoacyl-*t*RNAs, since each *t*RNA carries the amino acid which corresponds to it. Each *t*RNA contains a particular nucleotide triplet, situated in a single-stranded loop (see fig. 6-44) and whose bases are therefore not paired; this triplet is called *anticodon*, precisely because it is susceptible to pair with a nucleotide triplet of complementary sequence (or codon) present in the *m*RNA. Thus, ***each codon of the mRNA can be "read" (or translated) correctly because it is recognized only by the anticodon of the corresponding aminoacyl-tRNA***. It is this specificity of pairing, between successive codons of the *m*RNA and anticodons of the corresponding aminoacyl-*t*RNAs which ensures the ordered incorporation of amino acids into the protein. The *t*RNA therefore plays the role of a molecular *adaptor*, i.e. it binds an amino acid and brings it to the growing peptide chain, in order that it is incorporated at the correct place specified by the corresponding codon.

The *direction of synthesis of polypeptide chains* was determined by Dintzis who followed as a function of time, the course of incorporation of tritiated leucine in the α and β chains of the hemoglobin synthesized by reticulocytes in suspension (at low temperature, to slow down protein synthesis). At regular intervals (between 4 and 60 minutes), he isolated the hemoglobin, separated the α and β chains, subjected them to a hydrolysis by trypsin, fractionated the peptides obtained and determined their radioactivity. Knowing the position of peptides in the globin chains, he could observe that the peptides near the C-terminal end are the first to be labeled (after 4 minutes exposure to ^{3}H leucine), and that in course of time the number of radioactive peptides increases from the C-terminal end towards the N-terminal end of the polypeptide chains, with the result that after 60 minutes, the radioactivity of all peptides containing leucine is somewhat equivalent. The fact that the peptides near the C-terminal end are the first to be labeled indicates that they are synthesized last; for short exposures to ^{3}H leucine, the labeled amino acid is incorporated only at the end of the chain of almost completed globin molecules and the only complete molecules which are labeled are those which contain the radioactive amino acid in the last peptides formed (during these short exposures some ^{3}H-leucine is also incorporated at the beginning or in the middle of the chain, but these molecules have no time to be completed and are lost during the isolation of the complete hemoglobin). From these experiments it could be deduced that the peptide chains are synthesized from the N-terminal end towards the C-terminal end.

A. — Elongation of the polypeptide chain

Figure 6-46*a* represents a polypeptide chain in formation with the first two amino acids already linked to one another by a peptide linkage. The second amino acid is still bound by its carboxyl to the *t*RNA no. 2 responsible for its transfer. This *t*RNA no. 2 is also linked, on the one hand to codon no. 2 of the *m*RNA (by its anticodon), and on the other hand, to the ribosome in a site called "*peptidyl*" because it is the site normally reserved for the peptidyl-*t*RNA (by opposition to the "*aminoacyl*" site meant to receive the aminoacyl-*t*RNA). The A site (*aminoacyl*) is also called acceptor site, while the P site (*peptidyl*) is also called D site (donor). It will be noted that the NH_2 group of the first amino acid is not involved in the first peptide bond (it is its carboxyl group which is linked); actually, (as seen above) the protein chain is synthesized from its N-terminal end to its C-terminal end. For the chain to lengthen by one unit (i.e. by one amino acid), a number of reactions must take place, and 3 steps may be distinguished:

a) **Attachment of the aminoacyl-*t*RNA.** – The next aminoacyl-*t*RNA is selected thanks to its anticodon which must pair with codon no. 3 of the *m*RNA. Besides, it binds to the "aminoacyl" site of the ribosome (see fig 6-46*b*). The binding requires on one hand, GTP and on the other hand, an elongation factor EF-T which in reality, consists of 2 factors differing by their stability to heat: EF-Ts which is stable and EF-Tu which is unstable.

The factor EF-Tu forms with GTP a binary complex EF-Tu-GTP which binds the aminoacyl-*t*RNA to give an aminoacyl-*t*RNA-EF-Tu-GTP complex. This complex will link with the ribosome, permitting the binding of the aminoacyl-*t*RNA to the A site and the hydrolysis of GTP, which liberates a binary complex EF-Tu-GDP. The factor EF-Ts displaces the GDP from this complex, thus forming EF-Tu-EF-Ts (i.e. the complete EF-T factor), then EF-Ts is in its turn displaced by GTP, which regenerates the active form EF-Tu-GTP (see diagram).

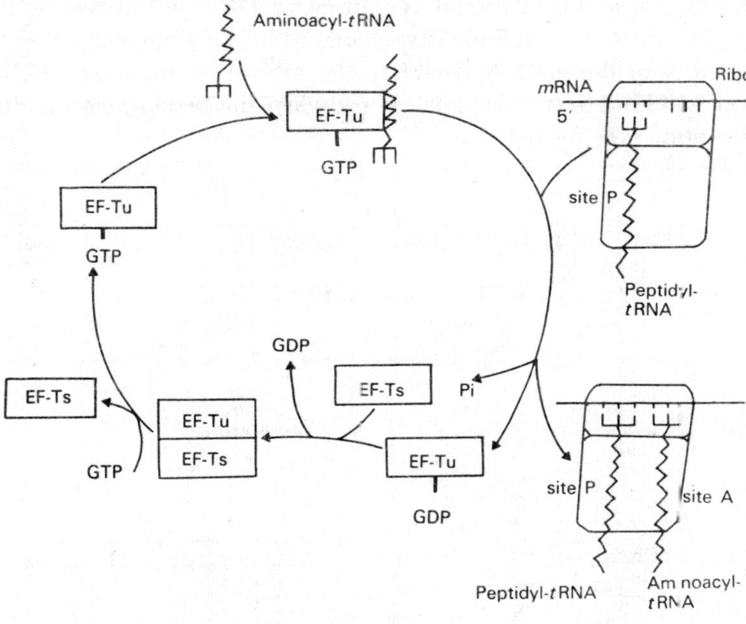

Diagram of the mechanisms of the attachment of the amino-acyl-*t*RNA.

Fidelity of translation implies the correction of errors if any (proof-reading). If an incorrect aminoacyl *t*RNA is bound to A site of the ribosome, it will be eliminated when GTP is hydrolyzed and the EF-Tu-GDP complex leaves the ribosome (the peptide linkage can form only after EF-Tu leaves aminoacyl-*t*RNA). It is in fact the GTPase activity of factor EF-Tu which determines the velocity of elongation of the protein chain and the fidelity of the process. The frequency of insertion of an erroneous amino acid is of the order of 10^{-4} per residue, which is compatible with the correct synthesis of a protein of about 1 000 amino acids (i.e. approximately 110 kdaltons). If EF-Tu-GDP remained longer on the ribosome (normally it remains on it a few milliseconds), more time would be available for the elimination of an erroneous aminoacyl-*t*RNA (i.e. an aminoacyl-*t*RNA whose anticodon is not paired in a stable manner to the codon which must be read) and the fidelity would be increased, but at the cost of the velocity of protein synthesis.

b) **Formation of the peptide linkage.** – The formation of the peptide linkage is catalyzed by a *peptidyltransferase* which is one of the proteins of the 50 S ribosomal particle (it is also called *peptide-synthetase*); it involves on one hand, the NH_2 of amino acid no. 3 and on the other hand the carboxyl group of amino acid no. 2 which is still linked to *t*RNA no. 2 by an ester linkage; this ester linkage will therefore break and the peptide formed will leave *t*RNA no. 2 and, lengthened by one unit, it will be attached to the *t*RNA no. 3, (see fig. 6-46*c*).

c) **Translocation.** – After the formation of the new peptide linkage, the tripeptide is carried by *t*RNA no. 3 which is still bound to the "aminoacyl" site. This peptidyl-*t*RNA must leave the "aminoacyl" site to occupy the "peptidyl" site and displace from it the deacylated *t*RNA no. 2 which binds transitorily to a third site on the ribosome (E site) before being ejected from the ribosome (it can then bind a new amino acid molecule). This translocation requires a protein factor – the EF-G factor ("Elongation Factor G") – and GTP (which is split into GDP + P_i) providing the energy required. A 3 nucleotide displacement of the ribosome brings it to codon no. 4 of the *m*RNA, enabling the arrival of aminoacyl-*t*RNA no. 4 which will bind to the "aminoacyl" site which has become vacant after the translocation (see fig. 6-46*d*).

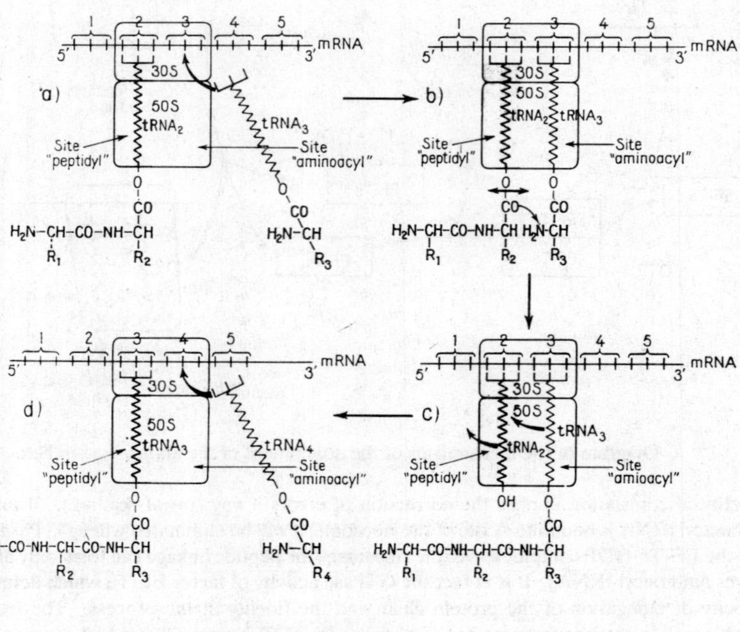

FIG. 6-46. — Elongation of the polypeptide chain.

The mechanisms of elongation of the peptide chain are almost the same in the eucaryotic cell, where one finds the eucaryotic elongation factors eEF-1 and eEF-2 which correspond respectively to the bacterial factors EF-T and EF-G. The eEF-2 factor required for translocation is inactivated by the diphteric toxin, which transfers the ADPR (adenosine-diphosphate ribosyl) group from NAD to eEF-2, and this ADP ribosylation is responsible for the lethal effect of the toxin.

B. – Cycle of factors required for protein synthesis

The number of factors of ribonucleic nature and protein nature, required for the mechanism of protein biosynthesis is very large (much greater than 100). We have already seen that in *E.coli* the translation of procaryotic *m*RNAs requires *t*RNAs (more than 70), aminoacyl-*t*RNA synthetases (about 20), ribosomes comprising three types of *r*RNAs and over fifty proteins; moreover, in addition to the elongation factors EF-T and EF-G just mentioned, are also needed initiation factors and termination factors which will be briefly described in the following part of this chapter (see pages 367 and 372).

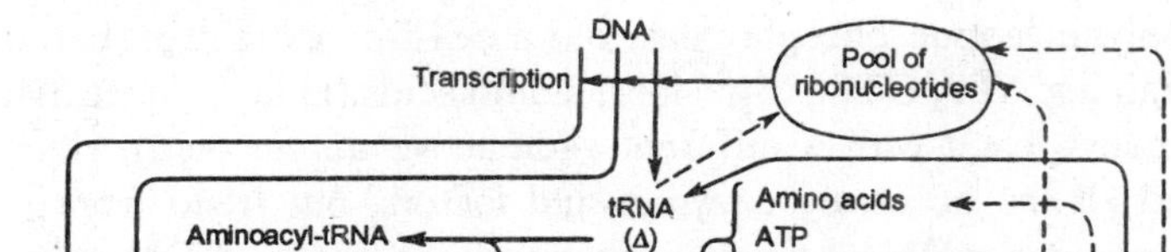

FIG. 6-47. — Cycle of the main factors necessary for protein biosynthesis. The degradation reactions (of ribonucleic acids and proteins) are indicated by dotted lines.

After their participation in the synthesis of a polypeptide chain these factors can be re-utilized a number of times before being degraded. This is illustrated in figure 6-47 which shows a polysome, i.e. a *m*RNA chain to which are bound several ribosomes, traveling from the 5′ to the 3′ end. The longer the distance traveled by the ribosome, in other words, the greater the region of the *m*RNA already translated, the longer the growing polypeptide chain. When the ribosome reaches the end of the message, the terminated protein chain is liberated; the ribosome detaches itself from the *m*RNA and dissociates into 30 S and 50 S sub-units which will join the pool of particles available for a new translation. As already mentioned the *t*RNAs are liberated, after delivering the amino acid they carried and after having for a moment, carried the growing peptide chain; they can then bind another amino acid. Lastly, after being translated a number of times, the messenger RNA will be enzymatically degraded into ribonucleotides; their longer life notwithstanding, the *r*RNAs and *t*RNAs will finally have the same fate, and these ribonucleotides can be used – with those synthesized *de novo* – for the synthesis of new RNA molecules.

C. – *Elucidation of the genetic code*

In 1961, Nirenberg and Matthaei carried out a decisive experiment: they added *polyuridylic acid or poly U* (a homopolynucleotide synthesized from UDP by polynucleotide phosphorylase) to a cell-free extract prepared from *E.coli* containing ATP, GTP_1, Mg^{2+} ions, amino acids (in fact, they carried out 18 experiments, each with a different radioactive amino acid), ribosomes, *t*RNAs and all the necessary enzymes and factors, but freed from the endogenous messenger RNA by pre-incubation in presence of DNase (thus by destruction of the template, *m*RNA can no longer be formed, and the pre-existing *m*RNA is rapidly catabolized). In these conditions they observed the formation of a particular polypeptide, consisting solely of phenylalanine residues (polyphenylalanine). If poly U codes for phenylalanine, it means that the *codon UUU is recognized by phenylalanyl-tRNA*. Subsequently, it was observed that poly C permits the formation of polyproline (CCC is therefore a codon of proline) and that poly A permits the formation of polylysine (AAA is therefore a codon specifying lysine). The use of copolymers, like poly UG, synthesized with the help of the polynucleotide phosphorylase could establish the probable composition of codons specifying a number of amino acids (e.g., 2U, 1G) but not their exact sequence.

A number of codons could be elucidated thanks to the use of *polyribonucleotides of alternate dinucleotide or trinucleotide sequences*, synthesized by Khorana and his team by means of DNA-polymerase and RNA-polymerase. Thus a polymer consisting of alternate cytidylic and adenylic residues, CpApCpApCpApCpApC... comprises only 2 codons, CAC and ACA, whatever the place where translation begins, and in a cell-free extract it directs the synthesis of a polypeptide comprising histidyl and threonyl residues in alternance. On the contrary, a polyribonucleotide of alternate trinucleotide sequence like *ApAp-GpApApG...ApApG* directs the synthesis of 3 different homopolypeptides: polylysine if the reading begins at the first A (codon AAG), polyarginine if the reading begins at the second A (codon AGA), polyglutamic acid if the reading begins at G (codon GAA). The fact that only homopolypeptides are obtained shows that translation, once started, proceeds sequentially, i.e. the nucleotides are read regularly in sets of three, till the end, without leaving out any nucleotide.

Codon sequence determination made great advances thanks to another method developed in Nirenberg's laboratory; it does not involve any protein synthesis, but only the formation of the specific complex between an aminoacyl-*t*RNA, the corresponding codon and the ribosomes. The formation of the specific complex takes place in presence of trinucleotide (e.g., UUU) and it is not necessary to use an oligo- or polynucleotide or to add GTP. The principle of the technique is very simple: a given triplet, a ^{14}C aminoacyl-*t*RNA and ribosomes are mixed and then filtered through a nitrocellulose membrane (millipore); the *^{14}C aminoacyl-tRNA is retained only if its anticodon has paired with the triplet*, but if it does not recognize this codon, it passes through the filter; therefore one simply measures the radioactivity present in the complex, i.e. retained on the filter. In this manner, one can determine the amino acid specified by a given codon, by determining – among the 20 aminoacyl-*t*RNAs – the one which can pair with the triplet studied.

We mentioned that the *m*RNAs are not only synthesized in the direction 5′⟶3′ but also translated in this direction. This latter point could be confirmed experimentally by introducing in a cell-free system the polynucleotide $_5$′*p*A (*p*A)$_n$ *p*A *p*C$_3$′ and studying the polypeptide formed in response to this message. A polylysine is formed (having therefore a lysine at the N-terminal end), but it has an asparagine residue in C-terminal position. This result is in conformity with a *translation of the mRNA in the direction 5′⟶3′*. The fact that *m*RNAs are translated in the direction 5′⟶3′ enables the translation to begin before the completion of the *m*RNA synthesis (i.e. a transcription-translation coupling) in procaryotes (see page 333). If the *m*RNAs were translated in the reverse direction only the terminated molecules could be translated.

All these results are summarized in the genetic code table (see fig. 6-48). This table calls for some remarks.

There are 61 codons having a meaning or sense (i.e. specifying an amino acid) and 3 non-sense codons (which will be studied in the following). Besides, it is observed that amino acids can be coded by several codons; there are two exceptions, tryptophan which is coded only by UGG and methionine coded only by AUG. Some amino acids like leucine, serine, arginine can be specified by 6 different triplets. It must be noted that the triplets YZU and YZC always correspond to the same amino acid; in most cases the same is true of codons YZA and YZG; there are only two exception: UG purine and AU purine; in many cases (8 boxes out of 16) the same amino acid is specified, whatever the

3rd letter of the codon (e.g., in the box UC: UCU, UCC, UCA and UCG, all code for serine). This suggests that the first two bases are of capital importance in the pairing. This is a factor of *genetic stability*, because a point mutation (replacement of a base by another) at the third position will often have no effect on the nature of the specified amino acid ("silent" mutation), especially if it is a transition, i.e. the replacement of a purine by another, or a pyrimidine by another (the replacement of a purine by a pyrimidine or vice versa is called a transversion).

1st ↓ / 2nd →	U	C	A	G	3rd ↓
U	PHE	SER	TYR	CYS	U
	PHE	SER	TYR	CYS	C
	LEU	SER	Non-sense	Non-sense	A
	LEU	SER	Non-sense	TRP	G
C	LEU	PRO	HIS	ARG	U
	LEU	PRO	HIS	ARG	C
	LEU	PRO	GLN	ARG	A
	LEU	PRO	GLN	ARG	G
A	ILE	THR	ASN	SER	U
	ILE	THR	ASN	SER	C
	ILE	THR	LYS	ARG	A
	MET (F-MET)	THR	LYS	ARG	G
G	VAL	ALA	ASP	GLY	U
	VAL	ALA	ASP	GLY	C
	VAL	ALA	GLU	GLY	A
	VAL (F-MET)	ALA	GLU	GLY	G

FIG. 6-48. — The genetic code.

Several triplets therefore correspond to one amino acid; it is said that the code is *degenerate*. One might wonder whether these diverse codons are recognized by a single *t*RNA or by several isoacceptor *t*RNAs. In fact, the same *t*RNA can often recognize several codons, by virtue of a certain flexibility, or wobble, in the pairing of the third nucleotide of the codon.

The wobble hypothesis, originally advanced by Crick, has been confirmed; it was indeed found that the first nucleotide of the anticodon can, in many cases, recognize several nucleotides at the 3rd position of the codon:

1st nucleotide of the anticodon		3rd nucleotide of the codon
C	can pair with	G
A	"	U
U	"	A,G
G	"	C,U
I	"	C,U,A

For example yeast *t*RNA^{ala} (see fig. 6-44) which has an I at the first position of the anticodon (IGC) can recognize the alanine codons GCU, GCC and GCA, and a *t*RNA^{phe} having an anticodon GAA can read the two phenylalanine codons UUU and UUC as may be seen in the diagram below. (it must be noted that the sequences are written in the direction 5'⟶3' and that the sequences of the corresponding codons and anticodons are antiparallel.) Other modified nucleotides are sometimes present at the first position of the anticodon of some *t*RNAs and can influence the recognition of codons. Lastly, in certain cases, it appears that the pairing of two nucleotides out of three could suffice, especially in the case of 2GC pairs (which are more stable than the other pairs). this mechanism is called "*two out of three*".

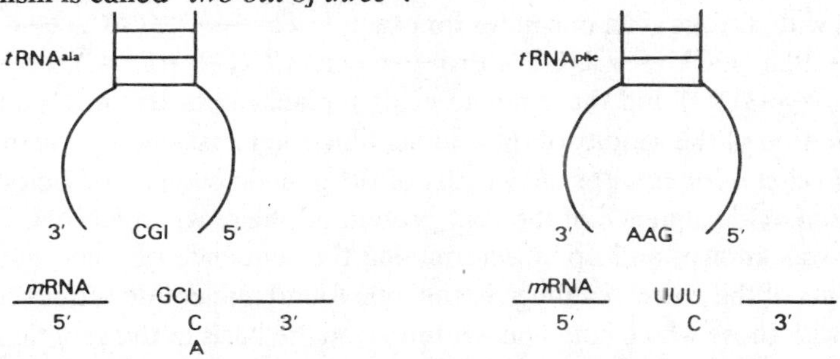

But the synonymous codons which differ at the level of one of their first two nucleotides must be read by different anticodons, therefore, by different *t*RNAs; this is the case for example for the leucine codons UUA and CUA, the arginine codons CGA and AGA, and of course for the serine codons UCU and AGU etc. whose first two nucleotides are different.

The genetic code was therefore elucidated by experiments carried out *in vitro*, with cell-free systems or ribosomes from *E.coli*. But it was later found that this code is *universal*, from viruses and bacteria up to higher plants and animals (including man). It was however noted recently that mitochondria (in yeast, mammals) use a genetic code which differs in some cases from the code called "universal" (see page 379).

Indirect confirmations of the validity of the genetic code thus determined *in vitro* were obtained by the study of mutations – spontaneous or induced – and of their consequences in the amino acid sequence of proteins corresponding to the mutated genes.

Among the *spontaneous mutations*, were especially studied those affecting the genes coding for the globin chains in man. These mutations are generally point mutations, i.e. they consist of a modification of a single nucleotide (which is replaced by another) with the result that the initial codon is transformed into a new codon which differs from the former by only one nucleotide. While studying the substitutions of amino acids resulting from these mutations, it was observed that they were indeed compatible with the replacement of a single nucleotide (in most cases, it is one of the four possible transitions, or A→U or

U→A transversions): thus in the abnormal human hemoglobins were noted the substitutions Gly⟶ Asp (which would result from a GGU⟶ GAU transition), Glu⟶ Lys (GAA⟶ AAA transition), His⟶ Tyr (CAU⟶ UAU transition), Asn⟶ Lys (AAU⟶ AAA transversion), etc. and an indirect confirmation of these codons was thus obtained.

Among the *induced mutations* were studied – particularly in the case of tobacco mosaic virus – those caused by nitrous acid which is known to act by causing a deamination of bases: cytosine is deaminated into uracil, adenine is deaminated into hypoxanthine (which pairs with C, with the result that this modification is actually equivalent to a A⟶ G transition). While studying the substitutions of amino acids in the capsid protein in the progeny of viruses treated with nitrous acid, one notes for example Thr⟶ Ile (ACA⟶ AUA), Ser⟶ Phe (UCU⟶ UUU), Pro⟶ Leu (CCC⟶ CUC), Ile⟶ Val (AUU⟶ GUU) and these amino acids replacements also give an indirect confirmation of the validity of the codons which were assigned to them.

A *direct confirmation* of the validity of the genetic code could be obtained: the amino acid sequence of the coat protein of phage R_{17}, a RNA-containing phage, was known, and upon determining the sequence of some nucleotide fragments of the corresponding cistron, one found nucleotide sequences coinciding with those which could be written – on the basis of the genetic code – from the amino acid sequences. It is interesting to note that the determination of the nucleotide sequence of the RNA of phage MS_2 has shown that in such simple genomes all the 61 sense codons are utilized.

With the development of the DNA sequencing techniques it then became possible to verify, in the case of numerous genes, that *the nucleotide sequence of the coding strand of the DNA corresponded to the amino acid sequence of the protein coded by the gene concerned.*

The RNAs of RNA-containing phages and viruses which can be prepared from viral particles, are pure messenger RNAs and it is not surprising that the *first protein biosynthesis experiments in vitro directed by natural RNAs were carried out with phage RNAs*. They allowed, in a cell-free system of bacterial origin, the synthesis of proteins specific of the phage. It is on the contrary much more difficult to isolate a particular messenger RNA of a higher organism because – as seen in the foregoing (page 355) – the totality of *m*RNAs represents less than 5% of the cellular RNAs and besides, they consist of a mixture of several hundreds (or even more) of different types of molecules.

However, using specialized cells or tissues which produce only a very limited number of different proteins (for example, reticulocytes for globin, some tumours for immunoglobulins, the eye-lense for crystallins, the oviduct for ovalbumin), one could isolate and purify mRNAs and obtain their translation into the corresponding proteins, either by introducing them in a cell-free system (prepared generally from reticulocytes or wheat germ), or by injecting them into *Xenopus* oocytes. In the case of cells containing a large number of different *m*RNAs, when one wants to obtain one type of *m*RNA present in small quantities, one can selectively precipitate the polysomes containing this *m*RNA with the help of antibodies raised against the corresponding protein.

Lastly, it may be noted that ambiguities and errors can be observed during translation when organic solvents, an excess of Mg^{2+} ions or streptomycin, is added to the cell-free system; in presence of streptomycin in particular, poly U stimulates not only the incorporation of phenylalanine, but also that of isoleucine, leucine, serine and tyrosine, when the cell-free system is prepared from streptomycine-sensitive bacteria. It was possible to determine the difference between bacteria sensitive and resistant to this antibiotic; this difference resides in one of the proteins of the 30 S ribosomal particle, protein S_{12} whose sequence is modified in the resistant mutants.

D. – Initiation of the peptide chain

It was observed that in *E.coli* there are 2 *t*RNAs specific of methionine, $tRNA_M^{met}$ and $tRNA_F^{met}$, which can be purified by conventional methods of fractionation of isoacceptor *t*RNAs (bidimensional electrophoresis on polyacrylamide gel, column chromatography). Both these *t*RNAs can be aminoacylated into methionyl-*t*RNA; but only methionyl – $tRNA_F^{met}$ can – as indicated by the letter "F" – be formylated (on the amino group of methionine) by N_{10}-formyl-tetrahydrofolic acid in presence of a *transformylase* and it incorporates methionine *at the beginning of the chain*, in response to the codon AUG (or GUG). As for the methionyl – $tRNA_M^{met}$, it incorporates methionine *inside the chains*, and only in response to the codon AUG.

Proteins synthesized in a cell-free system prepared from E.coli begin with N-formyl-methionine while methionine (non formylated) represents only 40% of the N-terminal residues of ribosomal proteins isolated from *E.coli* (it is however the dominant amino acid at this position); this suggests the existence *in vivo* of an enzyme which deformylates methionine (deformylase) and of an enzyme capable of eliminating methionine (exopeptidase).

N-formyl-methionyl-tRNA was also shown to be present in other bacteria and also in mitochondria and chloroplasts; this represents one more similarity between organelles and bacteria (we have already seen that organelles have 70 S ribosomes, like bacteria). On the contrary, no N-formyl-methionyl-*t*RNA is found in the cytoplasm of eucaryotic cells; the initiator *t*RNA is a *methionyl-*$tRNA_M^{met}$ (non-formylated but formylatable *in vitro* with the help of a bacterial transformylase), different from the methionyl-$tRNA_M$ responsible for the incorporation of methionine inside the chains. Here again, an enzyme can cleave off methionine which is then no longer found in N-terminal position in the completed protein.

In bacteria, there are *3 initiation factors* IF-1, IF-2 and IF-3 which participate in the formation of the initiation complex by binding to the 30 S particle. The binding of GTP to IF-2 enables the *m*RNA and the initiation *t*RNA to join the complex while IF-3 is liberated. We then have binding of the 50 S particle, with hydrolysis of GTP and liberation of IF-1 and IF-2. This hydrolysis of GTP takes place thanks to proteins L7 and L12 (see page 354) which also participate in the hydrolysis of GTP during the elongation of the protein chain. The 30 S particle participates in the formation of the initiation complex by binding to the *m*RNA on a ribosome binding site situated near the initiation codon AUG. The formation of this complex implies pairing between an oligonucleotide sequence situated at the 3′ end of the 16 S RNA in the 30 S particle and a complementary sequence situated on the *m*RNA upstream the initiation codon AUG (thus differentiating this initiation AUG from an internal AUG). Formyl-methionyl-$t\mathrm{RNA}_F^{met}$ joins the complex by binding at the P site (while other aminoacyl-*t*RNAs can only bind at the A site of the ribosome), with the result that the A site will be able to receive aminoacyl-*t*RNA no. 2 in order to permit the formation of the first peptide linkage.

If the ribosomes are allowed to bind to the *m*RNA but the elongation is blocked, the ribosomes remain in the initiation site, and if a ribonuclease is then made to act, it will spare the *m*RNA region involved in the initiation complex. This region can be isolated and studied, and it is observed that the bacterial ribosome protects a region of about 35 to 40 nucleotides, comprising two sequences common to all *m*RNAs: the initiation codon AUG (or GUG) and a sequence which has a complementarity with a hexameric sequence present near the 3′ end of the 16 S RNA:

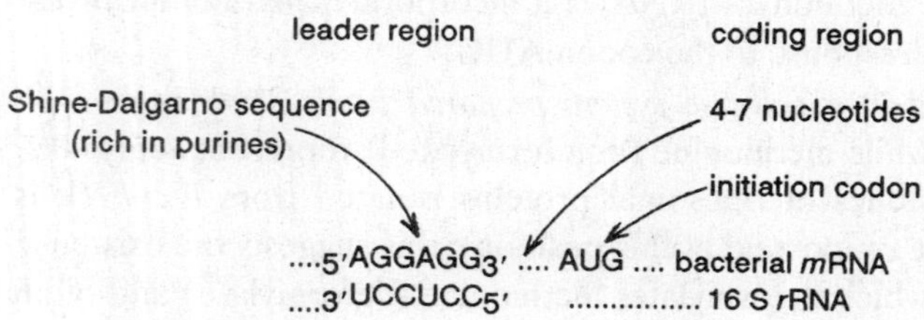

The sequence identified by Shine and Dalgarno is constantly present upstream the initiation codon in the procaryotic *m*RNAs; this is an argument in favour of its participation in the initiation process of protein biosynthesis. Besides, one knows a mutation in a gene of phage T7, where the terminal triplet AGG of this sequence is changed into AAG, and the binding of ribosomes to the *m*RNA is affected in this mutant. The study of a large number of Shine-Dalgarno sequences showed that the number of base pairs between the *m*RNA and the 16 S *r*RNA can vary from 3 to 9.

The 3′ region of 16 S *r*RNA seems to be involved in the initiation because it is very well conserved in bacteria, it is the target of kasugamycin (an inhibitor of initiation), and it is cross-linked with the initiation factors. This region can form a hairpin by intrachain base pairing. The hexameric sequence (UCCUCC) is involved, either in this intrachain pairing, or in the pairing with the *m*RNA (this is a temporary pairing, broken after the initiation step, which then allows the ribosome to move along the *m*RNA).

In most cases, the ribosomes can bind independently at the beginning of each cistron of a polycistronic mRNA. Thus, the tryptophan operon of *E.coli* for example, is transcribed into a *m*RNA of about 7 000 nucleotides coding for 5 enzymes which catalyze various steps of the pathway of biosynthesis of this amino acid and for each of these 5 proteins there is on the *m*RNA an initiation signal (and a termination signal) of translation. But since a ribosome covers a *m*RNA length of about thirty nucleotides (as shown by experiments of protection of the RNA in the *m*RNA-ribosome initiation complex, against the attack by the pancreatic RNase), one can imagine that when the intercistronic region is short (and all the more so when there is overlapping) the 30 S particle of a ribosome terminating the translation of a cistron does not leave the *m*RNA and can immediately initiate the translation of the next cistron. In some cases, the blocking of the translation, after the appearance of a non-sense codon in a polycistronic *m*RNA (see page 370) affects not only the expression of the mutated gene, but also that of the next (distal) cistrons; this phenomenon called polarity (or polar effect) is mainly due to an indirect effect at the level of transcription which is interrupted downstream the mutation, but there can be a polar effect due to the impossibility for the ribosomes to re-initiate the translation of the next cistrons.

In eucaryotes, near the initiation codon, there is apparently no sequence permitting, by base complementarity, the attachment of the small 40 S ribosomal sub-unit; the latter binds to the 5′ end of the *m*RNA and then moves up to the initiation site of protein synthesis. On the other hand, the eucaryotic initiation factors (called *e*IF) are more numerous (nine have been identified in red blood cells), and some are oligomeric, particularly *e*IF-2 and *e*IF-3. The sub-unit α of the *e*IF-2 factor can be phosphorylated by a kinase (which inhibits the initiation of protein synthesis), either when there is heme deficiency in the red blood cells, or due to the effect of a double stranded RNA (or of interferon, which is itself induced by double stranded RNAs and inhibits the synthesis of viral proteins). This is an example of the control of gene expression at the level of translation (see page 468).

The 3′ end of the small eucaryotic *r*RNA (18 S) presents a great similarity with that of the small procaryotic *r*RNA (16 S): in both cases the 3′ region can form, by intrachain pairing, a hairpin which comprises two adjacent N_6 dimethyl A. On comparing the twenty nucleotides separating these two methylated residues from the 3′ OH end of the RNA, an almost complete homology is observed with only two differences: a dinucleotide sequence UU in bacteria is replaced by a AU or AA sequence in eucaryotes, and in eucaryotes there is a deletion of the pentanucleotide CCUCC (i.e. the quasi totality of the sequence complementary to the Shine-Dalgarno hexanucleotide sequence).

The eucaryotic ribosomes do not bind to the *m*RNAs (which are, it may be recalled, mostly monocistronic) at an initiation site located just upstream the coding region, but they recognize the methylated cap, situated at the 5′ end of the non-translated leader region (whose length is generally less than 100 nucleotides), thanks to factors called cap binding proteins. The cap is therefore necessary, and the uncapped *m*RNAs (obtained by blocking cap formation or by enzymatic cleavage of the cap) are not translated efficiently but for very rare exceptions. Some viral *m*RNAs (in the case of poliovirus, for example) have no cap and inhibit the translation of the *m*RNAs of the host cell by blocking the cap binding proteins.

When the leader region is short (less than 40 nucleotides) it may be imagined that the ribosome covers both the cap and the initiation AUG codon. But when the distance is greater (e.g., it can reach 200-300 nucleotides), it is believed that the 40 S particle, after recognizing the cap, migrates along the *m*RNA till it meets the initiation codon. The internal initiation sites, for example in the case of some viral *m*RNAs containing more than one cistron, do not seem to be recognized until a cleavage creates a new 5′ end near the initiation codon.

E. – Termination of the peptide chain

When poly U is used as messenger in a cell-free system, the polyphenylalanine chains synthesized remain bound to the last *t*RNA and are not liberated from the polysomes. Poly U has no termination signal, a signal which natural messengers must have in order to permit the liberation of polypeptide chains.

Our knowledge on the termination mechanisms benefited in a large measure from studies carried out on the premature, accidental, termination of the synthesis of a protein taking place because of a *non-sense mutation*; this is a mutation which – after transcription of the cistron where it occurred – brings about the appearance in the *m*RNA of one of the 3 non-sense codons, UAG (amber), UAA (Ochre) or UGA (see fig. 6-48); these codons are called "non-sense", because normally, the cell has no *t*RNA with an anticodon able to pair with any one of these 3 codons; such a codon is therefore not translated, and the synthesis of the corresponding protein is blocked at this level. Thus, the RNA of a non-sense mutant of the bacteriophage f_2, having a UAG codon at the seventh position in the cistron coding for the coat protein, permits only – *in vivo* as well as *in vitro* – the synthesis of a hexapeptide (fmet – ala – ser – asn – phe – thr). The consequences, for the bacterial cell for example, of a non-sense mutation depend on the one hand, on the role – essential or otherwise – of the protein coded by the mutated gene, and on the other hand, on the position of the mutation in this gene. For example, a non-sense mutation in the gene of β-galactosidase will have no consequence if the bacterium grows in a glucose-containing medium, because in these conditions the gene is not transcribed (see fig. 8-4). A mutation in the *i* gene of the lactose operon (see page 445) will block the synthesis of the repressor with the result that the enzymes of the lactose operon will always be synthesized even in the absence of the product to be metabolized (lactose). On the contrary a non-sense mutation in the gene of the RNA polymerase will be lethal. The position of the mutation also has some importance, because one can imagine a non-sense codon near the 3′ end of the cistron, permitting the synthesis of a protein lacking only a few amino acids at its C-terminal end which may not be essential for the biological activity of this protein.

In any case, a non-sense mutation can be corrected in two ways:

– by a second mutation in the *same gene*, more precisely in the same triplet, transforming the non-sense codon into a sense codon (see fig. 6-49a). This is a *reversion*; there can be a true reversion if the second mutation brings back the codon as it existed before the first mutation and in that case the protein will be necessarily functional; but a new sense codon may also appear, different from the one which was present before the first mutation, and in that case the protein will be biologically active only if the amino acid coded by this new triplet is compatible with biological activity.

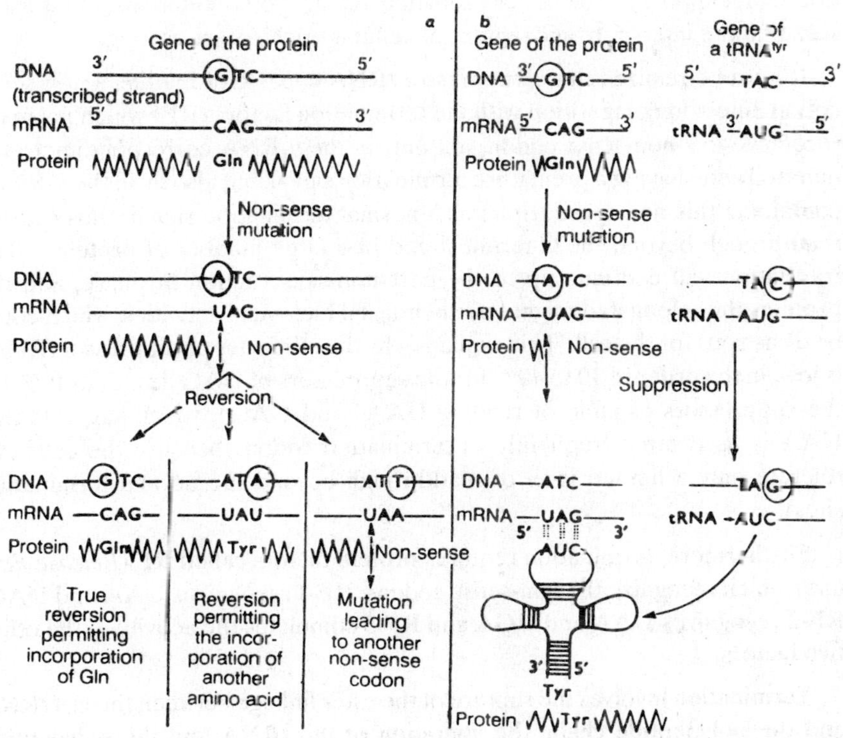

FIG. 6-49. — Non-sense mutation taking place in the gene of a protein. Possibility of correction by two mechanisms: reversion (a) and suppression (b).

– by a second mutation in an altogether different gene, the *gene of a tRNA*, the effect of which is more precisely a modification of the anticodon of this *t*RNA which becomes capable of pairing with the non-sense codon (fig. 6-49b). The non-sense codon therefore remains in the *m*RNA of the protein, but now it can be read. This is a *suppression*; in this case the protein will be active only if the amino acid brought by the suppressor *t*RNA is compatible with biological activity.

The natural termination of protein synthesis uses non-sense codons but it appears that the signal is not always limited to a single non-sense triplet. It is logical that a more elaborate signal was selected during evolution, because a single triplet presents the serious disadvantage of easy conversion into a sense codon by a simple point mutation (replacement of one base by another), which would bring about the appearance of a chain consisting of two proteins fused together, and which would perhaps have the properties of neither (so that 2 enzymes would be missing in the cell). It has been possible to determine the sequence of an inter-cistronic region in the RNA of phage R_{17}. It comprises the 2 non-sense codons UAA and UAG, then an initiation codon AUG followed by 6 sense codons, and lastly the third non-sense codon UGA. Such a termination signal appears to be better protected against point mutations, but it is not sure that the inter-cistronic regions of cellular *m*RNAs are as long.

It must be realized that a suppressor *t*RNA capable of reading a non-sense codon enters in competition with the termination factors (RF) which normally recognize this non-sense codon, not only in the *m*RNA corresponding to the mutated gene having a premature termination signal, but also in all the *m*RNAs containing this non-sense triplet as a normal termination signal, thus causing readthrough beyond the C-terminal end of a large number of proteins. The translation will continue up to the next non-sense codon in phase, and the proteins thus elongated might lose their usual biological activities, which could be disastrous for the cell. This explains why the efficiency of suppressor *t*RNAs is low, in the order of 10 to 40% for the suppressors of UAG, less than 10% for the suppressors capable of reading UAA (and UAG), which suggests that UAA is used more frequently as termination codon (because the cells can tolerate only a limited level of readthrough beyond the normal termination sites).

Furthermore, termination requires protein factors called RF (*Release Factors*) which recognize the non-sense codons: RF-1 recognizes UAA and UAG, RF-2 recognizes UAA and UGA and RF-3 stimulates the activity of the other two factors.

Termination involves the rupture of the ester linkage between the last *t*RNA and the polypeptide chain, the liberation of this *t*RNA and the polypeptide chain and the dissociation of the 70 S ribosome into 50 S and 30 S particles which necessitates factor IF-3.

F. — Inhibition of protein synthesis

In the binding of amino acids to the corresponding *t*RNAs, an antibiotic is known, *borrelidine*, which specifically inhibits the aminoacylation of *t*RNA^{thr}.

Puromycin is an antibiotic which presents a great structural similarity with the 3′ end of a tyrosyl-*t*RNA (see fig. 6-50); this enables it to react with a peptidyl-tRNA during the elongation of the peptide chain and to displace the *t*RNA to form a peptidyl-puromycin. But since puromycin comprises an amide linkage and not an ester linkage between the carboxylic group of para-methoxy-phenylalanine (or methylether of tyrosine) and the amino group of the pentosamine, this peptidyl-puromycin can no longer allow elongation to continue and is released from the ribosome.

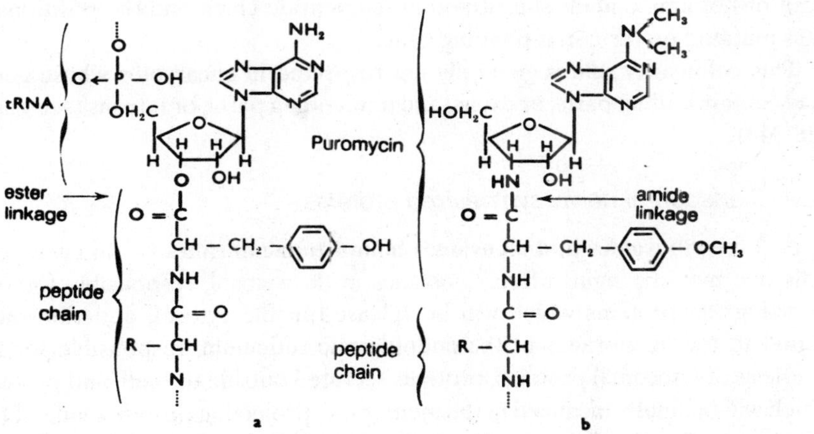

FIG. 6-50. — Structures of a peptidyl-*t*RNA (a) and a peptidyl-puromycin (b).

Chloramphenicol inhibits protein synthesis on the 70 S ribosomes of procaryotes, mitochondria and chloroplasts, but not that taking place on the 80 S ribosomes of the cytoplasm of eucaryotes. On the contrary, *cycloheximide* inhibits protein synthesis on 80 S ribosomes, but not that taking place on 70 S ribosomes (procaryotes, mitochondria, chloroplasts). As regards sensitivity to antibiotics as well as size of ribosomes (see pages 353-354) and use of a formyl-methionyl-*t*RNA$_{M}^{met}$ for the initiation process (see page 367), protein biosynthesis in mitochondria and chloroplasts resembles more the process taking place in procaryotic organisms than the one occurring in the neighbouring cytoplasm, which supports the hypothesis of an endo-symbiotic origin of these organelles; a number of arguments do suggest that mitochondria and chloroplasts could have derived from procaryotic organisms.

G. — Colinearity of the gene and the peptide chain

Studies carried out on the mutations consisting of an addition or deletion of nucleotides, and particularly, the observation that three mutations of the same sign taking place in the same gene restore a translation in phase (see page 357) had already provided arguments in favour of the colinearity of the gene and the corresponding polypeptide chain.

This colinearity was confirmed by studies carried out on non-sense mutations in the gene coding for the protein of the head of phage T_4; Brenner did show that there is a correlation between the length of the peptide chains formed and the position on the gene of codons mutated into non-sense codons. Yanofsky and co-workers carried out a similar study on mis-sense mutations (sense codon transformed into another sense codon specifying a different amino acid) that occurred in one of the chains of *E.coli* tryptophan synthetase which is 267 amino acids long and whose sequence was determined; they observed, in about 20 different mutations, a correspondence between the positions of amino acids substituted in the peptide chain and the positions of sites mutated on the corresponding gene.

This colinearity rule is generally not respected in eucaryotes whose genes may contain coding parts, or exons, and non-coding parts, or introns (see pages 338-343).

H. – *Transport of newly synthesized proteins*

a) **Free ribosomes and ribosomes bound to membranes.** – In eucaryotic cells one may distinguish free ribosomes in the cytosol, responsible for synthesizing the proteins which will be released in the cytosol, and ribosomes bound to the membranes of the endoplasmic reticulum, responsible for the synthesis of lysosomal proteins, proteins secreted outside the cell, and proteins which will be finally localized in the membrane, projecting on either side of this membrane. But the ribosomes themselves do not differ, they are free or bound to membranes depending on the protein they synthesize.

b) **Signal sequence or signal peptide.** – A protein destined to cross the membrane of the endoplasmic reticulum binds to this membrane thanks to a *signal sequence* of fifteen to thirty amino acids, present at the N-terminal end of the nascent chain, but which is then cut by a *signal peptidase* and is therefore absent in the protein once secreted. One knows at present, the signal sequence of more than hundred secreted proteins originating from diverse eucaryotic organisms; some common characteristics were observed: there is at least one residue having a positive charge at the N-terminal end, a hydrophobic part of 10 to 15 residues at the centre of the sequence (comprising leu, ile, vale, phe residues), and a more polar sequence of about 5 residues upstream the cleavage site. However, some secreted proteins have an internal signal sequence (e.g., ovalbumin) not situated at the N-terminal end. The addition of a signal sequence to a normally cytosolic protein enables the latter to be transported through the membrane of the endoplasmic reticulum, as shown by experiments where the signal sequence of β-lactamase of *E.coli* (an enzyme secreted in the periplasmic space between the external membrane and the inner membrane of the bacterium) was grafted at the N-terminal end of α-globin.

c) **Signal sequence recognition particle.** – This particle (called *SRP*: *Signal Recognition Particle*) consisting of a 7S RNA (305 nucleotides) and 6 different proteins binds to a ribosome comprising a nascent protein chain having a signal sequence and will recognize, on the membrane of the endoplasmic reticulum, a receptor of SRP. The ribosome carrying the nascent protein chain is thus brought to the level of the translocation system comprising especially 2 membrane proteins, the ribophorins I and II, while the SRP having done its work, returns to the cytosol. The elongation of the protein chain then continues with the ribosome fixed to the membrane of the endoplasmic reticulum.

The proteins to be secreted and the lysosomal proteins completely pass through the membrane of the endoplasmic reticulum. On the contrary, other proteins must form part of a membrane: some have only one helical region across the membrane, others have several; some have their N-terminal end projecting out of the membrane on the cytosolic side and their C-terminal end projecting on the extracellular side, for other proteins it is the reverse.

d) **Glycosylation of proteins at the level of the endoplasmic reticulum.** – At the level of the endoplasmic reticulum, the signal sequences are eliminated by the signal peptidase and a large number of proteins undergo glycosylations to become glycoproteins. As indicated in the chapter on Carbohydrates (see pages 162-164) glycoproteins contain oligosaccharide units bound, either by O-glycosidic bonds to the hydroxyl of a serine or threonine, or by N-glycosidic bonds to the amide group of asparagine. We have also seen how these units are transferred to the polypeptide chain thanks to an activated lipidic carrier, dolichol phosphate (see pages 208-211). Tunicamycin is an analogue of UDP-N-acetylglucosamine which blocks the fixation of N-acetylglucosamine on dolichol phosphate and therefore prevents the glycosylation of proteins.

e) **Transport of proteins to the Golgi apparatus.** – Proteins are transported by transfer vesicles from the cytoplasmic reticulum to the Golgi apparatus where modifications of oligosaccharide units takes place (see pages 208-211), as also a sorting of proteins which will be sent either to the plasmic membrane, or to the lysosomes, or to the secretory granules (or vesicles). Various vesicles perform the transfer of proteins from one compartment of the Golgi apparatus to the other (compartments cis, median and trans are distinguished) and from the trans compartment of the Golgi apparatus to various destinations (plasmic membrane, lysosomes, secretory granules).

f) **Transport of proteins from the Golgi apparatus to their final destination.** – Some enzymatic proteins (particularly hydrolases) undergo phosphorylation at the position 6 of several of their mannose residues; this forms a signal enabling their transport from the Golgi apparatus to the lysosomes (see page 211). A mucolipidosis, characterized by an accumulation of glycolipids not digested in the lysosomes, is due to the absence of mannose-6-phosphate in several hydrolases which are then exported instead of being sent to the lysosomes. In these patients, there is a lack of a specific phosphotransferase catalyzing the attack of the hydroxyl at position 6 of the mannose present in glycoprotein on the pyrophosphate bond of UDP-N-acetylglucosamine. The glycoproteins comprising mannose-6-phosphate bind to a protein receptor of the membrane of the Golgi apparatus, and vesicles containing the glycoprotein-receptor complex, detach themselves by budding of the trans compartment of the Golgi apparatus. After the glycoprotein has left the receptor and a phosphatase has dephosphorylated the mannose-6-phosphate, the transport vesicle can supply to the lysosomes, by fusion, the glycoproteins it contains.

On the contrary, it does not seem that for transport of other proteins, namely those which must be either secreted, or integrated in the plasmic membrane, particular monosaccharides serve as markers, and it is rather believed that elements of three-dimensional structure play a role in directing these glycoproteins to their correct destination.

g) **Transport of proteins to mitochondria and plasts.** – As will be seen in the following (see page 378), the mitochondrial genome codes only for a very small number of mitochondrial proteins (about twelve), with the result that the great majority is coded by the nuclear genome, synthesized in the cytosol and then imported into the mitochondria. The import of a protein into the mitochondrion requires the presence of a particular sequence (*pre-sequence*) at its N-terminal end, which must be recognized by receptors situated on the external membrane of the mitochondrion, but it appears that a certain destabilization of the three-dimensional structure of the protein is also necessary.

The passage through the external membrane therefore requires a pre- sequence, different from the signal peptide permitting entry in the endoplasmic reticulum (see above), and bringing residues having a positive charge and hydroxylated residues whose organization in space perhaps plays a decisive role.

The imported protein can remain integrated in the external membrane if the pre-sequence is followed by an anchoring sequence and a second sequence charged positively. But there are three other possible destinations, namely *the intermembrane space, the inner membrane and the matrix*, and for a protein to reach one of these three destinations there must be, on one hand, a protomotive force (see pages 137-139), because the addition of an uncoupling agent, like dinitrophenol, blocks the transport to these destinations and, on the other hand, a proteolytic cleavage of the amino-terminal pre-sequence. It appears that transport takes place at locations where outer membrane and inner membrane adhere to one another.

Techniques of genetic engineering have made possible the production of chimeric proteins; thus, when dihydrofolate reductase (a cytosolic enzyme) is preceded at its N-terminal end by the pre-sequence (61 amino acids) of cytochrome c_1, it is imported into the intermembrane space of the mitochondrion, i.e. at the place where the cytochrome c_1 is normally localized. But if this enzyme is preceded by the pre-sequence (27 amino acids) of alcohol dehydrogenase (an enzyme localized in the matrix), dihydrofolate reductase is found in the mitochondrial matrix.

The import of proteins into mitochondria (and plasts) is a post-translational process which may be performed *in vitro* by incubating organelles in presence of the protein precursor, and therefore differs from the transport of proteins through the membrane of the endoplasmic reticulum which takes place, as already mentioned, concomitantly with the translation (while the protein chain is being synthesized).

Most proteins of chloroplasts are also coded by the nuclear genome, synthesized in the cytosol and imported in the plast. But there are more possible localizations in the plast (6) than in the mitochondrion: outer membrane, inter-membrane space, inner membrane, stroma, membrane of the thylakoid, lumen of the thylakoid. As in the case of mitochondria, the import of proteins requires a pre-sequence, also called *transit peptide*, at the N-terminal end, comprising residues with positive charge and hydroxylated residues. This pre-sequence is cleaved during the transport which requires the hydrolysis of ATP rather than a protomotive force. In the case of chloroplasts also, it has been possible to import a cytosolic protein by grafting to it (at the level of the DNA) the transit peptide of a protein normally imported in the chloroplast (e.g., that of the small sub-unit of ribulose 1,5 bisphosphate carboxylase). It must be noted that the situation is more complex in the case of plasts, not only because of the existence of 6 different compartments in the chloroplast, but also owing to the presence, in plants, of plasts which are different with regard to their roles and therefore with regard to their proteins: chloroplasts (photosynthesis), chromoplasts (synthesis of carotenoids), amyloplasts (synthesis of starch).

h) **Transport of proteins to the nucleus.** – All nuclear proteins (especially histones, DNA polymerases, RNA polymerases and all proteins participating in the replication of DNA and transcription) are synthesized in the cytosol and must pass through the nuclear envelope of eucaryotes comprising an outer membrane and an inner membrane. This transport seems to take place thanks to nuclear pores of 70 Å in diameter, for the small proteins (for example, histones), but for larger proteins a short peptide sequence appears to be necessary. Thus, the transport to the cell nucleus of antigen T of SV 40 virus, a protein of molecular weight 92 000 daltons which controls the replication and transcription of viral DNA, requires a heptapeptide comprising 5 consecutive positively charged residues (Pro-Lys-Lys-Lys-Arg-Lys-Val). The replacement of a Lys residue by Thr or Asn prevents the import of antigen T in the nucleus. It was also possible to bring about the transport of proteins to the nucleus by grafting (at the DNA level) this heptapeptide sequence on pyruvate kinase or on other cytosolic proteins.

I. – Organization and expression of mitochondrial and chloroplastic genomes

The mitochondrial genome of various eucaryotic cells, especially yeast cells and human cells, has been extensively studied in recent years; the results obtained are interesting and – for some – unexpected.

The human mitochondrial genome, which is a circular double-stranded DNA, comprises 16 569 base pairs and its sequence has been entirely determined (the mitochondrial genome of the yeast *Saccharomyces cerevisiae* is five times larger and contains about 78 000 base pairs). This genome is very compact and characterized by a great economy. It codes for:

– 2 *r*RNAs, which are particularly short: 1 559 and 954 nucleotides respectively (3 200 and 1 660 nucleotides in yeast mitochondria).

– 22 *t*RNAs (24 in yeast). This number of *t*RNAs is clearly smaller than the minimum number of *t*RNAs needed to read the 61 sense codons, taking into consideration the wobble hypothesis (see page 364) i.e. 31 *t*RNAs (plus the initiation *t*RNA). The translation of codons by a reduced number of tRNAs in the mitochondria is however accomplished thanks to different possibilities of codon-anticodon pairing. Thus, some mitochondrial *t*RNAs, having a U at the first position of the anticodon, can read 4 codons differing only by their third nucleotide, i.e. the 4 codons of the same family (or the same box, in fig. 6-48), while procaryotic and eucaryotic *t*RNAs generally have a modified U at the first position of the anticodon, which restricts their possibilities of pairing.

– About thirteen polypeptides which are part of oligomeric membrane proteins: cytochrome oxidase, apocytochrome *b*, ATPase, NADH dehydrogenase (NADH ubiquinone oxidoreductase, sensitive to rotenone). It is of interest to note that the sub-unit no. 9 of ATPase is coded by the mitochondrial genome in yeast, but by the nuclear genome in *Neurospora crassa*, which suggests that transfers of genes took place during the evolution.

It was observed that human mitochondrial genes are not separated by intercistronic sequences (punctuation is carried out by the *t*RNA genes) and that the *m*RNAs contain neither leader sequence nor tail sequence. More than half of them do not even have, after transcription, any termination codon, the latter being formed only upon polyadenylation of the *m*RNA (the termination codon UAA is formed by addition of one or two adenylic residues to UA or U).

There seems to be only one promoter on each strand of the human mitochondrial DNA, which suggests that the genome is transcribed completely and symmetrically and that the two long precursors are cleaved (generally at the sites where the *t*RNAs are situated). In the yeast mitochondrion, on the contrary, transcription seems to take place from several different promoters.

Comparing the genes of apocytochrome *b* in human and yeast mitochondria, it is observed that the coding sequences are of the same size, but while there is no intron in the human mitochondrial gene, there are 5 introns in yeast (the intron no. 2 and the preceding exon seem to code for a maturase, the enzyme responsible for the excision of this intron). Besides, while in man, there is neither leader sequence in 5′, nor tail sequence in 3′ (and not even any complete termination codon), in yeast there is a leader sequence of about 1 000 nucleotides and a tail sequence of about 50 nucleotides.

Lastly it came as a surprise to find out that the genetic code used by the mitochondria is different from the genetic code which was thought to be universal (see fig. 6-48). For instance, UGA is a tryptophan coding (instead of non-sense) and AUA a methionine codon (instead of coding for isoleucine). AGA and AGG can be used as termination codons (instead of being codons for arginine), AUA and AUU can be used as initiation codons (instead of being codons for isoleucine) and the 4 CUN codons can be recognized by threonyl-*t*RNA (instead of leucyl-*t*RNA), but as regards these peculiarities of the mitochondrial genetic code, it appears that there are variations according to the organisms.

It must be noted that the mitochondrial genome of plants is also circular but that it is clearly larger than that of mammals (about 16 kpb) or yeast (about 78 kpb) and that its size varies, according to the species, from 200 to more than 2 000 kpb, while it is not known presently whether these large mitochondrial genomes contain more genes. Indeed, only about fifteen genes of proteins have been identified and sequenced at present; these are, on one hand, genes already identified in mitochondrial genomes of mammals or yeast, coding for sub-units I, II and III of cytochrome oxidase, sub-units of NADH-dehydrogenase, some sub-units of ATPase and cytochrome b (see above) and, on the other hand, some genes coding for proteins of the small ribosomal particle (S_{12}, S_{13}, S_{14}). As regards genes coding for *r*RNAs, one finds genes coding for the 2 large *r*RNAs and also – and this is characteristic of the mitochondrial DNA of plants – the gene of mitochondrial 5 S *r*RNA; another characteristic of plant mitochondria: a number of mitochondrial *t*RNAs are coded by the mitochondrial genome, but the other mitochondrial *t*RNAs are coded by the nuclear genome and are imported into the mitochondrion. Moreover, a heterogeneity is observed in the size of mitochondrial DNA molecules of the same plant; this is due to the presence, in the mitochondrial genome, of repeated sequences (either in the same direction or in opposite direction) which can generate sub-genomic DNA molecules by recombination.

Lastly, it was recently shown that in plant mitochondria, there exists a phenomenon of correction of genetic information at the level of *m*RNAs, called *RNA editing*. It consists in modifying some Cs into Us and thus changing some codons. This results in the synthesis of a protein whose sequence in amino acids is different from the sequence which could be deduced from that of the gene, but has more similarity to the amino acid sequence of the corresponding mitochondrial protein in other organisms (mammals, yeast).

On the contrary, the size of chloroplast genomes, which are also circular, does not vary considerably from one species to another and is about 120 to 190 kpb. In most of the species studied, the chloroplast DNA comprises a region (of about 10 to 25 kpb) present as two copies in opposite orientations (inverted repeats) which contain the genes of the *r*RNAs, and are separated by a large single copy region and a small single copy region.

The complete sequence of the chloroplast genome of two plants namely tobacco (155 kpb) and *Marchantia* (120 kpb), was recently determined. The chloroplast DNA codes for the following products:

– the chloroplast 23 S, 16 S, 5 S and 4.5 S *r*RNAs;

– thirty chloroplast *t*RNAs;

– 19 chloroplast ribosomal proteins, i.e. 11 proteins of the 30 S particle and 8 of the 50 S particle (the others are imported from the cytoplasm);

– the translation initiation factor IF-1 (and probably, the elongation factor EF-Tu in some species only);

– three sub-units of the RNA polymerase;

– the large sub-unit of ribulose 1,5 bisphosphate carboxylase (the small sub-units is coded in the nucleus, synthesized in the cytoplasm in the form of a slightly longer precursor which is imported into the chloroplast);

– two proteins of photosystem I, five proteins of photosystem II, six polypeptides forming part of the ATP-synthase complex, and three proteins involved in electron transport (in the cytochrome b_6/f complex).

In addition to the genes for these forty or so identified proteins, the chloroplast genome contains some forty open reading frames (ORF) which could also code for polypeptides, but most of the chloroplast proteins are coded by the nuclear genome.

Contrary to the mitochondria, it appears that chloroplasts use the universal genetic code.

GENETIC RECOMBINATION *IN VITRO* AND CLONING OF GENES[1]

I. – METHODS OF CLONING

In the past few years, an increasing number of laboratories have undertaken experiments of genetic recombination *in vitro*. These experiments consist in incorporating heterologous DNA in a stable manner in a bacterial cell (generally the strain of *E.coli* K_{12}), through a vector which can be, either a plasmid, or a phage such as phage λ, i.e. genetic material capable of existing independently of the bacterial chromosome and of replicating autonomously.

For example, if a plasmid is used (i.e. a circular double-stranded covalently closed DNA molecule present in the bacterial cell independently of the chromosome), the principle of the experiments will consist in isolating the

(1) Revised by C. Kedinger.

plasmid from the bacterial cell, splitting this DNA molecule by means of a very specific enzyme (a restriction endonuclease, see page 272), joining the cleaved plasmid DNA to the heterologous DNA to be cloned and finally re-introducing – thanks to the transformation process – in the bacterial cell, the plasmid carrying the heterologous DNA.

Among the most frequently used plasmids, one can mention those made *in vitro* using the DNA recombination techniques, and which often belong to the *Col* series, because they contain genes coding for the production of a *colicine* (an extracellular protein, having antibiotic properties, produced by *E.coli*). The plasmids used for the cloning of genes generally have the following properties: their size is comparatively small; they possess at least one single site sensitive to a restriction endoculease (see page 272); they contain one (or several) gene(s) which allow the selection of the cells transformed by the plasmid; they are present at the concentration of about 20 molecules per cell, but in presence of chloromphenicol their replication leads to the accumulation

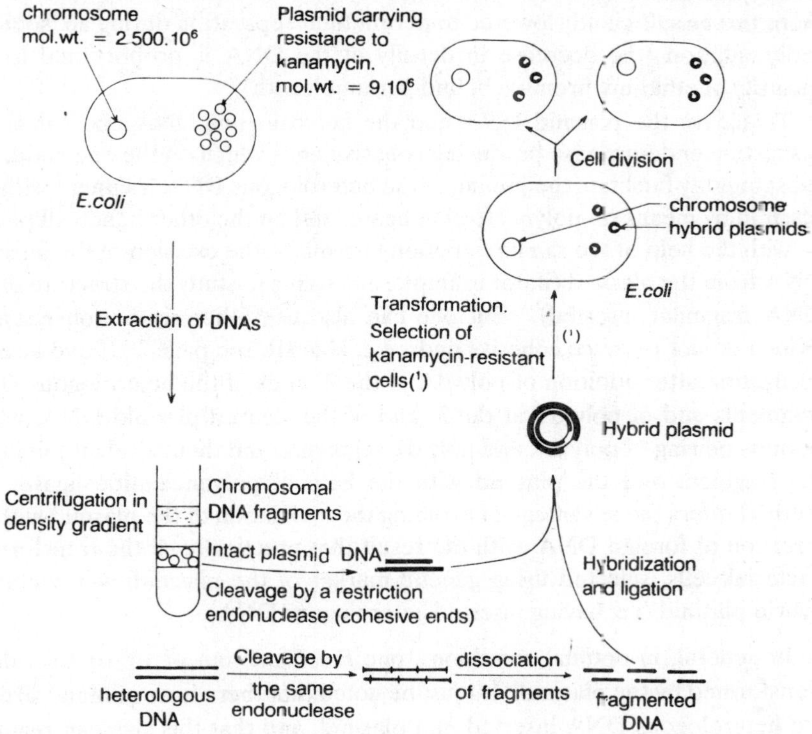

FIG. 6-51. — Diagram of the important steps permitting the cloning of heterologous genes in a *E.coli* cell.

---

(1) When the single site of cleavage by the restriction endonuclease is located in the gene responsible for the resistance to a second antibiotic, the insertion of heterologous DNA results in the loss of this resistance by the hybrid plasmid, with the result that this second antibiotic enables one to distinguish between the cells containing a hybrid plasmid (sensitive) and those containing plasmid which closed again without insertion of heterologous DNA (resistant).

of 1 000 to 2 000 circular plasmid DNA molecules per cell; they can be easily isolated from bacterial cells and can also be easily reintroduced in them.

The diagram of fig. 6-51 shows the main steps allowing the cloning of heterologous genes in a bacterial cell. Owing to its large size, the bacterial DNA is very sensitive to shearing forces, with the result that it is cut into fragments during the extraction of DNAs, while the plasmid DNA much smaller in size, present in the form of circular (covalently closed) double-stranded super-coiled molecules, remains intact and can be easily separated from the bacterial DNA by centrifugation in a density gradient, in presence of a molecule which intercalates between the bases of the DNA, ethidium bromide. The physical constraints existing in a covalently closed circular DNA molecule are such that the quantity of ethidium bromide bound will be limited. These constraints do not exist in a linear molecule and the quantity or ethidium bromide bound per unit DNA length will be much greater. Compared to the density of the circular complexes, the density of the linear complexes will therefore be sufficiently lowered to permit their separation during an isopicnic sedimentation (the decrease in density of the DNA is proportional to the quantity of ethidium bromide bound per unit length).

To cleave the plasmid DNA and the heterologous DNA one can use a restriction endonuclease producing cohesive ends which on the one hand, will subsequently facilitate the joining of the heterologous DNA fragment with the plasmid by means of polynucleotide ligase, and on the other hand will permit – with the help of the same restriction enzyme – the excision of the inserted DNA from the plasmid (if, for example, one wants to study the structure of the DNA fragment inserted). But one can also use other restriction enzymes (which do not produce cohesive ends, e.g. Hae III, see page 272) and achieve cyclization after addition of poly dA at the 3′ ends of the heterologous DNA fragments and of poly dT at the 3′ end of the cleaved plasmid DNA, which permits pairing of poly dA and poly dT segments, and then covalent joining of the fragment and the plasmid with the help of polynucleotide ligase; this method offers the advantage of avoiding the cyclization of the plasmid without insertion of foreign DNA, with the result that practically all the transformed bacterial cells, selected using genetic marker of the plasmid, will contain a hybrid plasmid (i.e. having inserted heterologous DNA).

In general, in optimal conditions, one *E.coli* cell out of 10^5 or 10^6 will be transformed by the plasmid. It must be noted that there is no minimal size for the heterologous DNA inserted in a plasmid, and that this size can reach at least 40×10^6 daltons, i.e. about 60 000 base pairs (= 60 kb). Moreover, the insertion of foreign DNA in plasmids of the *Col* series does not affect the number of plasmid molecules which can be present in the cell, so that the heterologous DNA inserted will be present also at the concentration of several copies in the transformed bacterial cell. Lastly, under the effect of chloramphenicol (mentioned above) one can have an amplification such that the circular DNA of the hybrid plasmid represents 40 to 50% of the totality of the DNA of the cell.

The DNA of the bacteriophage λ (47 000 base pairs = 47 kb) is often used as cloning vector because it offers some advantages over the bacterial plasmids. About 15 kb of the central part of the genome of this phage are not essential for its survival and can be replaced by heterologous DNA to be cloned. One can carry out *in vitro* the encapsidation of the recombinant DNA obtained and thus considerably increase the infection efficiency of this DNA. The encapsidation mechanism of the phage is such that only DNA molecules having a final size of about 47 kb will be encapsidated, which brings about a spontaneous selection of DNA molecules, having inserted a heterologous fragment of about 15 kb.

Cosmids are small plasmids which contain an encapsidation signal (the "cos" site of phage λ), an origin of replication and a gene of resistance to an antibiotic. They therefore combine the advantages of the bacteriophage (efficiency of infection of the encapsidated form) and those of bacterial plasmids (autonomous replication and ease of selection). Furthermore, one can insert in cosmids, fragments of heterologous DNA of larger size (up to 40 kb).

These two vectors of cloning (phage λ and cosmids) are often used for constructing genomic "banks" or "libraries" which are collections of recombinant phages or plasmids that together comprise the totality of the sequences of a given genome. It is clear that the greater the size of genomic DNA fragments inserted, the smaller the number of clones needed to cover the totality of the genome: this is the special advantage of cosmids.

II. – CLONING OF A SPECIFIC GENE

The problem is to select from a vast population of transformed cells, a cell which contains a hybrid plasmid carrying a particular heterologous gene. This is a serious problem, especially if the starting heterologous DNA is complex and if the fragments generated by the restriction enzyme are numerous. It is clear that either there must be a phenotypical property associated with the gene cloned, or one must have an appropriate probe.

If one clones a gene coding for an enzyme for example, one can transform a bacterial mutant lacking this enzyme and thus select the appropriate transformed cells on the basis of the restoration of the corresponding enzymatic activity. It was thus possible to clone the genes of the tryptophan operon, or the gene of the ligase of *E.coli*, starting from DNA fragments resulting from the hydrolysis of the whole chromosome of *E.coli*. But if one wants to clone a eucaryotic gene, it is often preferable to clone first the DNA copy (cDNA) of the *m*RNA corresponding to the gene sought. The size of these cDNAs is indeed generally much smaller than that of their genomic counterpart due to the absence of introns. One therefore constructs a cDNA "library" by inserting, in a plasmid or in a vector derived from bacteriophage λ (e.g., λgt 10), *c*DNA copies of the entire population of cytoplasmic *m*RNAs extracted from the cells expressing the gene concerned. These copies are made with the help of the reverse transcriptase (see page 323) which synthesizes the *c*DNA strands complementary to the *m*RNAs starting from oligo dT primers previously hybridized to the poly A tails of *m*RNAs. After alkaline hydrolysis of the RNAs, double-stranded *c*DNA molecules are then obtained by action of DNA polymerase I of *E.coli* (see page 308) which makes strands complementary to those synthesized during the first step. The synthesis is "self-primed" at the 3′ end of the single-stranded *c*DNA which folds back upon itself (see fig. 6-31*a*). A digestion by nuclease S1 (specific of single stranded nucleic acids) finally eliminates the loops at this end and gives perfectly double-stranded *c*DNAs. A *c*DNA library is said to be complete if it contains at least one bacterial clone for each starting *m*RNA.

There are several methods to identify the clones sought. If the *c*DNAs have been inserted downstream a bacterial promoter (as in the case of λgt 11), one can analyse (or "screen") the "library" by the expression of the *c*DNAs. This is done by immunological identification of the producer clones; antibodies specific of the corresponding protein will react selectively with the relevant clones which will thus be labeled. The success of this "screening" technique however depends on several factors: the *c*DNA must be inserted in the correct orientation with respect to the bacterial promoter of the vector, in order to permit the transcription of its coding strand; the translation of the messenger thus produced must take place in the correct reading frame; lastly, the tertiary structure adopted by the synthesized polypeptide must reconstitute the antigenic determinants (epitopes) recognized by the antibodies. A second approach, independent of these prerequisites, is based on recent developments in the micro-sequencing techniques of proteins and in the production of synthetic oligodeoxynucleotides. With the knowledge of the partial amino acid sequence of the protein one can, on the basis of the genetic code, deduce (taking into consideration the fact that the code is degenerate) the corresponding nucleotide sequence and carry out its chemical synthesis. These oligonucleotides, labeled by ^{32}P, are then used as radioactive probes to identify, by hybridization with the complementary sequences, the bacterial clones containing the *c*DNA sought.

The *c*DNAs thus isolated can in their turn serve as probes to look for the corresponding genes in a genomic "bank".

III. — EXPRESSION OF CLONED GENES

These hybrid plasmids represent not only an abundant source of inserted heterologous DNA, but also allow the production — by the bacterial cell — of large quantities of products specified by the genes inserted, particularly when the cloned genes are from the same bacterium or another procaryote: for example, the cloning of the tryptophan operon of *E.coli* in a plasmid of the *Col* series has allowed the induced synthesis of such large quantities of the five enzymes of this operon that they represented 40% of the total proteins of the cell. But when genes of eucaryotes are cloned in *E.coli* plasmids, it is observed that the synthesis – by the bacterium – of the products corresponding to these genes is very small, or even nil, most probably because of differences between the systems controlling transcription and translation in procaryotes and eucaryotes (for example in the initiation and termination of transcription and translation) or because of the absence of intron splicing systems in procaryotes (see page 339). Some of these problems can be solved by the insertion of a *c*DNA (DNA complementary to the mature *m*RNA) directly downstream a functional procaryotic promoter in the plasmid. In this manner one has obtained in bacteria, the synthesis of proteins of different origins: animal (growth hormone, somatostatin, insulin, interferon), plant (large sub-unit of ribulose 1,5 bisphosphate carboxylase) or viral (antigen of the hepatitis B virus, or of the foot and mouth disease virus). It must be emphasized however that the biological activity of some of these proteins can depend on subsequent modifications (specific proteolytic cleavages in the case of precursors or glycosylation) which the procaryotic host cannot perform.

IV. – CLONING OF GENES IN EUCARYOTIC CELLS

At the beginning, the cloning of genes was mostly carried out using *E.coli*, but the principles on which this process is based are also applicable to animal and plant cells. Viruses or plasmids capable of introducing genes in the chromosomes of eucaryotic cells were needed; it was therefore envisaged to use viruses derived from Simian virus 40 (SV 40). As far as the plant kingdom is concerned, it was shown that a fragment of the DNA (T-DNA) of a plasmid (Ti plasmid) of the bacterium *Agrobacterium tumefaciens* (responsible for the formation of a tumour called Crown-Gall) could be transferred to plant cell chromosomes. Moreover, one could insert, in this T-DNA, genes of bacterial origin (e.g., gene of resistance to an antibiotic) or plant origin (e.g., gene of the small sub-unit of ribulose 1,5 bisphosphate carboxylase) and thus obtain their transfer to a tobacco protoplast; since in various species (including tobacco) one can regenerate an entire plant from a protoplast, one could observe, starting from transformed protoplasts, the presence of the gene inserted in diverse tissues of the transformed plant, observe its expression and note that this expression is controlled by factors (such as light) which exert their regulatory effects at the transcriptional level in plants.

Some recent techniques enable the direct introduction of DNA into plant or animal cells without the help of any plasmid or viral vector (direct transfer of genes).

V. – ADVANTAGES OF THE CLONING OF GENES

Cloning of genes provided for the first time, large quantities of specific DNA fragments, in other words genes, especially of higher organisms. These DNA fragments can be used as *probe* and permit, by hybridization, the detection and dosage of the corresponding *m*RNAs in cells which are at different stages of their development or placed in diverse physiological conditions. On the other hand one can, thanks to the cloning of genes, undertake the study of the structure of genes and chromosomes of eucaryotes, and the study of the control mechanisms of gene expression in higher organisms.

But, beside the great advances that may be expected in the domain of fundamental knowledge in molecular biology, the cloning of genes can lead to extremely important applications.

In the field of agronomy, one can envisage transferring to plants:

– a gene responsible for resistance to a herbicide (isolated for example from a microorganism) thus allowing the protection of the cultivated plant when weeds are eliminated by the herbicide;

– a gene responsible for resistance to a pathogen (e.g., fungus, virus, insect). It has been possible to obtain plants protected from infection by a virus (e.g., Tobacco Mosaic Virus) by introducing into these plants the gene coding for the coat protein of this virus, or plants protected from attack by insects by introducing into these plants the gene of a bacterial toxin having insecticide properties;

– genes improving the growth of plants, for example, enabling a more efficient photosynthesis, or permitting atmospheric nitrogen fixation (although, in the microorganisms studied, this process implies the coordinated expression of 17 genes);

– genes improving the nutritional value of plants, for example, providing seed storage proteins richer in some essential amino acids, like lysine.

This assumes that the gene is not only introduced in the plant, but also expressed in the right tissue (e.g., the seed) and at the right time.

In medicine, the production of therapeutically important substances, by (bacterial, animal, human) cells in culture after introduction of the corresponding gene, is a major application of genetic engineering. Among substances which are already being produced or under study, one may cite: insulin (used in the treatment of diabetes), growth hormone (used against some forms of dwarfism), interferons (anti-viral action), the plasminogen activator (active against thromboses, because plasmin is an enzyme which degrades the fibrin of clots), α-anti-trypsin (active against emphysema, because it inhibits elastase which digests the elastin of the pulmonary conjunctive tissue), the blood factors VIII and IX (against hemophilia A and B, respectively), various viral antigens, like those of hepatitis B or foot and mouth disease (for the preparation of vaccines), etc. It must be noted that the advantage lies not only in the production of large quantities of human proteins, which will not cause allergies (unlike for example, pig insulin used till now), but also in the fact that the products are free from contaminants, especially of viral origin, present particularly in substances prepared from human blood, particularly when it becomes necessary to pool the blood of many donors to prepare appreciable quantities of these products (for example in the case of antihemophilic factors), which increases risks (several hemophilic patients were thus contaminated by the hepatitis or Aids virus).

Genetic engineering techniques have also led to major advances in the diagnosis of hereditary diseases. After action of a restriction enzyme on the DNA of a patient, one visualizes the DNA fragments produced at the genetic locus of the disease by hybridization with an appropriate radioactive probe (DNA segment or synthetic oligonucleotide corresponding to this locus or to a portion of it). The comparison of the size of DNA fragments thus revealed, with those produced in the same manner from a healthy subject permits the detection of anomalies if any: deletions or even, if the hybridization conditions are sufficiently selective (or "stringent"), point mutations. When the gene responsible for the disease in question is not known, one may utilize the existence in the genome, of zones whose sequence may vary from one individual to another (polymorphic zones or mutation hot points). These zones, characterized by the appearance or disappearance of cleavage sites for a given restriction enzyme, therefore present a restriction fragment length polymorphism (RFLP). Thus, if one has a probe which can reveal such a polymorphism near the presumed locus of the disease, one can associate the appearance of the disease in an individual with a restriction profile characteristic of the DNA of this individual. It is then possible, by analysis of the RFLP of the DNA of parents and descendants of this individual, to follow the segregation of the gene responsible for the disease, with the help of this probe.

It has also been envisaged to introduce the intact gene, into animal or human cells deprived of a particular enzymatic activity because of a mutation in the corresponding gene, and thus restore the activity lacking in these cells.

Important applications of genetic engineering are also possible:

– in the field of environment, for example for the biodegradation of toxic products, or for microbiological extraction of various metals from low grade ore.

– in the field of energy, for example, for the production of methane and methanol, or for the recovery of crude oil not easily accessible by conventional extraction methods.

VI. – PRECAUTIONS TO BE TAKEN DURING EXPERIMENTS OF GENETIC RECOMBINATION

It has been said that living species could preserve their identity for a multitude of generations thanks to the presence of natural barriers, and that cloning of genes, by violating these barriers, could create new forms of life, dangerous to man and his environment. Experiments on genetic recombination *in vitro* and cloning of genes, as well as the possible risks which could endanger humanity, have been the object of many controversies. Scientists carrying out such experiments in the United States as also in Europe have decided to lay down a number of precautionary measures. While we will not discuss them here, we will just mention that such measures concern on the one hand, the laboratories where the experiments are carried out and on the other hand the plasmids and bacteria used. As regards the laboratories, one must naturally work in conditions which exclude risks of dissemination as in the case of experiments with pathogenic bacteria and viruses. As far as the plasmids are concerned, one must preferably use plasmids lacking genes which permit bacterial conjugation; this will reduce the risk of dissemination of hybrid plasmid molecules in our environment. Regarding bacteria, strains of *E.coli* K12 were developed, having a number of mutations and necessitating in particular, diaminopimelic acid (a constituent of the bacterial cell wall). If this compound which is indispensable to the mutant is not provided in the culture medium, the cells die and their propagation outside the laboratory is ruled out.

CHAPTER 7[1]

METABOLISM OF NITROGEN COMPOUNDS

Living organisms contain a wide variety of nitrogen compounds. In Chapter 1, we studied the structure of amino acids and proteins, but there are other compounds which also contain nitrogen: nucleotides (co-enzymes or constituents of nucleic acids) porphyrins, alkaloids, osamines, phospholipids, etc. We have seen, in Chapter 6, how amino acids are assembled to form proteins, but amino acids are also, very often, the precursors of other nitrogen compounds, to which they can donate not only their nitrogen but also – in numerous cases – their carbon atoms; in the preceding chapter we examined how glutamine or aspartic acid can donate a nitrogen atom during the synthesis of purine nucleotides (see figs. 6-19 and 6-21) and how the whole molecule of glycine (see fig. 6-19) or aspartic acid (see fig. 6-22) can be incorporated during the *de novo* biosynthesis of purine or pyrimidine nucleotides; in this chapter we will see how amino acids participate in the synthesis of other nitrogen compounds like porphyrins, some hormones, pigments, etc. Amino acids therefore occupy a key position in the metabolism of nitrogen compounds and that is why we will study first the general reactions in which all amino acids may participate because these reactions (deamination-amination, transamination, decarboxylation) play a role, either in the formation of amino acids, or in the transformations the latter may undergo, or in their catabolism.

I. – GENERAL REACTIONS OF AMINO ACIDS

1. – Enzymatic reactions using pyridoxal phosphate as the co-enzyme

A number of reactions involving amino acids, like transamination, decarboxylation, elimination of the side chain with maintenance of the α-amino-carboxylic group (e.g., transformation glycine ←→ serine), elimination of H_2O or H_2S (for example, under the influence of serine-dehydratase or cysteine desulphydrase), *racemization* (e.g., L-alanine ←→ D-alanine) require the same coenzyme: *pyridoxal phosphate*. The fact that the reactions catalyzed are very

---

[1] Revised by Y. BOULANGER.

different, while the coenzyme is identical, shows the decisive influence of the apoenzyme (the protein part) on the evolution of the reaction. However, as will be seen for the first three reactions mentioned above, the same intermediate product is formed (a Schiff base), then the reactions proceed in different directions (see fig. 7-1). One must indeed imagine pyridoxal phosphate in space as a plane molecule (aromatic cycle). During the formation of the intermediate Schiff base, the α-carbon and nitrogen of the amino acid are placed in the plane of the coenzyme. Then the enzyme will force one of the three other bonds of the α-carbon of the amino acid to take a position perpendicular to this plane and will use the electron attractive power of the protonated coenzyme ($>N^{+}H$) to attract on the α-carbon the electronic doublet of this bond. In all cases, the α-carbon of the amino acid will become

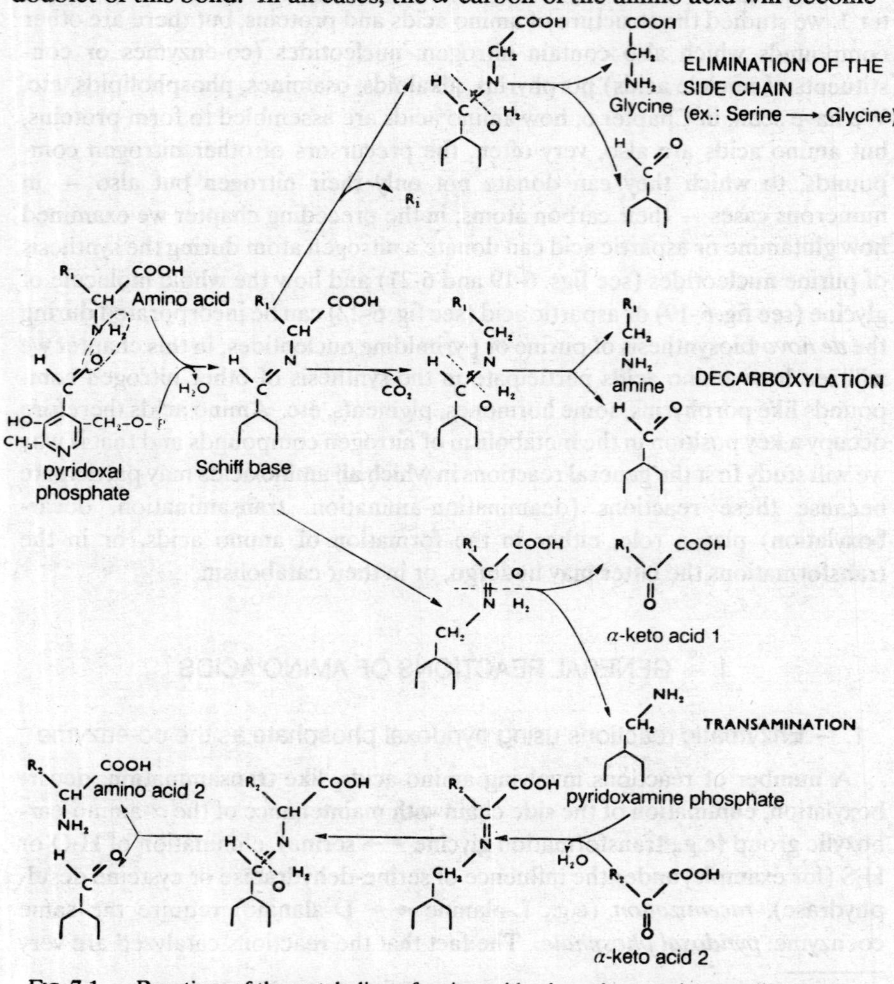

FIG. 7-1. — Reactions of the metabolism of amino acids where the same intermediate compound is formed in presence of pyridoxal phosphate.

a transitory carbanion ($C\bar{\alpha}$) and one of the three substituents, H, R or COOH will be pulled out in the form of cation (H^+, R^+ or $COOH^+ = CO_2 + H^+$).

It must be noted that pyridoxal phosphate cannot be synthesized in our organism, which explains the vitamin B_6 activity shown by pyridoxal, pyridoxine (the corresponding primary alcohol) and pyridoxamine (the corresponding primary amine which is formed intermediately during the transamination process).

A. – Decarboxylation

We have already studied the mechanism of this reaction (fig. 7-1) which leads to an amine. Some of these amines have an important physiological or pharmacological role and are therefore sometimes called *biogenic amines* The table below indicates – for some amino acids – the corresponding amine and its localization or role.

Amino acid	*Amine*	*Localization or role*
aspartic acid............	α–alanine	in proteins
	β–alanine	in coenzyme A
cysteine............	mercaptoethylamine	in coenzyme A
glutamic acid............	γ-aminobutyric acid	mediator of the central nervous system
histidine............	histamine	hypotensive effect
3,4-dihydroxy-phenyl-alanine (DOPA)..........	dopamine	precursor of adrenaline (hormone)
serine............	ethanolamine	in phosphatides
5 hydroxy-tryptophan............	serotonine	vasoconstrictive tissue hormone

Decarboxylations are catalyzed by *decarboxylases*, pyridoxal phosphate-containing enzymes, present in microorganisms and animal tissues. Intestinal bacteria especially, have enzymes which can decarboxylate lysine and ornithine respectively into cadaverine and putrescine, amines which are always present in small quantities in the intestine, but can cause intoxications if their concentration increases due to abnormal intestinal fermentations.

Bacterial decarboxylases have an extremely narrow specificity, with the result that a purified decarboxylase can permit the quantitative titration – by measuring the CO_2 liberated – of the corresponding amino acid in a complex mixture.

B. – Aldolization

By analogy with the cleavages of cis-diol bonds of carbohydrates this is an *elimination* of the side chain of serine and threonine transformed into glycine by cleavage of $C\alpha - C\beta$ bond:

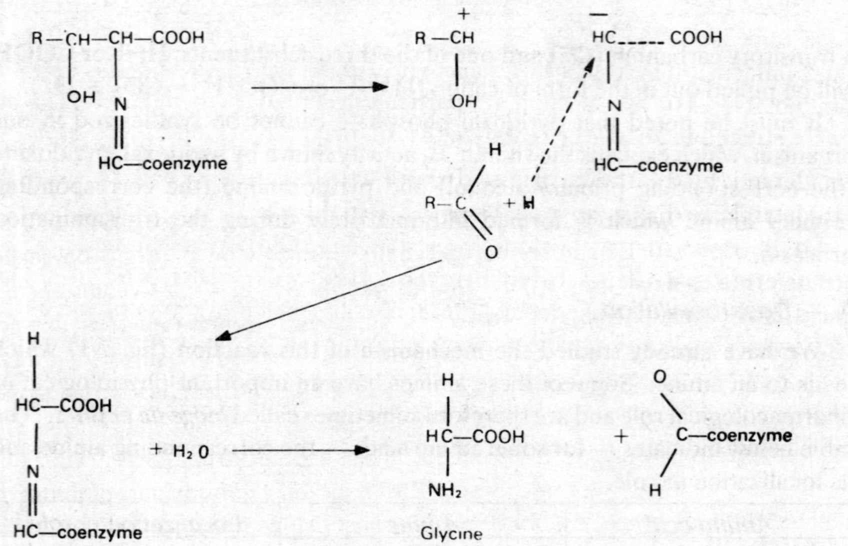

In the case of threonine (R = CH_3) the enzyme is called threonine-aldolase. It releases glycine and acetaldehyde which is transferred to the coenzyme thiamine-pyrophosphate (see oxidative decarboxylation of pyruvic acid). In the case of serine (R = H) it is serine-aldolase which, in addition to glycine, releases formaldehyde transferred to tetrahydrofolate (FH_4).

C. — Transamination

This term denotes *the reversible transfer of the amino group from an amino acid to a α-keto acid; there is no liberation of NH_3* unlike during deamination which will be studied in the following.

The reactive mechanism is illustrated at the bottom of figure 7-1; in fact, there are two steps, of which one is the reverse of the other; they can be summarized as follows:

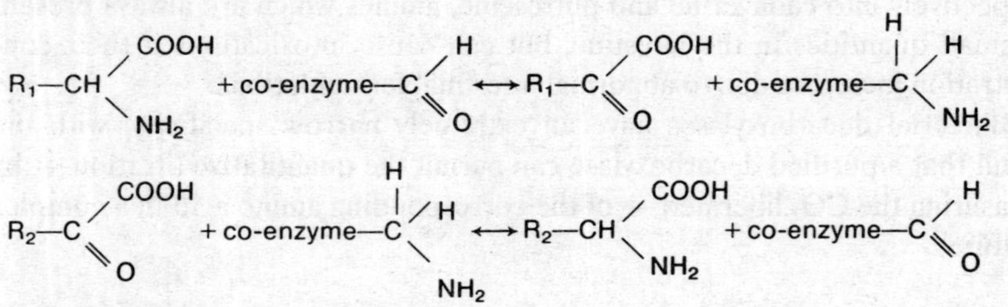

The sum may be written:

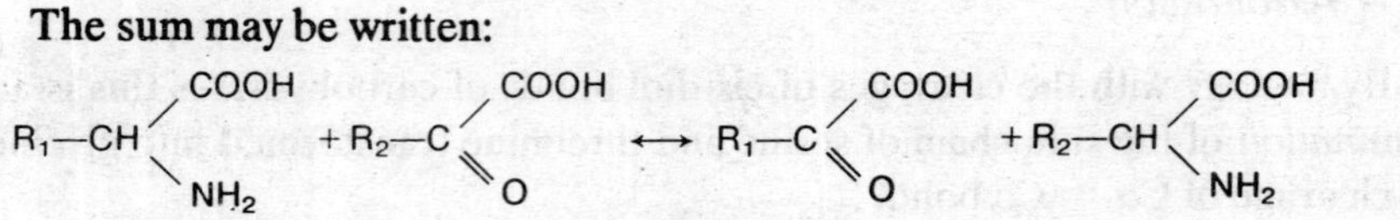

As we will note in the following, transamination is a very important reaction, for the formation of amino acids as well as the transformations they undergo. *Transaminases* are universally distributed: they are found in bacteria, plants and animals. There is a large number of transaminases specific of various amino acids, but two of them are particularly abundant in animal tissues and were studied in detail; they catalyze the following 2 reactions:

L-glutamic acid + oxalo acetic acid $\leftrightarrow$ α-ketoglutaric acid + L-aspartic acid
L-glutamic acid + pyruvic acid $\leftrightarrow$ α-ketoglutaric acid + L-alanine

It is observed that in both cases, the amino group of glutamic acid is transferred to a α-keto acid to form the corresponding amino acid. Glutamic acid is a major participant in transamination reactions; we will see in the following that it is very easily formed by binding of NH_3 to α-ketoglutaric acid (see fig. 7-4); its amino group can then be rapidly transferred by transamination and found in the other amino acids.

It must be noted that transamination is a transfer of primary amino group (NH_2) of a donor α-amino acid to an acceptor α-keto acid. This process cannot therefore apply to proline whose nitrogen group is a secondary amine. On the contrary, transamination can take place on a primary amino group situated at the end of a side chain of amino acid. An exchange, primary amine $\leftrightarrow$ aldehyde will then take place (see fig. 7-21).

D. – Racemization

The mechanism resembles that of transamination in the sense that proton H^+ is pulled out from α-carbon by the effect of a proton attracting amino acid belonging to the active site of the racemase. The latter can then add another proton H^+ through a proton donor amino acid *but on the other side of the plane of the coenzyme*. There is therefore inversion of the configuration of α-carbon of the substrate amino acid.

E. – Reactions of α-β and α-γ elimination

Here again the enzymatic mechanism resembles that of transamination and also begins by the extraction of proton H^+ from α-carbon of the substrate amino acid. These reactions are found in alcohol and sulphur-containing amino acids. The enzyme then pulls out the alcohol or sulphur group situated on β or γ carbon, which gave the name of reactions of α-β or α-γ *elimination*, and produces an unsaturated, unstable intermediate susceptible of other transformations (spontaneous or enzymatic). We will illustrate these reactions by two simple examples: the deamination of serine and cysteine by the action of serine-dehydratase and cysteine-desulphydrase. The action of these two enzymes containing pyridoxal phosphate results in both cases in the release of an α-β ethylenic amino acid (see fig. 7-2) which in aqueous solution exists in tautomeric imino acid form and the latter is spontaneously hydrolyzed by water into ammonia and pyruvic acid.

Similar mechanisms and enzymes exist for threonine (dehydration) and methionine (removal of CH_3-SH group) as well as for homoserine and homocysteine. We will find them especially in connection with reactions called

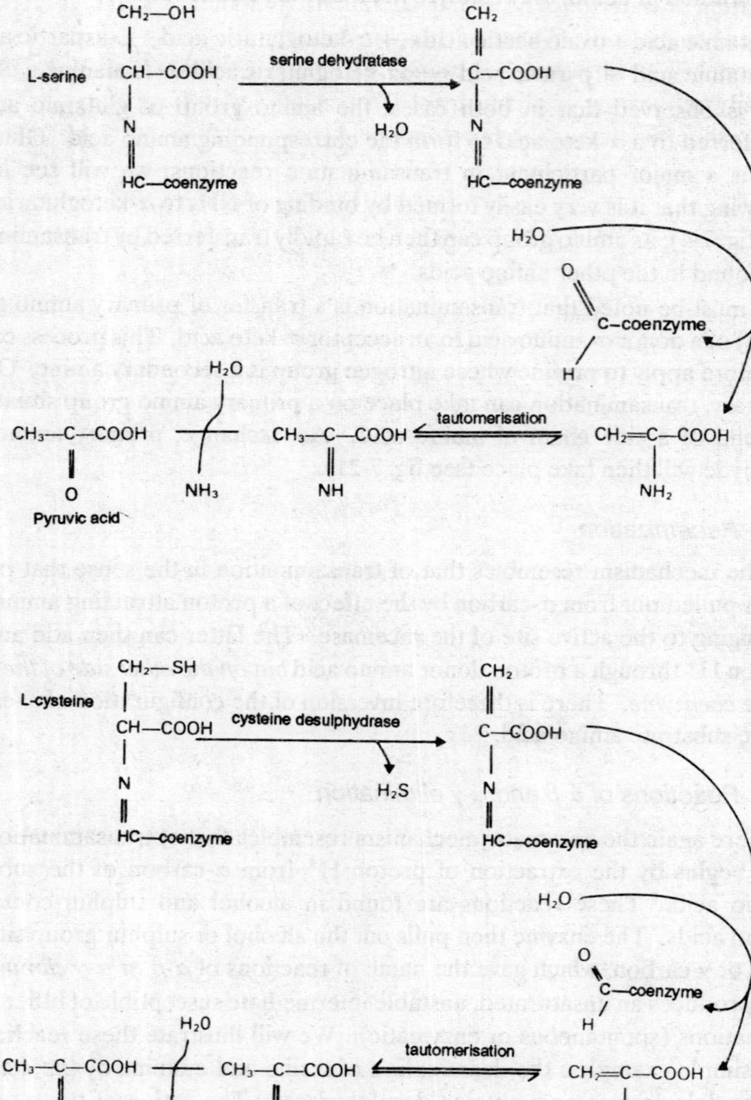

FIG. 7-2. — **Deamination of serine and cysteine.**

transulphuration reactions (see metabolism of sulphur-containing amino acids). It is noted that in these examples the actual enzymatic reaction is an elimination of water or of a sulphur-containing compound, deamination being a non-enzymatic consequence of this elimination.

In conclusion, whatever the example, an enzyme containing pyridoxal phosphate always acts by mechanisms of transfer of electrons and protons (plus other groups, alcohol or sulphur-containing groups in some particular cases).

2. – Deamination

A. – Oxidative deamination

As shown by figure 7-3, this essentially irreversible process takes place in two steps. In a first phase, the enzyme catalyzes a dehydrogenation of the amino acid into an imino acid; this enzyme is a flavoprotein and the 2 H atoms removed from the substrate reduce FAD to $FADH_2$ (or FMN to $FMNH_2$). Deamination is therefore a non-enzymatic consequence of the enzymatic process of dehydrogenation. The imino acid is hydrolyzed spontaneously into α-keto acid + ammonia. As regards the flavin coenzyme, it is reoxidized, generally by molecular oxygen; this leads to the formation of hydrogen peroxide which can then be decomposed by a catalase; but other electron acceptors can also permit the reoxidation of $FADH_2$. Amino acid oxidases are stereospecific enzymes: in higher animals one has thus identified a L-amino acid oxidase (liver and kidney) not very abundant and not very active, which acts neither on glycine, nor on alcohol-containing, sulphur-containing, basic and acid amino acids. One has also characterized a D-amino acid oxidase (liver and kidney) very abundant and very active on most D-amino acids and on glycine; this abundance is paradoxical at first sight considering the absence of corresponding substrates (D-amino acids) in higher animals.

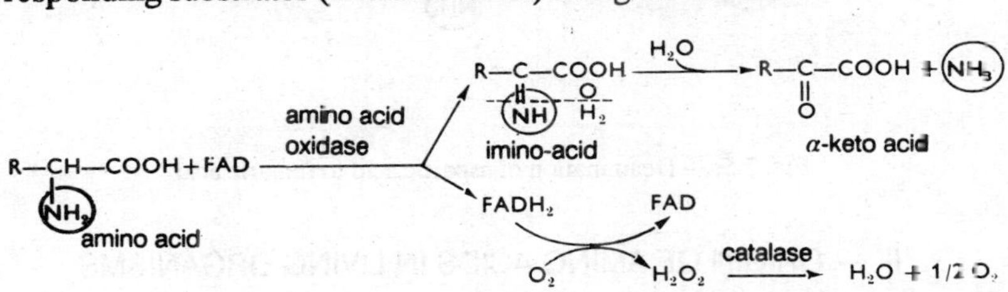

FIG. 7-3. — Oxidative deamination of amino acids.

Glutamic acid is deaminated by a particular enzyme, *L-glutamate-dehydrogenase*, an enzyme which contains NAD or NADP depending on the organisms, and catalyzes the reaction represented in figure 7-4. Contrary to other oxidative deaminations, this reaction is reversible and it is particularly important as a reducing amination; this is indeed one of the main processes of fixation of ammonia in organic compounds and the amino group thus formed

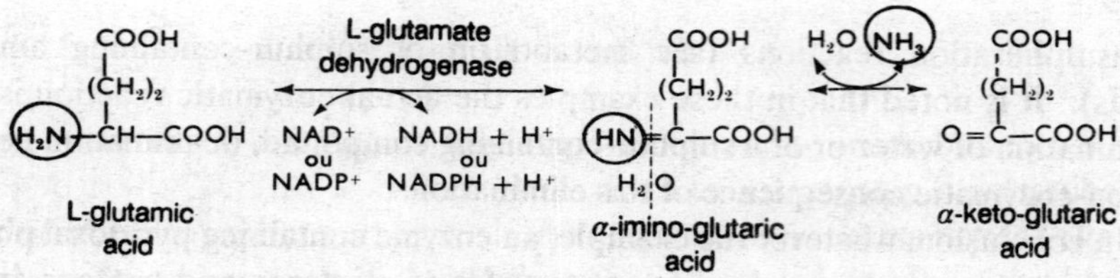

FIG. 7-4. — Interconversion between glutamic acid and α-ketoglutaric acid by oxidative deamination or reducing amination.

can easily be transferred from glutamic acid to other amino acids by transamination, as mentioned in the previous section. Moreover, this conversion from glutamic acid to α-keto glutaric acid (which is one of the compounds of the Krebs cycle) and vice-versa constitutes one of the contact points between the carbohydrate and protein metabolisms.

B. – Desaturating deamination

It concerns only a small number of amino acids. In this process the enzyme catalyzes the removal of a molecule of ammonia with formation of a double bond between the α and β carbons. We will cite the deamination of L-aspartic acid into fumaric acid (fig. 7-5), a reversible reaction which permits the fixation of HN_3 in a compound of the Krebs cycle. This reaction is catalyzed by aspartate-ammonium lyase, an enzyme present in microorganisms and plants.

In the same manner are deaminated histidine in animals and phenylalanine in plants.

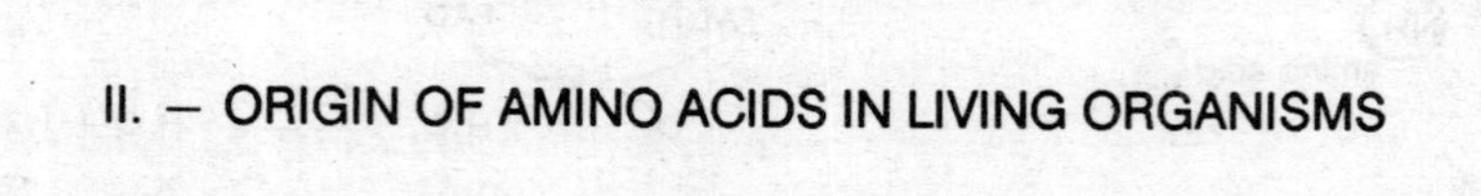

FIG. 7-5. — Deamination of aspartic acid to fumaric acid.

II. – ORIGIN OF AMINO ACIDS IN LIVING ORGANISMS

With relation to the origin of amino acids, living organisms are divided in two groups: there are organisms capable of synthesizing all the 20 natural amino acids; here the amino acids are therefore of *endogenous* origin. On the other hand there are organisms which are incapable of synthesizing some amino acids; these *essential* amino acids must be supplied by the medium (or through diet); their origin is *exogenous*. We must therefore study successively the synthesis of amino acids and the absorption of preformed amino acids.

1. – Synthesis of amino acids

E.coli and many other micro-organisms can be grown in an inorganic medium with glucose added as the only source of carbon, which definitely shows that all amino acids can be synthesized by these cells from glucose[1]. It may be considered that plants and other photosynthetic organisms synthesize the 20 amino acids from inorganic substances, because, as explained in Chapters 3 and 4, they form carbohydrates and thereby all organic substances by reduction of CO_2. The synthesis of amino acids can be divided in several steps: formation of ammonia, incorporation of ammonia in an organic compound (generally in the form of glutamic acid), synthesis of the carbon skeleton of amino acids (i.e. formation of the corresponding α-keto acids) and lastly, transfer of the amino group from the glutamic acid to these various α-keto acids by transamination.

A. – *Formation of ammonia*

Most microorganisms and plants use the ammonia formed during the degradation of nitrogenous organic substances. But we must mention here the formation of ammonia from inorganic substances: atmospheric nitrogen and the nitrite and nitrate ions:

a) **Fixation of atmospheric nitrogen in the form of ammonia.** – This is a reduction of N_2 to $2NH_3$, in other words, an addition of electrons to the two nitrogen atoms, a process whose mechanisms is not yet fully understood. This formation of NH_3 takes place:

– on the one hand, in microorganisms living independently: aerobic soil bacteria (of the genus *Azotobacter*) or anaerobic soil bacteria (of the genus *Clostridium*), photosynthetic bacteria, etc.

– on the other hand, in some bacteria of the genus *Rhizobium* living in *symbiosis* with higher plants (for example, of the Leguminosae family) and forming on the roots of these plants, *nodules* where nitrogen fixation takes place.

b) **Reduction of nitrates and nitrites to ammonia.** – FORMATION OF NITRITES AND NITRATES – They can be formed directly from atmospheric nitrogen oxidized by lightenings during storms. But they mainly originate from the oxidation of NH_3; actually, as mentioned in the previous paragraph, the ammonia formed either by decomposition of organic material, or by nitrogen

[1] However, after the action of mutagenic agents and selection, one can isolate mutants which have lost the capacity of synthesizing one (or several) amino acids; these mutants are called *auxotrophic* vis-a-vis the amino acid concerned; this amino acid has become essential and must be supplied to enable their growth. The study of such mutants has led to great advances in the study of metabolic processes and regulation mechanisms.

fixation, undergoes *nitrification*, a process that can be summarized by the following 2 reactions:

– in nitrifying bacteria of the genus *Nitrosomonas*:

$$NH_3 + 3/2\ O_2 \longrightarrow NO_2^- + H_2O + H^+ \ (\Delta G_0 = -66.5 \text{ kcal})$$

– in nitrifying bacteria of the genus *Nitrobacter*:

$$NO_2^- + 1/2\ O_2 \longrightarrow NO_3^- \ (\Delta G_0 = -17.5 \text{ kcal})$$

These 2 groups of bacteria are *autotrophic* and synthesize all their organic compounds from CO_2 using, not light energy like the photosynthetic organisms, but the energy produced by the oxidation of inorganic compounds during the exergonic reactions written above, hence the name *chemo-autotrophs* given to these bacteria.

ASSIMILATION OF NITRITES AND NITRATES. – This is the name given to the process used by diverse organisms – especially plants – to reduce nitrates and nitrites to NH_3. The first step is catalyzed by a *nitrate-reductase*, an enzyme present in various organisms (bacteria, fungi, plants) and which catalyzes the reduction of NO_3^- to NO_2^- with NADH, NADPH or $FADH_2$ as the source of electrons. Then the nitrite formed is itself reduced by a nitrite-reductase and finally NH_3 is obtained, after passing through a number of intermediates (apparently including hydroxylamine), and thanks to flavoproteins containing molybdenum indispensable for their activity.

B. – Incorporation of ammonia into an organic compound

The four most important processes permitting the "fixation" of NH_3 into organic compounds have already been studied; they are:

– reducing amination of α-ketoglutaric acid to glutamic acid by glutamate-dehydrogenase (see fig. 7-4)
– amination of fumaric acid to aspartic acid (see fig. 7-5)
– formation of carbamyl-phosphate (see fig. 6-22)
– formation of an amide like glutamine (see fig. 6-22)

The diverse modes of formation of ammonia and its incorporation into organic compounds can be considered within the general scope of the *nitrogen cycle*, which can be represented in a very simplified form (see fig. 7-6).

C. – Synthesis of the carbon skeleton of amino acids

The α-keto acids corresponding to a number of amino acids are common products of the intermediate metabolism, especially of glycolysis and Krebs cycle; this is particularly the case of pyruvic acid (α-keto acid corresponding to alanine), α-keto glutaric acid (corresponding to glutamic acid), oxaloacetic acid (corresponding to aspartic acid). Some other α-keto acids derive more or less directly from compounds of the intermediate metabolism, for example, 3-phospho-hydroxy-pyruvic acid (corresponding to phospho-serine) which

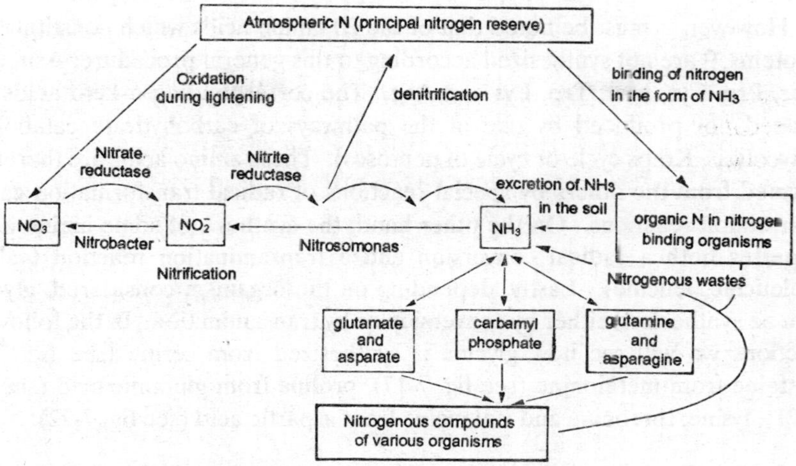

FIG. 7-6. — Simplified diagram of the nitrogen cycle showing particularly the formation of NH_3 and its incorporation into organic compounds.

results from 3-phospho-glyceric acid by dehydrogenation (see fig. 7-8). But while microorganisms and plants are capable of forming all the α-keto acids, higher animals and particularly man, are incapable of achieving the synthesis of some of them; these α-keto acids or the corresponding amino acids must be supplied to them through diet; we will study in the following these amino acids called "*essential*". As regards the organisms capable of synthesizing all the α-keto acids, the reactions required are sometimes numerous and complex – especially for the aromatic amino acids – and we will not describe in this chapter the biosynthesis of all amino acids.

D. – Transfer of the amino group to the α-keto acids

Generally, amino acids are formed form the corresponding α-keto acids, by *transamination*. Since *glutamic acid* is the principal compound resulting from the incorporation of NH_3 in organic form, it will be the *main donor of the amino group* in transamination reactions, as mentioned earlier while studying this process (see page 392). As for α-keto acids, they are produced by the pathways of carbohydrate catabolism (pyruvic acid, oxaloacetic acid) or derive from these pathways.

The formation of various amino acids from ammonia and α-keto acids can be summarized in a simple diagram (see fig. 7-7):

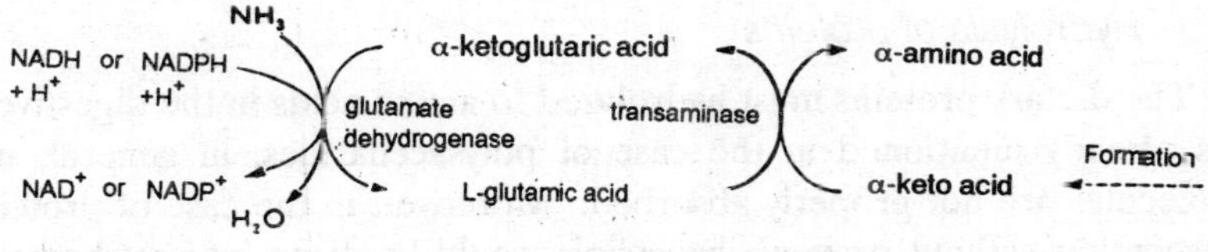

FIG. 7-7. — Diagram summarizing the synthesis of an amino acid from the corresponding α-keto acid and NH_3.

However, it must be noted that of the 20 amino acids which constitute the proteins, 9 are not synthesized according to this general procedure: Asn, Gln, Thr, Pro, Cys, Met, Trp, Lys and Arg. The corresponding α-keto acids are indeed not produced by one of the pathways of carbohydrate catabolism (glycolysis, Krebs cycle or cycle of pentoses). These amino acids are therefore formed from the others by special reactions of radical transformation called conversion reactions. On the other hand, the synthesis of some amino acids requires both a radical conversion and a transamination reaction (valine, isoleucine, leucine). Lastly, depending on the organism considered, glycine can be synthesized either by conversion or by transamination. In the following sections we will see how glycine is synthesized from serine (see fig. 7-9), cysteine from methionine (see fig. 7-17), proline from glutamic acid (see fig. 7-21), lysine, threonine and isoleucine from aspartic acid (see fig. 7-22).

2. – Absorption of preformed amino acids

A. – *Essential amino acids*

In mammals – and especially in man – a number of amino acids cannot be formed by the mechanisms we have just described (see fig. 7-7), either because the corresponding α-keto acids are not present (and cannot be synthesized), or because they can undergo neither amination nor transamination. In man, these amino acids are 8 in number: three amino acids with hydrocarbon side chain (Leu, Ile, Val), one basic amino acid (Lys), one hydroxylated amino acid (Thr), one sulphur-containing amino acid (Met) and 2 aromatic amino acids (Phe and Trp). They are called "*indispensable*" or "*essential*" amino acids and must be supplied by diet. There are 9 amino acids essential for the growth of rat – all the 8 already cited for man and additionally, histidine – as may be proved by dietary experiments in which the animal is fed, not with proteins but with an artificial mixture of free amino acids, omitting by turn each amino acid: if the amino acid omitted is essential, the biosynthesis of proteins is perturbed. Theoretically, normal life and growth can be ensured by providing the essential amino acids and ammonium salts to allow the formation of other amino acids. But in reality, animal as well as human diet includes proteins which yield the essential amino acids as well as those which could be synthesized: one can thus avoid consuming the corresponding precursors like the α-keto acids participating in the glycolysis or Krebs cycle (they can thus continue to function in the oxidation of carbohydrates or fatty acids which produces ATP).

B. – *Hydrolysis of proteins*

The dietary proteins must hydrolyzed to amino acids in the digestive tract. As already mentioned in the case of polysaccharides, in general, macromolecules are not properly absorbed. Moreover, in the case of proteins, an absorption without previous hydrolysis would be of no interest because the proteins are specific: we do not need dietary plant or animal proteins; we need proteins of our own with very different sequences, resulting from the expression of our genes. Lastly, the introduction of foreign proteins in our organism may be fraught with real dangers and can be followed by reactions ranging from relatively minor allergic manifestations (skin reactions in some persons following ingestion of fish, crustacea, or hay fever in presence of some pollens), to serious conditions of anaphylactic shock (observed for example during serotherapy when the patient is injected with rather large and repeated quantities of horse serum containing serum proteins of this animal and especially the antibody desired). For all these reasons the proteins must be hydrolyzed to amino acids and the latter can then be absorbed, not by simple diffusion, but thanks to specific proteins called permeases.

We have seen in Chapter 1 that, in proteins, amino acids are linked to one another by the peptide linkage and one might think that a single enzyme would suffice for the complete hydrolysis of all the proteins as there is only one type of bond to be split. In reality, the *proteolytic enzymes*, or *proteases*, or *peptidases*, reveal a more or less pronounced specificity vis-a-vis the *position of the peptide bond in the chain* and the *nature of the amino acids* involved in this linkage. As a result, the rapid and complete hydrolysis of a protein requires the participation of several enzymes capable of attacking the polypeptide chain at diverse points.

a) **Endopeptidases.** – They are the enzymes which catalyze the hydrolysis of peptide bonds within the chains. There are 3 important endopeptidases in our digestive tract:

– *pepsin* is secreted by the gastric mucosa in the form of inactive *pepsinogen*, which is transformed – by removal of a number of small peptides – into an active enzyme whose optimum pH is very low (around 2). As mentioned in Chapter 2 (see fig. 2-26), pepsin preferentially hydrolyzes the peptide bonds involving the amino group of an aromatic amino acid (Phe, Tyr);

– *trypsin* is contained in the pancreatic juice and continues in the intestine the digestion of proteins started in the stomach by pepsin. It is secreted by the pancreas, also in the form of an inactive precursor – *trypsinogen* – whose transformation into trypsin is an activation which is either *autocatalytic* (under the influence of trypsin itself), or catalyzed by another proteolytic enzyme such as *enterokinase* secreted by the duodenal mucosa, and which consists in the removal of a hexapeptide. Trypsin has an optimum pH around 8; it preferentially hydrolyzes the peptide bonds in which a basic amino acid (Arg, Lys) participates by its carboxyl (see fig. 2-26).

– *chymotrypsin* is also secreted by the pancreas in the form of a zymogen, the *chymotrypsinogen*, transformed into chymotrypsin by the action of trypsin. (see page 89). It also has an optimum pH near 8 and preferentially splits the peptide bonds in which aromatic amino acids (Trp, Tyr, Phe) are involved; but contrary to pepsin which splits the bonds in which amino groups of aromatic acids are involved, chymotrypsin splits those in which carboxylic groups are involved (see fig. 2-26).

The action of these 3 endopeptidases on the dietary proteins leads to a mixture of peptides of varied sizes and some free amino acids. These peptides will then be attacked by exopeptidases.

b) **Exopeptidases.** – They are the enzymes which catalyzes the hydrolysis of peptide bonds at the end of chains:

– the pancreatic *carboxypeptidases* hydrolyze the peptide bond in which the C-terminal amino acid is involved;

– the intestinal *aminopeptidases* hydrolyze the peptide bond in which the N-terminal amino acid is involved. As indicated by its name, the *leucine-aminopeptidases* has a preference for leucine, but this is not an absolute specificity and we have seen in Chapter 1 (see page 22) how this enzyme, as also the carboxypeptidases,permit the determination of the nature of amino acids situated at the ends of the peptide chains;

– the intestinal *dipeptidases* can split the dipeptides; but a particular enzyme, the prolidase, is required for hydrolyzing the dipeptides in which proline is the amino acid having the free carboxyl because this is not a conventional peptide bond.

We only mentioned the proteolytic enzymes enabling the hydrolysis of dietary proteins in our digestive tract. But many animal cells also contain endopeptidases called *cathepsins A, B and C* whose specificity is analogous respectively to that of pepsin, trypsin and chymotrypsin; these cathepsins are responsible for the post-mortem autolysis but their role in living cells is not yet fully understood. Lastly, it must be said that there are other proteolytic enzymes in living organisms: subtilisin (in *Bacillus subtilis*), papain (in the papaya tree, a higher plant), etc.

III. – METABOLISM OF AMINO ACIDS

We have seen how ammonia can be formed and then be found in the form of *amino group* of glutamic acid. We have studied the reactions – transamination and deamination – by which amino groups can be added to or removed from amino acids. At the end of this chapter we will see the forms in which ammonia resulting from the amino groups of amino acids, can be eliminated by living organisms.

As for the *carbon skeleton* of the amino acid, it can be used for regenerating the corresponding amino acid (e.g, by transamination). It can also be used, either as such, or after some modifications for *synthesizing other amino acids* or for participating in reactions leading to the *synthesis of compounds important* for the organism (nucleotides, pigments, hormones, porphyrins, etc[1]). Lastly, it can be degraded and thus join the carbohydrate or lipid metabolisms whence the notion of *glucogenic and ketogenic* amino acids.

---

(1) As will be seen in the following, these transformations may be relative to either the carbon skeleton, or the amino acid which has conserved its amino group.

The first experiments on this point were carried out in rather abnormal physiological conditions as they consisted in administering the amino acids to be studied, to animals made diabetic, either by removal of the pancreas, or by the action of phlorizin (which prevents the reabsorption of the glucose in the tubules by lowering the threshold of renal elimination); it was thus observed that in these conditions some amino acids cause an increase of glucosuria (glucogenic amino acids), and other amino acids, an increase of ketonuria (ketogenic amino acids). The glucogenic character of some amino acids was also tested by studying the accumulation of hepatic glycogen, after administration of these amino acids to fasting rats. These experiments were confirmed, in normal physiological conditions, by the use of amino acids containing a radioactive isotope of carbon (^{14}C). There are *glucogenic* amino acids whose carbon atoms are found in glucose and glycogen upon neoglucogenesis. (see pages 185-187). The reactions we have already studied at the beginning of this chapter (in connection with transamination and deamination) explain the fact that among the above-mentioned glucogenic amino acids, one finds for example, glutamic acid (as it gives α-keto glutaric acid), aspartic acid (which gives oxalo-acetic acid), alanine, serine and cysteine (which all three give pyruvic acid); other amino acids are also glucogenic, particularly glycine and proline and we will understand the reason after studying the metabolism of these amino acids. Among the *ketogenetic* amino acids whose catabolism joins that of fatty acids we may cite phenylalanine and tyrosine, the metabolism of which leads to the formation of acetoacetic acid as we shall see in the following (fig. 7-24). In fact these two aromatic amino acids can also be considered as glucogenic amino acids since their catabolism also yields fumaric acid which is an intermediate of Krebs cycle and can thus lead to glucose by neoglucogenesis. In Chapter 5, (while studying the oxidation of a branched fatty acid) we have seen that isoleucine is also a glucogenic and ketogenic amino acid since its catabolism (see fig. 5-15) leads to, on the one hand, succinyl-coenzyme A which can permit the formation of glucose by neoglucogenesis and on the other hand, acetyl coenzyme A which can be used either for the ketogenesis or the biosynthesis of lipids.

But the metabolism of the carbon skeleton, because of the wide variety of the structures of amino acids, is extremely different from one amino acid to another. And just as we had to exclude from this chapter a description of the biosynthesis pathways of all amino acids, we will not be able to study here the metabolism of all amino acids. We will examine the metabolism of glycine and serine (two closely related amino acids), sulphur-containing amino acids, glutamic and aspartic acids and the derived amino acids, phenylalanine and tyrosine (also closely related).

1. – Metabolism of glycine and serine

They are two non-essential amino acids; as we shall see in the following they are interconvertible, so that for a long time, one did not know which one was formed first and gave rise to the other. It is now known that it is not the simpler of the two (glycine), but serine which is synthesized first and we will therefore study it in the first place.

A. – *Synthesis of serine*

Serine is formed from 3-phosphoglyceric acid (a compound of the intermediate metabolism which appears during glycolysis) by a series of 3 reactions represented in figure 7-8.

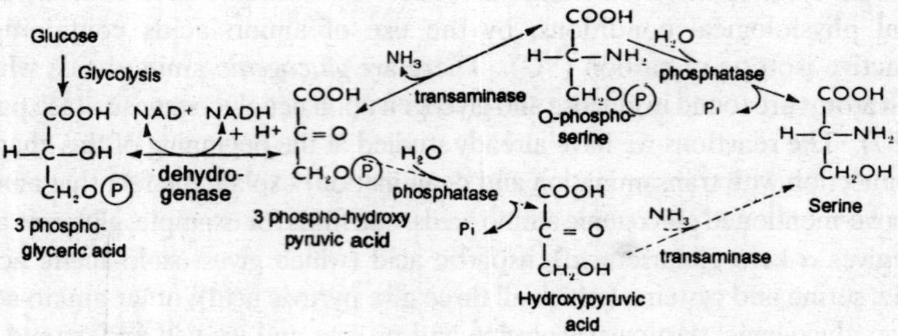

FIG. 7-8. — Formation of serine. (It appears that the transaminase acts before the phosphatase and that the process passes through the intermediary of phosphoserine, but the last two reactions could also take place in the reverse order).

B. – *Metabolism of serine*

Under the action of serine dehydratase, serine is deaminated to pyruvic acid (see fig. 7-2). We shall see in a later section that the metabolism of serine is linked with that of sulphur-containing amino acids (see fig. 7-17). The decarboxylation of serine gives ethanolamine which will be referred to in connection with glycine (see fig. 7-10).

C. – *Serine-glycine interconversion*

This interconversion is catalyzed by a specific enzyme (its name, serine transhydroxymethylase, is a misnomer, because it is not a hydroxymethyl but a formyl which is transferred) and requires two coenzymes: on the one hand, pyridoxal phosphate which forms intermediately a Schiff base with the amino acid, as already seen (see fig. 7-1), and on the other hand a tetrahydrofolic derivative (FH_4). As observed in Chapter 2 (see fig. 2-19), FH_4 is one of the two carriers of one-carbon units (the other carrier is S-adenosyl-methionine which we will study in a subsequent section of this chapter). Some authors suggested that a hydroxymethylated derivative of FH_4 is formed (on N_5 or N_{10}), but in fact it rather appears that it is N_5-N_{10}-methylene FH_4 which is formed by the removal of a molecule of water as shown by figure 7-9.

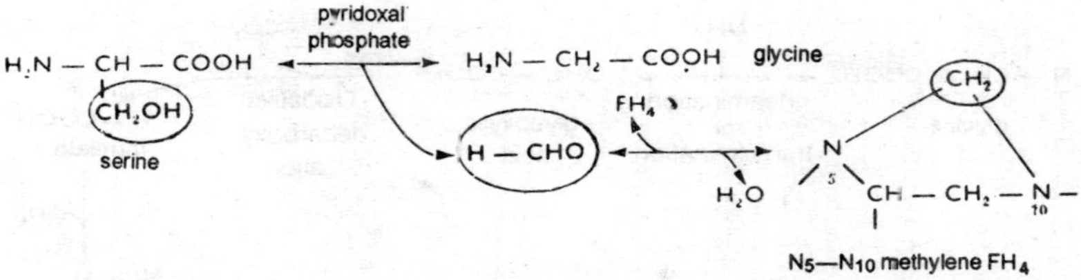

FIG. 7-9. — Serine-glycine interconversion.

D. – Metabolism of glycine

a) **Formation of glycine from ethanolamine.** – To begin with, *ethanolamine* must be transformed into *choline* through 3 successive methylations at the cost of S-adenosylmethionine: this takes place while ethanolamine is linked to a phosphatidic acid; in fact, there is therefore conversion of phosphatidyl-ethanolamine into phosphatidyl-choline (see structures in figure 5-3). The liberated choline can be oxidized to betain and then lose successively its 3 methyl groups to give glycine (see fig. 7-10). FH_4 picks up these methyl groups.

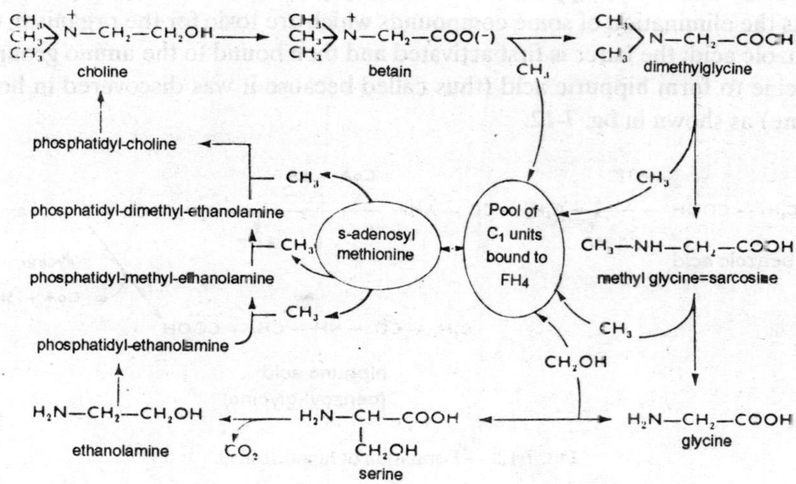

FIG. 7-10. — Formation of glycine from ethanolamine through choline.

b) **Deamination and transamination of glycine.** – Glycine is transformed into *glyoxylic acid* by oxidative deamination (under the influence of a specific glycine oxidase) or by reversible transamination. This compound undergoes an oxidative decarboxylation to CO_2 + formiate which is not liberated in the mixture but picked up by FH_4 to form N_5-formyl FH_4 or N_{10}-formyl FH_4 (see fig. 7-11).

Glycine can also be formed by transamination – as the process is reversible – from glyoxylic acid. The latter, as we have seen, can arise from isocitric

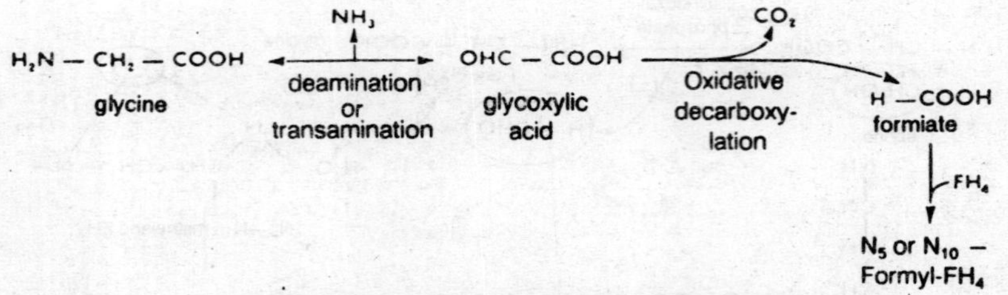

FIG. 7-11. — Catabolism of glycine.

acid (see fig. 5-26) and the glyoxylic acid cycle also represents a contact point of carbohydrate and lipid metabolism (see fig. 5-27).

It is of interest to note that during the 3 processes we have just studied (serine-glycine interconversion, transformation of betain into glycine, catabolism of glycine), one-carbon atom units are detached and taken over by FH_4; they join the pool of one-carbon units which can be used in various reactions like *de novo* biosynthesis of purine nucleotides (see fig. 6-19), methylation of *d*UMP to *d*TMP (see page 299). etc.

c) **Formation of hippuric acid and conjugated bile acids.** – Glycine permits the elimination of some compounds which are toxic for the organism, e.g. benzoic acid; the latter is first activated and then bound to the amino group of glycine to form hippuric acid (thus called because it was discovered in horse urine) as shown in fig. 7-12.

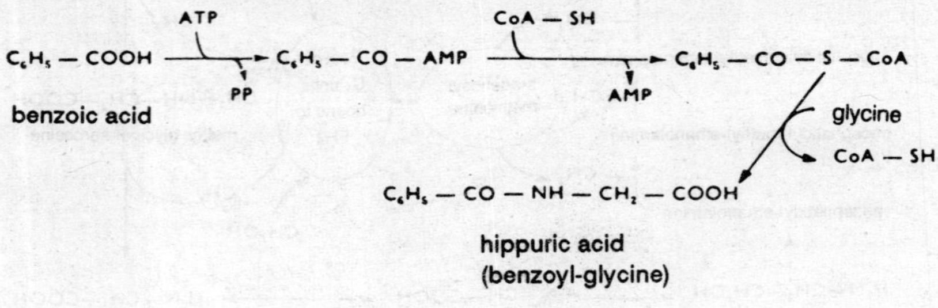

FIG. 7-12. — Formation of hippuric acid.

By a similar mechanism, cholic and deoxycholic acids are first activated and then conjugated with glycine to form glycocholic and glycodeoxycholic acids (see fig. 5-7).

d) **Synthesis of glutathione.** – The synthesis of glutathione (see fig. 1-12) takes place in 2 steps:

Glutamic acid + cysteine + ATP ⟶ γ-glutamyl-cysteine + ADP + P_i

γ-glutamyl-cysteine + glycine + ATP ⟶ glutathione + ADP + P_i

e) **Synthesis of creatine.** – This synthesis begins with a transfer of the guanidic group from arginine to glycine; the product formed is then methylated to creatine by S-adenosyl-methionine (see fig. 7-13). It is observed that the

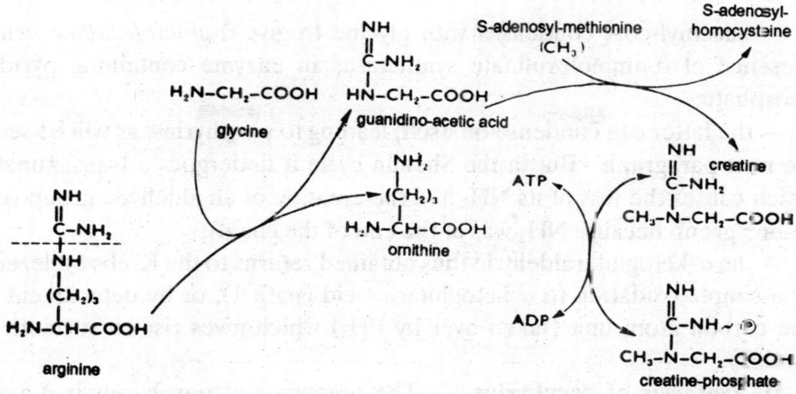

FIG. 7-13. — Synthesis of creatine and storage of the ATP energy in the form of creatine-phosphate.

carbon atoms of creatine originate from 3 different amino acids. We have already emphasized in Chapter 3 (see page 110-111) the importance of creatine: when ATP concentration is high, creatine is converted in the muscle into creatine-phosphate which can once again decompose with formation of ATP when the muscle needs it (see fig. 7-13); creatine-phosphate thus constitutes an immediately usable energy reserve, supplying the ATP required for muscular contraction, until the glycolysis stimulation brings about an increased formation of ATP.

***f)* Shemin cycle. – This cycle is a *derivation of the Krebs cycle* at the succinyl-coA level. Its 3 steps are shown in fig. 7-14.**

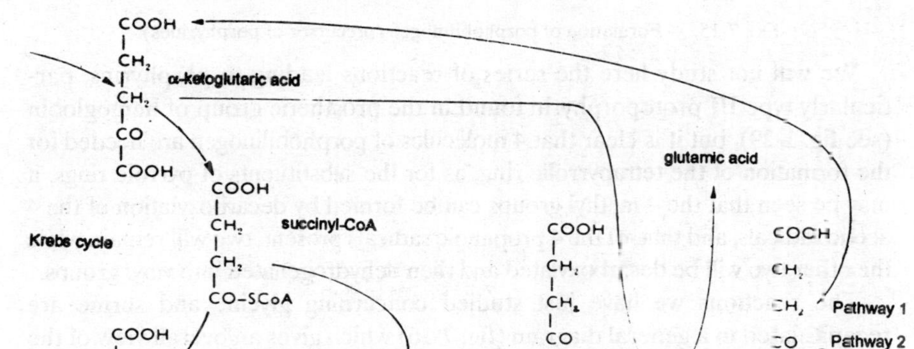

FIG. 7-14. — Shemin cycle.

– succinyl-coA condenses with glycine to give *δ-aminolevulinic acid*, in presence of δ-aminolevulinate synthetase, an enzyme containing pyridoxal phosphate;

– the latter can condense on itself, leading to porphyrins, as will be seen in the next paragraph. But in the Shemin cycle it undergoes a transamination, which causes the loss of its NH_2 and the creation of an aldehyde group (not a ketone group because NH_2 was at the end of the chain);

– the α-ketoglutaraldehyde thus obtained returns to the Krebs cycle, either by a simple oxidation to α-ketoglutaric acid (path 1), or by detachment of a one-carbon atom unit (taken over by FH_4) which gives rise to succinic acid (path 2).

g) **Synthesis of porphyrins**. – The precursor of porphyrins is δ-aminolevulinic acid; its mode of formation was explained in the preceding paragraph. As shown by figure 7-15, the condensation of 2 molecules of this compound leads to *porphobilinogen*.

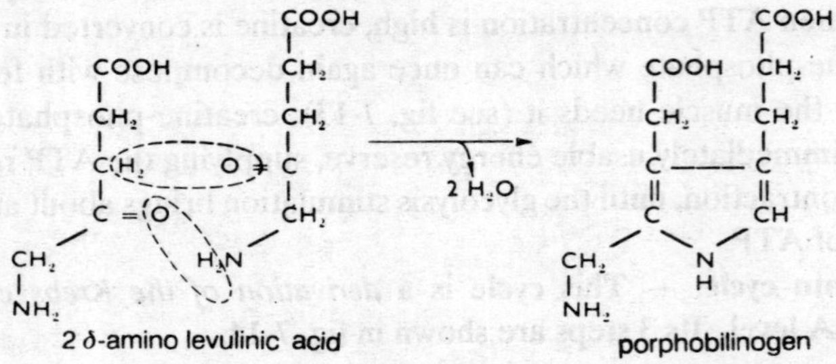

FIG. 7-15. — Formation of porphobilinogen (precursor of porphyrines).

We will not study here the series of reactions leading to porphyrins, particularly type III protoporphyrin found in the prosthetic group of hemoglobin (see fig. 1-29), but it is clear that 4 molecules of porphobilinogen are needed for the formation of the tetrapyrrolic ring; as for the substituents of pyrrole rings, it may be seen that the 4 methyl groups can be formed by decarboxylation of the 4 acetic radicals, and that, of the 4 propanoic radicals present, two will remain while the other two will be decarboxylated and then dehydrogenated into vinyl groups.

The reactions we have just studied concerning glycine and serine are recapitulated in a general diagram (fig. 7-16) which gives an overall view of the metabolism of these two amino acids. It will be noted that a number of reactions yield fragments in C_1; this is why glycine and serine are often counted among amino acids yielding one-carbon units.

2. – Metabolism of sulphur-containing amino acids

There are 3 sulphur-containing amino acids, *methionine, cysteine* and *cystine*, but since the last two are very easily interconvertible by oxidation-reduction,

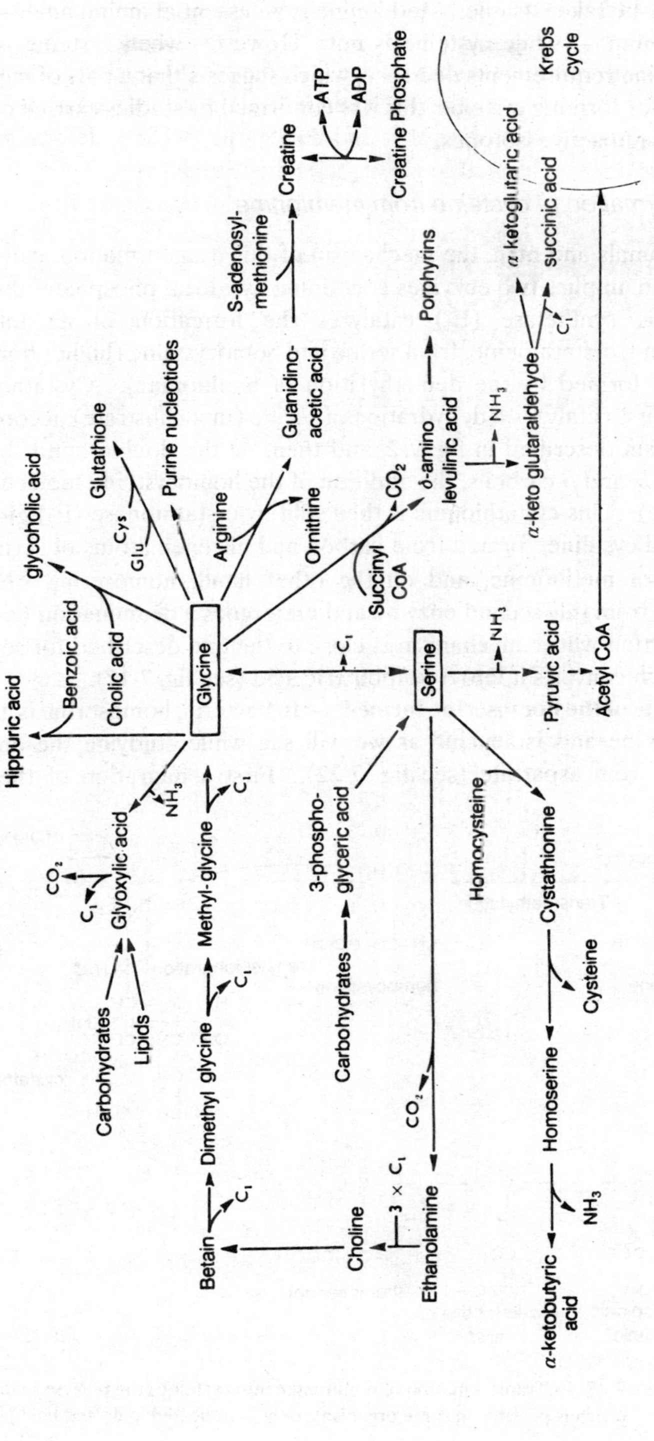

FIG. 7-16. — Overall diagram of the metabolism of glycine and serine.

they may be taken as one. Methionine is an essential amino acid – for rat as well as man – while cysteine is not. However, when cysteine is supplied, methionine requirements decrease, which suggests that a part of methionine is utilized for forming cysteine; this was confirmed by studies carried out with the help of radioactive isotopes.

A. – Formation of cysteine from methionine

In animals and man, the mechanism of this transformation, called transulphuration, implies two enzymes containing pyridoxal phosphate: the first, cystathionine synthetase (E_1) catalyzes the formation of an intermediate compound, cystathionine, from serine and homocysteine (higher homologue of cysteine, formed by the demethylation of methionine). Cystathionine synthetase first catalyzes a dehydration of serine (first substrate) according to the mechanism described in fig. 7-2, and then, on the double bond thus formed between α and β carbons, the addition of the homocysteine molecule (second substrate). This cystathionine is then split by cystathionase (E_2) giving on the one hand cysteine, formed from carbon and nitrogen atoms of serine and the sulphur of methionine, and on the other hand, homoserine which is not released from this second enzyme and undergoes a deamination (α-γ elimination reaction whose mechanism is close to the one described for serine in fig. 7-2) which converts it into α-ketobutyric acid (see fig. 7-17).

a) **Fate of the homoserine formed.** – In bacteria, homoserine is a precursor of threonine and isoleucine as we will see while studying the amino acids deriving from aspartate (see fig. 7-22). First a migration of the alcoholic

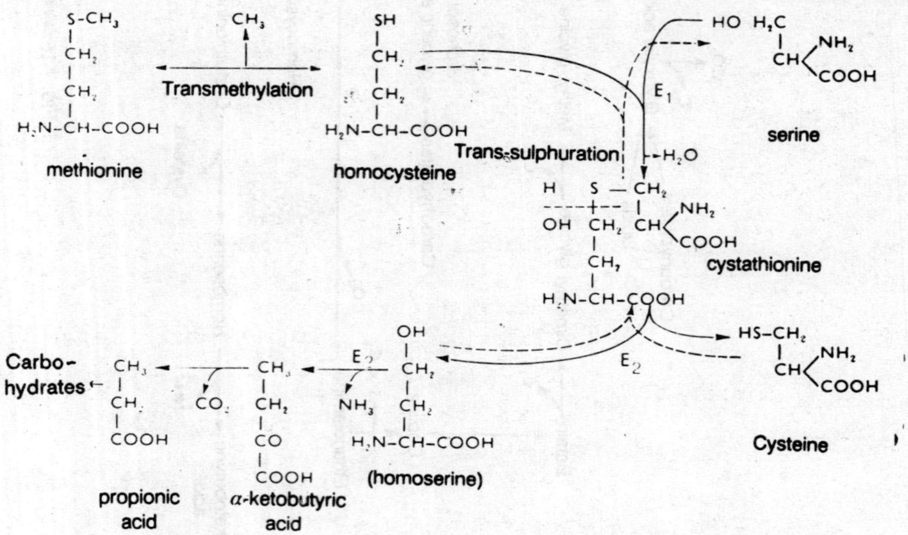

FIG. 7-17. — Transformation of methionine into cysteine (the reverse reaction which is possible in some organisms only, is indicated in dotted lines).

hydroxyl leads from homoserine (or more exactly from its derivative phosphorylated on the alcohol group) to threonine, then a deamination gives α-acetobutyric acid (see fig. 7-22). There are now two possibilities: either the formation of isoleucine (see fig. 7-22), or the decarboxylation of α-ketobutyric acid to *propionic acid* (see fig. 7-17). But we have seen (see fig. 5-13) that propionyl-coA can be transformed into succinyl-coA; the carbon atoms of propionic acid can then follow this pathway and be present for example in glucose after neoglucogenesis.

b) **Reversibility of *trans*-sulphuration.** – In animals and man, trans-sulphuration is not reversible, but it is so in various organisms which can therefore form cystathionine from homoserine and cysteine under the action of a cystathionine synthetase (E_1') and split this cystathionine by a cystathionase (E_2') into homocysteine, pyruvate and ammonia (see fig. 7-22); in other words these organisms can (by this process) form homocysteine from cysteine and are thus capable of synthesizing methionine.

B. – *Methylation of homocysteine into methionine*

If the methyl groups of choline or betain are labeled (see structures in figure 7-10), radioactivity is observed in CH_3 of methionine. Methylation is therefore possible (it takes place thanks to N_5-methyl-FH_4 which brings the one-carbon unit); it is possible even in animals. Therefore, it is homocysteine rather than methionine, which is essential, and if it is fed to the animal it is methylated into methionine.

C. – *Metabolism of methionine*

In addition to its role of amino acid constitutive of proteins and its part in the initiation of protein biosynthesis (see page 367), methionine is mostly a supplier of methyl groups. It must be first *activated* by ATP to give *S-adenosyl-methionine* which is the real agent participating in the *transmethylation* processes. It can then yield its methyl group to very diverse compounds and be converted into S-adenosyl-homocysteine, as shown by figure 7-18; this is the mechanism involved for example in the biosynthesis of choline (see fig. 7-10), creatine (see fig. 7-13), adrenaline (see fig. 7-28), in the methylation of some bases (called abnormal or rare) of *t*RNAs (see page 335), etc.

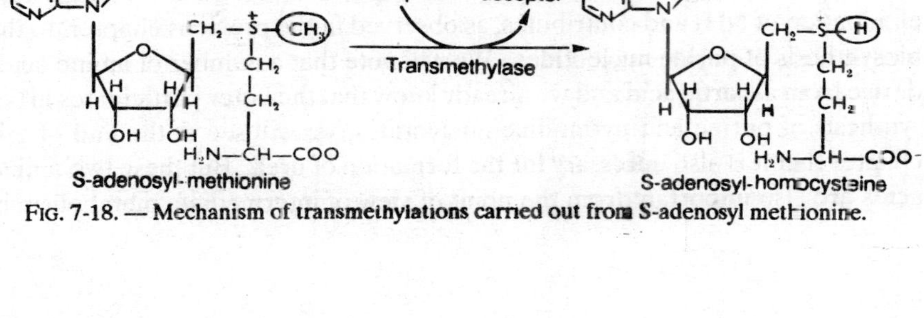

FIG. 7-18. — Mechanism of transmethylations carried out from S-adenosyl methionine.

D. – Metabolism of cysteine

Let us first recall that cysteine is a constituent of *glutathione* (see fig. 1-12) whose mode of formation we discussed while studying the metabolism of glycine (page 406). Moreover, one must remember the *reversible oxidation cysteine* ←→ *cystine* (thiol ←→ disulphide, i.e. 2R–SH ←→ R–S–S–R) which can take place in the free amino acid, in glutathione or in proteins. As for the *decarboxylation* of cysteine, it leads to mercaptoethylamine or cysteamine ($H_2N-CH_2-CH_2-SH$) which is a constituent of coenzyme A (see fig. 2-18).

a) **Transformation of cysteine into pyruvic acid.** – Two pathways are possible:

– the first, which we already studied (see fig. 7-2), consists of a deamination catalyzed by cysteine desulphydrase;

– the second consists of an oxidation of cysteine to cysteine-sulphinic acid which loses its amino group by transamination and gives sulphinyl-pyruvic acid. The latter gives pyruvic acid by liberating sulphur which is found in the form of sulphite and then *sulphate* (see fig. 7-19).

This sulphate is activated in the form of adenosine 3′-phosphate-5′-phospho-sulphate which can react with diverse compounds like phenols or steroids which are then eliminated in urine in the form of sulpho-conjugated derivatives (detoxication).

b) **Formation of taurine.** – Cysteine-sulphinic acid whose formation by oxidation of cysteine we mentioned above can – instead of being transaminated – be decarboxylated to hypotaurine, which is then oxidized to taurine (there can also be, first oxidation to cysteic acid, then decarboxylation to taurine) as shown by figure 7-19. Taurine can – like glycine – be conjugated with the bile acids (see fig. 5-7).

Figure 7-20 presents a recaptulative diagram of reactions of the metabolism of sulphur-containing amino acids.

3. – Metabolism of dicarboxylic amino acids and their amides

Glutamic acid, aspartic acid and their amides occupy an important place in the metabolism of nitrogenous compounds. We have already seen that glutamic acid was the main entry point of ammonia into biological compounds and we will see that its metabolism is linked to that of several other amino acids; as for its amide, glutamine, it plays an important role in the storage and elimination of NH_3 and contributes, as observed in the previous chapter, to the biosynthesis of purine nucleotides. We will note that a number of amino acids derive from aspartic acid and we already know that the latter participates in the synthesis of purine and pyrimidine nucleotides; we will see at the end of this chapter that it is also necessary for the formation of urea. But these two amino acids are also important from the point of view of intermediate metabolism in

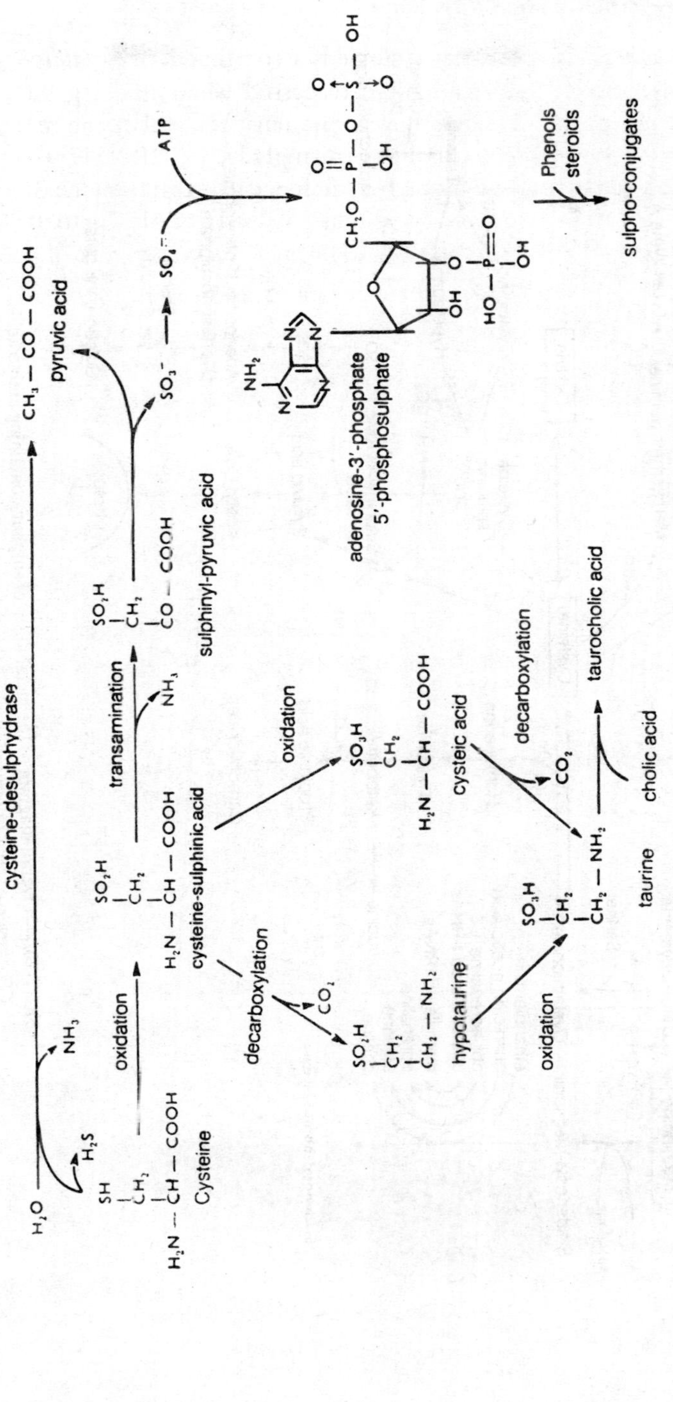

FIG. 7-19. — Metabolism of cysteine.

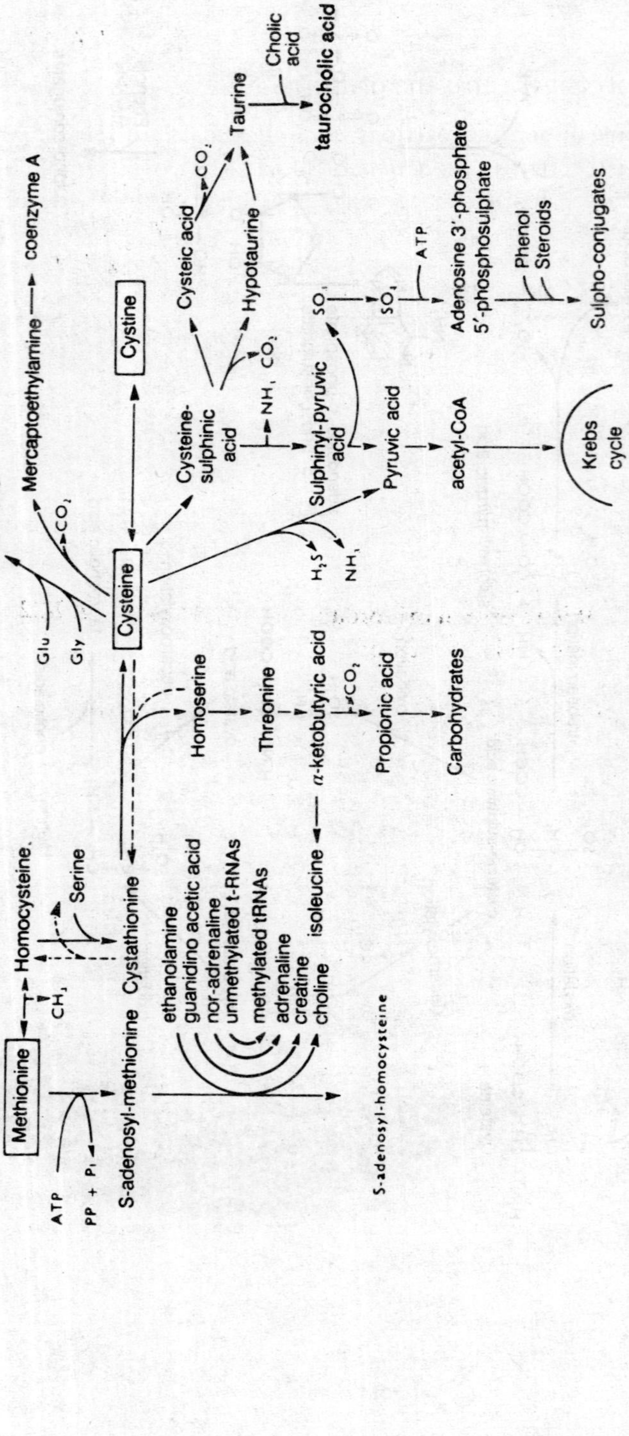

FIG. 7-20. — Overall diagram of the metabolism of sulphur-containing amino acids. (The reverse trans-sulphuration which does not take place in animals and man is indicated in dotted lines).

general; actually, their corresponding α-keto acids are intermediates of Krebs cycle and therefore constitute contact points between carbohydrate and protein metabolisms.

A. – Metabolism of glutamic acid and glutamin

a) **Review of some important reactions.** – *Glutamic acid is formed from α-ketoglutaric acid*, either by *reducing amination* (see fig. 7-3), or by transamination. As observed in the foregoing, (see page 393), glutamic acid very often participates in transamination processes during which it parts with its amino group to permit the formation of other amino acids. By *decarboxylation* it gives *γ-aminobutyric acid*, an inhibition mediator of synapses mainly found in the brain. It is also a constituent of *glutathione* (see fig. 1-12), the synthesis of which we studied in connection with glycine (page 406). Besides, it is a constituent of pteroylglutamic acid or *folic acid* (see fig. 2-19), whose tetrahydrogenated derivative (FH_4) plays a capital role in the transport of one-carbon units.

b) **Metabolic inter-relations between glutamic acid, ornithine and proline.** – As shown by figure 7-21, *glutamic γ-semialdehyde* (resulting from glutamic acid by reduction of its γ-carboxyl) can either cyclize itself and give *proline* (and hydroxyproline, but hydroxylation takes place on the proline already incorporated in proteins), or undergo a transamination which converts it into *ornithine*; ornithine easily leads to arginine by the ureogenesis cycle which will be studied later in this chapter (see fig. 7-30). These reactions are reversible.

c) **Metabolism of glutamine.** – Glutamine enables the *storage* of NH_3 which is toxic for animal tissues, and it is an intermediate in the elimination of

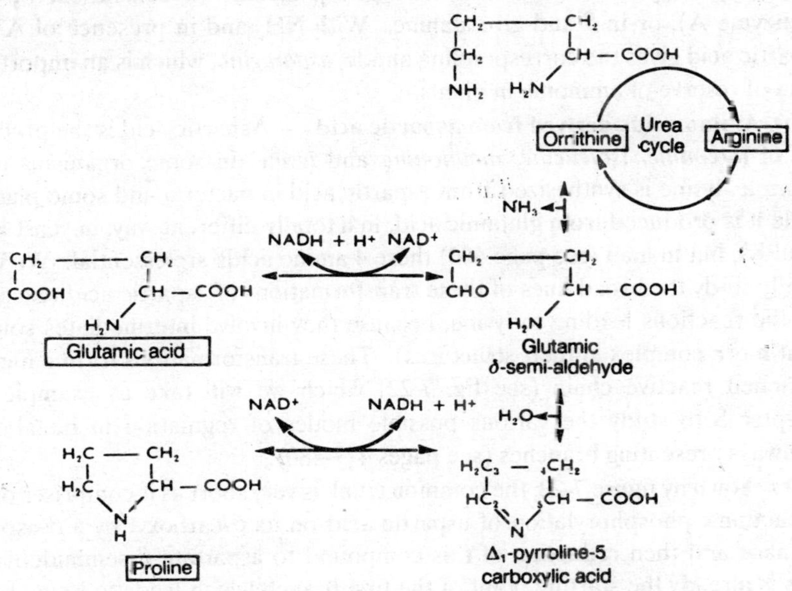

FIG. 7-21. — Interconversions between glutamic acid, ornithine and proline.

NH_3; the major part of urine ammonia in mammals results from the hydrolysis of blood glutamine in the kidney (see page 427).

We have seen how glutamine is formed from glutamic acid (see fig. 6-22) and how it yields its amidic NH_2 during the synthesis of purine nucleotides (see figs. 6-19 and 6-21), the formation of carbamyl-phosphate (see fig. 6-22) and the amination of UTP to CTP (see fig. 6-23). It also permits the formation of osamines.

Lastly, like glycine, glutamine participates in *detoxication* processes, and phenyl-acetic acid for example (see fig. 7-25), is excreted combined with the α-amino group of glutamine.

B. – Metabolism of aspartic acid and asparagine

a) **Review of some important reactions.** – Aspartic acid is formed by transamination from oxalo-acetic acid (see page 393). It can yield its amino group by transamination to permit the formation of other amino acids and it gives back oxaloacetic acid. *Oxalo-acetic acid* occupies a key position, because on the one hand it is so to speak the "starting point" of the Krebs cycle, since it takes over the *acetyl-coA* molecules from the carbohydrates and lipids, and on the other hand, it is the starting point of *neoglucogenesis* (see figs. 4-34 and 4-38). Besides, aspartic acid can also join the Krebs cycle at the level of fumaric acid by deamination (see fig. 7-5). We have seen that aspartic acid is a precursor of *purine nucleotides* (see figs. 6-19 and 6-21) and *pyrimidine* nucleotides (see fig. 6-22). We will see in this chapter that aspartic acid also plays an important role in *ureogenesis* (see fig. 7-30). Aspartic acid can undergo a decarboxylation either in α and give β-alanine (a constituent of the coenzyme A), or in β and give alanine. With NH_3 and in presence of ATP, aspartic acid gives the corresponding amide, *asparagine*, which is an important form of reserve of ammonia in plants.

b) **Amino acids derived from aspartic acid**. – Aspartic acid is the precursor of *threonine, isoleucine, methionine* and *lysine*, in some organisms (for example, lysine is synthesized from aspartic acid in bacteria and some plants, while it is produced from glutamic acid, in a totally different way, in yeast and moulds), but in man (see page 400) these 4 amino acids are essential. We will briefly study the broad lines of these transformations of aspartic acid (excepting the reactions leading to lysine, because they involve intermediates somewhat more complex in their structures). These transformations form a highly branched reactive chain (see fig. 7-22) which we will take as example in Chapter 8 to study the various possible modes of regulation in metabolic pathways presenting branches (see pages 477-480).

As shown by figure 7-22, the common trunk is very short as it comprises only 2 reactions: phosphorylation of aspartic acid on its β-carboxyl by a β-aspartokinase and then reduction of this compound to aspartate-β-semialdehyde. This is already the starting point of the first branch which leads to lysine by 7 reactions. On the other hand, aspartate-β-semialdehyde can be reduced to homoserine where a second branch is found: one of the pathways leads to threonine and then isoleucine; the second leads to methionine thanks to a trans-sulphuration which we already studied in connection with the transformation methionine ⟶ cysteine (see fig. 7-17) indicating that it takes place in the direction cysteine ⟶ methionine only in some organisms.

Figure 7-23 is an overall diagram of the metabolism of aspartic and glutamic acids and their amides.

4. – Metabolism of phenylalanine and tyrosine

It is not surprising that the metabolism of these two aromatic amino acids is closely linked because they differ only by the presence of a phenolic OH in the tyrosine molecule. We will not study here their biosynthesis which – in organisms where it takes place, especially bacteria – starts from 2 compounds of the carbohydrate metabolism (erythrose-4-Ⓟ and phosphoenolpyruvate) and passes through shikimic acid which is also a precursor of tryptophan. In man, phenylalanine is an essential amino acid while tyrosine is not because, as we will see in the following, it can be easily formed from phenylalanine. The study of the metabolism of these two amino acids is interesting on more than one score, because among the transformations they can undergo in the organism, some lead to hormones or pigments; but more importantly, some hereditary diseases are characterized by disorders of the catabolism of these two amino acids, which will enable us to examine how, in some cases, one can determine exactly which enzyme (i.e. which gene) is affected by the lesion which causes a hereditary disease.

A. – Normal catabolism of phenylalanine and tyrosine

As shown by figure 7-24, phenylalanine is hydroxylated to tyrosine; this reaction is not reversible which explains that the phenylalanine requirements cannot be covered by administration of tyrosine. One molecule of oxygen participates in this reaction: one atom is incorporated to the substrate, the other is reduced to H_2O by FH_4 which is oxidized to FH_2 (it is then again reduced to FH_4 by NADPH). Tyrosine then loses its amino group by transamination and gives *para*-hydroxy-phenyl-pyruvic acid which is oxidized to quinol. After transposition, an oxidative decarboxylation leads to alcaptone or homogentisic acid which is normally catabolized, on one hand, to fumaric acid (which can join the Krebs cycle and give glucose by neoglucogenesis which confers on the two amino acids a glucogenic character), and on the other hand, to aceto-acetic acid (thence the ketogenic character of phenylalanine and tyrosine).

B. – Disorders of the catabolism of phenylalanine and tyrosine

a) **At the level of phenylalanine-hydroxylase.** – This enzyme is absent in persons suffering from *phenylketonuria*, a congenital disease also called *phenyl-*

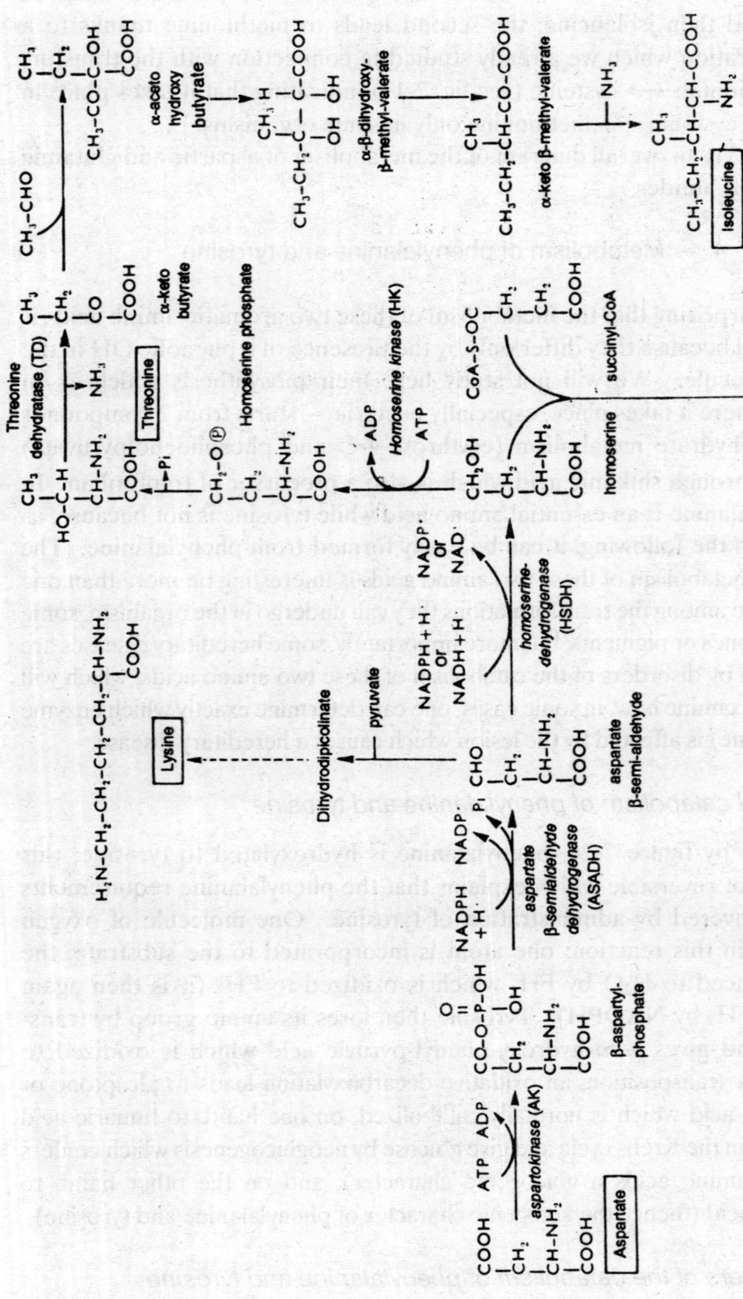
Aspartate
aspartokinase (AK)
β-aspartyl phosphate
aspartate β-semialdehyde dehydrogenase (ASADH)
aspartate β-semi-aldehyde
homoserine-dehydrogenase (HSDH)
homoserine
succinyl CoA
Homoserine kinase (HK)
Homoserine phosphate
Threonine dehydratase (TD)
Threonine
α-keto butyrate
α-aceto hydroxy butyrate
α-β-dihydroxy β-methyl-valerate
α-keto β-methyl-valerate
Isoleucine
Dihydrodipicolinate
pyruvate
Lysine

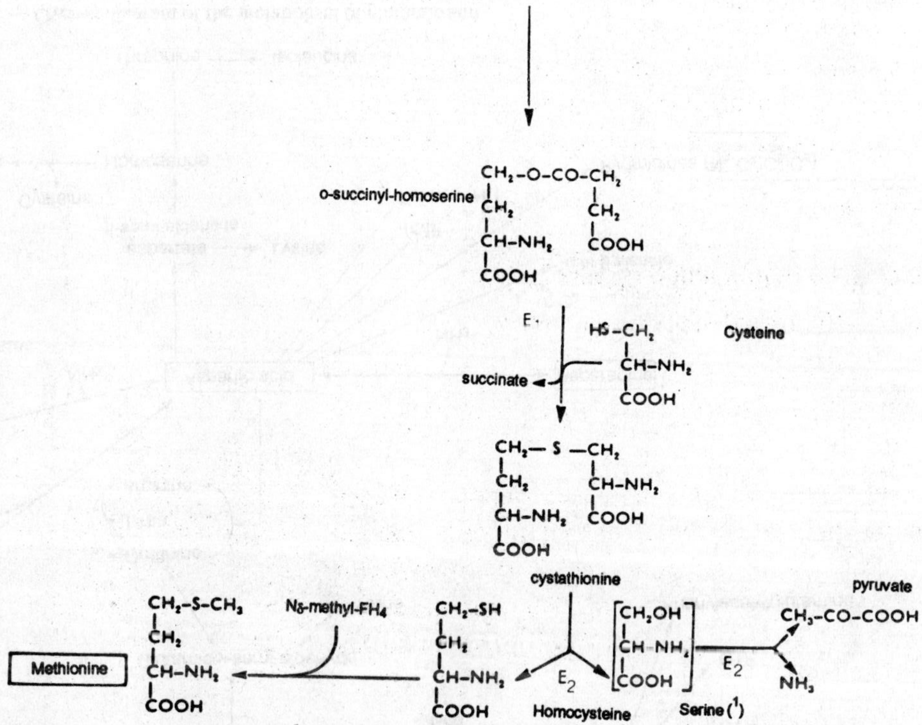

FIG. 7-22. — Branched reactive chain leading from aspartate to lysine, threonine, isoleucine and methionine.

---

(1) Serine is not liberated from the enzyme E'2 (cystathionase) and undergoes a dehydrating deamination which converts it into pyruvic acid.

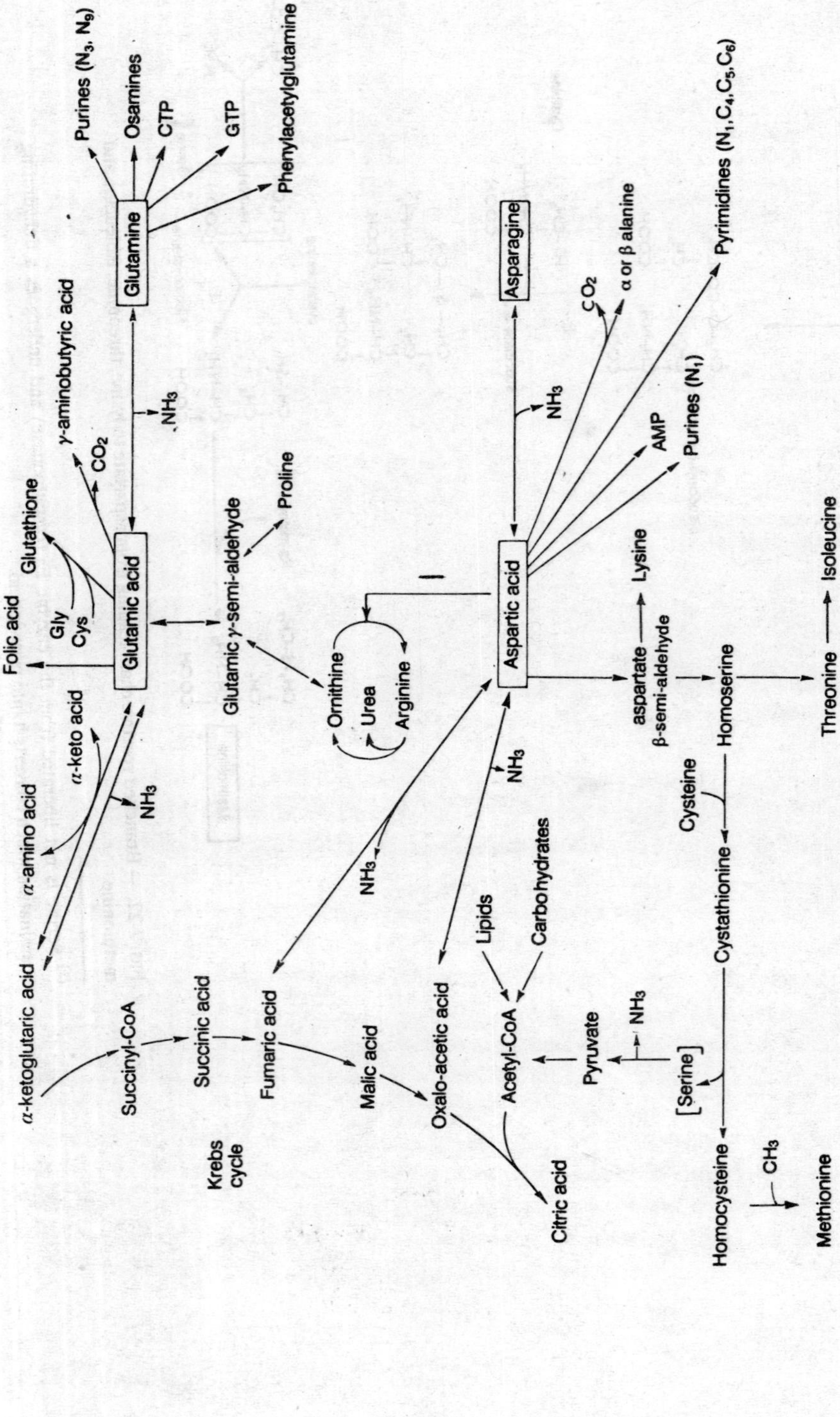

FIG. 7-23. — Overall diagram of the metabolism of glutamic and aspartic acids and their amides.

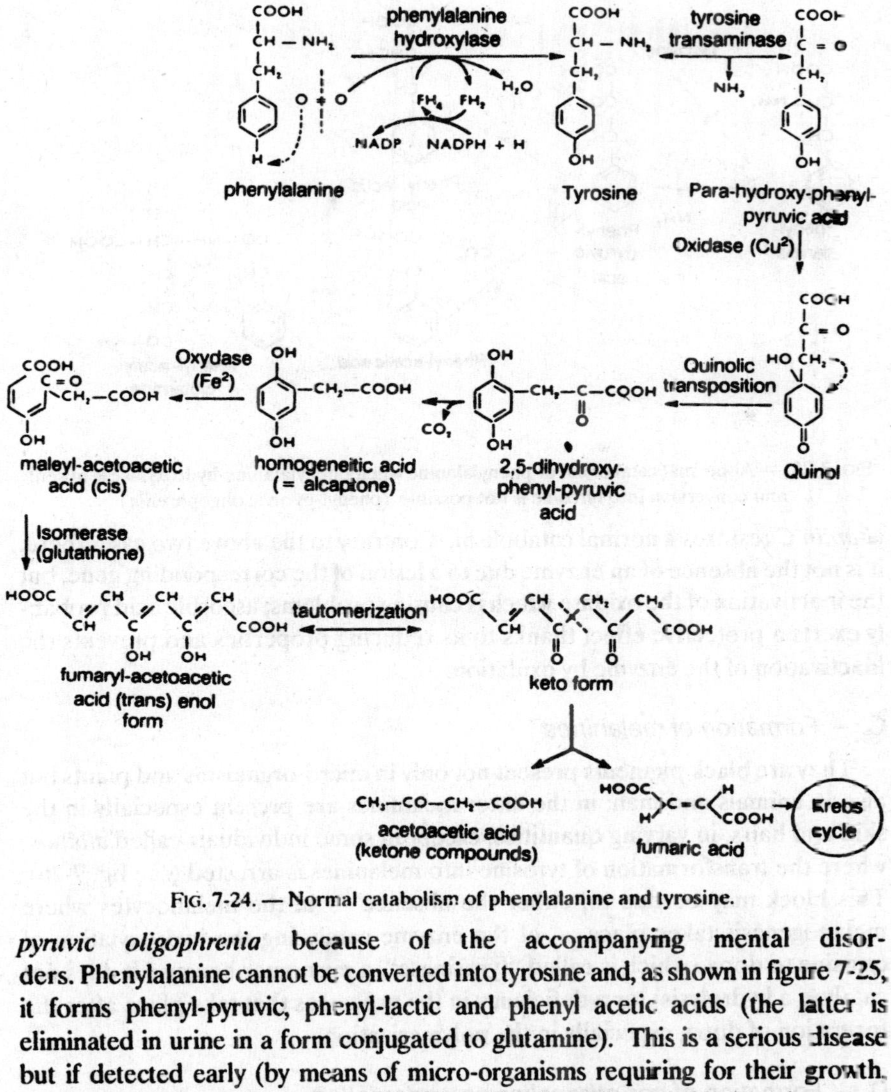

FIG. 7-24. — Normal catabolism of phenylalanine and tyrosine.

pyruvic oligophrenia because of the accompanying mental disorders. Phenylalanine cannot be converted into tyrosine and, as shown in figure 7-25, it forms phenyl-pyruvic, phenyl-lactic and phenyl acetic acids (the latter is eliminated in urine in a form conjugated to glutamine). This is a serious disease but if detected early (by means of micro-organisms requiring for their growth phenylalanine, phenyl-pyruvic acid or phenyl-lactic acid), the infants can be put on a rigorous diet (without excess phenylalanine) which will prevent the appearance of pathological symptoms.

b) **At the level of homogentisic acid oxidase.** — In the absence of this enzyme, homogentisic acid can no longer be catabolized and passes into the urine which takes up a brown coloration after alkalinization, either spontaneous, by formation of NH_3, or induced (alcaptone gets oxidized and forms brown pigments). *Alcaptonuria* is not a serious hereditary disease.

c) **At the level of *para*-hydroxy-phenyl-pyruvate-oxidase.** — In humans and animals suffering from scurvy, one observes an elimination of *para*-hydroxy-phenyl-pyruvic acid and *para*-hydroxy-phenyl lactic acid; administration of

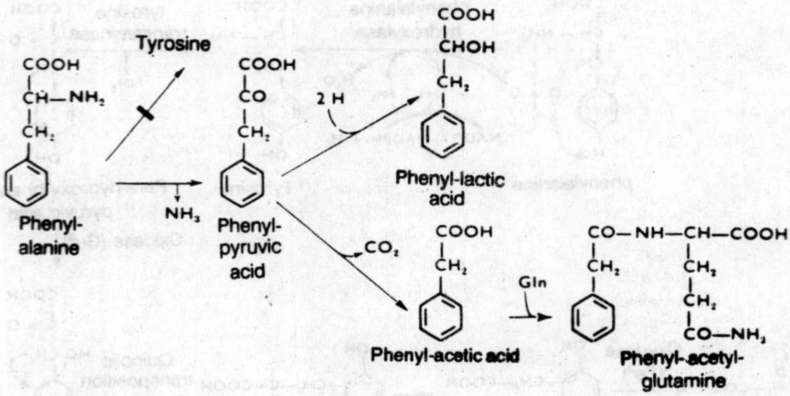

FIG. 7-25. — **Abnormal catabolism of phenylalanine when phenylalanine-hydroxylase is absent and conversion into tyrosine is not possible (phenyl-pyruvic oligophrenia).**

vitamin C restores a normal catabolism. Contrary to the above two cases, here, it is not the absence of an enzyme due to a lesion of the corresponding gene, but the inactivation of the oxidase which is causing problems; ascorbic acid probably exerts a protective effect thanks to its reducing properties and prevents the inactivation of the enzyme by oxidation.

C. – *Formation of melanines*

They are black pigments present not only in micro-organisms and plants but also in animals and man; in the latter melanines are present especially in the skin and hairs, in varying quantities, except in some individuals called *albinos*, where the transformation of tyrosine into melanines is arrested (see fig. 7-26). This block may be due to, either the absence – in the melanocytes where melanogenesis takes place – of the enzyme catalyzing the hydroxylation of tyrosine to dopa (which is called "tyrosinase", a misnomer because it does not catalyze a hydrolysis), or a deficiency in the processes that take place after the formation of dopa, especially in the polymerization.

D. – *Formation of nor-adrenaline and adrenaline*

As shown by figure 7-27, phenylalanine and tyrosine are also the precursors of nor-adrenaline and adrenaline. This series of reactions comprise first 2 successive hydroxylations (of phenylalanine to tyrosine and tyrosine to dopa), a decarboxylation leading to dopamine, then another hydroxylation giving nor-adrenaline and finally a methylation carried out at the cost of S-adenosyl-methionine and giving adrenaline.

E. – *Formation of thyroid hormones*

Iodides (I^-) circulating in blood are picked up by the thyroid gland where they are oxidized (to I° or I^+). The H atoms which are in *ortho* with respect to

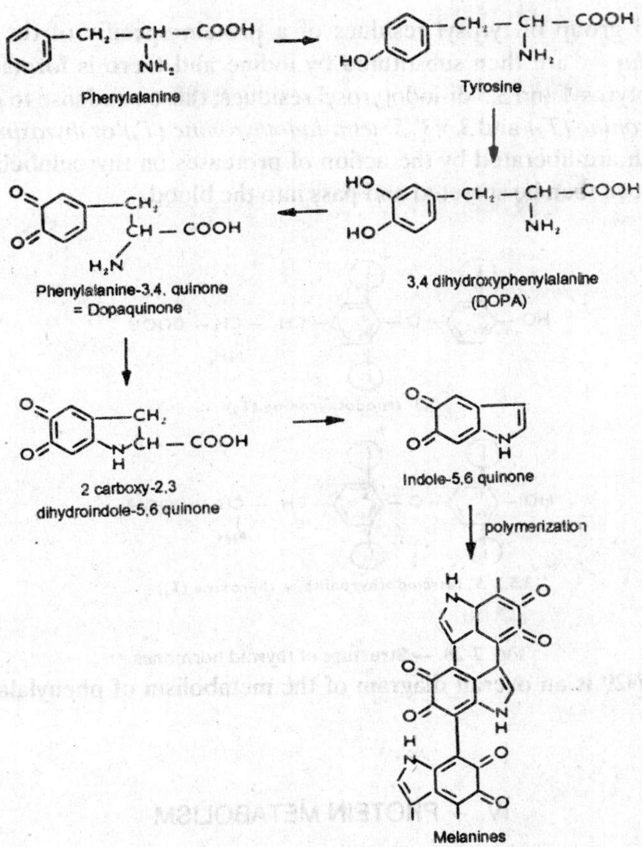

FIG. 7-26. — **Formation of melanines (principal steps).**

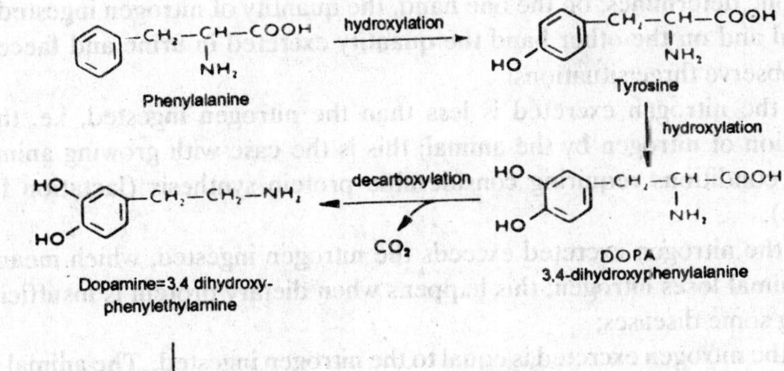

FIG. 7-27. — **Formation of adrenaline.**

the phenol group of tyrosyl residues of a protein specific of the gland – *thyroglobulin* – are then substituted by iodine and there is formation of 3-mono-iodotyrosyl and 3,5 di-iodotyrosyl residues; these condense to give *3,5,3′ tri-iodothyronine (T_3)* and *3,5,3′,5′ tetra-iodothyronine (T_4)* or *thyroxine* (see fig. 7-28) which are liberated by the action of proteases on thyroglobulin. The 2 hormones can then be secreted and pass into the blood.

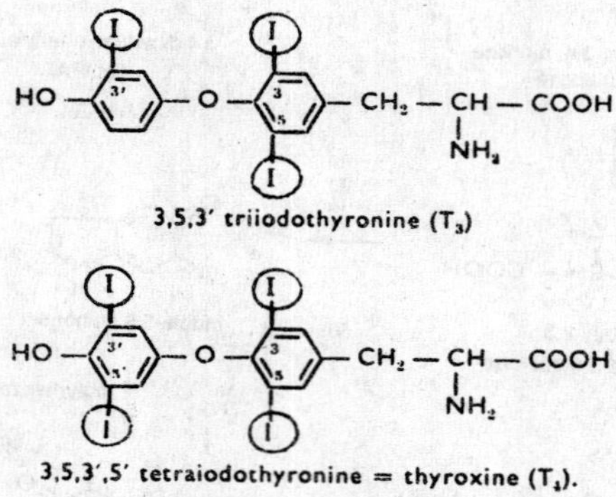

FIG. 7-28. — Structure of thyroid hormones.

Figure 7-29 is an overall diagram of the metabolism of phenylalanine and tyrosine.

IV. – PROTEIN METABOLISM

1. – Notion of nitrogen balance

If one determines, on the one hand, the quantity of nitrogen ingested by an animal and on the other hand the quantity excreted in urine and faeces, one may observe three situations:

– the nitrogen excreted is less than the nitrogen ingested, i.e. there is retention of nitrogen by the animal; this is the case with growing animals or other conditions requiring considerable protein synthesis (lactation for example);

– the nitrogen excreted exceeds the nitrogen ingested, which means that the animal loses nitrogen; this happens when dietary protein is insufficient or during some diseases;

– the nitrogen excreted is equal to the nitrogen ingested. The animal is said to be in *nitrogen balance*; one can define the quantity of nitrogen to be supplied through diet to cover exactly the metabolic requirements of the organism and maintain the (adult) animal in nitrogen balance; this is called the "*protein minimum*" (in man, about 1 g proteins/kg body weight/24 h). But this protein

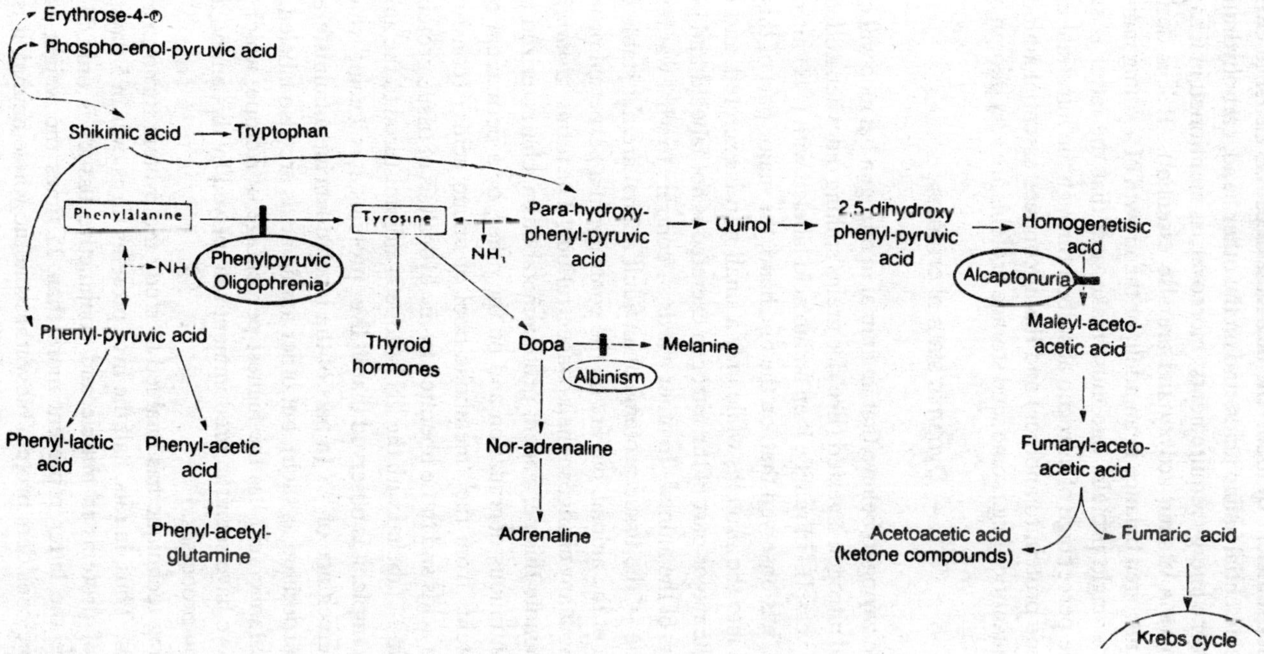

FIG. 7-29. — Metabolism of phenylalanine and tyrosine.
(The bars indicate the reactions which cannot take place — due to the absence of the corresponding enzyme — in some hereditary diseases).

minimum is not sufficient by itself: the diet must of course also include, on the one hand, the essential amino acids, and on the other hand, carbohydrates and lipids to meet the energy requirements (moreover, as mentioned in Chapter 5, some lipids have a vitamin activity and are also essential). If it is desired to maintain the nitrogen balance in an adult, or the growth of a young individual, by means of a single protein, one must make sure that the latter is suitably digestible (the percentage of nitrogen absorbed through the intestinal mucosa varies from one protein to another) and that it contains essential amino acids in sufficient quantities (e.g., some seed storage proteins are very poor in lysine).

2. – Dynamic state of proteins

It would be wrong to believe that an animal in nitrogen balance will excrete any additional nitrogen ingested (like rice grains spilling out when added to an already filled vessel). In reality, if an amino acid labeled with (^{15}N) nitrogen is administered, it is observed that, on the one hand, the major part of this ^{15}N is incorporated into the proteins while only a small part is excreted, and on the other hand, the major part of the nitrogen excreted is not labeled and belongs to the proteins of the animal. In other words – contrary to what was believed for a long time – the tissue proteins of an adult animal are not stable during the entire life of the animal, but are on the contrary, in a perpetual process of renewal (without any change in their concentration). The terms "*dynamic state of proteins*" denote this permanent protein metabolism, characterized on one hand by a continuous degradation and on the other, by a permanent biosynthesis of proteins from the "metabolic pool of amino acids" (consisting of dietary amino acids and those liberated by the hydrolysis of tissue proteins). If the amino acid – labeled with the ^{15}N isotope – administered to the animal is leucine for example, it is observed that in liver proteins, it is leucine which has the highest percentage of ^{15}N (a part of the labeled leucine administered was therefore incorporated as such), but other amino acids are also labeled. Next to leucine, glutamic acid has the highest percentage of isotope, which is not surprising as we have seen the predominant role played by this amino acid in transamination processes.

This labeling technique has enabled to study the protein turnover rate in various tissues. Thus, in rats, half the liver proteins turns over in 5 days, while the half-life of proteins in muscle and conjunctive tissues is only 21 days; however since the latter represent more than 25 times the weight of liver proteins, if one considers protein biosynthesis in the whole animal, it is clear that much more proteins is formed per unit time in the muscle and conjunctive tissues than in the liver. This permanent turnover of proteins (from dietary amino acids or amino acids produced by the hydrolysis of tissue proteins), the fact that mammals cannot build large reserves of proteins or amino acids (contrary to our observation concerning carbohydrates and lipids) and particularly the fact that the ordered incorporation of amino acids into proteins is controlled by the genetic information as mentioned in the previous chapter, constitute some remarkable characteristics of protein metabolism. We will see in Chapter 8 – in the case of bacteria, especially – the mechanisms which enable the cell to control the synthesis of some enzymatic proteins and adapt it to its requirements.

V. – ELIMINATION PRODUCTS OF NITROGEN METABOLISM

We have seen that some amino acids are utilized by the animal organism as detoxication agents and that products like benzoyl-glycine (hippuric acid) or phenyl-acetyl-glutamine can be found in urine. We will also see that urobilin and stercobilin, the degradation products of hemoglobin (see fig. 7-33), are present in faeces. But generally, nitrogen is eliminated in 3 forms:

– *ammonia and ammonium salts* in ammoniotelic organisms (aquatic invertebrates, fishes);

– *uric acid*, in uricotelic species (terrestrial invertebrates, birds, reptiles);

– *urea* in ureotelic animals (especially mammals including man).

These various nitrogeneous wastes can co-exist in the same organism; for example, man excretes nitrogen mostly in the form of urea (80%), but ammonium salts and uric acid are also found in urine.

Ammonia is a highly toxic product; the organism must be able to eliminate it fast and that is probably why this form of elimination was maintained in aquatic animals where NH_3 can be expelled rapidly by diffusion into the external medium. In terrestrial animals less toxic products are used: uric acid and urea. This idea is confirmed by the fact that some animals (e.g., the tadpole) excrete their nitrogen in the form of NH_3 during the aquatic period of their life, but acquire the property of forming urea when they leave water (frog).

1. – Ammonia and ammonium salts

As observed above (see figs. 7-2 to 7-5), in most tissues, the deamination of amino acids leads to the formation of NH_3. This toxic ammonia is taken over by glutamic acid to form *glutamine* (a reaction which requires ATP, see fig. 6-22) and it is in this form that NH_3 is transported by blood up to the kidneys. The glutamic acid-glutamine system therefore plays a capital role in the transport of NH_3, and it must be mentioned that these 2 amino acids represent a very high percentage of the non-protein nitrogen of animal tissues (in human plasma they together form 1/3 of the total nitrogen of amino acids).

In the kidneys, glutaminase catalyzes the reverse reaction and *decomposes glutamine which thus provides the major part of urine NH_3.* The rest is produced by the deamination of amino acids in the renal parenchyma. In the liver, the ammonia produced by glutamate dehydrogenase (see fig. 7-4) is directly detoxified in the urea cycle by a mechanism we will study in the next paragraph. In the case of *acidosis*, for example during prolonged physical exercise (production of lactic acid), or in diabetes (ketonemia), or after injection of acid to an animal, excretion of NH_3 increases – at the cost of urea formation – because it is used for neutralizing acidity, which permits an economy of other ions required by the organism (Na^+, Ca^{2+}, Mg^{2+}); therefore, in acidosis, more NH_3 is eliminated and less urea. Exactly the reverse takes place in the case of *alkalosis*. It may be said that in general, in an animal receiving a fixed diet, the total nitrogen excreted ($N_{NH_3} + N_{urea}$) does not vary much, but the respective proportions of the two forms of elimination can vary. As we will see now, NH_3 is a precursor of urea, but it must remembered that the reverse is not true (urea is not a precursor of urine NH_3, the precursor is mostly blood glutamine, and – to a lesser extent – all the amino acids whose deamination in the kidney produces NH_3 which is immediately excreted).

2. – Ureogenesis

Urea formation is localized in the liver; this was shown by various observations and experiments and especially the fact that if an isolated liver is perfused with a solution containing NH_3, the liquid flowing out is free from NH_3 but contains urea. Krebs and Henseleit succeeded in bringing about urea formation by slices of rat liver and showed that arginase catalyzes urea formation from *arginine*. But there was not enough free arginine to explain urea production by the hydrolysis of this amino acid, and this led to the idea that arginine acts catalytically; it was later found that two other amino acids, *ornithine* and *citrulline* have the same action and Krebs proposed the mechanism represented in figure 7-30, known as *ornithine cycle, urea cycle*, or Krebs-Henseleit cycle, the various steps of which we will examine now.

A. – Synthesis of carbamyl-phosphate and transformation of ornithine into citrulline

The formation of carbamyl phosphate is catalyzed by a mitochondrial enzyme using NH_3 as substrate (the latter results from the deamination of glutamate by glutamate dehydrogenase). A derivative of glutamate, acetyl-glutamate, is the allosteric activator of this carbamyl-phosphate synthetase (the carbamyl-phosphate required for the synthesis of pyrimidines is synthesized by a cytosolic enzyme, using glutamine as substrate and being regulated in a totally different manner, see figure 6-22 and page 296). Carbamyl-phosphate reacts with the δ-amino group of ornithine to give *citrulline*, under the influence of ornithine-carbamyl-transferase. It must be noted that during this step, carbamyl phosphate supplies the carbon atom and one of the nitrogen atoms of the future urea molecule.

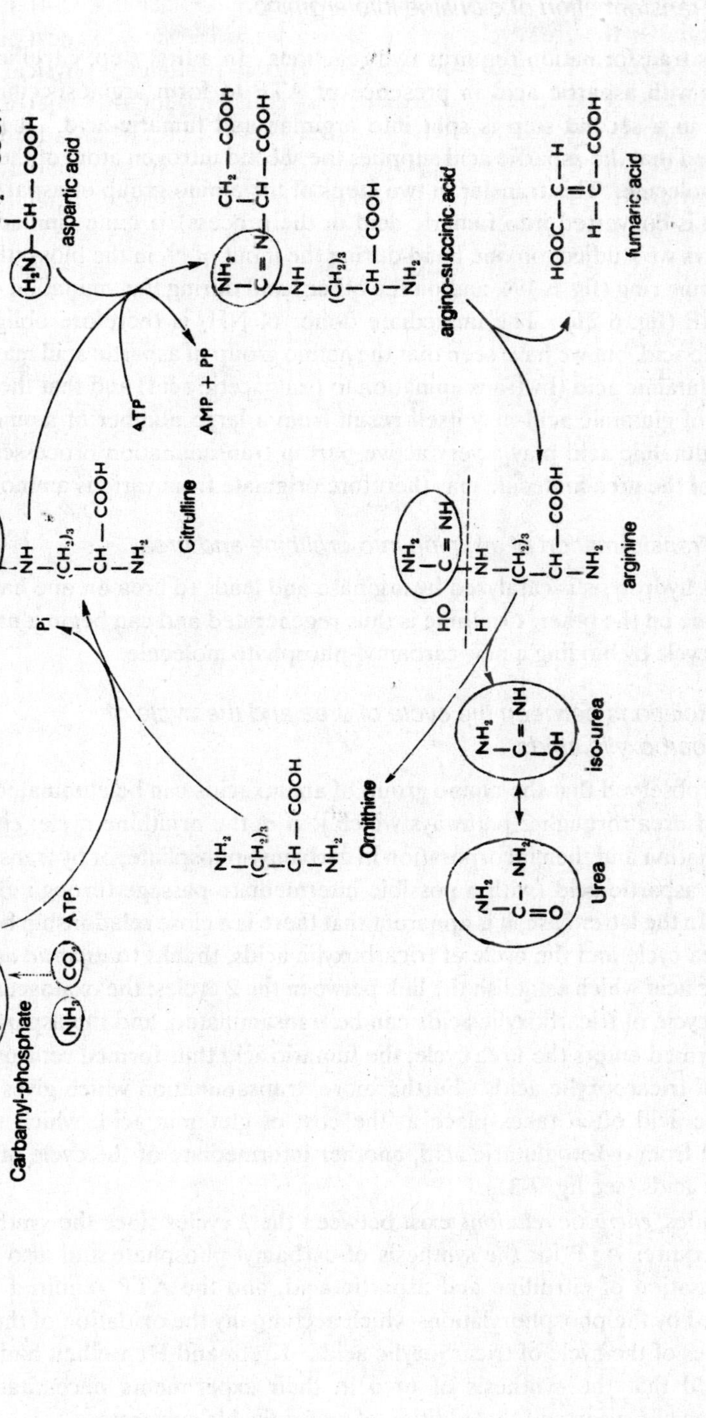

FIG. 7-30. — Urea cycle.

B. – Transformation of citrulline into arginine

This transformation requires two reactions. In a first step, citrulline condenses with aspartic acid in presence of ATP to form arginosuccinic acid, which, in a second step is split into arginine and fumaric acid. It may be observed that the aspartic acid supplies the second nitrogen atom of the future urea molecule. This transfer in two steps of the amino group of aspartic acid (which is converted into fumaric acid in the process) is quite similar to the transfers we studied, on one hand during the input of N_1 in the biosynthesis of the purine ring (fig. 6-19), and on the other hand during the amination of IMP to AMP (fig. 6-21). The immediate donor of NH_2 is therefore obligatorily aspartic acid, but we have seen that the amino group of aspartic acid may result from glutamic acid (by transamination to oxaloacetic acid) and that the amino group of glutamic acid may itself result from a large number of amino acids since glutamic acid play a very active part in transamination processes. The $2NH_2$ of the urea molecule may therefore originate from various amino acids.

C. – Transformation of arginine into ornithine and urea

This hydrolysis is catalyzed by arginase and leads to urea on one hand and ornithine on the other; ornithine is thus regenerated and can begin a new turn of the cycle by binding a new carbamyl-phosphate molecule.

D. – Relations between the cycle of urea and the cycle of tricarboxylic acids

It is observed that the amino group of amino acids can be eliminated in the form of urea through 2 pathways which join in the ornithine cycle: either by deamination and then incorporation in carbamyl-phosphate, or by transamination to aspartic acid (with a possible intermediate passage through glutamic acid). In the latter case, it is apparent that there is a close relationship between the urea cycle and the cycle of tricarboxylic acids, thanks to *aspartic acid* and *fumaric* acid which establish the link between the 2 cycles: the oxaloacetic acid of the cycle of tricarboxylic acids can be transaminated, and the aspartic acid thus formed enters the urea cycle; the fumaric acid thus formed returns to the cycle of tricarboxylic acids. Furthermore, transamination which gives rise to aspartic acid often takes place at the cost of glutamic acid, which is itself formed from α-ketoglutaric acid, another intermediate of the cycle of tricarboxylic acids (see fig. 7-31).

Besides, *energetic relations* exist between the 2 cycles since the synthesis of urea requires ATP for the synthesis of carbamyl-phosphate and also for the condensation of citrulline and aspartic acid, and the ATP required can be supplied by the phosphorylations which accompany the oxidation of the intermediates of the cycle of tricarboxylic acids. Krebs and Henselheit had in fact observed that the synthesis of urea in their experiments necessitated the presence of oxygen and the addition of an oxidizable substrate.

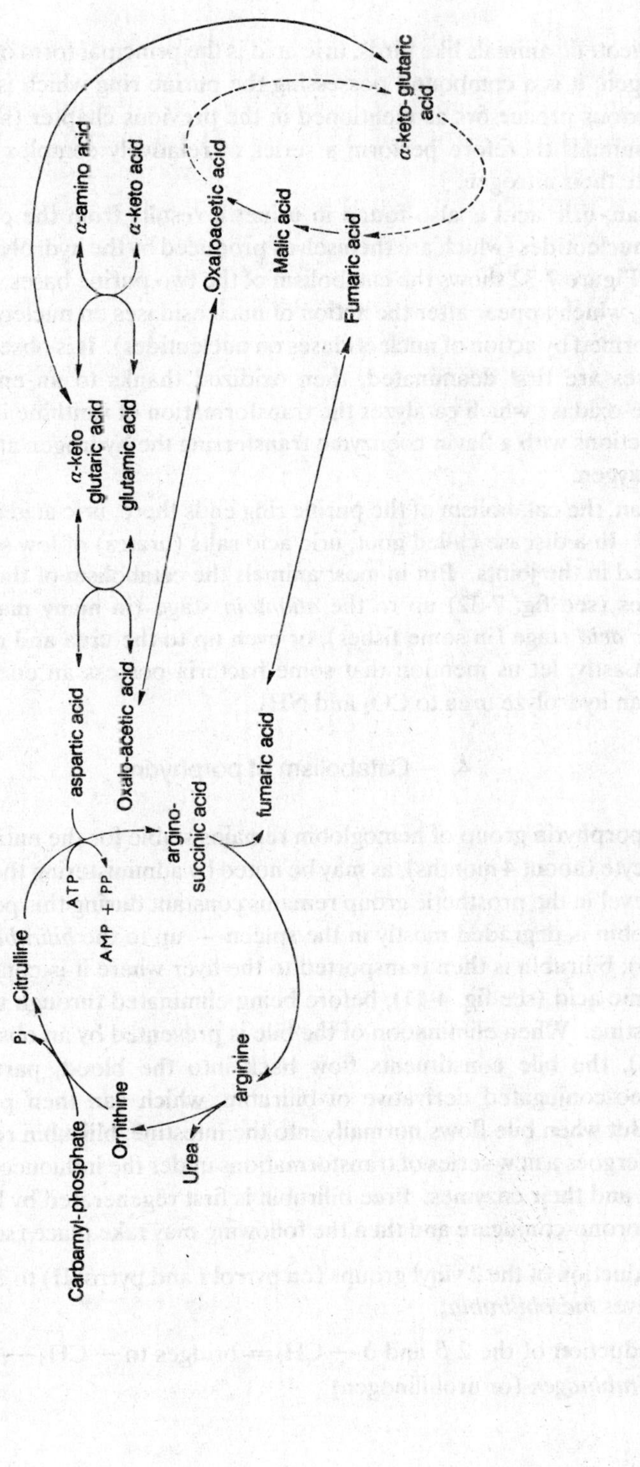

FIG. 7-31.— Relations between the urea cycle and the cycle of tricarboxylic acids.

3. – Uric acid

In uricotelic animals like birds, uric acid is the principal form of elimination of nitrogen; it is a compound possessing the purine ring which is synthesized from various precursors as mentioned in the previous chapter (see fig. 6-19). These animals therefore perform a series of relatively complex reactions to eliminate their nitrogen.

In man, uric acid is also found in urine; it results from the catabolism of purine nucleotides (which are themselves produced by the hydrolysis of nucleic acids). Figure 7-32 shows the catabolism of the two purine bases, adenine and guanine, which appear after the action of nucleosidases on nucleosides (themselves formed by action of nucleotidases on nucleotides). It is observed that the two bases are first deaminated, then oxidized thanks to an enzyme called xanthine-oxidase which catalyzes the transformation of xanthine into uric acid and functions with a flavin coenzyme transferring the hydrogen atoms directly to the oxygen.

In man, the catabolism of the purine ring ends there, uric acid is eliminated in urine. In a disease called gout, uric acid salts (urates) of low solubility, are deposited in the joints. But in most animals the catabolism of the purine ring continues (see fig. 7-32) up to the *allantoin* stage (in many mammals), the *allantoic acid* stage (in some fishes), or even up to the urea and *glyoxylic acid* stage. Lastly, let us mention that some bacteria possess an enzyme, *urease*, which can hydrolyze urea to CO_2 and NH_3.

4. – Catabolism of porphyrins

The porphyrin group of hemoglobin remains stable for the entire life of the erythrocyte (about 4 months), as may be noted by administering the ^{15}N isotope whose level in the prosthetic group remains constant during this period. Then, hemoglobin is degraded mostly in the spleen – up to the *bilirubin* stage (see fig. 7-33); bilirubin is then transported to the liver where it is conjugated with glucuronic acid (see fig. 4-11), before being eliminated through the bile, into the intestine. When elimination of the bile is prevented by an obstacle (e.g., a calculus), the bile constituents flow back into the blood, particularly the glucorono-conjugated derivative of bilirubin, which can then pass into the urine. But when bile flows normally into the intestine, bilirubin reaches there and undergoes a new series of transformations under the influence of intestinal bacteria and their enzymes. Free bilirubin is first regenerated by hydrolysis of the glucorono-conjugate and then the following may take place (see fig. 7-33):

– reduction of the 2 vinyl groups (on pyrrol I and pyrrol II) to ethyl groups, which gives *mesobilirubin*;

– reduction of the 2 β and δ $-CH_2=$ bridges to $-CH_3-$, which gives *mesobilirubinogen* (or urobilinogen).

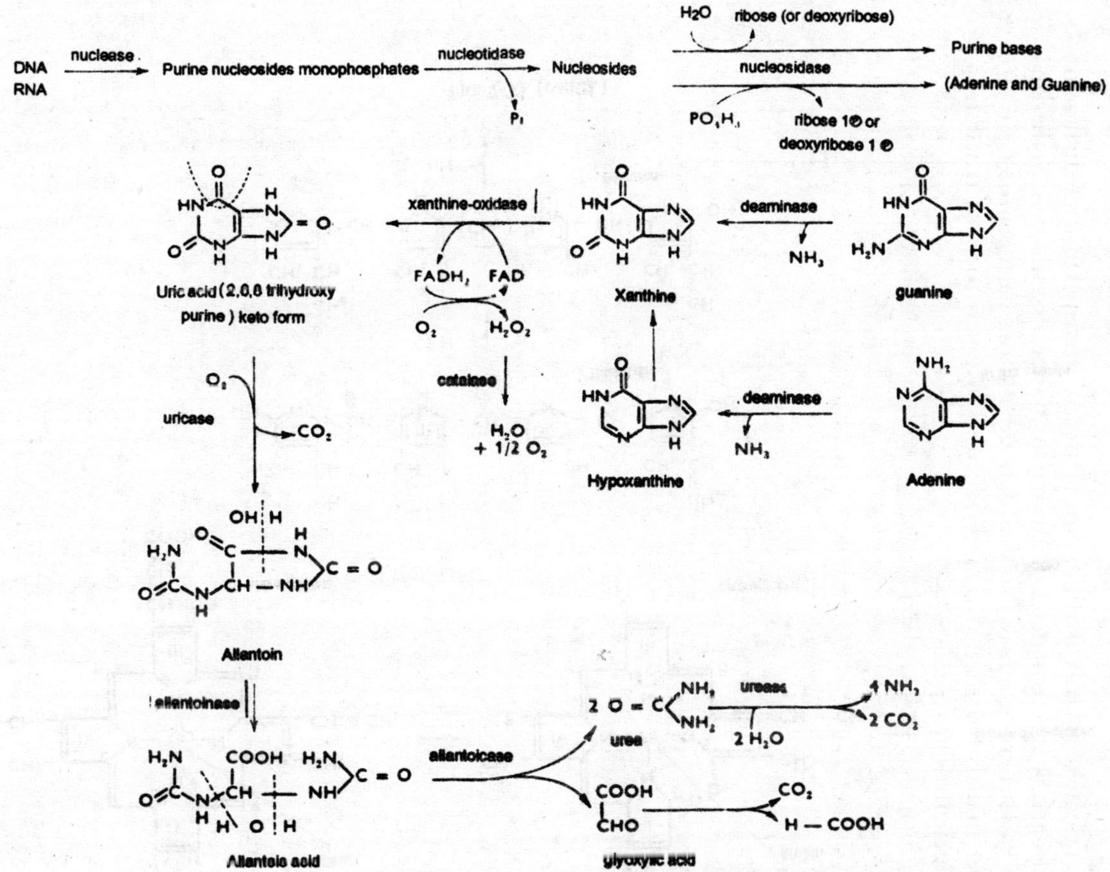

FIG. 7-32. — Catabolism of purine nucleotides: formation of uric acid and subsequent degradations possible in some organisms.

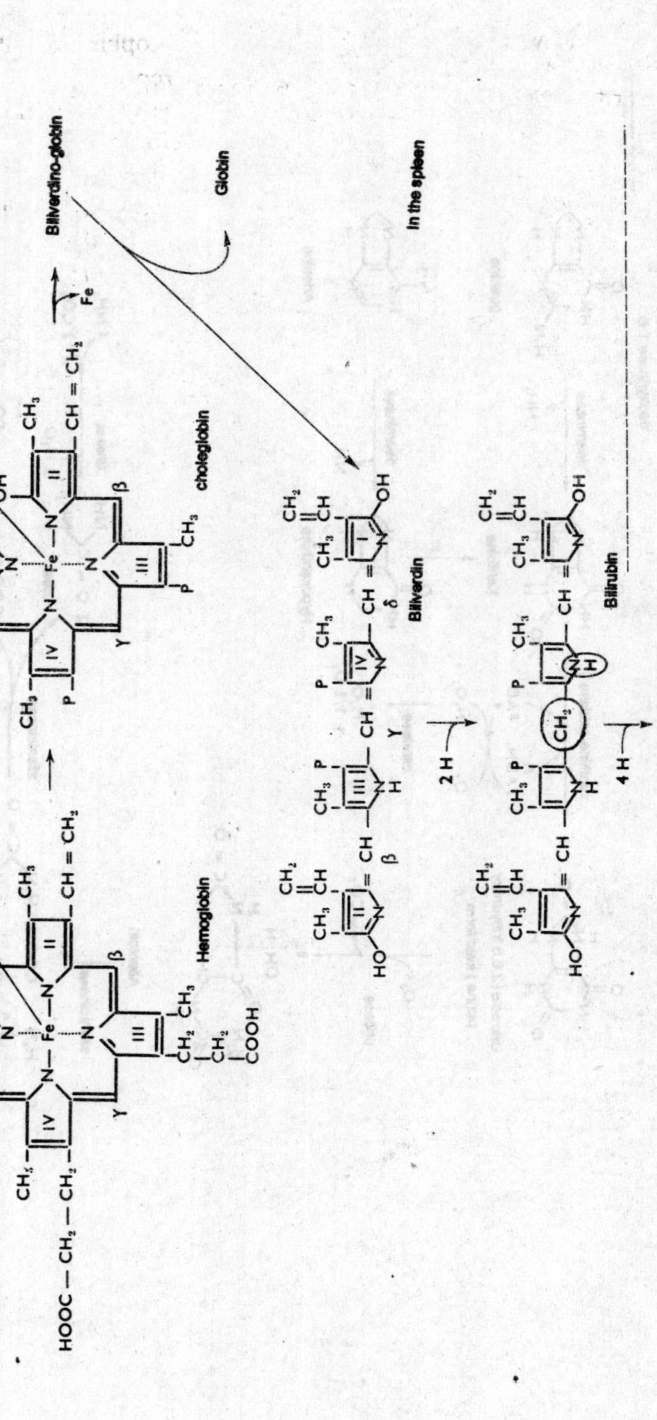

FIG. 7-33. (contd.)

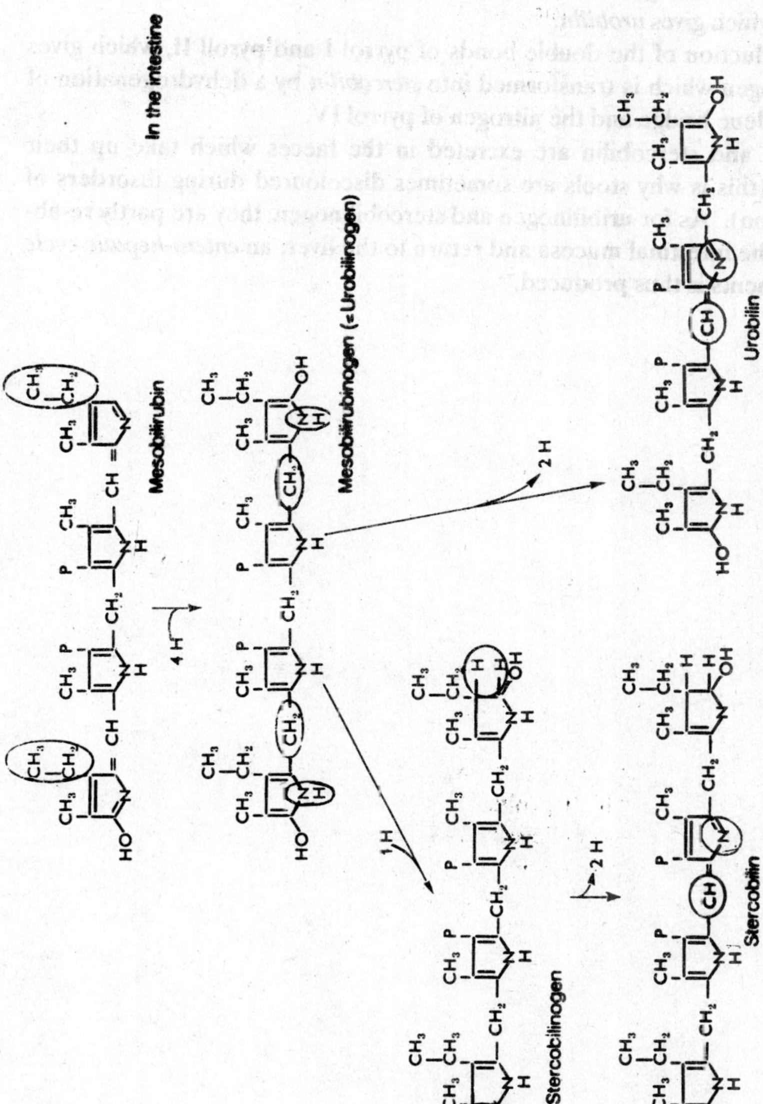

FIG. 7-33. — Catabolism of hemoglobin into bile pigments.

There are then 2 possibilities:

– either dehydrogenation of the γ methylene bridge and the nitrogen of pyrrol IV, which gives *urobilin*,

– or reduction of the double bonds of pyrrol I and pyroll II, which gives stercobilinogen which is transformed into *stercobilin* by a dehydrogenation of the γ methylene bridge and the nitrogen of pyrrol IV.

Urobilin and stercobilin are excreted in the faeces which take up their coloration (this is why stools are sometimes discoloured during disorders of bile secretion). As for uribilinogen and stercobilinogen, they are partly re-absorbed by the intestinal mucosa and return to the liver; an *entero-hepatic cycle* of bile pigments is thus produced.

CHAPTER 8[1]

REGULATION OF CELLULAR METABOLISM

INTEGRATION OF CARBOHYDRATE, LIPID, NUCLEIC ACID AND PROTEIN METABOLISMS

We have successively studied the metabolism of carbohydrates, lipids, nucleic acids and proteins, but we have already underlined the fact that this division is rather arbitrary; it is justified mainly because it facilitates the presentation. In fact – as observed repeatedly – inter-relations between various types of compounds are numerous and the entire cellular metabolism must be regarded as a set of harmoniously integrated reactions. These inter-relations ought to be presented in a complete overall diagram; this could indeed be done, but the diagram would be extremely complex considering that even partial diagrams comprise a large number of reactions; we will therefore give only a simplified diagram showing the broad lines of the intermediate metabolism (fig. 8-1).

From the energy point of view it is seen that *formation of ATP* takes place during glycolysis and especially during the electron transfers which accompany photosynthesis (photophosphorylation) and respiration (oxidative phosphorylation), processes we studied in Chapter 3. Oxidative phosphorylation follows the Krebs cycle where *acetyl-coA* resulting from *glucose, fatty acids* or *some amino acids*, is oxidized. We have seen that other amino acids can also join the Krebs cycle at various points.

Considering the main possibilities of interconversion, it is observed that:

– *carbohydrates can be converted into fatty acids* through *acetyl-coA* which is the real pivot of intermediate metabolism;

– *carbohydrates can be converted into some amino acids*; this merely requires the formation of the corresponding α-keto acid which can then be aminated: pyruvic acid $\longrightarrow$ alanine, oxaloacetic acid $\longrightarrow$ aspartic acid, α-ketoglutaric acid $\longrightarrow$ glutamic acid, 3-phosphoglyceric acid $\longrightarrow$ serine, etc.,

– *some amino acids can be the source of carbon atoms of glucose*, when their metabolism leads to an intermediate of the Krebs cycle, because oxaloacetate

(1) Revised by C. Kedinger and J.-C. Patte.

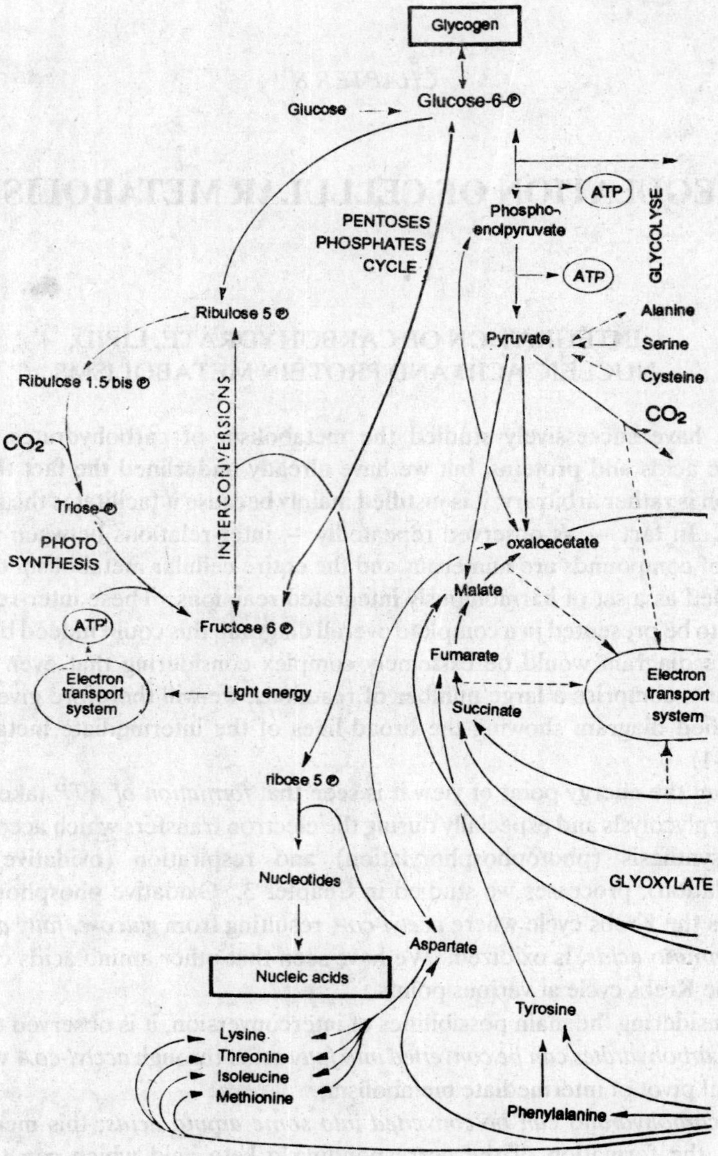
Glycogen
Glucose
Glucose-6-Ⓟ
ATP
Phospho-
enolpyruvate
GLYCOLYSE
PENTOSES
PHOSPHATES
CYCLE
ATP
Ribulose 5 Ⓟ
Alanine
Pyruvate
Serine
Ribulose 1.5 bis Ⓟ
Cysteine
INTERCONVERSIONS
CO_2
CO_2
Triose-Ⓟ
oxaloacetate
PHOTO
SYNTHESIS
Malate
ATP
Fructose 6 Ⓟ
Fumarate
Electron
transport
system
Light energy
Succinate
Electron
transport
system
ribose 5 Ⓟ
Nucleotides
GLYOXYLATE
Aspartate
Nucleic acids
Tyrosine
Lysine
Threonine
Isoleucine
Methionine
Phenylalanine

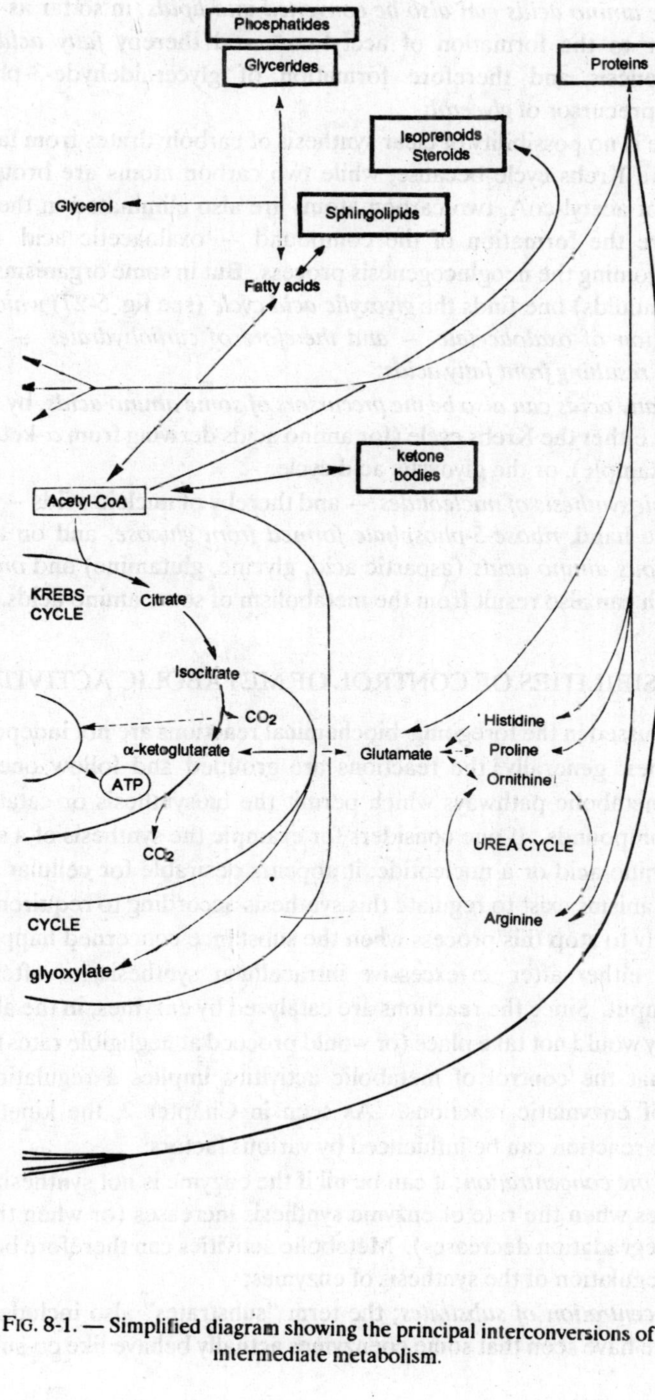

FIG. 8-1. — **Simplified diagram showing the principal interconversions of intermediate metabolism.**

can give phosphoenolpyruvate and the latter leads to glucose by nucleogenesis (glucogenic amino acids);

– *some amino acids can also be converted into lipids*, in so far as they can lead either to the formation of acetyl-coA and thereby *fatty acids*, or to neoglucogenesis and therefore formation of glyceraldehyde-3-phosphate which is a precursor of *glycerol*;

– there is no possibility of clear synthesis of carbohydrates from fatty acids through the Krebs cycle because, while two carbon atoms are brought by a molecule of acetyl-coA, two carbon atoms are also eliminated in the form of CO_2 before the formation of the compound – oxaloacetic acid – which permits rejoining the neoglucogenesis process. But in some organisms (plants, bacteria, moulds) one finds the *glyoxylic acid cycle* (see fig. 5-27) *which allows the formation of oxaloacetate – and therefore of carbohydrates – from the acetyl-coA resulting from fatty acids*;

– *the fatty acids can also be the precursors of some amino acids*, by the same processes: either the Krebs cycle (for amino acids deriving from α-ketoglutaric acid for example), or the glyoxylic acid cycle.

– *the biosynthesis of nucleotides* – and thereby of nucleic acids – requires on the one hand, *ribose-5-phosphate formed from glucose*, and on the other hand, *various amino acids* (aspartic acid, glycine, glutamine) and *one-carbon units* which can also result from the metabolism of some amino acids.

POSSIBILITIES OF CONTROL OF METABOLIC ACTIVITIES

As discussed in the foregoing, biochemical reactions are not independent of one another: generally, the reactions are grouped and follow one another forming metabolic pathways which permit the biosynthesis or catabolism of various compounds. If one considers for example the synthesis of a substance like an amino acid or a nucleotide, it appears desirable for cellular economy that mechanisms exist to regulate this synthesis according to requirement, and particularly to stop this process when the substance concerned happens to be in excess, either after an excessive intracellular synthesis, or after a large external input. Since the reactions are catalyzed by enzymes, in the absence of which they would not take place (or would proceed at negligible rates), it would appear that the control of metabolic activities implies a regulation of the kinetics of enzymatic reactions. As seen in Chapter 2, the kinetics of an enzymatic reaction can be influenced by various factors:

– *enzyme concentration*; it can be nil if the enzyme is not synthesized at all; it increases when the rate of enzyme synthesis increases (or when the rate of enzyme degradation decreases). Metabolic activities can therefore be controlled by a regulation of the synthesis of enzymes;

– *concentration of substrates*; the term "substrates" also includes co-substrates (we have seen that some coenzymes actually behave like co-substrates);

– *concentration of activators and inhibitors*; in fact they act on the concentration of active enzyme. Cellular metabolism can therefore be *controlled by a regulation of enzymatic activities.*

– *variation of the specific molecular activity of the enzyme* by covalent modification of the protein.

We will briefly examine these diverse possibilities. What follows must therefore be considered as an introduction to the mechanisms of regulation.

I. – REGULATION OF THE SYNTHESIS OF ENZYMES

The synthesis of enzymes – or more generally, the synthesis of proteins – implies the incorporation of amino acids in an order dictated by the genetic information. We have seen in Chapter 6 that the passage of this information from the DNA to the proteins takes place in two steps: transcription and translation. We will now consider the possibilities of control of the expression of genes at these two levels.

1. – Control of the expression of genes at the level of transcription [1)]

In addition to the adaptation mechanisms operating in the cells in response to changes in environmental conditions, cellular *differentiation* and *development imply the sequential expression of genes or groups of genes*. It is presently established that control of gene expression takes place mainly at the level of transcription.

This transcriptional regulation operates through protein factors which interact with particular nucleotide sequences located generally upstream of genes (promoter sites) and with the RNA polymerase itself. Various methods applicable to procaryotes and eucaryotes can reveal this mode of control, either by a direct quantification of the RNA transcribed from a given gene, or by an estimation of the accumulation of the corresponding protein.

A technique called "analysis by nuclease S1" or S1 mapping (see fig. 8-2) *allows to quantify with precision the RNA molecules whose transcription has started at the correct site*. In this technique, the RNAs synthesized either in *in vitro* systems or in transformed cells, are hybridized with a single-stranded DNA fragment labeled with ^{32}P at its 5′ end and including the site of transcription initiation. This DNA fragment, constituting a radioactive probe, will be placed in the hybridization medium in large excess with respect to the specific RNA. After hybridization, the unpaired molecules or portions of molecules are degraded by a treatment with nuclease S1 of *Aspergillus*, which hydrolyses selectively the single-stranded nucleic acids. The fragments of the probe resis-

---

(1) Revised by C. KEDINGER.

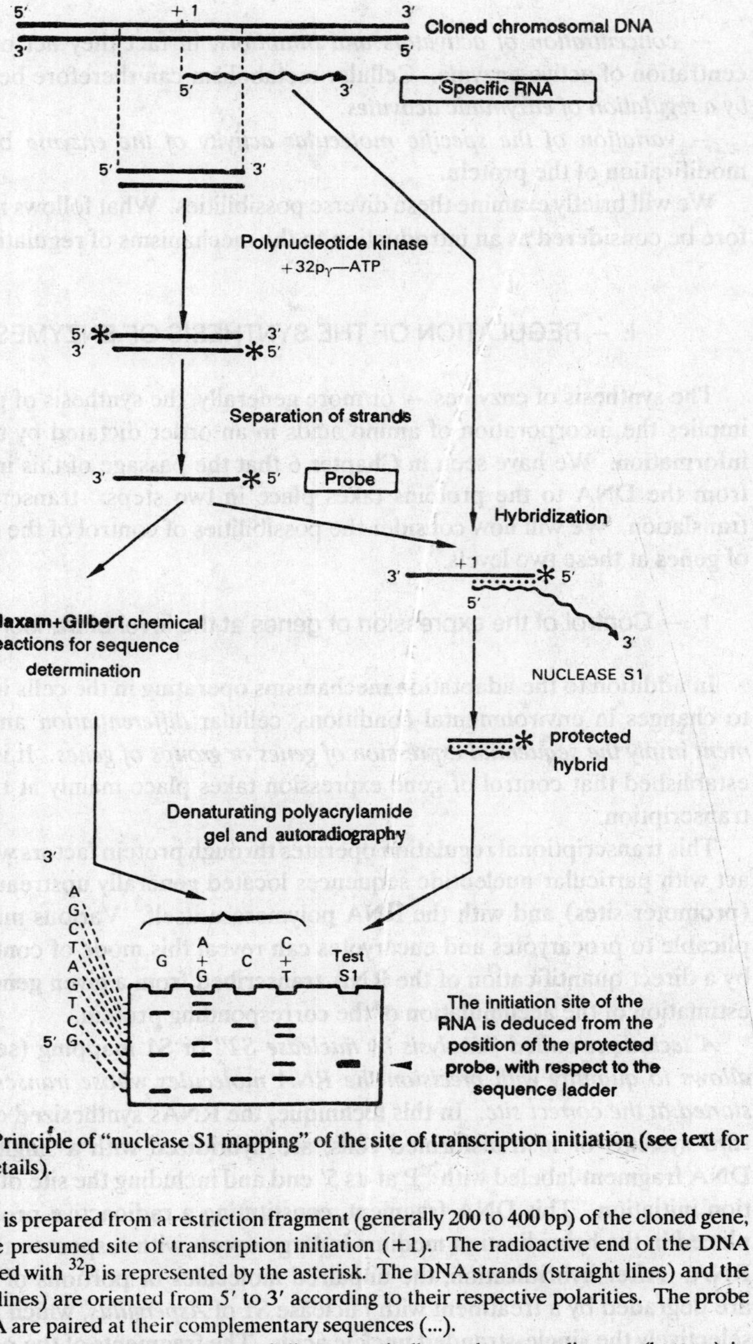

FIG. 8-2. — **Principle of "nuclease S1 mapping" of the site of transcription initiation (see text for additional details).**

The probe is prepared from a restriction fragment (generally 200 to 400 bp) of the cloned gene, including the presumed site of transcription initiation (+1). The radioactive end of the DNA strands labeled with ^{32}P is represented by the asterisk. The DNA strands (straight lines) and the RNA (wavy lines) are oriented from 5′ to 3′ according to their respective polarities. The probe and the RNA are paired at their complementary sequences (...).

The portion of the sequence represented diagrammatically on the autoradiogram (bottom) is: 5′—GCTGATCGA—3′; the probe fragment resistant to nuclease S1 migrates at the level of the band marked "T" (third nucleotide from the 5′ end of the above sequence). Knowing that the chemical cleavages take place on the 5′ side of the bases modified by the specific reactions, this band corresponds in our example to a probe fragment terminating at the 3′ end by a C. This C is therefore the nucleotide +1 of the coding strand of the gene analyzed.

tant to S1 nuclease are separated by electrophoresis on polyacrylamide gel in denaturing conditions and revealed by autoradiography. The probe fragment protected by the pairing with the specific RNA will be of a precise size equal to the distance between the initiator nucleotide (+1) and the labeling site (5′ end) of the probe. The intensity of the corresponding radioactive band on the autoradiogram will be directly proportional to the quantity of specific RNA present.

When the product of the gene studied is easily detectable (either by its enzymatic activity or by its immunological properties) one can estimate the activity of this gene by determining the quantity of protein synthesized. If the detection of this protein is difficult, one can "fuse" by recombination *in vitro*, the promoter of the corresponding gene with the nucleotide sequence coding for a "reporter" protein easy to estimate, like β-galactosidase or chloramphenicol-acetyl-transferase (CAT).

These regulation mechanisms involve the action of *regulator proteins* which modify the expression of genes by binding in a specific manner to the corresponding promoter sequences. The study of these interactions is now possible thanks to conventional methods.

1) **Method of "gel retardation".** – The DNA fragment studied, made radioactive by labeling *in vitro* is incubated in presence of proteins which can bind to the DNA. After the formation of specific complexes, the mixture is subjected to electrophoresis on polyacrylamide gel in non denaturing conditions. The DNA-protein complex will migrate more slowly than the fragment of free DNA and its position can be visualized on the gel by autoradiography.

2) **Foot-printing method.** – It consists in incubating the protein fractions with a fragment of double-stranded DNA labeled at one end and corresponding to the promoter region of a gene (see fig. 8-3). After the formation of specific complexes the DNA is subjected to the controlled action of DNase which will cleave the fragment in the accessible zones not protected by interaction with a factor. Here, it is important to choose digestion conditions leading on average, to only one cleavage per DNA molecule. The position of the protected regions is then revealed by the study of the size of radioactive fragments of the residual DNA and comparison with the size of fragments obtained in the same conditions by action of DNase on the naked DNA, in absence of protein fractions.

A. – Procaryotic systems

It is certainly in these systems (bacterial cells and bacteriophages) that the transcription regulation mechanisms are best understood at the molecular level. We will see that the modulation of the transcription of a gene is performed by the joint action of RNA polymerase and a series of regulatory proteins (activators, repressor, terminators, anti-terminators) which, in presence of appropriate co-factors, interact with the DNA (or *m*RNA) at the level of particular nucleotide sequences (targets). Two steps of the transcription

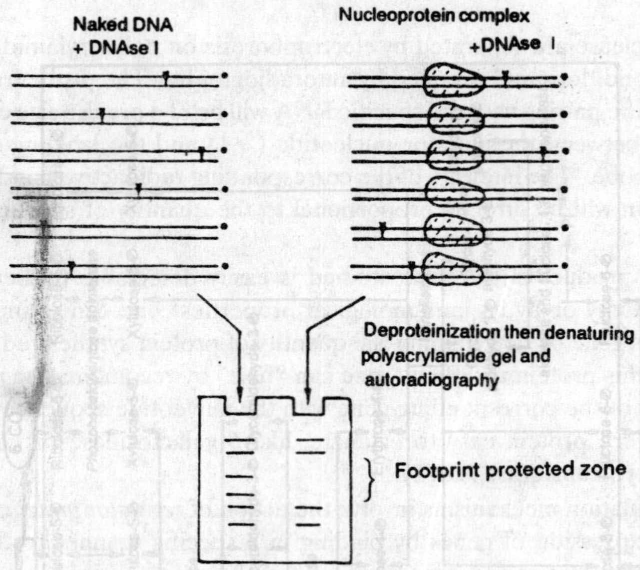

FIG. 8-3. — Principle of the demarcation of the DNA region protected by the interaction with a protein.

The radioactive end of the DNA fragment analyzed is represented by the asterisk. Arrow heads indicate sites of cleavage by DNase 1 on a series of naked DNA molecules (left) and on a series of identical DNA molecules complexed with the protein (right): digestion conditions are adjusted to produce statistically only one cleavage per molecule. Autoradiography will reveal only the radioactive part of molecules thus cleaved. The cleavage points are the same on the 2 types of molecules except in the interaction region where they are masked by the protein (right). The absence of cleavages in this region is reflected by the absence of corresponding bands on the autoradiography, thus determining the footprint of the protein on the DNA.

are thus subjected to regulation: initiation (or start of transcription) and termination.

a) **Initiation of transcription.** – During this step the RNA polymerase recognizes particular sequences on the double-stranded DNA and binds to it. The sub-unit (or factor) σ confers on the core enzyme the capacity to bind selectively to these sequences and start the transcription. These sequences form the promoter sites. The study of the sequence of numerous promoters of *E. coli* revealed the presence of two regions of homology (called "consensus sequences") localized at about 10 and 35 base pairs upstream the initiation point (see fig. 8-7) and respectively called Pribnow box or TATA box (because it is rich in nucleotides A and T) and sequence – 35 (owing to its position).

The efficiency of the initiation of transcription is mainly modulated by regulatory proteins which also recognize and bind to specific target nucleotide sequences near the promoter site. Generally, these proteins act in the dimeric or tetrameric form and bind to a region of the DNA presenting a point of symmetry at the sequence level (palindromic sequence): the sequence of nucleotides along one strand is similar to that of the complementary strand read in the other direction (see fig. 8-7).

There are generally two types of control at the transcription initiation level:

– *the negative control*: transcription which normally requires the presence of RNA polymerase alone is *prevented* by the presence of a protein factor (e.g., like the repressor of the lactose system);

– *the positive control*: transcription cannot take place in presence of RNA polymerase alone; other protein factors are needed to *permit* this transcription, e.g. the activator of the maltose system, or the CRP-protein required for a number of operons (see page 451). This positive control is comparable with the action of specific σ factors, recognizing specific promoters (see page 453);

– ***Control of transcription by induction-repression.***

α) **Lactose operon.** – JACOB-MONOD MODEL. – In 1961, these authors proposed a model for the control of the expression of genes, on the basis of the observations made concerning the functioning of the *lactose operon* in *E. coli*. This operon contains the structural and regulator genes enabling the synthesis of enzymes involved in the metabolism of lactose, i.e.: a β-galactosidase (coded by the gene *z*), a permease (coded by the gene *y*) and a transacetylase (coded by the gene *a*).

The 3 enzymes of the lactose operon are called *inducible* (the term *adaptative* is also used): if *E.coli* is grown in a conventional glucose-based medium, they are present only in negligible quantities in the cell, but if *E.coli* is made to grow in a medium which contains nothing but lactose as the only source of carbon, the bacterium adapts the expression of its genome and the concentrations of these 3 enzymes increase by about 1 000 times. The problem is therefore to identify the control mechanisms enabling the 3 structural genes of the lactose operon to be expressed only when needed and not the rest of the time.

Figure 8-4 shows the lactose operon repressed, a situation which exists in *the absence of lactose*. The regulatory gene *i* controls the synthesis of a *repressor*, a protein capable of diffusing in the cytoplasm and which has an affinity for the region *o* (operator site) of the lactose operon. This *binding of the repressor to the operator prevents the transcription of the structural genes of the 3 enzymes by*

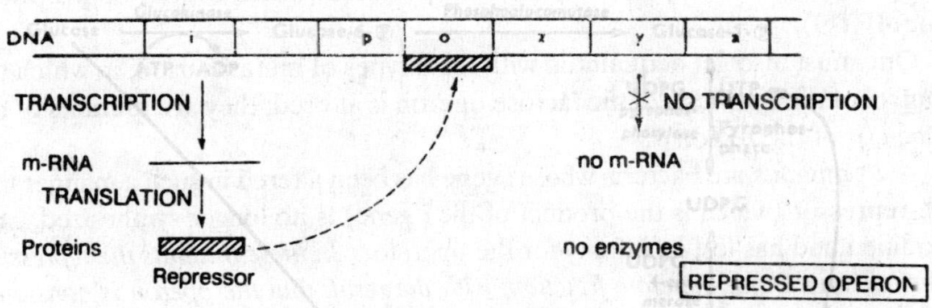

FIG. 8-4. — Lactose operon in repressed state (in absence of lactose or another inducer).

the RNA polymerase and as a result of the absence of messenger RNA, the 3 enzymes are not synthesized.

In presence of lactose (or generally, of an appropriate inducer), *the repressor forms with the inducer a complex whose structure is modified by allosteric transconformation in such a manner that it can no longer bind to the operator site* (see fig. 8-5). Genes *z*, *y* and *a* can then be transcribed into a polycistronic *m*RNA, which permits the coordinated synthesis (in a constant quantitative ratio) of the 3 corresponding enzymes. The *operon is induced*. The initiation site recognized by the RNA polymerase is the *promoter* site (*p*).

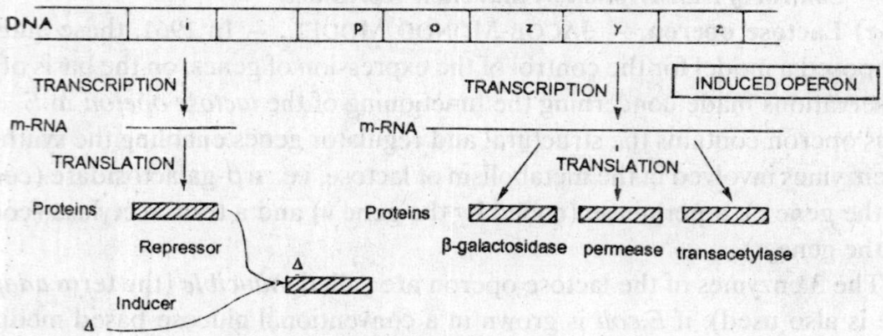

FIG. 8-5. — Lactose operon in induced state (in presence of lactose or another inducer like IPTG).

EXPERIMENTAL CONFIRMATION OF THE PREDICTIONS OF THIS MODEL. – Before presenting the experimental arguments supporting the Jacob-Monod model, a word must be said about the inducer which was used in these studies: *iso*propyl*thio*galactoside (IPTG), a synthetic galactoside capable of inducing the lactose operon like the natural galactoside that is lactose(1), but which differs from lactose by the fact that it cannot be metabolized; in other words it can "cheat" the regulating mechanisms the functioning of the operon, but not β-galactosidase. Such an inducer, which makes the cell produce enzymes of which it cannot be the substrate is called *gratuitous inducer*. The advantage of using a gratuitous inducer lies in the fact that one can thus eliminate – for the study of phenomena – the metabolites of lactose (which cannot be produced from IPTG).

One must also get acquainted with the 3 types of mutants used, in which the control of the synthesis of the lactose operon is altered; they are mutants of the gene *i* or of the region *o*:

– i^- *mutants* are bacteria whose *i* gene has been altered in such a manner that the repressor (which is the product of the *i* gene) is no longer synthesized, or is modified and has lost its affinity for the operator. *In these mutants the repression mechanism cannot therefore function, with the result that the operon is constantly*

---

(1) In fact it has been shown that the true natural inducer is not lactose, but alloctactose, a derivative formed in the cell from lactose.

induced and the 3 enzymes of the lactose operon are synthesized continuously — even in the absence of inducer — in maximum quantities.

— *i^s mutants* are super-repressed, because mutation of the *i* gene resulted in the synthesis of a repressor which has no affinity for the galactosides and is therefore bound permanently to the operator even in presence of the inducer. *The operon cannot therefore be induced and the 3 enzymes are never synthesized*;

— *o^c mutants* ("constitutive operator") are characterized by a deletion (more or less important) of the DNA of the *o* region, which can no longer bind the repressor, with the result that *the 3 enzymes are synthesized continuously, even in the absence of inducer.* In o^c mutants as also in i^- mutants, the operator-repressor combination cannot take place; in the former case the operator site is defective, in the latter, the repressor.

More than 6 years lapsed before the existence of the repressor postulated by the Jacob-Monod model could be experimentally confirmed, thanks to the experiments of Gilbert and Muller-Hill. These authors observed that a bacterial extract placed in a dialysis bag binds IPTG (after allowing the cell-free extract to come into equilibrium with a radioactive IPTG solution, they observed that the IPTG concentration was slightly higher inside the bag), and using this binding of IPTG as a test to follow the repressor, they purified the repressor. It is a *protein* — which was expected — because "suppressible" non-sense mutations of the *i* gene were known; as described in Chapter 5, suppression is due to a mutation modifying a *t*RNA which becomes capable of reading a non-sense codon (see fig. 6-45), it is therefore a "correction" which takes place at the time of protein synthesis; since a mutation of the *i* gene could be "suppressed", i.e. corrected by a second mutation whose effects manifest themselves at the protein synthesis level, it was clear that the product of the *i* gene had to be a protein. The purified protein obtained by following the binding of IPTG is indeed the repressor, in other words the product of the *i* gene, *because the extracts of a i^s mutant* (super-repressed, non-inducible) *do not bind IPTG.*

The other predictions of the model were then verified by the use of repressor labeled with ^{35}S (after culture of *E.coli* in presence of ^{35}S-sulphate whose ^{35}S is incorporated into the sulphur-containing amino acids, and therefore into proteins). This labeled repressor was mixed with DNA containing the lactose operon, and the mixture was centrifuged in a sucrose gradient: the DNA sediments faster than the proteins and it was observed that ^{35}S was present in the DNA. It was shown that this radioactivity does reflect the *binding of the repressor to the operator* by using, in another experiment, the *DNA of an o^c mutant* — whose operator site is defective — and observing that in this case the quantity of ^{35}S recovered after centrifugation in the DNA band was either nil, or much smaller (depending on the extent of the lesion in the *o* region). Moreover, it must be indicated that the repressor binds only to native DNA, and not to denatured DNA. On the other hand, by mixing the DNA extracted from a wild strain with the *radioactive proteins obtained from an* i^- *mutant* (which synthesizes no repressor or a repressor without affinity for the operator) no ^{35}S is observed in the DNA band after centrifugation, which shows that in the first experiment the radioactivity found in the DNA band does represent the *binding of the repressor*.

Lastly, by mixing an intact repressor with DNA containing an intact operator site, *the binding of the labeled repressor to the DNA does not take place in presence of IPTG*, which confirms the prediction of the model concerning the mechanism of induction (see fig. 8-5): in presence of an inducer, there is no binding of the repressor to the operator.

It was therefore verified that the repressor, product of the *i* gene, binds the inducer (except when it originates from a non-inducible i^s mutant), that it binds to the operator site (except when this site has undergone a deletion), but that the binding of the repressor to the operator does not take place in presence of an inducer and this is all in conformity with the predictions of the model.

With the help of cell-free extracts containing all the required factors, it was possible to obtain *in vitro* the transcription of the DNA comprising the lactose operon into *m*RNA specific of this operon, and then the translation of this *m*RNA into the corresponding specific proteins. The addition of the purified repressor to the incubation mixture blocks the synthesis of *m*RNA, and therefore of the specific proteins, but addition of the inducer lifts this repression.

β) **Tryptophan operon.** – In the case of the lactose operon we have seen that the repressor binds to the operator, except in presence of lactose (or another inducer) which lifts the repression of the operon. Lactose induces the synthesis of enzymes which will enable the cell to metabolize this disaccharide. In the case of the tryptophan operon which contains the genes coding for the enzymes required for the synthesis of the amino acid by the bacterium, one has the opposite situation: when the amino acid is absent in the cell, the operon is functional (it is said to be *derepressed*), so that the amino acid is synthesized; on the contrary when the amino acid is present in excess, the operon is repressed. In spite of this difference which is apparently fundamental, the model proposed by Jacob and Monod is also applicable: one only has to consider the repressor as a protein molecule unable to bind to the operator when it is alone and which requires – in order to bind – a molecule of co-repressor present in the cell when the amino acid is in excess. Hence, *in the absence of amino acid, the operator is not blocked by the repressor, and the transcription of the structural genes corresponding to the enzymes can take place*. On the contrary, *when the amino acid is present in the excess, the co-repressor combines with the repressor, which is then capable of binding to the operator and blocking the transcription of the structural genes*. This model was also confirmed experimentally by the isolation of the tryptophan repressor, and by *in vitro* experiments of transcription and translation of DNA containing the tryptophan operon, experiments which enabled the verification of the role of the repressor.

Let us also indicate a variant in terminology: when the amino acid concentration decreases in the cell (and when therefore there is a decrease in the concentration of co-repressor needed by the repressor to repress the operon), one uses the term *derepression* which is in fact equivalent to *induction* in the sense that both denote a situation characterized by the fact that the RNA polymerase will be able to resume the transcription of the structural genes of the operon (in derepression due to the disappearance of the co-repressor, in induction due to the appearance of the inducer).

γ) **Prophage λ.** – When the bacteriophage λ infects *E.coli*, there are two possibilities:

– either the phage multiplies in the host cell leading to the liberation of a number of new phage particles and death of the bacterial cell. This is the infectious cycle, which will not be studied here;

– or, *the deoxyribonucleic genome of phage λ is integrated in the* bacterial chromosome; it is then observed that the genes of λ (which is then called prophage and not phage) are not expressed; there is no infectious process, the bacteria continue to grow and divide, apparently normally. In fact, the genes of λ are replicated as those of *E.coli* during the duplication of the chromosome: they are thus transmitted to the daughter-cells and the infectious process can start at any moment (even after numerous generations). This is lysogeny.

The problem which arises is therefore to know how is the expression of the prophage λ genes regulated. It was found that there is a *repressor*, coded by the cI gene of the prophage, which prevents the expression of all the genes of the prophage except that of cI (therefore, the cI gene is to the prophage λ what the *i* gene is to the lactose operon). A cell of *E.coli*, carrying the prophage therefore contains molecules of the λ repressor which, not only repress the expression of genes of the prophage genes, but also prevent the expression of genomes of infectious λ phages which could penetrate into the cell (whence the name "immunising substance" sometimes given to the repressor).

Ptashne succeeded in isolating the repressor of the λ phage. He started with 2 cultures of *E.coli* containing prophage λ, previously irradiated by UV to decrease to the minimum the synthesis of bacterial proteins, and he infected them, one by λ phages having an intact cI (in presence of ^{14}C-leucine), the other by λ phages having an altered cI gene (in presence of ^{3}H-leucine). After one hour, he stopped the 2 cultures, extracted the proteins and compared the fractionation profiles of proteins on a DEAE cellulose column: he observed a number of peaks where the ratio of the 2 isotopes was constant, but he found that an important peak (labeled with ^{14}C), resulting from cells infected by λ having cI intact, was absent in the extract (labeled with ^{3}H) resulting from cells infected by λ having the cI gene altered. This protein peak therefore had to be the product of cI gene, which was confirmed later.

Ptashne then showed that the *radioactive repressor is found bound to the DNA of λ* after mixing the two and centrifuging in a sucrose gradient; he verified that *this binding does involve the repressor* (no radioactivity was found in the DNA band when the radioactive proteins originated from a mutant having a non-sense codon in the cI gene) and *requires a homologous operator site* (there is no radioactivity in the DNA band, if this DNA is extracted from a phage having an operator different from that of the wild type λ phage which was used for preparing the labeled repressor). It can be seen that the repression of the genes of the prophage λ also uses the mechanisms predicted by the Jacob-Monod model. However, a difference to be noted is the fact that the affinity of protein cI for its operator sites is not modulated, as in the case of the repressor of the lactose operon, by the fixation of a ligand. It is the fact that cI is (or is not) synthesized from the genome of the phage which will (or not) lead to a repression of the expression of this genome. This is actually a highly simplified representation of the regulation of phage λ, because there is in fact a complex system of regulations in cascade permitting the sequential expression of the phage genes in the host cell once the repression is lifted.

δ) **Remarks on the control of transcription by induction-repression.** – Repression blocks the synthesis of new enzyme molecules by preventing the production of the corresponding *m*RNA, but it does not affect the enzyme molecules existing in the cell, which will continue to function until they disappear by the usual degradation processes. It also does not prevent the already synthesized *m*RNA molecules from continuing to be translated. *The effects of repression are therefore less rapid than those of feedback inhibition* (see page 473). On the contrary induction, owing to the rates at which transcription and translation take place, produces effects which can be observed very early and one may see enzymes appearing in bacteria a few minutes after the addition of an inducer.

We have seen that an excess of tryptophan – due for example, to an addition of this amino acid into the medium where the bacterium is grown – causes a repression of the tryptophan operon. Neither the synthesis of the corresponding polycistronic *m*RNA, nor the synthesis of the various enzymes required for the formation of the amino acid will take place, and since these processes require much energy, it is clear *that repression enables the cell to make a significant economy of energy.*

The systems described above (lactose operon, tryphophan operon) were taken as examples because they are in perfect agreement with the Jacob-Monod model. But variant models have been established. The best known is the positive regulation (see page 445). The standard example is the *maltose* system. In this system, the enzymes required for the catabolism of this disaccharide are synthesized only when a protein activator coded by a regulatory gene, is activated by maltose and will interact with an initiator site, thus permitting the transcription of the structural genes. This target site is located upstream the binding site of RNA-polymerase (and not downstream as in the case of lac operon), and the interaction of the activator protein with its target will make more efficient the starting of the transcription. The same molecule, according to the position of its site of fixation with respect to the poly, play the role of repressor or activator; this is in fact the case of ... in the λ phage.

... *arabinose* operon, the product of the regulatory gene behaves as a ...or in the absence of arabinose, but in presence of arabinose it behaves ...activator and permits – as in the maltose system – the transcription of ...ructural genes of the arabinose operon.

On the other hand, the target sites of regulatory proteins are often more ...mplex than in the examples described above. They are often numerous and their interaction with the proteins can lead to major modifications (loops) of the DNA double helix.

Interactions between regulatory proteins and DNA are now being better understood. There are various biochemical, physico-chemical and genetic methods, allowing to determine the nucleotide residues which are in relation with the protein. In most cases the sequences involved are in palindromic organization (see fig. 8-7), in relation with the symmetrical polymeric structure of the protein. It has been possible (especially for the lactose and λ repressors, and the CRP protein) to identify in the protein structure, several rather independent domains; only one of them binds strongly to the DNA (another domain acting mainly in the association between promoters). The study of the three-dimensional structure of purified proteins and protein-DNA complexes deduced from crystallographic examinations revealed, in the domain which binds to the DNA, the presence of a preserved motif consisting of 2 α helices separated by a small β-turn (see fig. 8-6). In this motif called helix-turn-helix,

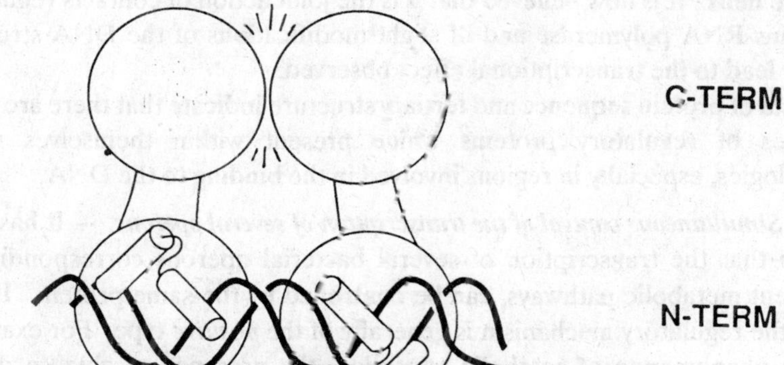

FIG. 8-6.—Diagrammatic representation of the complex between the dimer of the λ repressor cI and the operator site (after M. Ptashne).

The two domains of each of the protomers of cI represented by the spheres correspond to the carboxy (C) and amino (N)-terminal parts of the peptide chain. Proein-protein interactions are established between the C-terminal regions. The structural motif in helix-turn-helix of the N-terminal region interacting with the DNA is represented : the parts of the peptide chain adopting a conformation in α-helix are represented by cylinders and the β-turn by a curved line. The double helix of the DNA is also shown.

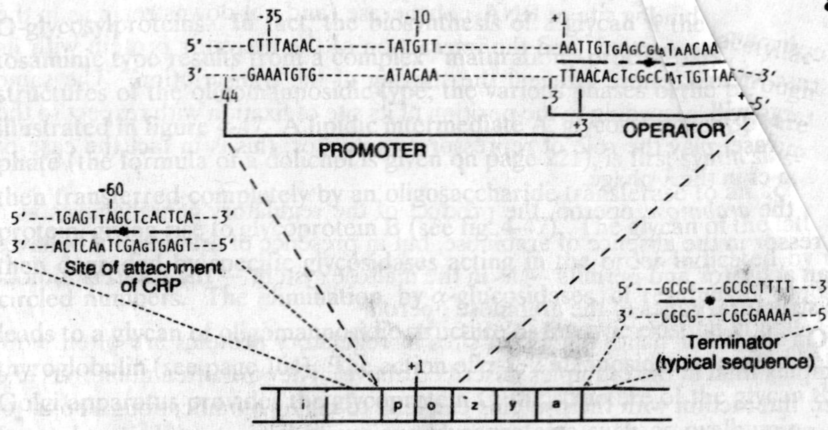

FIG. 8-7. — Diagrammatic representation of signals controlling the transcription of the lactose operon.

Transcription starts at the nucleotide +1 and proceeds towards the right in this diagram (the coding strand of the DNA is here the lower strand). The nucleotides are numbered negatively upstream and positively downstream this point, with respect to the direction of transcription. The palindromic sequences are indicated by a horizontal line between the two strands of the DNA; the symmetry point of these sequences is represented by ——*——; nucleotides not conforming to this symmetry are represented in smaller letters.

one of the 2 α helices of the protein is inserted in the large groove of the DNA, establishing in it specific contacts with the atoms of the underlying nucleotides. The second helix positioned across the large groove, stabilizes this structure by interacting simultaneously with the DNA and the first helix. Furthermore, the close interaction between the DNA and these proteins produces a slight curvature of the DNA which probably has repercussions on the stability of the double helix. It is now believed that it is the joint action of contacts regulatory proteins-RNA polymerase and of slight modifications of the DNA structure which lead to the transcriptional effect observed.

Data of protein sequence and tertiary structure indicate that there are other families of regulatory proteins which present within themselves major homologies, especially in regions involved in the binding to the DNA.

a) ***Simultaneous control of the transcription of several operons***. — It has been shown that the transcription of several bacterial operons corresponding to different metabolic pathways, can be controlled by the same protein. In this case, the regulatory mechanism is generally of the *positive* type. For example, in the phenomenon of *catabolic repression*, the presence of glucose during bacterial growth inhibits the syntheses of a very large number of inducible enzymes pertaining to diverse catabolic pathways. This phenomenon is due to an arrest of the transcription of the corresponding operons, transcription which requires not only the appropriate inducer (lactose for example, in the case of the lactose operon), but also a specific protein, having bound cyclic AMP (*c*AMP), which is called CAP (*C*atabolite gene *A*ctivator *P*rotein) or

CRP (*C*yclic AMP *R*eceptor *P*rotein) and which binds to the DNA upstream of the promoter (see fig. 8-7). In presence of glucose, the *c*AMP content decreases and the deficiency in CRP-*c*AMP complex causes a block of the transcription.

Multiple regulation models were also established in the case of regulations by the nitrogen source, phosphate, ionic strength, etc.

On the other hand, the *specificity of transcription* can also be under a *positive* type of control by modifications at the RNA polymerase level itself. As mentioned earlier (see page 445), positive control implies the existence of protein substances which permit the expression of some genes (which are not transcribed in the absence of these substances). The sub-unit (or factor) σ of the RNA polymerase of *E.coli* belongs to this group of substances. Thus the synthesis of particular σ factors, presenting new specificities, permits the expression of some operons in *E.coli*; this is the case for example of the σ^{32} factor (thus named because it has a molecular weight of 32 kdaltons) which recognizes the promoters of heat-shock genes coding for a series of proteins synthesized in response to a thermal shock; these promoters have a −10 sequence different from the usual −10 sequence:

Usual promoter: 5′——TTGACA———TATAAT ———3′

Heat shock promoter: CNCNCTTGAA———CCCATNT———3′.

A similar mechanisms operates in the sporulation of *Bacillus subtilis*, which represents a simple model of cell differentiation. It also plays a role in the control of transcription of the viral genome, at different stages of the infection of *E.coli* by some bacteriophages of the T_{even} series (one knows for example, a factor σ T_4 which acts during the infection of *E.coli* by this phage). Other mechanisms can also play a role in the positive control of transcription: for instance appearance of new RNA polymerase (in the case of phages T_3 and T_7), or the modification of the core enzyme leading to new specificities.

b) **Termination of transcription.** – The transcription of an operon ends with the liberation of the RNA synthesized and of the RNA polymerase which leaves its template.

The termination of transcription confers on each transcription unit its individuality with respect to the neighbouring operons but permits also, in certain conditions, the fine regulation of the expression of a given operon (as we shall see while studying the mechanism of attenuation).

In *E.coli*, termination takes place generally at specific sites on the DNA; these sites are of two types: some, called ρ (rho)-dependent, require the action of a termination protein, the ρ factor. Others are ρ-independent; in this case a comparison of the 3′ terminal sequences of numerous messenger RNAs reveals the existence of palindromic sequence, rich in G + C, which permit a folding back of the RNA by pairing of complementary bases (see an example in the typical sequence of the terminator of fig. 8-7); these sequences are followed by a series of uridylic residues (poly U) at the 3′ end of the RNAs. These sequences appear in the form of a series of thymidylic residues in the typical sequences of the terminator of figure 8-7.

The exact mechanism of termination is not yet fully understood but it appears probable that multiple and complex interactions take place, involving the RNA polymerase (core enzyme), the secondary hair-pin structure of the RNA, a destabilization of the RNA : DNA hybrid at the poly U residues, the ρ factor, or other cytoplasmic proteins (e.g., protein Nus A).

Mutations introducing non-sense codons in the nucleotide sequence of an operon can be the cause of a *premature termination* of the transcription of distal genes of this operon (the proximal genes are nearer the operator, the distal genes are more distant from the operator). These mutations are called "polar" because they bring about not only the arrest of translation in the mutated gene but also a reduction of the expression of distal genes. These non-sense mutations (where translation is interrupted by liberation of the ribosomes) probably unmask transcription termination sites situated further downstream on the RNA and which are normally inactivated by the progression of ribosomes.

The majority of these termination sites, called weak, located inside the multicistronic operons, do not play an important physiological rule (see however, the case of complex operons, page 456). It has however been shown that some internal sites participate in the regulation of transcription by being sensitive to particular components having no effect at the level of the initiation of transcription. Regulatory proteins will then form specific complexes with RNA (and not DNA). This regulatory mechanism functions by selective reduction of the transcription of the distal portions of an operon. Such a system operates during the lytic cycle of the bacteriophage λ where a particular protein, an *antiterminator* protein, is required to prevent or lift the premature termination of transcription at precise sites.

Another regulatory mechanism, called *attenuation*, comes into play in most bacterial operons grouping the genes involved in the biosynthesis of amino acids (tryptophan, histidine, leucine, phenylalanine...). In this type of operon, the cellular concentration of the corresponding amino acid (whose synthesis is catalyzed by the products of the operon) indirectly influences its own biosynthesis by modulating the efficiency of the premature termination of transcription at the level of a signal situated between the promoter and the first structural gene of the operon. This attenuation mechanism implies a close coupling between the transcription of this proximal portion of the operon (leader region) and its translation. The RNA sequence transcribed from this region (leader RNA) being particularly rich in codons of the amino acid corresponding to the operon, it is finally the cellular concentration of the corresponding amino acyl *t*RNA which will regulate the translation of the leader RNA. The arrest of transcription or its continuation downstream the leader RNA is controlled by a process involving two types of alternative secondary structures at the level of the 3′-terminal end of the leader RNA[(1)].

The model presently accepted for the attenuation mechanism is the following *when the cellular concentration of the amino acid (and therefore that of the amino acyl-tRNA) corresponding to the operon is low, the ribosomes moving along the leader RNA take a pause at the level of the codons corresponding to this amino acid. The free part of the leader RNA*, situated between the first ribosome and the RNA polymerase, can then adopt only one of the two possible conformations, here, a structure which does not prevent the continuation of the transcription and expression of the structural genes of the operon (fig. 8-8a). However, when the cellular concentration of the amino acid increases, the ribosomes stop their progression only at the level of the termination codon of the leader peptide. In this case, the only possible conformation for the free

The intracellular concentration of the amino acid (and therefore of the aminoacyl-*t*RNA) synthesized by the enzymes coded by this operon is low, with the result that the translation of the leader RNA, rich in codons corresponding to this amino acid, is slow and the ribosomes pause at the codons specifying this amino acid. Sequence 2 can pair with sequence 3, and the conformation taken by the leader RNA permits the continuation of transcription by the RNA polymerase.

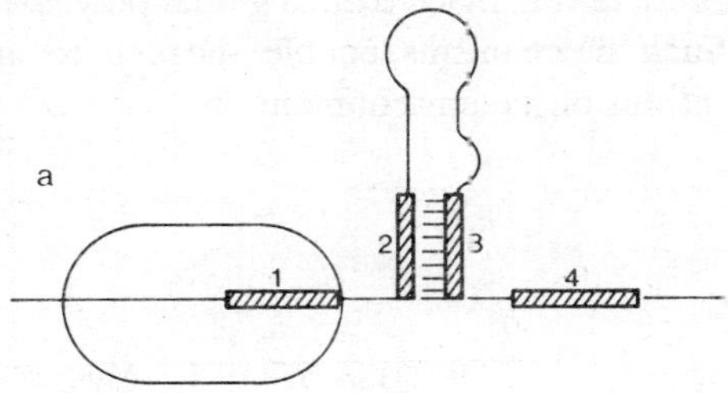

The concentration of the amino acid (and therefore of the aminoacyl-*t*RNA) is high with the result that the translation of the leader RNA is rapid. Sequence 3, which cannot pair with sequence 2 pairs with sequence 4 and the resulting conformation of the leader RNA brings about the termination of transcription.

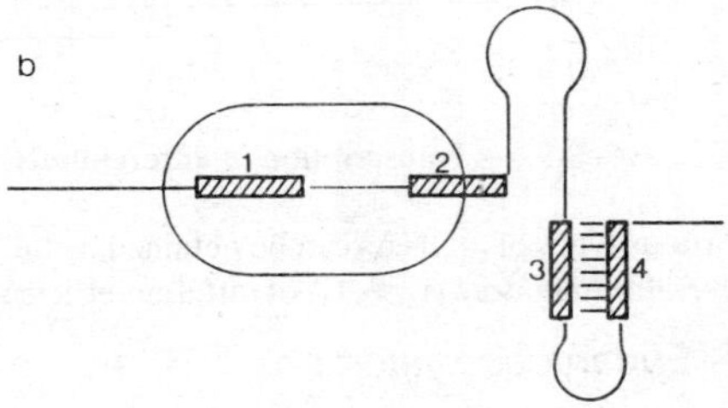

FIG. 8-8. — Control by attenuation in the case of an operon regrouping the genes of enzymes catalyzing the biosynthesis of an amino acid.

---

(1) The attenuation mechanism differs in this respect from the mechanism proposed for the polar effect of non-sense mutations (see above).

part of the leader RNA is a structure typical of the termination signal for the RNA polymerase and the transcription stops at the end of the leader sequence (see fig. 8-8b).

For the tryptophan operon there are therefore two successive 'locks', one at the level of the operator (see page 448), the other at the level of the attenuation site, to decrease the quantity of RNA synthesized in response to increasing cellular concentrations of tryptophan and tryptophanyl-*t*RNA respectively.

— COMPLEX OPERONS. — There are now numerous examples in *E.coli* where a number of genes are adjacent on the chromosome, genes whose expression can be obtained from several types of poly (or mono) cistronic messengers. For this purpose there are several promoters and several terminators, localized at various positions along the DNA (see fig. 8-9) and of different efficiencies. Furthermore, the efficiency of these signals can depend on various external regulatory factors leading to the preferential increase (or decrease) of the expression of one, or several, of these genes in particular physiological conditions, while in most cases, the expression will be coordinated, taking place from a global polycistronic *m*RNA.

Such mechanisms enable the cell to appreciably refine its response to variations of the environment.

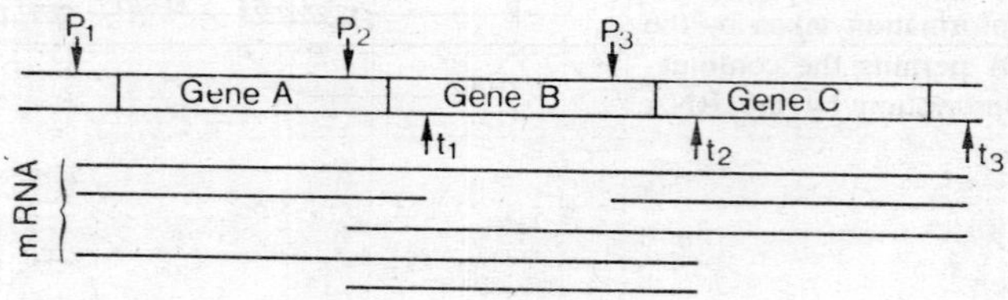

FIG. 8-9. — Transcription of different *m*RNAs from the same group of genes.

Various types of *m*RNAs can be obtained by the use of several promoters (P_1, P_2, P_3) and several terminators (t_1, t_2, t_3) of different efficiencies and subjected to diverse regulations.

B. — *Eucaryotic systems*

The control of eucaryotic genes is indeed far from being as fully understood as that of some procaryotic genes, but it is now increasingly clear that the expression of eucaryotic genes is controlled by a variety of different genetic mechanisms, unknown in procaryotes.

a) **Modifications at the genome level.** — In some protozoans, nematodes, crustaceans and insects, the cellular differentiation of somatic lineages may be accompanied by a decrease of the quantity of chromatin or the loss of particular chromosomes. But it is not very likely that this phenomenon of *programmed elimination* of a part of the genetic material is universally widespread.

The increase of the number of copies per genome of a class of genes is a means used by some cells to produce massive quantities of RNAs or specific proteins. This mechanisms of *gene amplification* operates for example in the oocytes during oogenesis in some batrachians for the massive synthesis of ribosomal RNA or in the egg of some insects to enhance the production of chorionic proteins (proteins constituting the outer envelope of the embryo).

Examples are known of *genetic rearrangements* where several coding regions, scattered on the same chromosome, fuse together during the development of a cell to form a new transcription unit. This phenomenon is called "somatic recombination". The best known example is the formation of immunoglobin genes during the differentiation of lymphocytes B, cells specialized in the production of antibodies.[(1)]

A molecule of antibody or immunoglobulin consists of 4 sub-units: 2 identical, light polypeptide chains (220 amino acids) and 2 identical heavy chains (330-400 amino acids) joined to one another by disulphide bridges and non-covalent bonds to constitute a Y-shaped molecule (see box in fig. 8-10). The amino acid sequence of the NH_2-terminal end of these chains determines the specificity of the antibody vis-a-vis a given antigen and differs from one antibody to another (variable domains). This region represents the idiotype of the antibody or specific binding site of the antigen. The remaining parts (constant domains) are not all identical, but exist in a limited number of forms: 2 types for the light chains (λ and κ) and 5 types for the heavy chains (μ, δ, γ, ε and α), the latter determining the belonging of the immunoglobulins to each of the 5 classes (or isotypes) IgM, IgD, IgG, IgE and IgA present in mammals. The organization of nucleotide sequences coding for the various regions and their assembly during the differentiation of lymphocytes accounts for the extreme diversity of antibodies which can be produced by an individual from a relatively small number of genes. Thus, the variable domain of the κ light chains for example is coded by about three hundred different V1 elements and 4 different J1 elements, while the constant domain is coded by a single Ck element. These elements, all present on the same chromosome, are arranged in the order $V1_n$ – $J1_4$ – Ck indicated in figure 8-10. The arrangement of elements coding for the heavy chains is similar but between the VL and JL elements, one finds about twenty different D elements which contribute to an increase in the diversity of the variable domains of these chains. The sequences corresponding to the constant domains Cm Cd Cl, Ce and Ca are localized on the same chromosome (different from the one programming the light chains) after the VL_n D_{20} and JL_4 elements (see fig. 8-10).

---

(1) Such rearrangements exist also in procaryotes (*Bacillus subtilis*, Cyanobacteria). It must be emphasized that the event being irreversible, can only take place in the case of some cells which have differentiated and do not reproduce any more, as their genome has become different from the parental genome. On the contrary, some reversible rearrangements have been described, which lead to the possibility of expressing (or not) some bacterial or viral genes (for example variation of antigenic phase in *Salmonella*).

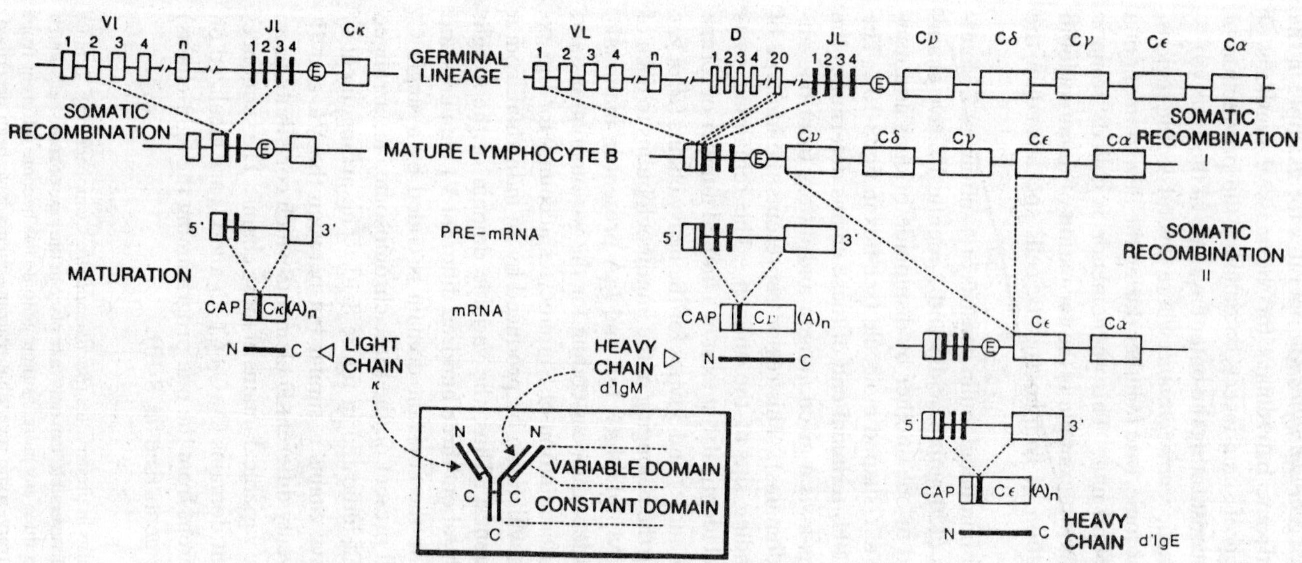

FIG. 8-10. — Structure and expression of genes of immunoglobulins.

The regions coding for the chains of immunoglobulins are represented by rectangles of different sizes, corresponding to the elements V, D, J and C. E denotes the enhancer sequence present upstream the elements C. During the maturation of a lymphocyte B, a recombination occurs at the level of elements coding for the light chain of immunoglobulins, while several successive recombinations (two are shown here), at the level of elements coding for the heavy chain, are responsible for the changes of isotypes (e.g., IgM→IgE). The maturation of the pre-*m*RNA leads to the binding of the cap, the polyadenylation A(*n*) and the splicing of intermediate sequences, leaving in the *m*RNA only one representative of each of the elements J and D. The translation of *m*RNAs leads respectively to the synthesis of light and heavy chains and then their association by disulphide bonds (.) to constitute the active molecule (see box): N and C correspond to the —NH2 and —COOH terminal ends of the peptides synthesized.

The assembly of a functional immunoglobulin gene takes place by the recombination of some coding elements at the level of the chromosomal DNA of lymphocytes of some of the coding elements. In the case of light chains, anyone of the V1 elements fuses at random with one of the 4 J1 elements leading to the sequence $V1_i\ J1_i - C_{\lambda \text{ or } \kappa}$. For the heavy chains this first step includes an additional recombination at the level of a D element and leads to the structure $VL_i\ D_i\ JL_i - C_{\mu-\alpha}$. The mechanism of this somatic recombination is not yet clear, but it involves particular "signal" sequences located near each of the V,D and J elements. This recombination has two important consequences. The first is the activation of transcription from the V element of the unit thus constituted. This stimulation of the constitutive promoter present in the V element is related to its coming closer to an enhancer (see pages 461) situated between J and C. The effect of this enhancer acts only on the nearest juxtaposed V element. The primary transcript resulting from the transcription of this unit then undergoes a maturation which eliminates the intermediate sequences according to the splicing mechanism of introns (see pages 338-343). The second consequence of this recombination is that it allows, thanks to the multiple possible combinations, the synthesis of a large number of chains differing by the sequences of the V, D and J elements. The combination of a light chain and a heavy chain can thus probably provide nearly 10^7 different sites of binding of antigens. There is an additional factor of diversity which results from the lack of precision of the mechanism of juxtaposition of the V,D and J elements: the junction is established with a difference of a few nucleotides more or less and furthermore, insertions of nucleotides catalyzed by a nucleotidyl-terminal-transferase, can take place during the recombination. These two phenomena explain that antibodies coded by identical elements are however able to present slightly different binding sites for antigens. Lastly, recent indications suggest that in V, D and J elements, spontaneous somatic mutations are more frequent than in the rest of the genome. The joint effect of combinative mechanisms, junction errors, insertions of nucleotides and somatic mutations probably takes the potential number of different antibodies to nearly one billion.

Another example of diversification of the genetic expression from the genomic sequences is provided by the variability of the surface antigens of the trypanosome (a flagellate protozoan, blood parasite of various animals, responsible for sleeping sickness) during its development in the animal organism. This variability which enables it to escape from the immune system of the host, is due to the successive expression of genes coding for related but appreciably different glycoproteins, which bind to the envelope of the trypanosome. In the genome of this protozoan there are therefore some hundred "basic versions" of this gene. These genes are generally not transcribed, but one of them can be duplicated and the copy made ("expression copy") is expressed if it is transferred by translocation at the end of a chromosome (telomer). Such a mechanism is applied by the parasite about once in 15 days and the activation of a new copy leads, in most cases, to the destruction of the previous one. In this manner, the parasite modifies its surface antigens before the immune response of the host organism can neutralize it.

Besides these important modifications of the genome there are others which are more discrete, especially modifications of some nucleotide residues of the DNA. A cytosine can for instance be methylated on carbon 5 of the pyrimidine heterocycle (5-methyl-cytosine), if it is adjacent to a guanine (CpG) in the nucleotide sequence of the DNA of eucaryotic cells. It seems, at least in vertebrates, that the proportion of methylated CpG in a gene or in its immediate neighbourhood reflects the state of activity of this gene: a given gene is generally under-methylated when it is actively transcribed and over-methylated when it is not expressed.

Changes in the rate of methylation seem to involve the replication of the DNA, but the mechanisms operating in this type of control are not fully understood. It is possible for example that methylation exerts its effect by modifying the interaction of some regulatory protein factors with their binding sites on the DNA. Lastly, it must be emphasized that except for vertebrates, the relation between genetic expression and degree of methylation of the DNA is much less clear: for instance in insects, cytosine is practically never methylated, while the quasi totality of CpG remains constantly modified in plants. It is however possible that in these cases, only a limited number of strategic sites undergo a controlled methylation, playing the same role as in vertebrate but having so far escaped the detection methods which are not sufficiently sensitive.

Finally, let us not forget that the DNA of eucaryotic cells is arranged in a *nucleosomal structure* (see pages 282 and 318). Studies on the digestion of chromatin by some nucleases suggest that the genes being transcribed are generally more sensitive to the action of these nucleases than the genes which are not expressed. Moreover, active genes possess a hypersensitive region situated immediately upstream the initiation site and probably free from nucleosomes. It is likely that such domains constitute sites accessible to the regulatory proteins in particular, and to the transcription machinery in general.

b) **Specialization of RNA polymerases.** – While in procaryotes only one enzyme catalyzes the transcription of all kinds of RNAs (*m*RNAs, *r*RNAs and *t*RNAs), three different classes of RNA polymerases share this task in eucaryotes (see page 333).

c) **Particular nucleotide sequences (or particular signals).** – Conventional genetics has been particularly useful for the study of the expression of procaryotic genes (see page 445) but due to the material itself, it was not widely applied to eucaryotic systems. It is only with the development of the technology of gene cloning (see page 380) that rapid progress could be made. With these techniques it became possible to isolate a large number of eucaryotic genes of varied functions and origins and to study their structure. The comparison of their nucleotide sequences revealed the existence of structural motifs common to most of them and the study of the role of these highly conserved sequences was undertaken by "*in vitro* genetics" experiments. Purified genes were thus modified *in vitro* in these regions (deletions or point modifications) and the effect of these "mutations" on the control of the expression of these genes was tested either in systems of transcription *in vitro*, or after reintroduction into live cells (microinjection in the nucleus of cells in culture or transfer by endocytosis of a precipitate of calcium phosphate-DNA deposited on the cells). The level of specific transcription (i.e. starting at the correct site on the DNA) is then determined by S1 mapping (see fig. 8-2).

These studies have shown that the sequences involved in the control of transcription in eucaryotes differ notably from those of procaryotes. Thus, the promoter of eucaryotic genes transcribed by RNA polymerase B and coding for proteins seems to cover a DNA region much larger than in procaryotes, comprising the transcription initiation site, and extending to more than 100 base pairs upstream. One finds a TATA sequence similar to that of procaryotic promoters but localized at about 30 base pairs from the initiation site (instead of 10 in procaryotes). This TATA box, present in the large majority of eucaryotic genes coding for proteins, seems to be involved in the selection of the site of transcription initiation. Other elements are required for an effective transcription by RNA polymerase B (see fig. 8-11). They are localized upstream the TATA box, generally between 40 and 100 base pairs upstream the initiation site. Their sequence is less conserved but it is often richer in G and C. The elements present in the 100 base pairs which precede the sequences coding for the RNA thus constitute the structure of the "minimum promoter" and provide the base (or constitutive) level of transcription by RNA polymerase B.

While the relative positions of the elements of the constitutive promoter are not very flexible, another type of promoter element called *enhancer*, stimulates the transcription of genes (between 50 and more than 1 000 times), independently of its orientation and position with respect to the initiation site, whether it is situated upstream the latter, in an intron, or even downstream the gene. This activity, two-directional, and relatively independent of the distance, is characteristics of this type of promoter elements.

Originally identified in viral genomes, these enhancers have since been found in numerous cellular genes. Of variable length, they are generally formed by an association of distinct, often repeated motifs (enhansons). Each of these motifs being the binding site of a particular protein factor, it appears that the enhancer effect results from the combined (often synergic) action of these factors on the initiation complex formed by the RNA polymerase and the general factors of transcription. To permit the interaction between proteins, sometimes separated by considerable distances on the genome, the folding back in loop of the DNA has been postulated. It is the promoter nearest the enhancer which however appears to benefit most from its effect. Thus, in the genes of immunoglobulins, the enhancer is localized between the elements J and C of the genes coding for the light chains κ and the heavy chains. This particular position of the enhancer, downstream the initiation site, explains the fact that only the promoter present in the nearest V element, fused with the J or DJ elements during the differentiation of lymphocytes B, is activated. The promoter elements of the non-rearranged V elements, situated very far upstream, are probably outside the field of the enhancer and remain inactive. Most of these enhancer elements present, in addition, the interesting characteristic of being active only in a particular tissue thus conferring on the genes to which they are linked a tissue-specific expression. The enhancer of immunoglobulins genes is thus functional only in lymphoid cells. It is therefore probable that the enhancers are required for the intensive expression of genes coding for a major product of cells having reached (or passing through) a particular stage of differentiation. The presence in a tissue of the adequate population of factors recognizing all the enhansons constituting a enhancer and the absence or inactivation of some of these factors in other tissues, may explain the cellular specificity of enhancers.

An example of more flexible regulation is provided by the hormonal control of gene expression. The steroid hormones (glucocorticoids, estrogens, progesterone and testosterone in vertebrates and ecdysteroids in insects) can thus selectively induce or stimulate the transcription of some genes in particular tissues.

These hormones penetrate into the target-cells and bind to protein receptors localized in the cytoplasm (receptors of glucocorticoids) or in the nucleus (e.g., receptors of estrogens and progesterone). Binding of the hormone induces the dimerization of the hormone-receptor complex, the activation of the inducer domain (possibly by dissociation of a inhibitor protein bound to the inactive receptor) and the translocation of the hormone-receptor complex into the nucleus where it combines with particular sequences on the DNA. These binding sites, generally large in number, are localized at variable distances upstream the TATA box (or sometimes downstream, in an intron) of controlled genes. The properties of these sites were thus found comparable with those of the enhancer elements described above and it has been proposed that they are in fact conditional enhancers functioning only in presence of the hormone-receptor complex.

DNA sequences having properties opposite to those of enhancers were discovered and called *silencers*. Such elements which bind particular protein factors and inhibit transcription independently of their orientation and position, were identified near a large number of genes (genes of interferons α and β, gene of ovalbumin...) and sometimes juxtaposed to enhancers (enhancer of the gene of heavy chains of immunoglobulins). There is no doubt that the simultaneous presence of silencers and enhancers must contribute to a finer control of tissue-specific expression.

The salivary glands of the Drosophila larva represent a choice system for the study of the mechanisms of hormonal induction because the stimulation of transcriptional activity of the chromosome can be visualized directly. A polytenization of the DNA (cycle of 9 tó 10 successive replications without separation of chromatids and without division of the nucleus) has taken place in these cells leading to the formation of giant chromosomes where the actively transcribed zones appear like swellings called "puffs". Thus, in the 30 minutes following a treatment with ecdysteroids, one can observe the formation of half a dozen specific puffs whose induction is independent of protein synthesis. On the contrary other puffs, appearing later after this treatment, seem to depend on neo-synthesized proteins. The study of the mechanism of action of these proteins will reveal whether, as may be imagined, they are initiation factors of transcription presenting new specificities (of the bacterial σ factors type for example).

As for the genes transcribed by RNA polymerase A (genes coding for the ribosomal RNAs, or *r*DNAs), the promoter sequences extend approximately from –150 to +10 (see fig. 8-11). The sequences which include the initiation site (+1) up to the position –40 however seem to play a dominant role in the selectivity of the transcription. No obvious consensus sequence could be found from the comparison of genes of different species. In the case of the ribosomal transcription units of *Xenopus laevis*, sequences possessing properties of enhancer elements were identified in the spacers (about 4 Kb) which separate the copies of *r*DNA. The structure of these enhancers is however original in that they correspond to more or less perfect repetitions of a sequence of 42 base pairs present in the promoter region, between the positions –114 and –72. It is therefore likely that these repeated elements play an attractive role for the protein factors which will then bind to the promoter and activate the transcription.

The situation is again different for the genes transcribed by RNA polymerase C where the specificity of initiation is controlled by nucleotide sequences situated inside the coding region (see fig. 8-11). In the case of genes coding for the ribosomal 5S RNA (which has about 120 nucleotides), the responsible sequences are situated between positions +50 and +80. In the case of *t*RNA genes (which have about 80 nucleotides), called *t*DNAs, the essential elements are constituted by two distinct regions generally between +10 and +30 (*block A*) and between +50 and +70 (*block B*) separated by a central element not essential for the promoter activity, comprising frequently one or several introns. These two promoter elements contain each a different sequence of about 10 base pairs found in the genes of all the *t*RNAs as well as in those of the 5S *r*RNA. In the latter however, only the second seems to be involved in the promoter. A series of recent observations indicate that some genes transcribed by RNA polymerase C (e.g., gene coding for RNA U6) possess control elements situated upstream the transcription initiation site and especially a sequence of TATA type. In some conditions, these elements outside the transcribed sequences are sufficient to bring about an efficient transcription by RNA polymerase C and can even be recognized by RNA polymerase B. These observations which suggest that different polymerases can utilize common transcription factors, must be viewed together with the fact that some polypeptide sub-units which constitute the RNA polymerases are present in the 3 classes of enzymes.

Reconstitution studies of *in vitro* specific transcription systems from cell extracts or partially purified fractions could show that the promoter elements defined above are binding sites of particular transcription factors required in addition to the RNA polymerase. These specific nucleoprotein interactions were revealed by "footprinting" experiments. (see fig. 8-3).

These experiments revealed the existence of multiple factors, specific of different promoter elements, whose combined action seems to be required for optimal promoter activity. The purification of an increasing number of these factors from eucaryotic cells and the cloning of the corresponding genes enabled the determination of some of their structural and functional characteristics. Concerning the domain involved in the binding of factors to the DNA, the existence of 3 types of structural motifs (see fig. 8-12) could be deduced from the study of the amino acid sequence of these factors. The *helix-turn-helix* motif (similar to the one identified by crystallographic study in some procaryotic regulation factors like cI and CRP) was thus found in the homeodomain of most proteins which are involved in the control of the early steps of development (see below). The *zinc finger* motif corresponds to a folding back of the peptide chain into a loop of 12 to 14 amino acids, stabilized by coordination bonds between a Zn atom and pairs of cysteine and/or histidine. This type of structure was initially discovered in a transcription factor of RNA polymerase C, the factor TFIIIA (which binds to the intragenic promoter of sequences coding for 5S RNA) where this motif is repeated 9 times, consecutively. Similar structures have since been identified in regulator proteins of polymerase B, like factor Spl (3 fingers) binding to the motif GGGCGG (see fig. 8-11), or receptors of steroid hormones (2 fingers). A third type of motifs was found in molecules which bind efficiently to the DNA only in the form of dimers. This motif, generally situated near the C-terminal end of the protein, corresponds to a portion of the polypeptide chain in α-helix, comprising 4 or 5 leucines, each spaced by 6 amino acids. Thanks to this periodicity (1 leucine every 7 amino acids), the hydrophobic side chains of these leucines are all localized on the same face of the α-helix. Two molecules having this structure can thus combine together at these helices by interdigitation of the side chains of leucines, the interaction being possibly stabilized by hydrophobic bonds between the leucines themselves, or between these residues and other hydrophobic amino acids positioned between the leucines. This structure called "*leucine zipper*", is believed to be involved in the binding to the DNA, in an indirect manner, by enabling the coupling of proteins having only a low affinity for the DNA separately. Leucine zippers were found in transcription factors binding to the DNA in the form of homodimers (like factor C/EBP which binds to consensus sequences GTGG$^{AA}_{TT}$ AG – see fig. 8-11 – or CAAT present in some promoter elements) or heterodimers (like the proteins coded by the proto-oncogenes c-jun and c-fos which assemble together before binding to their recognition site on the DNA). Let us mention lastly, that leucine zipper structures were also found in proteins comprising helix-turn-helix or zinc finger motifs. This observation suggests that leucine zippers have a more general role in the field of protein-protein interactions.

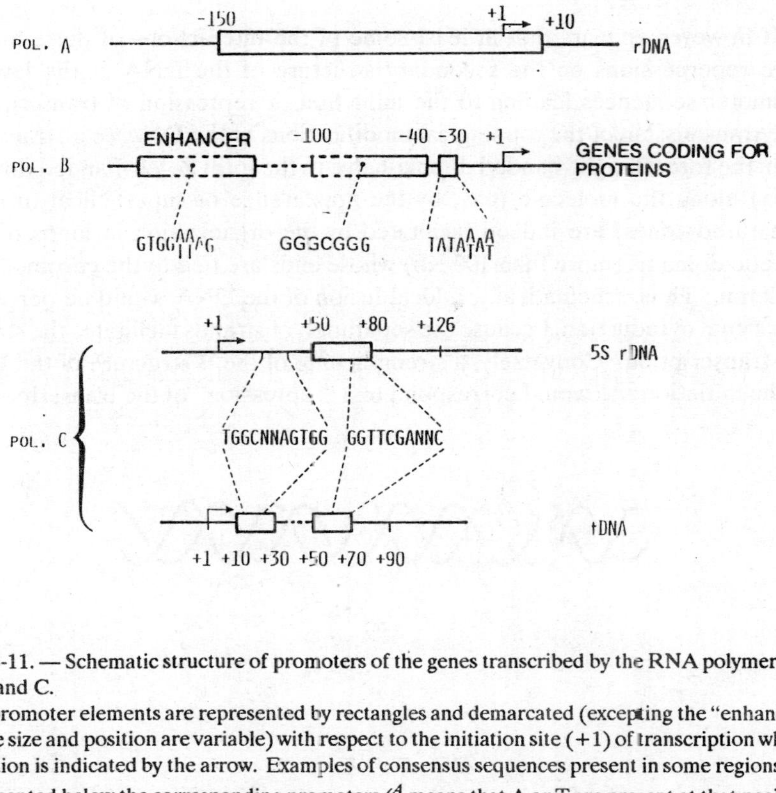

FIG. 8-11. — Schematic structure of promoters of the genes transcribed by the RNA polymerases A, B and C.
The promoter elements are represented by rectangles and demarcated (excepting the "enhancer" whose size and position are variable) with respect to the initiation site (+1) of transcription whose direction is indicated by the arrow. Examples of consensus sequences present in some regions are represented below the corresponding promoters ($\frac{A}{T}$ means that A or T are present at that position at equivalent frequencies; N corresponds to any one of the 4 nucleotides).

While the DNA binding domain of these transcription factors is responsible (thanks to the structural motifs described above and their immediate environment) for the high specificity of protein-DNA interactions, a different domain is involved in transcriptional activation. It appears that these enhancer regions often consist of α-helices of amphipathic character, i.e., presenting groups of negative charge on one face of the helix and groups of positive charge on the other. Experiments of construction of chimerical factors, by genetic recombination *in vitro*, confirmed that the DNA binding domains were responsible for the specificity for a particular promoter element, but showed that the enhancer regions were generally interchangeable (between factors of diverse functions and origins). These observations emphasize not only the relative independence of the two domains, but also the high degree of conservation of the transactivation mechanism, from yeast to the human cell. At the molecular level, this mechanism is not yet known and its deciphering will necessarily involve the comparative study of the tertiary structure of these specific nucleoprotein complexes.

It however appears that at least some of the interactions of these factors have repercussions on the secondary structure of the DNA at the level of promoter sequences leading to the induction or repression of transcription. The transmission of the topological modifications of the DNA (e.g., transition from the form B, right-handed double-helix, to the form Z, left-handed double-helix) along the molecule (e.g., by the appearance of superhelical turns or denatured zones) are indeed facilitated by the organization in loops of vast genetic domains (more than 100 Kb) whose ends are tied to the chromosomal skeleton. Thus, schematically, a local fusion of the DNA would be perceived as a signal of induction, because the separation of strands facilitates the start of the transcription. Conversely, a strengthening of the B structure of the DNA at the initiation site would correspond to a "repression" of the transcription.

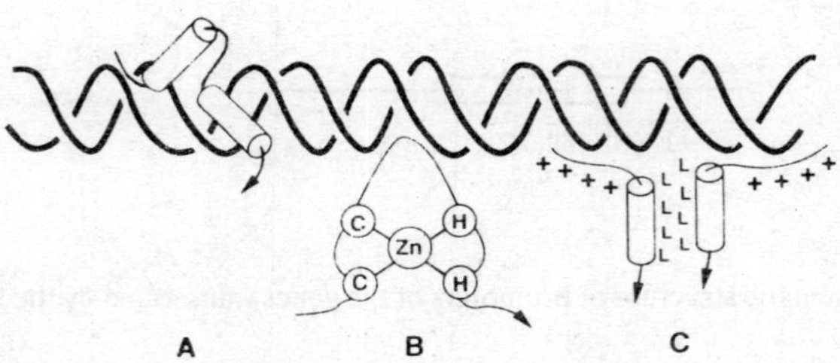

FIG. 8-12. — Diagrammatic representation of the three types of structural motifs involved in the binding of transcriptional regulation factors with the DNA, in eucaryotic cells.

A. Helix-turn-helix motif. Parts of the peptide chain adopting a α-helix conformation are represented by cylinders; the β-turn connecting the 2 portions in α-helix is shown by a curved line.

B. Zinc finger motif. A finger shaped structure (which may comprise a α-helix and β folds) is stabilized by coordination bonds between a zinc atom, two cysteines (C) and two histidines (H) (or just 4 cysteines).

C. Leucine zipper motif. The dimerization of two polypeptide chains is stabilized by hydrophobic interactions between the side chains of leucines (L) suitably positioned on regions in α-helices. The neighbouring regions, charged positively (+), interact with the DNA.

The figure shows the double helix of the DNA and the peptide chains with the arrows pointing towards the carboxy-terminal end.

In this perspective, it is quite remarkable that genes whose products are involved in the same function are very frequently grouped together, forming real specialized genetic domains. This is especially the case with ribosomal genes, genes of histones, globins or interferons. This organization in battery is clearly favourable to a coordinated regulation of the genes concerned. Other families of genes can be controlled simultaneously without being necessarily clustered. In this connection one may cite genes under hormonal control (see above), heat-shock genes, genes induced by light (whose products are involved in photosynthesis systems), genes coding for at least a part of the translation system (proteins constituting ribosomes, elongation factors, sub-units of the RNA polymerase A). In these examples a promoter sequence specific of each family is recognized by diffusible factors (activated by binding of hormones, by thermal or photonic excitation, or by other stimuli) controlling the level of expression of the corresponding genes.

Lastly, one can mention the particular case of a series of genes (homeotic genes) controlling morphogenesis in the Drosophila. Little is known yet on the functioning of these genes which are expressed each at a precise stage of the embryonic development. It was however found that they present a strong homology in a sequence of 180 base pairs (homeobox), situated in one of their exons. The homology is even more striking when one compares the sequences of the 60 amino acids coded by these regions. The basic character of this peptide sequence (homeo-region) is remarkable because at least 1 amino acid out of 4 is, either an arginine or a lysine. This observation suggests that the homeotic genes code for protein factors which would bind to well defined sequences on the DNA through basic regions corresponding to the homeobox. These factors would thus control the activity of other genes involved in the embryogenesis. The fact that the homeobox has been conserved in the course of evolution (since it was found in several higher organisms including man) underlines the importance of its function.

In addition to numerous factors which, in association with precise sites on the DNA, ensure the selectivity of the transcription initiation and control its efficiency, there are also other factors which modulate the rate of elongation of RNA chains.

Little is known at present on the processes of termination of transcription in eucaryotes. While the RNA polymerase C finds, at the 3′ end of all the genes it transcribes, a precise termination signal comparable with that of the ρ-independent procaryotic genes (a short palindromic sequence, followed by a series of T on the non-coding strand), no typical sequence of termination could be identified for the genes transcribed by the polymerases A and B. It however appears that the transcription of ribosomal genes generally stops upstream 3 consecutive thymidines. The primary transcripts coding for proteins often possess heterogenous 3′ ends, due to terminations in multiple sites occurring in a DNA zone, of size varying with the genes but always containing sequences rich in AT.

2. – Post-transcriptional control of gene expression

The physical separation of the genetic material (nucleus) and the machinery responsible for its translation (cytoplasm) involve in eucaryotes, mechanisms of maturation and transport of the transcripts. Such mechanisms do not exist in procaryotes where transcription and translation are coupled. In the latter, the post-transcriptional regulation will therefore operate mostly at the translation level, whereas in eucaryotes it can also take place during the steps of maturation and transport of RNAs to the cytoplasm.

A. – Maturation and transport of mRNAs

In procaryotes, the preferential destruction of distal parts of messengers enables, in some cases, the decoordination of the expression of genes belonging to the same multicistronic operon.

In eucaryotes, each step of the maturation of *m*RNAs (see pages 335-343 and fig. 6-41) is a potential site of post-transcriptional regulation of gene expression. A particularly striking example illustrating this mode of control is that of alternating splicing. In many cases a given pre-*m*RNA can lead to two or several different mature *m*RNAs, corresponding each to the combination of a different series of exons. This phenomenon is mostly observed in viruses where it permits maximal exploitation of the coding capacity of the DNA by the combined use of the three reading frames. But there are also examples of alternating splicing in animals where, in this manner, the same gene can code for different proteins according to the tissue where it is expressed.

Lastly, another way of modifying the expression of a gene is by controlling the stability of its messengers. The binding of a protein (*"poly A-binding-protein" or PABP*) on the poly A sequence at the 3′ end of *m*RNAs increases the life span of these *m*RNAs, probably by protecting them against an exonucleolytic degradation from this end. Conversely, the presence of a zone rich in AU in the non-translated 3′ region of some *m*RNAs seems to be the cause of their instability by preventing the binding of the PABP. On the other hand it was shown that some hormones could have an effect on the stability of some *m*RNAs : for example, prolactin prolongs the life of casein *m*RNAs in mammary glands.

B. – Translation

We are giving below some examples illustrating various possibilities of regulation at the translational level :

a) **Modification of the activity of translation initiation factors.** – It has been shown that the phosphorylation of one of the eucaryotic initiation factors (*e*IF-2) leads to the inhibition of this step and therefore to an arrest of protein synthesis in the cell considered (see page 369). Such a phosphorylation has been shown in cells treated with interferon (resulting in an arrest of translation of viral RNAs) as well as in cells synthesizing hemoglobin (where the action of hemin permits a regulation of the synthesis of globin).

b) **Negative self-control of translation.** – Some ribosomal proteins of *E.coli* seem capable, when present in excess in the cell, to bind to the corresponding polycistronic *m*RNA, in a site located near (upstream) the initiation site of the translation of the first cistron, and thus block their own production by preventing the binding of the ribosome. A control of the same type was recently proposed for the translation of threonyl-*t*RNA synthetase.

In some cases, the recognition site on the *m*RNA has some homology (at the level of the primary or secondary structure) with the region recognized by the protein on the RNA with which this protein normally interacts, for instance on the *r*RNA in the case of ribosomal proteins, or on *t*RNA^{Thr} in the case of threonyl-*t*RNA synthetase.

c) **Role of the secondary structure of the RNA.** – The best documented example is the expression (*in vivo* and *in vitro*) of RNA-containing phages (type Qβ and R17). The RNA of these phages, which acts as messenger, codes for 4 proteins (see page 345) whose expression varies in quantity and in time. Such a control disappears after treatment of the RNA with denaturing agents. It appears that it is the accessibility of the AUG initiation codons which is involved in this regulation, accessibility which can be modified by the intervention of proteins having a "regulatory effect", as well as by the translation of the message itself.

In fact the secondary structure of *m*RNAs can lead to the appearance of a specific binding site for proteins; the formation of such complexes will result in a variation of the efficiency of the translation and/or stability of the corresponding *m*RNA (e.g., regulation of the expression of ribosomal proteins in *E.coli*).

A very particular model of modification of gene expression was observed recently : some regions of the DNA can be transcribed in both directions. Beside the "normal" messenger (used for the synthesis of proteins) will be synthesized whole or part of a complementary RNA called "*anti-sense RNA*", the presence of which will perturb the expression of the "normal" messenger. The sequence complementarity of the two RNAs (messenger and anti-sense) leads to the formation of hybrids in the cytoplasm of the cell. In such hybrid molecules, the structuring of the *m*RNA is no longer possible. On the other hand, the accessibility of some sequences of the *m*RNA is no longer possible, like that of the Shine-Dalgarno sequence (indispensable for translation initiation in procaryotes; see page 368) to ribosomes. It has been shown that the injection of synthetic anti-sense RNA in cultured cells leads to an appreciable decrease of the translation of the corresponding *m*RNA. Active research is now in progress in view of the therapeutic or biotechnological utilization of this method; it is thus hoped to block the expression of pathogenic viral genomes for which it is difficult to prepare a vaccine (e.g., AIDS virus).

d) **Role of *t*RNAs and aminoacyl-*t*RNA synthetases.** – Some differentiation phenomena lead to the synthesis of very large quantities of specific proteins. This is the case with the silk gland of the silkworm where is synthesized fibroin a protein very rich in some amino acids (glycine, alanine and serine alone represent 90% of the amino acids). Protein synthesis may therefore be limited by the concentrations of *t*RNAs charged with these amino acids. It was shown that in these differentiated cells, the synthesis of aminoacyl-*t*RNA synthetases and *t*RNAs specific of these amino acids (and only these) was increased, thus fulfilling this requirement. This model can be compared with the one suggested for explaining the synthesis, from the genome of some phages, of specific *t*RNAs, which permit the translation of some phage messengers particularly rich in codons recognized by these *t*RNAs.

e) **Cleavage of a polyprotein precursor.** – In procaryotes the genes of the same operon are transcribed into only one *m*RNA, called "poly-cistronic". Translation starts independently at the signals preceding each cistron of this *m*RNA. In eucaryotes there is only one translation initiation site per molecule of *m*RNA, but several proteins can be cleaved from a single precursor polypeptide. The maturation of these 'polyprotein" precursors represents another possible control point of the expression of eucaryotic genes.

3. – Importance of the coordinated expression of genes. An example of perturbation: oncogenesis

The control mechanisms of gene expression not only maintain the "housekeeping" functions (metabolism, growth and cellular division) taking place in all living cells, but also participate in the processes of cellular differentiation and specialization implied by the ontogenic development (from the egg to the adult) of multicellular organisms. This closely coordinated set of regulations, where each gene or group of genes is expressed in a particular tissue for a particular time and at precise rates, thus adapt the functioning and multiplication of cells of each organ to the needs of the entire organism. However, it happens that some cells acquire new properties enabling them to free themselves from the constraints imposed on normal cells: such cells called "transformed" cells, are generally characterized by an active and unlimited multiplication, no longer needing their binding to a support. Within an organism, the appearance of transformed cells whose development escapes any control, is characterized by the formation of cancerous tumors (solid or liquid depending on whether the cells have a structural role or are circulating cells).

The molecular bases of cancerogenesis (or oncogenesis) are being actively investigated. They seem to be related to the perturbation of some cellular genes whose products normally participate in intra- or intercellular communications. These genes (called *proto-oncogenes* in normal cells) can code, for example, for transcription factors, growth factors or membrane receptors of growth factors. The activation of these receptors, due to the binding of the corresponding growth factors, produces within the cell a cascade of events (phosphorylation of particular proteins, increase of pH and of concentration of

Na^+ and Ca^{++}) whose effects contribute to the stimulation of replication and transcription of the genome (mitogenic action). Moreover, the products of other proto-oncogenes can modulate the intermediate reactions leading to these phenomena.

In this context, one can easily imagine that a modification of the activity of these proto-oncogenes can upset the functions controlling especially the cellular cycle and lead to malignant transformation. Thus, point mutations (caused by carcinogenic chemical substances or by ionizing radiations) can alter the coding sequences of proto-oncogenes (which are then called "c-onc", for *cellular oncogenes*) and can thus modify the properties of their expression products. The consequences of these modifications can be spectacular if they inactivate the corresponding protein or prevent the regulation of its activity. Similar consequences can result from an increase in the rate of expression of a c-onc either by stabilization of the corresponding *m*RNA, or by activation of its transcription (after translocation of the gene near a transcriptional activator or enhancer), or by gene amplification (i.e. multiplication of the number of copies of the gene). Lastly, some oncogenes are activated by their integration in the genome of a retrovirus. These viruses, whose genome is an RNA molecule, but whose replication involves a "retrotranscription" of this RNA into DNA (see reverse transcriptase, page 323) and possibly an insertion in the cellular chromosomes, may have, in the course of their history incorporated a proto-oncogene of the host cell. The cellular gene generally mutated, placed under the control of the viral promoter, can then behave like a *viral oncogene* (v-onc)[1], when the virus infects other cells.

It therefore appears that cancerous transformation can result from qualitative or quantitative alterations of different proto-oncogenes or varied combinations of proto-oncogenes. These multiple possibilities explain the great variety of cancers observed due to the tissue specificities of the expression of these genes.

II. – EFFECTS OF SUBSTRATE CONCENTRATION

We refer the reader to Chapter 2 where we studied the effects of variations of substrate concentration on the velocity of an enzymatic reaction (see fig. 2-4)

---

(1) As examples let us cite some viral oncogenes, (out of some twenty known) whose products have been partially characterized; *v-src* (carried by the Rous sarcoma virus which is responsible for the chicken sarcoma) codes for a membrane protein kinase; *v-erbB* (carried by the virus of avian erythroblastosis which induces leukemia in birds) codes for a protein related to the cytoplasmic part of the transmembrane receptor of EGF (epidermal growth factor); *v-sis* (carried by the virus of Simian sarcoma which causes sarcomas in the monkey) codes for a protein related to one of the chains of PDGF (growth factor of blood platelets); *v-fos* and *v-jun* (carried respectively by a virus of murine osteo-sarcoma and a virus of avian sarcoma) code for altered versions of cellular proteins coded by the *c-fos* and *c-jun* genes, which combine together into a heterodimer representing a transcription factor.

We may only add that in some cases, variations of substrate or co-substrate concentration [S] can exert important effects on the metabolism by permitting either the synthesis, or on the contrary the degradation of a substance X, or by determining the metabolic pathway in which a substance X will be directed.

Thus, for example – as observed in Chapter 4 – *the NAD^+ concentration* (which in fact depends on the oxygen content) *is a determinant factor of the fate of pyruvic acid*. If O_2 is present in sufficient quantities, the electron transport system will function properly, with the result that NAD^+ will be present in the cell in sufficient amount to enable pyruvic acid to undergo oxidative decarboxylation, process during which NAD^+ is required for reoxidizing the reduced lipoic acid (see fig. 4-36); if on the contrary little oxygen is available, the electron transport system will not permit the reoxidation of NADH to NAD^+, and in these conditions – as already mentioned – pyruvic acid is reduced by NADH + H^+ to lactic acid (see fig. 4-30). This clearly shows that oxygen concentration determines whether pyruvic acid is reduced to lactic acid or oxidized to acetyl-coenzyme A which enters the Krebs cycle.

One may also cite as example, *the effect of ATP concentration on the synthesis and degradation of glycogen*. An increase of ATP concentration favours the formation of UTP, then of UDP-glucose and thereby promotes the synthesis of glycogen; this effect is further enhanced by the fact that a high ATP concentration is established at the cost of ADP and P_i, whose concentrations decrease, which diminishes the degradation of glycogen as it requires P_i (see fig. 4-23); there is therefore an increase of the synthesis of glycogen and a decrease of its degradation. On the contrary a low concentration of ATP (and consequently, high concentration of ADP + P_i) is unfavourable for the synthesis of glycogen, but favourable for its degradation into glucose-1-phosphate which enters the pathway of glycolysis followed by the Krebs cycle (which will again raise the ATP concentration).

III. – REGULATION OF ENZYME ACTIVITY BY ACTIVATION OR INHIBITION

We have seen, in connection with the tryptophan operon, that an excess of tryptophan can cause a repression of the genes of this operon, leading to an arrest of the synthesis of the enzymes required for the formation of tryptophan. Beside this possibility of *repression*, there is very often *feedback inhibition*, i.e. the possibility for an *essential* metabolite (amino acid, nucleotide, etc.), which is the *final product of a series of biosynthetic reactions, of inhibiting the activity of an enzyme catalyzing one of the first reactions of this series*. As will be seen, the inhibited enzyme is generally the one which catalyzes the first reaction leading *specifically* to the final product, and not an enzyme which catalyzes a reaction common to several metabolic pathways; it is the enzyme situated at a *strategic junction* whose activity is inhibited by the final product.

This type of regulation is particularly characterized by the fact that the *effector* (the substance which activates or inhibits the enzyme) and the substrate of this enzyme are generally not isosteric., i.e. they have no structural analogy (contrary to the situation in competitive inhibition exerted by analogues of the substrate, see page 73). That is why they are called *allosteric effectors*, while the term *allosteric enzymes* denotes the enzymes in which this type of control is observed.

1. — General properties of allosteric enzymes

We have already indicated, in Chapter 2 (see pages 74-76) and in connection with aspartate transcarbamylase (see page 296), some important characteristics of the regulation at the level of allosteric enzymes, but it is of interest here to review the main properties.

A. — Kinetics of reactions catalyzed by allosteric enzymes

In general, allosteric enzymes have special *kinetic properties*, and when studying the variation of velocity as a function of substrate concentration one obtains, not a branch of an equilateral hyperbola as in the case of most enzymes (see fig. 2-4), but a *sigmoid* curve (see fig. 2-12). This S-shaped curve reflects a *cooperative effect* i.e. the fact that at least 2 substrate molecules interact with the enzyme and that the binding of the first molecule facilitates that of the second. Very often such a cooperative effect is also manifest in the binding of allosteric effectors (see fig. 8-13), which suggests that the binding of the first allosteric activator (or inhibitor) molecule favours the binding of the second. These results already suggest that there are *more than one catalytic site* and *more than one allosteric site* per molecule of enzyme and imply the *polymeric or oligomeric* nature of allosteric enzymes.

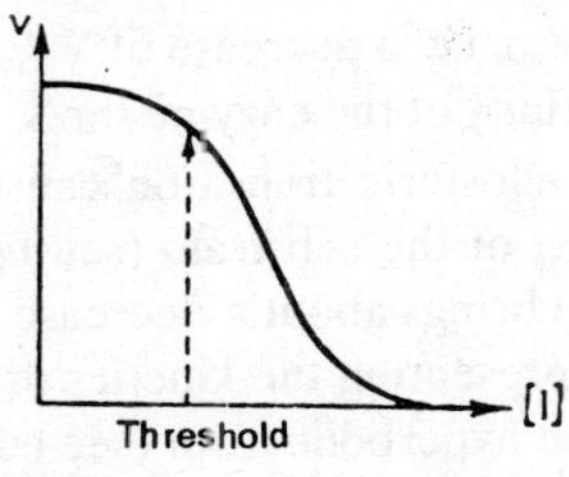

FIG. 8-13. — Variations of the velocity of a reaction catalyzed by an allosteric enzyme, as a function of the concentration of allosteric inhibitor.

We have not indicated where the allosteric effector binds, but our knowledge of the specificity of enzyme-substrate interaction (see page 90) and our observations on the absence of any structural similarity between substrate and effector suggest that the allosteric effectors do not bind to the active sites, but to different sites called *allosteric sites*.

Considering the sigmoid nature of the curves expressing enzymatic activity as a function of either the substrate (see fig. 2-12), or the allosteric inhibitor (see fig. 8-13), one therefore has a *threshold* effect when inhibitor concentration increases or when substrate concentration increases. Below the threshold, an increase of [S] (see fig. 2-12) or [I] (see fig. 8-13) does not cause a significant change of velocity; but beyond the threshold, velocity varies considerably for a relatively small increase of [S] or [I]. This enables the cell to adjust the enzymatic activity according to relatively small variations of [S] or [I], but occurring in a zone of critical concentration which corresponds to the intracellular concentrations of the metabolites involved.

B. – Action of allosteric effectors

There are *various types of allosteric inhibitors* and using the Lineweaver-Burk plot (see fig. 2-9), it is observed that some allosteric inhibitors are of the competitive type and others of the non-competitive type. But contrary to what we have seen while studying the competitive inhibition of conventional enzymes (see page 70), there is no competition – in the case of allosteric enzymes – between S and I for the active site of the enzyme (because they have no structural analogy). The two types of inhibitors bind to allosteric sites, different from the active sites, as shown by experiments of desensitization and fractionation of sub-units (see below). In the case of a *non-competitive allosteric inhibitor*, the binding of I to the allosteric site of an enzyme to give E – I can thus cause a change of conformation which still permits the binding of S to give E – S – I, but $1/V_{max}$ is increased (see fig. 2-9) therefore V_{max} is lower. During the binding of a *competitive inhibitor* to the allosteric site, there is a change of conformation, *an allosteric transition*, which causes repercussions on the active site to which S can no longer bind (see fig. 2-14). There is a decrease of $-1/K_m$ (see fig. 2-9), i.e. an increase of K_m, in other words a *decrease of the affinity of the enzyme for S*. There exists a type of *mixed inhibition*, characterized by an increase of $1/V_{max}$ i.e. a decrease of V_{max}, as well as an increase of K_m, i.e. a decrease of the affinity of the enzyme for S.

We have seen how the allosteric transition caused by the binding of an activator favours the binding of the substrate (see fig. 2-13). This change of conformation of the enzyme brings about a decrease of K_m, i.e., *an increase of affinity for S*. The curve representing the kinetics of the reaction can change from the sigmoid form to the hyperbolic form (see curve 2 of figure 2-12), but one must be very cautious because it has been shown in some cases that this change of order of the reaction was only apparent (it was due to an inaccuracy in the first part of the curve which did not show the sigmoid character).

C. – Desensitization and dissociation of allosteric enzymes

Allosteric enzymes can be made *insensitive to allosteric effectors*, either after a mutation, or *in vitro* by a physical or chemical treatment: variation of pH, temperature, ionic strength; action of urea, mercurial agents, proteolytic enzymes, etc.

This desensitization generally does not affect catalytic activity, which supports the hypothesis of separate catalytic and allosteric sites. It is often reflected by a modification of kinetics which changes from the sigmoid form to the hyperbolic "Michaelian" form. The fact that the enzyme conserves a catalytic activity but is no longer sensitive to the allosteric site was first interpreted as the sign of an alteration of the allosteric site (by the desensitizing agent) not affecting the catalytic site. In reality, *for the regulatory effects to manifest themselves* not only the allosteric sites must be intact and the effectors able to bind to these sites but also, *a conformation of the enzyme must be preserved which will enable the allosteric transition* and especially the repercussion – at the catalytic site – of an event affecting the allosteric site. In fact, it was observed in some cases, that after desensitization the effector can still bind to the allosteric site; on the contrary, due to a modification of the spatial structure of the enzyme, there is *disappearance of the cooperative interactions between the various catalytic sites of the same enzyme molecule, between its various allosteric sites and between its catalytic and allosteric sites,* thus preventing the allosteric transition. *This allosteric transition consists of a modification of the bonds joining the promoters to one another*, which permits the passage of the enzyme from a relaxed state to a constrained state, or vice-versa (see fig. 8-14).

The existence of separate sites for substrate and inhibitor, confirmed experimentally in numerous cases, is particularly evident when it is possible to dissociate an allosteric enzyme in distinct sub-units, some carrying the catalytic sites and others carrying the allosteric sites. This is the case for example with aspartate transcarbamylase, (already discussed in Chapter 6, page 296) which is inhibited by CTP and activated by ATP, can be desensitized by heat or urea, but can also be dissociated by mercurial agents: the native enzyme (molecular weight = 310 000) consists of 6 polypeptide chains with catalytic activity (molecular weight = 33 000) and having each a site for the binding of the substrate, and 6 regulatory chains (molecular weight = 17 000) enabling the binding of 6 CTP. On isolating the catalytic sub-units it is observed that their specific activity (quantity of substrate transformed per unit time, referred to the quantity of protein) is greater than that of the native enzyme, which is not surprising because the elimination of the regulatory sub-units – inactive in the catalysis process – is in a way a purification of the enzyme if one considers only the catalytic point of view.

D. – Model of Monod, Wyman and Changeux

To explain the phenomena observed during the study of allosteric enzymes, these authors proposed a model whose important characteristics are as follows:

– *the allosteric enzymes are oligomers*, whose protomers are associated so that the molecule comprises at least *one axis of symmetry* (the protomer is defined as the structure which has a binding site for each *ligand*, i.e. for each substance capable of *binding* – i.e. substrate, activator and inhibitor – and must not be mistaken for the sub-unit which results from the dissociation of the enzyme and can contain – as in the case of aspartate-transcarbamylase – only one site, catalytic or allosteric);

– *each protomer possesses only one site permitting the formation of one specific complex with each category of ligand;*

– *the allosteric enzyme may have different but interconvertible conformations. One often speaks of relaxed state and constrained state. These states are in equilibrium and differ either by the distribution and energy of bonds between the protomers* (which determine the constraints imposed on protomers), *or by the affinity of the various sites for the corresponding ligands.*

Figure 8-14 shows a simple diagram – with only 2 protomers – to illustrate the model. At first, there is equilibrium between the relaxed form and the constrained form; if one of the ligands (e.g., the substrate) has a greater affinity for one of these 2 forms, a relatively small concentration of this ligand will permit the binding of a substrate molecule to a protomer of the form considered, which will shift the equilibrium in favour of this form and will facilitate the binding of the substrate. But an increase of the concentration of an antagonistic ligand (here the inhibitor) is enough for the equilibrium to be shifted in the reverse direction. Allosteric phenomena are *reversible* and depend on the concentrations of the various ligands. Such a model explains the fact that a sigmoid curve is obtained when velocity is expressed as a function of [S] or [I].

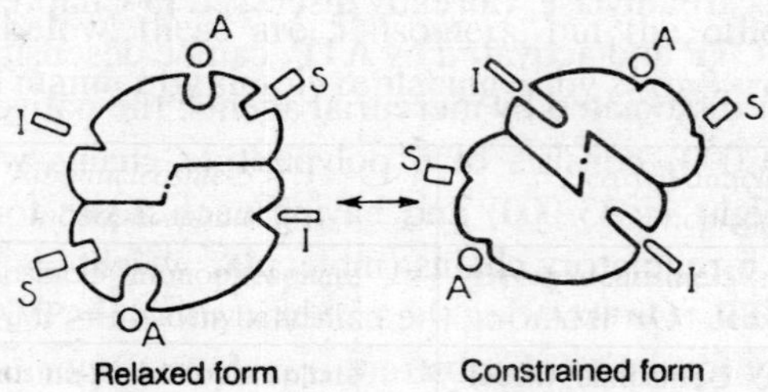

FIG. 8-14. — Equilibrium between the relaxed form and the constrained form of an allosteric enzyme of the K type. (The relaxed form has greater affinity for S and the constrained form greater affinity for I.)

The diagram of fig. 8-14 is valid for an allosteric enzyme of the K type. In this case, in the absence of substrate, the equilibrium is in favour of the form having a low affinity for the substrate. But as observed above, when [S] increases, the equilibrium is shifted in favour of the form having a greater affinity for S. Conversely, the equilibrium is shifted by the inhibitor in favour of the form having a low affinity for S and *the allosteric transition consists precisely in this change of equilibrium*. Therefore, the inhibitor decreases the affinity of the enzyme for S (K_S increases), and conversely the substrate decreases the affinity of the enzyme for the inhibitor (K_i increases), whence the name K type enzyme.

Other models were proposed to explain the kinetic properties of allosteric proteins. According to the model of *induced fit* proposed by Koshland, Nemethy and Filmer, there is only one configuration for the protein in the absence of ligand; it appears that the binding of the ligand *induces* a conformational modification of the protomer, which transforms the interactions between the sub-units and changes the catalytic properties.

It appears that the conformation of an enzymatic protein, which we called tertiary and quaternary structure in Chapter 1, is not exclusively determined by the primary structure (see page 33). Actually, it is observed that small molecules (substrates, activators, inhibitors), by binding to specific sites, are capable of causing slight modifications of the spatial structure of the protein.

2. – Main modes of regulation

– *Feedback inhibition* consists in the inhibition of the first enzyme of a reaction series by the metabolite which is the terminal product of this series. The intracellular concentration of this metabolite therefore controls the rate of its own biosynthesis. In the following we are considering feedback inhibition in straight and branched reaction series.

– *Activation* of an enzyme by a *precursor* of the substrate or by the substrate itself.

– *Activation by a degradation product of the terminal metabolite* causing a new increase of the concentration of this metabolite (which may be a high energy potential substance for example).

– *Activation of an enzyme of a metabolic series leading to a metabolite A by a metabolite B, which is synthesized by an independent series*, when A and B are both necessary for the synthesis of the same macromolecules, which permits a coordinated production of precursors (in the case of nucleotides for example, see especially page 290).

The activity of an allosteric enzyme can be controlled by several of these regulatory modes. Thus, *aspartate transcarbamylase*, the first enzyme of the pathway leading to the synthesis of pyrimidine nucleotides, is feedback-inhibited by a terminal product (CTP), activated by the substrate and also activated by ATP, a ribonucleoside triphosphate required – jointly with UTP and CTP – for the biosynthesis of RNAs.

3. – Feedback inhibition in straight and branched reaction chains

A. – Feedback inhibition in straight reaction chains

In *straight metabolic sequences*, it is generally the first enzyme (E_1) which is the regulatory enzyme, i.e. the enzyme subjected to a control of allosteric type (see fig. 2-15). By "first enzyme" one must generally understand the enzyme which catalyzes the first reaction specific of the metabolic pathways considered. For example, in the case of the biosynthesis of pyrimidine ribonucleotides, it is aspartate transcarbamylase which is subjected to feedback control and not an enzyme permitting the synthesis of carbamyl-phosphate or aspartate; these two compounds can also enter other metabolic pathways, while their combination to give carbamyl aspartate is really the first reaction leading specifically to pyrimidine nucleotides (see fig. 6-22). The first enzyme of the chain is generally the only one to be inhibited by the final product; its activity therefore determines the functioning of the whole sequence of reactions.

The inhibition of this enzyme by the final product of the chain of reactions is of obvious interest. When this final product is in excess, the inhibiting effect it exerts on the first enzyme decreases the rate of this first reaction and consequently restricts its own biosynthesis. Since the series of biosynthetic reactions usually require energy, this regulation process enables the cell *to save energy*. This economy is however smaller than the one made through the repression process: when a substance X is in excess, repression enables the cell to dispense with not only the biosynthesis reactions of X, but also the transcription of genes into *m*RNA and the translation of the polycistronic *m*RNA into the enzymes required for the biosynthesis of X, while in feedback inhibition the enzymes required are present but do not function.

On the contrary, feedback inhibition appears as a process *more rapid* than repression. An excess of a substance X can immediately inhibit the first enzyme of the chain of reactions leading to X, while the effects of repression are manifest only after the disappearance – through catabolism – of the molecules of enzymes and *m*RNAs existing in the cell (and which are not replaced because the expression of the corresponding genes is blocked). Feedback inhibition which is based – as mentioned above – on the phenomenon of allosteric transition, i.e. on the shift of a state of equilibrium in favour of one of the two conformations of the enzyme, is an easily reversible process, very sensitive to small variations of the concentrations of ligands beyond a particular threshold, and is therefore characterized by a *great flexibility*.

B. – Feedback inhibition in branched reaction chains

Feedback inhibition poses special problems in the case of *branched reactions chains* where one could *a priori* fear that the excess of the final product of one of the branchings would cause – if it inhibits the first enzyme of the chain – the arrest of the synthesis of substances produced by the other branchings, substances which are not necessarily in excess. To study these problems we will take the example of the biosynthesis of amino acids deriving from aspartate, a set of reactions we examined in Chapter 7 (see fig. 7-22); this will enable us to study their regulation with the help of a simplified diagram.

a) **Feedback inhibition limited to branchings.** – As shown by figure 8-15, *the amino acid which is the final product of a branching can inhibit the first step of the sequence of reactions leading only to its biosynthesis*. The biosynthesis of other amino acids is therefore not affected. Lysine inhibits dihydrodipicolinate synthetase, threonine inhibits homoserine kinase (HK), methionine inhibits the succinylation of homoserine and isoleucine inhibits threonine deaminase (TD).

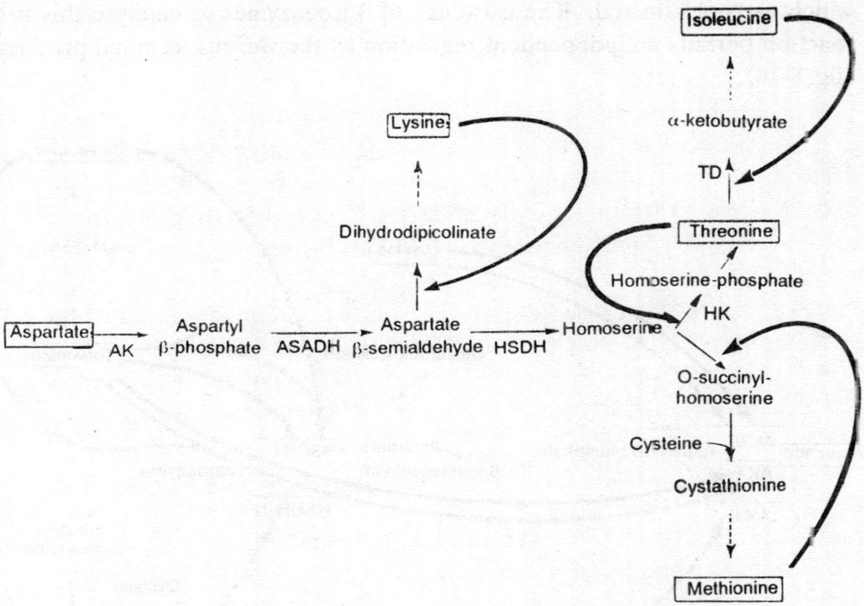

FIG. 8-15. — Biosynthesis of amino acids deriving from aspartate: examples of feedback inhibition limited to branchings.

b) **Iso-enzymatic control.** – Three aspartokinases (AK) have been identified in *E.coli*; each of them is subjected to a regulation by a specific repression mechanism and two are subjected to a feedback inhibition which is also specific. Moreover, there are also 2 homoserine dehydrogenases (HSDH) whose regulation is identical to that of the first two aspartokinases, as shown by the table below:

Enzyme	*Repressed by*	*Feedback-inhibited by*
AK I	Thr + Ile	Thr
AK II	Met	—
AK III	Lys	Lys
HSDH I	Thr + Ile	Thr
HSDH II	Met	—

It has been shown that the two catalytic activities AK I and HSDH I are carried by one and the same polypeptide chain; the same is true of the activities AK II and HSDH II. It is evident that the existence, for example in the case of aspartokinase, of 3 isoenzymes whose synthesis and activity are controlled by different terminal products, enables the cell – in case of repression of the biosynthesis or inhibition of the activity of one of the aspartokinases due to a high concentration of one amino acid – to continue to synthesize the other amino acids deriving from aspartate thanks to the other two aspartokinases which are not affected. The existence of 3 isoenzymes to catalyze this first reaction permits an independent regulation by the various terminal products (fig. 8-16).

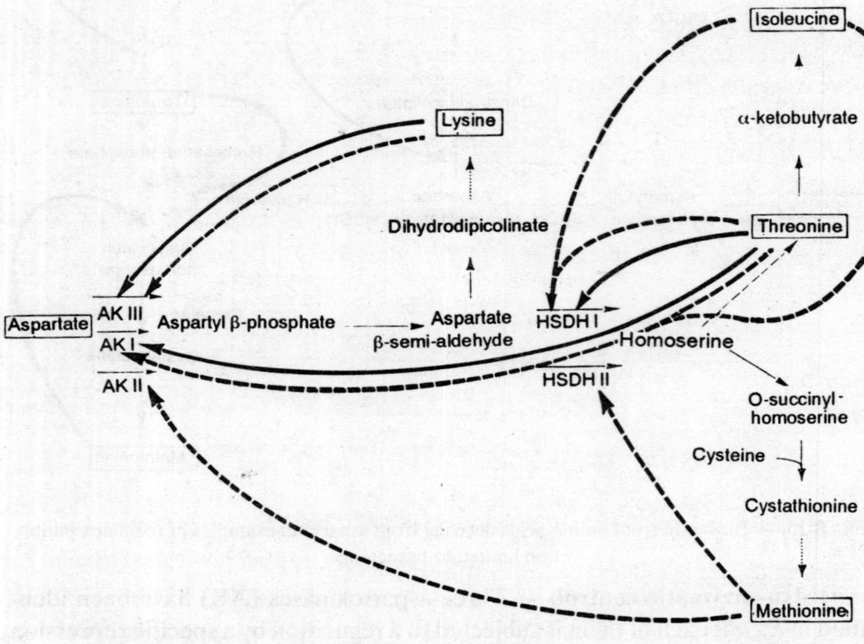

FIG. 8-16. — Biosynthesis of amino acids deriving from aspartate: examples of isoenzymatic control by feedback inhibition (solid lines) or repression (dotted lines).

c) **Concerted or multivalent feedback inhibition.** – In some organisms of the genus *Rhodopseudomonas* or *Bacillus* there is only one aspartokinase which is not affected by the excess of only one of the terminal products (Lys, Thr, Ile), but which is feedback inhibited when there is excess of both lysine and threonine. However this concerted feedback inhibition is not total, which permits the synthesis of methionine.

Other possibilities of control exist in some organisms in this branched chain of biosynthesis of amino acids deriving from aspartate. We cannot study all of them but the types of regulation we examined do show that varied mechanisms were selected by the living organisms in the course of evolution to solve the special problems posed by regulation in the metabolic pathways presenting branchings.

IV. – COVALENT MODIFICATION OF PROTEINS

In numerous cases, it has been shown that proteins can exist either in the native form or in a form modified by the covalent binding of a radical (for example, phosphate, methyl, acetyl, adenyl, etc.). These modifications can bring about major variations of the catalytic activity of the protein: for example phosphorylation of phosphorylase *b* (see page 483); adenylation of glutamine synthetase in *E.coli*. The reactions of binding and elimination of the radical are catalyzed by specific enzymes, whose activity and (or) synthesis can themselves be subjected to regulation. Thus, activators or inhibitors of these modification enzymes will lead to a variation of the ratio Native protein/Modified protein.

We may compare these phenomena with the modification of enzymatic activity by limited specific proteolysis (see for example the trypsinogen-trypsin activation, pages 89 and 401).

It appears that such mechanisms play a very important role in the variations of enzymatic activities in cells of higher organisms.

V. – EFFECTS OF HORMONES

Pluricellular organisms are characterized by the existence of tissues, i.e. sets of cells which have acquired specialized functions. This specialization is advantageous in the sense that it permits a distribution of work but it necessitates means of controlling and integrating the activities of various tissues. Hormones constitute precisely a means of communication between territories distant from one another. It was suggested that hormones act on the activity of enzyme (as activators or inhibitors) or on the repression and induction of the synthesis of enzymes (see page 462). In fact, although the physiological effects of hormones are generally well known, their primary site of action and their mechanism of action – in molecular terms – have not yet been clearly explained. Furthermore, the control of the synthesis of hormones, and of their secretion by endocrine glands, in response to a change of the concentration of a compound (food or waste) in the extra-cellular fluids, represents a domain which is not yet understood, though it is known that secretion can be stimulated either by a nervous impulse originating from the brain, or by a hormone secreted by another gland (e.g., an antepituitary stimuline).

1. – Control of the metabolism of glycogen

Glycogen is the principal source of energy for muscle contraction. Its degradation and synthesis are subjected to a control which modifies the degree of phosphorylation of *glycogen phosphorylase* and *glycogen synthetase* respectively, by the action of numerous proteins, especially 5 different protein kinases, 4 protein phosphatases and various regulatory proteins like calmoduline, troponine C, etc. Thanks to these proteins, the metabolism of glycogen can be controlled by Ca^{2+} ions and by cyclic AMP (*c*AMP), denoted "*second messengers*" because they relay the action of the "*first messengers*" which are respectively electric stimulation and β-adrenergic stimulation (by adrenaline).

The hormonal control of glycogen metabolism involves adrenaline (in the muscle and liver) and glucagon (in the liver).

Glucagon is a peptide hormone (29 amino acids) secreted by the α cells of Langerhans islets of the pancreas (the β cells secrete insulin). *Adrenaline* has a simpler structure and derives from tyrosine as mentioned in the previous chapter (see fig. 7-27). The physiological effect of these two hormones can be easily shown: they cause an increase of the blood glucose and of the glucose which enters the oxidation processes. This increase of available glucose takes place at the expenses of glycogen but, as we will see in the following, the primary site of action of these two hormones is a reaction rather distant from the degradation reaction of glycogen.

In the muscle, the adrenaline – called *first messenger* by some authors – brought by the circulating blood, binds to the membrane of the target-cell and activates *adenyl-cyclase*, an enzyme catalyzing the scission of ATP into PP + *cyclic AMP* (or adenosine 3′, 5′ cyclic monophosphate or *c*AMP). The latter formed on the internal surface of the cell membrane, will diffuse into the cell; it is the *second messenger* which has taken the relay from the first. It will act on a *protein kinase* which needs cyclic AMP for its activity. This protein kinase transforms an inactive non-phosphorylated *phosphorylase b kinase* into active phosphorylated phosphorylase *b* kinase. Phosphorylase *b* kinase can then catalyze the phosphorylation of *phosphorylase b* (which contains 2 subunits) into *phosphorylase a* (which contains 4 subunits); this phosphorylation modifies 4 serine residues (one per polypeptide chain) which are transformed into phosphoserine by the action of phosphorylase *b* kinase. *Phosphorylase a* thus formed constitutes the active form of *glycogen phosphorylase* which permits, as mentioned earlier (see page 172) , the glycogenolysis, i.e. the *phosphorolysis of glycogen*. All these reactions are schematically represented in figure 8-17. In the liver – under the influence of adrenaline or glucagon – there is a set of similar reactions, although there are some differences, especially at the level of the phosphorylases.

In addition to *this activating action on glycogenolysis, cyclic AMP also has an inhibiting action on the synthesis of glycogen*. As shown by figure 8-17, the protein kinase activated by cyclic AMP also has an action on glycogen synthetase and catalyzes – with other kinases – the transformation of active non-phosphorylated glycogen synthetase (form I) into inactive phosphorylated glycogen synthetase (form D).

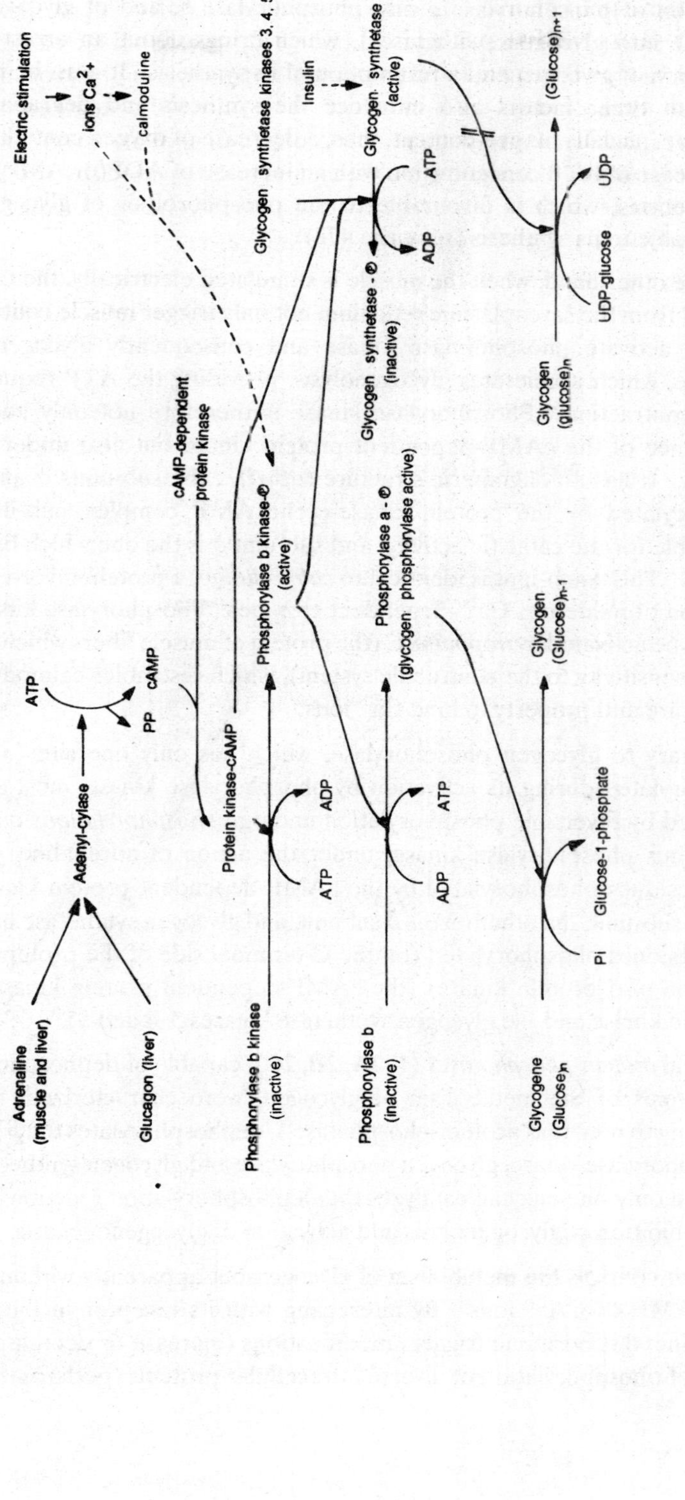

FIG. 8-17. — Hormonal and nervous control of the degradation and biosynthesis of glycogen.

Conversely, when cyclic AMP concentration decreases, there is dephosphorylation of phosphorylase *a* into phosphorylase *b*, and of glycogen synthetase D into glycogen synthetase I, which brings about an arrest of the degradation of glycogen and a resumption of its synthesis. It must be remembered that other factors also influence the synthesis and degradation of glycogen, especially oxygen content: thus, a decrease of oxygen content results in a decrease of ATP concentration with an increase of ADP(or AMP) and P_i concentrations, which is favourable to the phosphorolysis of glycogen and unfavourable to its synthesis (see page 472).

On the other hand, when the muscle is stimulated electrically, the Ca^{2+}ions liberated from the sarcoplasmic reticulum not only trigger muscle contraction, but also activate phosphorylase kinase and consequently glycogen phosphorylase, which accelerates glycogenolysis, providing the ATP required for muscle contraction. Phosphorylase kinase is therefore not only under the dependence of the *c*AMP-dependent protein kinase but also under that of Ca^{2+}ions. It has an oligomeric structure $(\alpha\beta\gamma\delta)_4$: the sub-units α and β are phosphorylated by the protein kinase-cyclic AMP complex, sub-unit γ is responsible for the catalytic activity, and sub-unit δ is the one which binds the Ca^{2+}ions. This δ sub-unit is identical to *calmoduline*, a protein involved in the regulation of numerous Ca^{2+}-dependent enzymes. Phosphorylase kinase also seems to be activated by *troponine C* (the protein of muscle fibers which confers calcium sensitivity to the contractile system), which resembles calmoduline by its structure and property to bind Ca^{2+}ions.

Contrary to glycogen phosphorylase, which has only one site (a serine) phosphorylated during its activation by phosphorylase kinase, most enzymes controlled by reversible phosphorylation undergo *phosphorylations on several sites*. Thus phosphorylase kinase, under the action of adrenaline, has two serine residues phosphorylated by the *c*AMP-dependent protein kinase, one on the α sub-unit, the other on the β subunit, and glycogen synthetase has seven serine residues phosphorylated (on the C-terminal side of the protein) under the action of 5 protein kinases (the *c*AMP-dependent protein kinase, phosphorylase kinase and the glycogen synthetase kinases 3, 4 and 5).

Several *protein-phosphatases* (1, 2A, 2B, 2C), capable of dephosphorylating the enzymes of the metabolism of glycogen, were characterized, but it is interesting to note that protein-phosphatase 1, dephosphorylates the β sub-unit of phosphorylase kinase, glycogen phosphorylase and glycogen synthetase, and therefore only one enzyme catalyzes the dephosphorylation reactions leading to the inhibition of glycogenolysis and activation of glycogenogenesis.

Insulin controls the metabolism of glycogen but apparently without calling upon *c*AMP or CA^{2+} ions. By interacting with its receptor in the plasmic membrane, this hormone triggers modifications (increase or decrease) of the degree of phosphorylation of diverse intracellular proteins (perhaps through a second messenger). It thus controls the metabolism of glycogen by stimulating the activity of glycogen synthetase.

2. – Control of other cellular processes

It soon became obvious that the *cAMP-dependent protein kinase* must have activities other than the one first shown (activation of the phosphorylase kinase) to explain the various effects of *c*AMP. According to a hypothesis that was advanced, the specificity of a hormone is determined by the presence of the corresponding receptor in the membrane of the target-cell and by the nature of the substrate (or substrates) of the *c*AMP-dependent protein kinase present in this cell. When these substrates are enzymes, the effect of phosphorylation is often to modify the constants of affinity for a substrate, an enhancer or an inhibitor (this is particularly the case with the enzymes of the metabolism of glycogen).

Numerous enzymes are controlled by the *c*AMP-dependent protein kinase, which explains the great diversity of action of hormones in various cells. Thus, triglyceride lipase is activated and glycerol-phosphate acyl transferase is inactivated, which permits the respective control of the *catabolism and synthesis of triglycerides in the adipose tissues* under the action of adrenaline in the same way as activation of phosphorylase kinase and inactivation of glycogen sythetase permit the control respectively of the catabolism and synthesis of glycogen under the action of adrenaline (in the muscle) or glucagon (in the liver).

The activation of *cholesterol esterase* increases the pool of cholesterol available for the *steroidogenesis* in the adrenal cortex under the action of ACTH. The phosphorylation of *acetyl-coenzyme A carboxylase* decreases its activity, which explains the *inhibition of the synthesis of fatty* acids in the adipose tissues under the action of adrenaline and in the liver under the action of glucagon. The inactivation of pyruvate kinase and the activation of fructose 1,6 bisphosphatase play an important role in the stimulation of *gluconeogenesis* by glucagon in the liver. But the hormonal control of many other processes seems to involve the phosphorylation of proteins, especially during growth, cellular differentiation, muscle contraction, transmembrane transports, etc., although the corresponding molecular mechanisms have not yet been explained.

It must be noted that the *c*AMP-dependent protein kinase is a very specific enzyme, phosphorylating at a significant rate only few proteins, and only on one or two sites among the large number of residues of serine and threonine. It appears that its physiological substrates always comprise two adjacent basic amino acids on the N-terminal side of the phosphorylated residue.

Numerous biological effects of Ca^{2+} ions involve a protein, *calmoduline*, which, after binding Ca^{2+} ions, undergoes conformational changes and thus acquires new possibilities of interaction. About twelve enzymes activated by calmoduline have already been identified, especially *proteins kinases* (particularly phosphorylase kinase) and *proteins phosphatases* which affect the activity of key-enzymes in various tissues. For example, in the brain a calmoduline-dependent protein kinase phosphorylates *tryptophan hydroxylase* and *tyrosine hydroxylase*, thus controlling the *synthesis of serotonine* and *dopamine* (two neuromediators).

A number of analogies are observed between the activation of proteins (by phosphorylation) under the influence of *c*AMP on the one hand, and calmoduline on the other. The same process can be activated by either of these control systems, depending on the tissue considered. Besides, the *c*AMP-dependent protein kinase and the calmoduline-Ca^{2+}-dependent proteins kinases often control the phosphorylation of the same proteins. We have seen that the *c*AMP-dependent protein kinase can phosphorylate the phosphorylase kinase, which is also a calmoduline-dependent Ca^{2+} enzyme.

The fact that the concentration of *c*AMP-dependent protein kinase is high in tissues where glycogen metabolism is not important was one of the first observations made in favour of the role of this enzyme in metabolisms other than that of glycogen. The same observation was made for various *protein phosphatases* (1, 2A, 2B, 2C), suggesting that they also play a role in the control of processes other than the metabolism of glycogen. These phosphatases act particularly in the liver on *pyruvate kinase, acetyl coenzyme A carboxylase, hydroxymethylglutaryl coenzyme A reductase*, and therefore seem to be involved in the control of *neoglucogenesis, synthesis of fatty acids* and *synthesis of cholesterol*. These phosphatases also act on the light chains of myosin, which suggests that they are involved in the control of *muscle contraction*. Protein phosphatases 1 and 2A act on the initiation factor eIF-2 and can therefore be considered as stimulating protein synthesis (since, as mentioned on page 468, the phosphorylation of this factor causes inhibition of translation).

Processes of phosphorylation and dephosphorylation of proteins occupy an important place in numerous control mechanisms of cellular activity. In addition to those already mentioned, one may cite the phosphorylation of nuclear proteins (especially the histones) or that of ribosomal proteins which probably act in the control of the organization and expression of genes, but whose mechanisms and role are not yet known exactly. We have mentioned that cyclic AMP plays a role in the control mechanisms in microorganisms, particularly in the phenomenon of catabolic repression (see page 452). In higher organisms, it constitutes, as just mentioned, the relay of various hormones and therefore acts in the hormonal control of numerous cellular processes.

INDEX

A

D

F

G

H

I

J

K

N

Q

R

S

U